Sea-level research

Sea-level research:

a manual for the collection and evaluation of data

EDITED BY

Orson van de Plassche

Free University, Amsterdam

a contribution to Projects 61 and 200

Cover design and three drawings of similar signature elsewhere in this book are by *eleonora tammes*. The drawings were selected from the series 'dō-gi' (motion).

British Library Cataloguing in Publication Data

Sea-level research : a manual for the
collection and evaluation of data.
1. Sea level
I. Van de Plassche, Orson
551.46 GC89

ISBN 0-86094-187-6

Published by Geo Books,
Regency House,
34 Duke Street,
Norwich.
NR3 3AP

Printed in Great Britain by Galliard (Printers) Ltd, Great Yarmouth

CONTENTS

List of figures

Chapter 11

List of tables

Preface

An editorial by Wanless (1982), entitled "Sea level is rising - so what?", tells the case of an executive editor of a major city newspaper, who, when confronted with evidence for a recent sea-level rise, replied: "That just means the ocean is six inches deeper, doesn't it?". Whether his "so what?" attitude was real or put on to dike a threat of sensation, there is at present a wide and deepening interest in ongoing and future global sea-level change. This interest has grown along with the concern over global warming due to increasing levels of CO2 and trace gases. A stage has been reached where investigators of climate - sea-level relationships call for long-term measurement programmes for ice-volume changes (using satellite altimetry) and changes in temperature and salinity of the oceans (thermal expansion).

This manual, however, is primarily concerned with sea-level changes in the past, mainly since the end of the last glaciation. Its major objective is to help answer the question: "how?", which, of course, is little else but to assist in the gathering of fuel for the burning question: "why?" Good fuel, hopefully, for the less smoke and ashes, and the more heat and light produced by that fire, the better scientists are enabled to develop a quantitative understanding of past, and hence of future, sea-level changes on different spatial and temporal scales.

Summarizing the "Paris symposium on world sea levels of the past 11 000 years" (VIII INQUA Congress, 1969), Gill (1971) pointed out the "continued inadequacy of the data" and the need for a "more sophisticated approach to our recording of sea levels in the past". One of the objectives of Project 61 (1974-1982) of the UNESCO/IUGS International Geological Correlation Programme, entitled "Sea-level movements during the last deglacial hemicycle (about 15 000 years)", was the preparation of guidelines for site evaluation and sample collection. These were considered necessary for improving the quality and comparability of time/altitude data and sea-level graphs. The Dutch Working Group decided to prepare guidelines for sample collection and data evaluation based on experience gained in subsiding, alluvial coastal areas such as occur along the southern North Sea.

An unfinished draft for a sea-level manual (by the present author) was distributed at the meeting of the International Working Group for IGCP Project 61 during the X INQUA Congress (Birmingham, 1977). The structure of that interim report was based on a discussion of the factors that determine the quality of sea-level data in general, which opened the way for dealing with all kinds of sea-level indicators, not just submerged peats. This inspired the large and active French Working Group NIVMER to organize a seminar on "Les indicateurs de niveaux marins" (Paris, December 1978). The many valuable communications presented at that meeting were published in the 1979-80 volume of OCEANIS.

A serious start for the production of a sea-level manual had finally been made by early 1980, when a considerable number of experienced sea-level researchers and other experts had accepted the invitation to contribute to the writing of the book. The approach adopted has been to provide and discuss, for a range of sea-level indicators and for aspects of age and height determination, basic information, diagnostic criteria, practical recommendations and examples for the purpose of aiding researchers, beginners and the more experienced alike, in collecting field data, sample treatment, deriving the paleo-tide level and/or nature of sea-level change from available evidence, and in determining the accuracy of the derivation. Eventually, eight of the solicited papers did not materialize for various reasons. A number of sea-level indicators, therefore, remain undiscussed, as does the important topic of changes in tidal range in time and space. These omissions and other deficiencies of the volume remind us that the task, to discuss and share knowledge and experience regarding the wide variety of phenomenological evidence of paleo-sea levels and to develop and apply a communal, uniform approach in reading and rating the data, is far from completed.

October 1985 O.v.d.P

REFERENCES

Gill, E.D., 1971. The Paris symposium on world sea levels of the past 11 000 years. In: Les niveaux marins Quaternaire, I.-Holocene. *Quaternaria*, XIV, 1-6

Groupe NIVMER, 1979-80. Les indicateurs de niveaux marins. *OCEANIS*, fascicule hors-série, 5, 145-360

Wanless, H.R., 1982. Editorial: Sea level is rising - so what? *Journal of Sedimentary Petrology*, 52(4), 1051-1054

IGCP Projects 61 and 200

IGCP PROJECT 61
(1974-82)

Title: Sea-level movements during the last deglacial hemicycle (about 15 000 years)

Project Leader: A.L. Bloom

Objectives: (as formulated in the original Project proposal, 1974): The primary objective of the Project is to establish a graph of the trend of mean sea level during the last deglaciation and continuing to the present time. This graph must be an expression of the changing hydrologic balance betw-en ice and water in response to climatic change. Future trends of sea-level movements will be predicted.

Lines of approach: Individual records of relative sea-level movements from localities all over the world will be used to compile the sea-level graph. In turn, the differences among the local records will be used to draw conclusions about local crustal movements along the coasts, and about fundamental parameters of strength and elasticity of the earth's outer layers. Operational phases: (i) compilation and computer storage of global data; (ii) determination of key areas with deficient data, and arranging for local experts to obtain the necessary data; (iii) mathematical modelling and prediction; and (iv) successive approximations of the final curve, to be tested against accumulating data.

IGCP PROJECT 200
(1983-87)

Title: Late Quaternary sea-level changes: measurement, correlation and future applications

Project Leader: P.A. Pirazzoli

Objectives: To identify and quantify the processes of sea-level change by producing detailed local histories that can be analysed and correlated for tectonic, climatic, tidal and oceanographic fluctuations. The ultimate purpose is to provide a basis for predicting near-future changes for application to a variety of coastal problems, with particular reference to densely populated low-lying coastal areas.

Sea-level variations result from a complex of local, regional and global processes. Sea-level data contain a wealth of information concerning internal and external effects and provide the only possibility for reconstructuring palaeogeoid surfaces and testing complex models. Project IGCP-200 intends to investigate these modulating factors and their interactions in order to attempt a definition of the temporal, areal, and altitudinal scales at which changes in sea level occur, with associated effects on coastal and shelf deposit evolution and the separation and quantification of the causes of these changes.

Lines of approach: (i) Collection, analysis, interpretation and correlation of new and existing sea-level data; (ii) survey and data analysis of coastal and shelf deposits, and (iii) analysis of tide-guage records and the modelling of other short-term sea-level fluctuations.

Acknowlegements

The compilation of this book has taken me an estimated one year, spread out over five. At least two too many, no matter how valid the reasons. My expression of gratitude towards each of the contributors is therefore accompanied by an apology for the lengthy preparation of the book; I thank them for their cooperation and patience. Special thanks are due to Dick van Harten and David Hopley for rescuing the contributions on ostracods and beachrock respectively, and to Douglas Grant and Andrew Palmer for editorial help and advice. I am no less grateful for the efforts of those whose contributions could not be included in this volume. The comments, suggestions and encouragements of unnamed reviewers are gratefully acknowledged. Furthermore, I wish to mention Ir. B.P. Hageman and Dr. A.J. Wiggers, who put me on the track to this manual, and Duncan Wood (Geo Books), who paved the way for its final publication. Pergamon Press Ltd. kindly permitted reproduction of C. Kidson's review paper on 'Sea-level changes in the Holocene' (Quaternary Science Review 1, 2, 1982, 121-151). The active and critical support I received from Eleonora Tammes, as well as her persistent artistic productivity have been and are a continuous source of inspiration. Support was provided by the Institute for Earth Sciences, Free University, Amsterdam and the Netherlands organisation for the advancement of pure research (full-grants 075-95 and H75-262).

1

Introduction

O. VAN DE PLASSCHE

You may find yourself
With this beautiful curve

You may ask yourself
Is it right? Is it wrong?
How did I get it?

You may say to yourself
What have I done!

(paraphrazed from Talking Heads'
'Once in a lifetime')

THE SEA-LEVEL MARKET

The SEASAT images of the highly unlevel ocean surface shown in Figures 1 and 2, with their inspiring visualisation of the gravity field as a potential cause for sea-level (SL) change (but at what scales?), nicely express the tension and fascination of the general objective of SL research: to know and understand past and future SL/sea-surface changes. Experts from a wide range of disciplines are involved or interested in answering one or both of the questions implied (how? why?) and/or in scientific or practical application of those answers. This multidisciplinary engagement in producing and consuming SL-study results relates primarily to three factors: (i) the wide spectrum and interrelationships of possible causes of SL change, (ii) the variation in the nature of SL change evidence (records), and (iii) the fact that SL itself is a variable in process-response systems that are studied for reasons of applied or academic interests.

With respect to the first of these factors, SL may be seen as the communal pen and ink, or rather broad brush and paint, with which astronomical, climatological, meteorological, oceanographical and geodynamical processes express

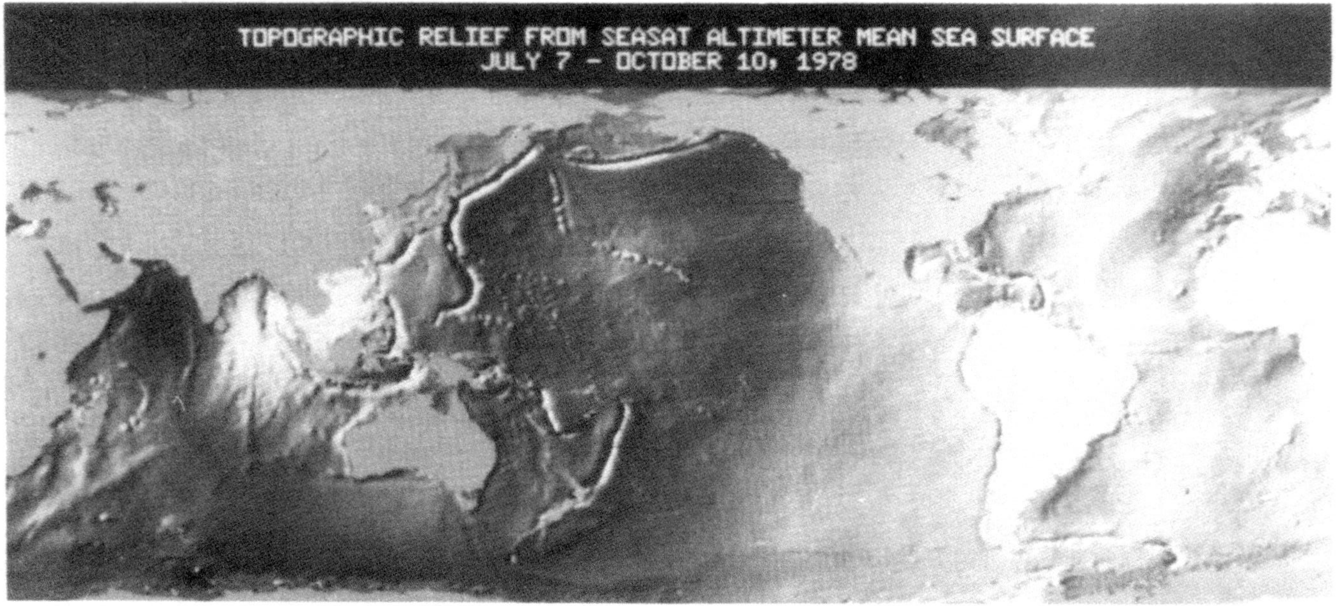

Figure 1 SEASAT relief map of ocean-surface topography (courtesy NASA Jet Propulsion Lab)

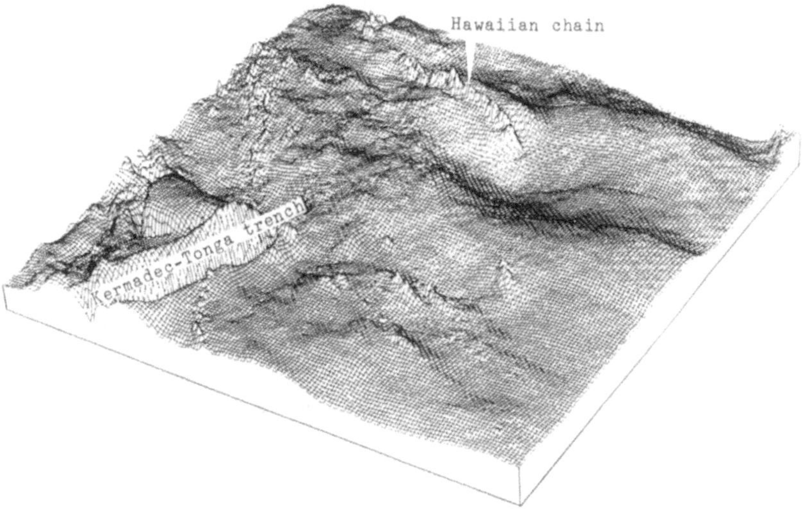

Figure 2 SEASAT geoid of the central Pacific Ocean (adapted from Cazenave & Dominh, 1985)

themselves, often interdependently and simultaneously, on and along the edges of continents and islands. It is this 'all in one' that makes SL change records such rich but complex sources of information on (interrelationships between) the factors and processes involved. The human factor (combustion of fuels, deforestation, dike building, subsurface gas and water extraction, dredging, ...), through its influence on climate and local oceanographic and hydrodynamic and hydrostatic conditions, has become increasingly important. As for SL change evidence, we can distinguish records based on tide-gauge measurements, phenomenological (field) evidence, and, in the near future, on satellite altimetry. Field-data records are principally derived from a variety of SL and SL change indicators of physical, physicochemical, biological or archaeological nature, such as raised beaches, sedimentary structures, carbonate diagenetic phenomena, decomposition horizons in peat beds, soil formation in subaquatic deposits, prehistoric occupation levels in tidal creek levees, together with geological, geomorphological, ecological and other context information. Finally, SL change as a factor plays an important role in studies concerned with, for example, the geological, paleogeographical and paleoecological development of alluvial coastal plains and estuaries and their occupation history, changes in tidal range due to changing water depth and/or coastal configuration, exploration of mineral resources, and (see Barth & Titus, 1984) with coastal management and protection.

The SL market thus shows a coming and going of experts (drawing) from such (sub)disciplines as marine biology, geodesy, sedimentology, stratigraphy, geophysics, archaeology, coastal engineering, climatology, urban planning, coastal geomorphology, oceanography, anthropology, seismology, historical geography, geochronology, glaciology, and others.

The production and academic or applied consumption of SL change records and of appropriate results obtained within the forementioned areas of investigation constitute the 'SL research economy', schematically illustrated in Figure 3. It developed rapidly in the 1970s when the demand for more, higher quality, and better global distribution of SL data increased strongly with the introduction of model-based predictions of past and future worldwide relative SL changes (Walcott, 1972; Clark, Farrell & Peltier, 1978 among others), the presentation of the 'geoidal challenge' (Mörner, 1976), an increasing attention for seismotectonic events and earthquake prediction (eg. Yoshikawa, 1970; Kaizuka *et al.*, 1973; Pirazzoli, 1979), and a greater awareness of the influence

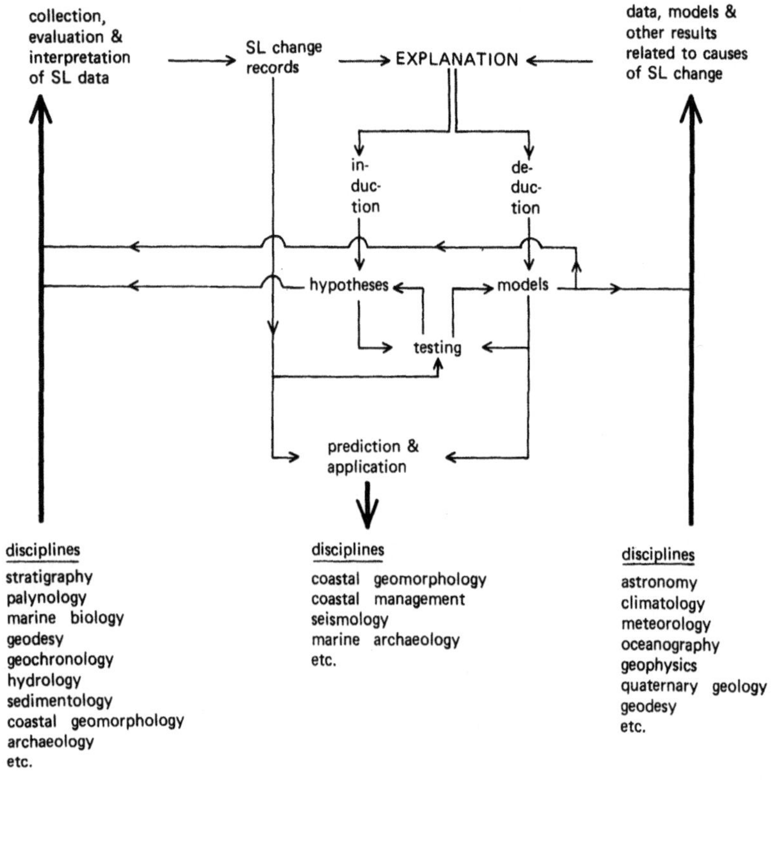

Figure 3 Scheme illustrating the 'sea-level research economy'

of local factors and error sources. The IGCP Projects 61 and 200, in cooperation with the Commission on Quaternary Shorelines of the International Union for Quaternary Research (INQUA), make (made) it part of their business to help meet this demand, to which Fairbridge (1961, p.137) had already drawn attention as '...the most urgent needs of Quaternary marine geology'.

This manual aims to stimulate the production of new, and reassessment of existing, SL data and graphs, and to do so on the basis of a developing body of knowledge regarding phenomenological evidence or indications of former SL positions and SL change and on systematic (methodic) consideration and evaluation of the factors that may reduce or seriously frustrate the comparability of results. Comparative analysis, accounting for differences and similarities between (SL or SL and other kinds of) data, is an important explanatory procedure in most stages of SL research. Hence, it is essential that such differences or similarities are real (true) and can be expected to pertain to the cause(s) in terms of which their explanation is sought. This introductory chapter deals with a number of aspects pertinent to the question of methodic comparability of SL data and graphs. The latter may not be sufficiently accurate or detailed (other important comparability requirements) to help solve certain problems, but if they lack definition, suffer from contamination with unrepresentative time- or space-data or have been inadequately evaluated with respect to errors of meaning, age and height determination, then the results have to be discarded as pieces of 'another' puzzle, or worse, the data or graphs may lead to erroneous conclusions and correlations, especially if they seem to 'fit'.

During the past thirty years, but particularly since the mid-seventies, many publications have appeared which, in part or in whole, implicitly or more explicitly, deal with the issue of method and quality of SL research: eg. Zeuner (1958), Fairbridge (1961), Fairbridge & Richards (1967), Gill (1968, 1971), Jardine (1975, 1981), Louwe Kooijmans (1976), Van de Plassche & Preuss (1979), Tooley (1978), Kidson & Heyworth (1979), Pinot (1979), Streif (1979), Behre & Streif (1980), Van de Plassche (1982), Devoy (1982), Tooley (1982) among others (see Kidson, this volume). However, the issue has been taken furthest by Heyworth & Kidson (1982), Kidson (1982, this volume) and Shennan (1982a,b, 1983), whose papers deserve careful study.

Comparative study of SL change is possible because: (i) SL can be defined as average tide levels *and* in terms of a vertical relationship between one of those levels and (a variety of) natural phenomena (SL indicators or markers), and (ii) SL change may lead to geological and other kinds of changes over larger or smaller areas (SL change indicators). Observations of both expressions of SL change (vertical change in position of SL indicators and areal/qualitative effects) can complement each other, provided the arguments are kept strictly separated. The reason for this requirement is that SL change is but one factor of geological, ecological or other changes and that vertical displacement of SL indicators may be due to environmental factors rather than a change in SL.

TERMS AND DEFINITIONS

1. Sea level

Records of past SL change are, until now, based either on regular measurements by means of tide gauges or on (interpreted) field evidence. Hence, a brief discussion of the nature of these records with respect to defining the term SL is in order.

a) Tide-gauge records

The simplest definition of SL is the height, relative to some datum, of the surface of the sea as measured *at a given place and time.* This so-called *instantaneous* SL is in continuous change, the most observable being those caused by surface waves and tides. Tide gauges, employing a damped flotation mechanism, record all but the very short-term (surface wave) variations of instantaneous SL.

'A continuous record of height of tide, accurately observed at fixed intervals of time, and referred to a stable bench mark' (Rossiter, 1962) provides the raw material for determining *average* SLs: mean tidal levels, such as mean high water, mean low water spring, etc, and MSL* (Table 1). An average SL is defined as the arithmetic mean of all (giving MSL), or of a set of specific (yielding MHW, MLWS, etc.) measurements at selected hours of wave-smoothed instantaneous SL, obtained at a *given place over a given period of time.* Thus, there are daily, monthly, yearly and x-annual records of MHW, MSL, etc. Clearly, a change of a given average SL can be determined only by comparing similar temporal averages.

A special average SL is mean or half tide level (MTL), which is the plane midway between MHW and MLW. MTL is of importance as a substitute for MSL (see below, and Jardine, this volume). Depending on the symmetry of the tidal curve, MTL occurs above or below MSL (Fig.4).

---

* World data on MSL are collected on behalf of the International Association of the Physical Sciences of the Ocean (IAPSO) and the International Union of Geodesy and Geophysics (IUGG) by a standing commission, the Permanent Service for Mean Sea Level. The data are collated and computerized at the Bidston Observatory of the British Institute of Oceanographic Sciences, situated near Liverpool (U.K.). The computer data bank of the PSMSL at present holds series from over 1000 stations. Data and other information are available free of charge to the scientific community.

Table 1 Tide levels

Tide level	Abbreviation	Exposure	Zone
highest astronomical tide	HAT		SUPRATIDAL ZONE
MEAN HIGH WATER SPRING	MHWS		
lowest high water spring	LHWS		
		wet at least once every monthly cycle	
mean higher high water	MHHW		
MEAN HIGH WATER	MHW		INTER-
MEAN HIGH WATER NEAP	MHWN		
lowest high water neap	LHWN		
		wet every tide of the year	
MEAN SEA LEVEL	MSL		
MEAN TIDE LEVEL	MTL		TIDAL
		exposed every tide of the year	
highest low water neap	HLWN		
MEAN LOW WATER NEAP	MLWN		
MEAN LOW WATER	MLW		ZONE
mean lower low water	MLLW		
		exposed at least once every monthly cycle	
highest low water spring	HLWS		
MEAN LOW WATER SPRING	MLWS		
Lowest low water spring	LLWS		
			SUBTIDAL ZONE

In areas with a small tidal range, water-level fluctuations resulting from meteorological conditions (pressure, wind) may far exceed in magnitude the tidal fluctuations between spring tides.

The first tide gauge was established in 1682 in Amsterdam. However, most tide-gauge records cover a much shorter time span. Only 112 stations have records from before 1900. For the study of SL changes in the more distant past and in areas where no tide gauges have been installed, one has to resort to evidence of naturally recorded SL positions, the kind of evidence dealt with in this manual. (Paleo-) SLs inferred from natural evidence may be called *derived;* ie. derived from (fossil) SL indicators and other field data.

b) Field-data records

Consider the following observations regarding the vertical relationships between average SLs and (i) upper limit of barnacles along the coast of New England, and (ii) the level of colonizing aquatic plants in SW Netherlands.

Barnacles. Observations on the New England coast (eastern USA) by Kaye (1964) have shown that the upper limit of barnacles varies between about local *annual* MHWN and 30

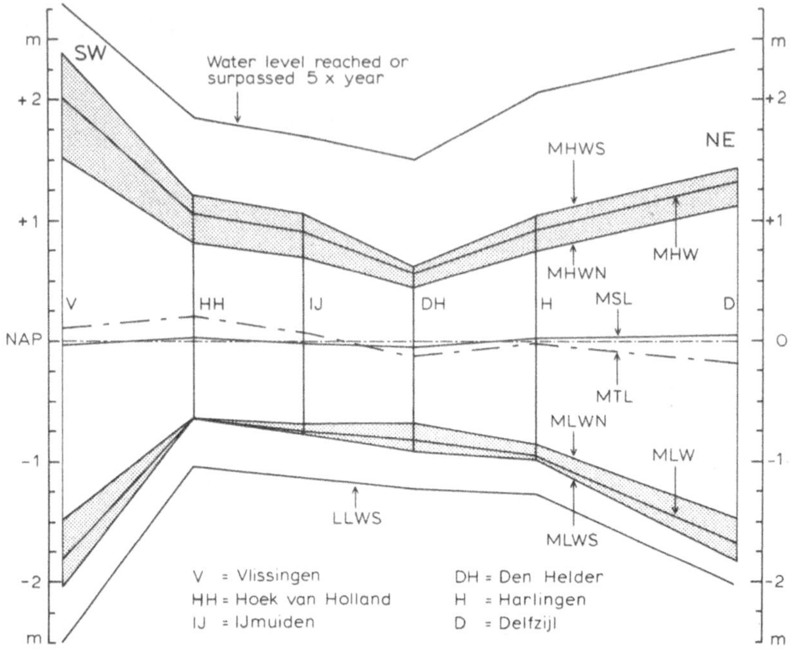

Figure 4 Differences in mean tidal levels (1971-1980) reached along the Netherlands coast (compiled from Getijtafels voor Nederland, 1985)

centimeters above local *annual* MHW, depending on (i) the mean calm-weather wave size, (ii) degree of protection from drying, and (iii) exposure to fresh water. The barnacle limit, in places that are well protected from waves, falls several centimeters short of annual MHW. In completely protected localities, where, during fair weather, waves are very small or non-existent, the level would be about annual MHWN. Where waves are unusually large, the barnacle limit has been noted at levels about 10 cm above MHW. Where the rock surface is in full shade and thus protected somewhat from drying, as in deep joints and crevices in the rock or at the bottom sides of rock overhangs, the barnacle limit has been observed to reach 30 cm above local annual MHW.

Colonizing vegetation. In the former brackish to fresh-water tidal area, mainly comprising the Haringvliet, the Hollands Diep and the Biesbosch (SW Netherlands), the level at which bare sand or mud areas were colonised by aquatic vegetation (mainly rushes) varied between 0.1 and 0.8 m below local *annual* MHW, depending on the salinity of the water. In the brackish environment this boundary was situated at 0.1 to 0.2 m below local annual MHW, whereas in the fresh-water environment of the Biesbosch it occurred at about 0.7 to 0.8 m below local annual MHW (De Glopper, 1973). Two classic studies on the ecology and water-level relationships of plant associations in tide influenced environments are those by Johnson & York (1915) and by Zonneveld (1960).

These two examples go to show that 'coastal' features (or some aspect thereof, eg. the upper limit ...) may have a more precise, site dependent and a less precise, overall vertical relationship to one or other local average SL. This is called the *reference water-level relationship*, or water-level relationship for short, of modern SL indicators. It is expressed in terms of an average SL or tide level (TL)-related groundwater level (the reference level), and a vertical range relative to that level, above, below or within which the indicator has been observed to occur, either in general or under specific local conditions. The vertical range may be called 'indicator range'.

The water-level relationship of physical, biological and other features originating in coastal areas imply that they possess *indicative meaning* with respect to present-day or former average SL or TL-related groundwater level (GWL) positions (hence the term SL indicator). The indicative meaning of a SL indicator is the 'reciprocal' of its water-level relationship. It is expressed in terms of the indicator, or a relevant or convenient aspect thereof, and a vertical range with respect to that indicator or aspect, above, below or within which the reference water level occurs or occurred given the water-level relationship of the indicator. The vertical range should be called 'reference water-level range', but is more conveniently referred to as the 'indicative range'. Thus, derived SL is defined by the indicative meaning of the indicator used. As with average SLs, a change in derived SL can only be established correctly if the reference water levels are of the same temporal mean.

In summary, direct measurement of SL yields instantaneous or average SL, while field evidence gives derived SL. All are place- and time-dependent. Average SL comprises the high and low TLs, MTL and MSL. These are the main reference water levels with respect to which the indicative range of indicators is to be expressed. Two other reference water levels are extreme TL and TL-related GWL. For the sake of discussion TL-related GWL is included in the general term SL.

2. Sea-level change

a) Space-time characteristics of SLs

Because definition and classification of SL change depend on the space-time characteristics of SLs, a brief and general discussion of these characteristics, which merit a separate chapter, is given first.

A fundamental requirement of quantitative comparison of SL change records from different sites or areas is, that the SL (reference water level) considered does not change its spatial configuration relative to the geoid with time.

As is generally known, all of the above mentioned reference water levels are in fact irregular surfaces with respect to the geoid, the global reference datum for height determination and itself highly irregular relative to the best fitting ellipsoid of rotation (Figs. 1 & 2) and none is stable over time (Montgomery, 1937; Pattullo *et al.*, 1955; Fairbridge, 1966; Pattullo, 1966; Wilson, 1966; Rossiter, 1967; Meade & Emery, 1971; Hamon, 1971; Wemelsfelder, 1971; Lisitzin, 1974; Mather, 1974, 1975; Lennon, 1975, Pugh & Alcock, 1976; Bray, 1977; Kjerve *et al.*, 1978; Figueiredo, 1980; Hamon & Godfrey, 1980; Molinari, 1980; Vanicek *et al.*, 1980; Gornitz *et al.*, 1982; Merry & Vanicek, 1983; Barnett, 1983; Ganeko, 1983; Blaha, 1984; Fu & Chelton, 1984; Maul *et al.*, 1985; Schott & Zantopp, 1985).

In terms of factor hierarchy, one can say that the geoid carries MSL which in turn carries the tidal wave (the TLs), from which characteristics MTL is derived, and which is one of the factors controlling TL-related GWL. The geoid, as a global reference surface for heights, is the equipotential surface of the earth's gravity field that best fits MSL in the spatial least square sense (Mather, 1975; Torge, 1980). The departure of MSL from the geoid is called sea-surface topography (SST).

The *geoid*, with its maximum relief difference of about 180 m (-105 and +75 m), is a function of the earth's rotation and mass anomalies. Changes in height occur over distances as large as thousands of kilometers (meters to tens of meters) to at least as small as 30 km (decimeters). Since the geoid is the reference surface for height determination, the spatial variation of geoid height is irrelevant for the question of comparability of SL data and graphs provided it is stable. The question 'on what temporal and spatial scales mass redistribution on and within the earth during the Quaternary, and earth rotational changes have affected the geoid?' remains to be answered.

MSL (or *SST*) is primarily a function of oceanographic, meteorologic and hydrologic factors: dynamic effects of currents (decimeters), temperature and salinity (density; decimeters), atmospheric pressure (centimeters), wind stress (decimeters), precipitation (centimeters-decimeters), and river discharge (centimeters-decimeters). Bathymetric configuration also is of influence. The effects of these factors is much larger near the coast than in the open ocean. The maximum relief difference of SST is ± 2m, established over thousands of kilometers distance. However, MSL differences of several decimeters can occur within several kilometers. The large scale features of SST are quasi-permanent. Otherwise, temporal scales of change vary from hours and days to hundreds or thousands of years. The short-term changes (say less than a year) are filtered out by tide gauges (average SLs), but should be taken into account when reference water levels and water-level relationships of indicators are to be established at places without a tide gauge nearby. For example, seasonal changes in wind stress, atmospheric pressure and precipitation can result in rather large differences in monthly MSL within a study area.

In addition to the factors governing MSL, *TLs* are a function of tidal range and river gradient. The spatial distribution of tidal range is controlled by a complex of factors of which coastal configuration, water depth and inclination of the moon play an important role. Tidal range is zero in amphidromic points; present-day maximum is reached at the head of the Bay of Fundy: about 15 m. It can change over rather short distances and the change in time can be gradual or sudden. The effect of the moon's nodal cycle (18.6 years) can be measured in decimeters (Fig. 5). Other cycles affect the tidal range without changing the spatial distribution. For example, the fortnightly spring-neap cycle, with neap tidal range usually being 10 to 30% less than mean tidal range, and the perigee-syzygy cycle (31.01 years), which may raise high TLs as much as 3 decimeters (Fairbridge, *pers. comm.*).

In addition to the factors influencing MSL, *MTL* is also a function of tide-curve symmetry (Fig. 4). This factor is of importance when, for the purpose of comparison, derived TL data have to be converted to derived MTL, by adding or subtracting half the derived, estimated or assumed former local mean tidal range. The overall difference between MSL and MTL within a study area may amount to a few decimeters.

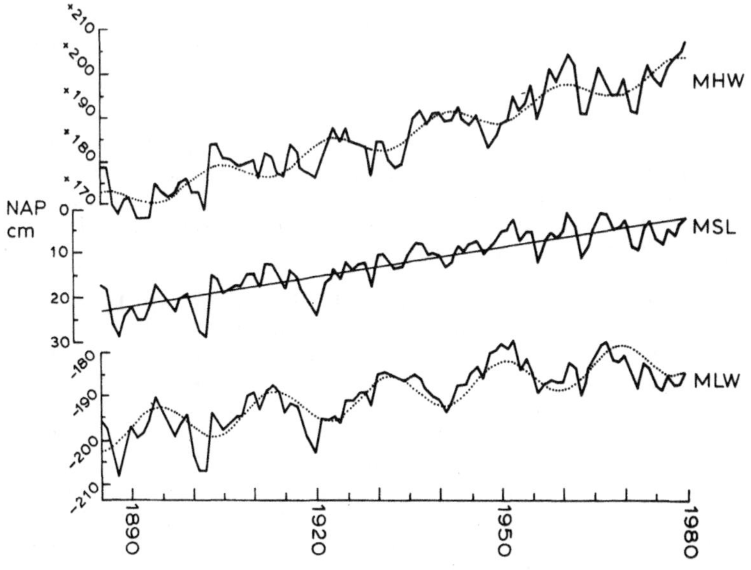

Figure 5 Moon's nodal cycle effect in annual MHW and MLW tide records for the period 1885-1980 at Flushing, S.W. Netherlands. This effect is negligible for MSL. The increase in tidal range at Flushing is probably due to dredging in the Wester Scheldt estuary. (After De Ronde, 1983)

TL-related GWLs are a function of a large number of factors such as tidal range, river gradient, permeability of subsurface sediments (seepage), morphology, size of sediment units, precipitation, drainage, etc., which can lead to water-table irregularities over short distances which, moreover, can be quite variable with time.

In conclusion, each of the SLs (surfaces) discussed above is characterized by *time-dependent topography*. Consequently, the definition of SL change based on data from one locality must include a conditional term for comparability with respect to time. Furthermore, because SL is necessarily defined with respect to location, the definition of SL change based on data from different sites must contain a conditional term for comparability of SL heights with respect to place. As for classification of SL change, a fundamental distinction is between whether the change is brought about by a change in SL topography or not.

b) SL change definitions

Instantaneous or average SL change, in the strict sense of the term, is the *net* change in height of the (wave-smoothed) surface of the sea or of MSL, MTL or a given TL(-related GWL), *averaged over comparable periods of observation* (time-representativity), between two 'moments' in time as measured at a given locality. (In the wider sense of the term, SL change refers to the approximated *course* of SL change with time and the spot(s) or area for which the change is considered to be representative.) A similar kind of definition applies for derived SL change, provided the SL indicators relate, or can be converted, to the same reference water level and have been obtained from the same locality. If the two samples come from different sites, they must (also) fulfill the requirement that the height of each (converted) derived SL is representative for the other sampling site (place-representativity). An SL graph based on a data set containing time- or place-unrepresentative derived SLs is *contaminated* by incomparable, though otherwise possibly correct, age-altitude information.

The above given definitions of SL change apply to situations in which the change can be quantified. Often, however, a change in SL is evident, but cannot be expressed quantitatively because derived SL heights are incomparable (due to, for example, compaction effects or because the derived SLs cannot be converted to the same reference water level) or there are no derived SL heights to be compared (eg. a decomposition horizon within a peat layer). In such a case derived SL change is expressed as the (positive or negative) *direction(s) of change* of a given reference water level prior to or since a point, or between two points, in time at or within a given locality or area.

c) Classification

The nature of SL change can be classified according to whether sea- and/or land-level changes are or have been involved. Recorded SL change resulting entirely from land-level movements is referred to as *apparent*, whereas it is called *real* (or absolute) in the case that no land-level changes are involved (at the recording site). The resultant of simultaneously occurring real and apparent SL changes is termed *relative* SL change, which term also applies to any (part of a) SL change record or graph for which the contribution of apparent SL change is unknown. The model studies by Walcott (1972), Chappell (1974), Clark *et al.* (1978) and others have convincingly demonstrated that even presumed stable coasts far away from glacial rebound areas experience(d) isostatic land-level movements. Hence, all 'raw' SL data or graphs must be considered as relative.

Furthermore, as alluded to above, real SL change can be subdivided into topographic and non-topographic, depending on whether or not the change is or was caused by a change in spatial configuration of the SL considered. Both kinds of real SL change can occur simultaneously at different spatial and temporal scales, ie. over greater or smaller distances, rapid or slow, continuous or intermittent, regular or irregular. Moreover, non-topographic SL change can lead to immediate or retarded topographic SL change.

A final remark concerns the term *eustasy*. The original meaning of eustasy is: world-wide change of SL. However, since it became clear that due to hydro-isostatic crustal movements and possible geoidal changes no SL graph could ever claim world-wide validity (representativity), the term has also been used synonymously with (local or regional) real SL change (eg. geoidal eustasy). This, however, prohibits the use of the term in an adjective sense. Geoidal SL change is a typical example of topographic SL change. The term eustasy exactly covers world-wide, non-topographic SL change. Therefore, the words eustasy and eustatic, even if 'only' conceptually useful, should be employed in their original sense. In physical-mathematical model studies of SL change, the term eustasy requires operational definition (eg. Walcott, 1972).

3. Land-level changes

The condition of place-representativity of derived SLs from different localities is inherent to topographic SL change, which, by definition, is differential in space. Rates of crustal movements (apparent SL change) also differ from place to place or area to area. Hence, the question of place-representativity is, in the choice of study area (size) or location of sampling site(s), one of the first to be considered in any SL research effort. In principle, differences in crustal movements between two areas or sites can only be established correctly from a comparison of relative SL graphs if each graph is based on data having the same place-representativity with respect to crustal movement history. Again, otherwise a data set becomes contaminated with incomparable apparent or relative age-altitude information.

DATA EVALUATION

Avoidable and unavoidable errors of indicative meaning, age and altitude determination are a problem for comparability of data when they remain undetected or unevaluated. If properly assessed, however, these errors do not formally affect comparability. But when it comes to establishing significant differences between (sets of) SL data, the after-taste of having correctly established the large, hence overlapping, error margins is bitter rather than sweet.

For some artificial flavour of error, take the case of coastal sites A and B which, during the past 5500 years, experienced a relative MSL rise of 1.27 m/ka and 1.63 m/ka respectively. Tidal amplitude (half the tidal range) has remained constant over time: 1 m. How accurately can we reconstruct the given MSL rise at A and B with 11 correctly dated samples for each site, one for every 500 years, and each sample representing a biological indicator known to occur only between MSL and MHW? To answer this question, we need to know where exactly between MSL and MHW each dated organism lived. Then, by considering only the depth of the samples, ie. pretending not to know the SL associated with each depth, one can draw error limits within which former MSL must have occurred.

We shall consider three cases (Fig. 6): (i) all samples from site A represent former MHW, those from site B represent former MSL (Figs. 6a & b), (ii) opposite to first case (Figs. 6c & d), and (iii) all samples from sites A and B represent various but regularly changing levels between MSL and MHW (Figs. 6e & f). Given the constant tidal range of 1 m, it is easy to see that former MSL must have occurred within a 1m band *below* each measured sample depth. In this way we obtain for each of the three cases two envelopes containing the (known) linear MSL rise at sites A and B.

Figures 1b and d show how the generated error bands tend to lead to an over- and underestimation respectively of the difference in rate of MSL rise at the two sites. Moreover, in the second case, the crossing of the error bands suggests the possibility of a sudden crustal movement that raised site B or lowered site A. The undulating error bands in the third case are very suggestive of opposing changes in rate of MSL rise: it takes objective judgement not to overlook the possibility that the fluctuating envelopes contain two linear MSL rise histories.

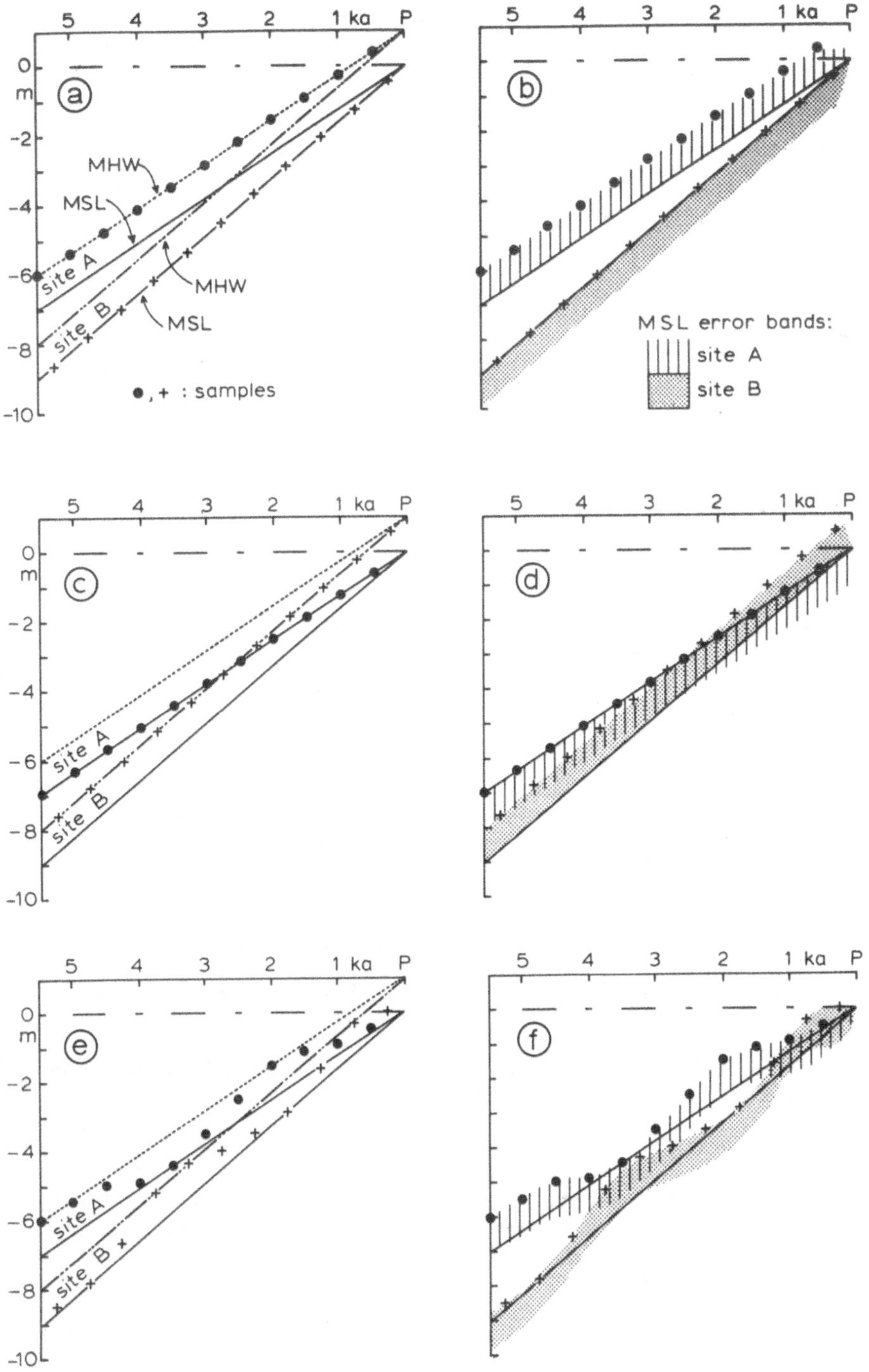

Figure 6. Exercise to reconstruct, under controlled error-band generation, given linear relative MSL rise at two coastal sites. Explanation in text.

This simple exercise of SL rise reconstruction under controlled error generation may serve to illustrate that even with the above given data it is not without problems to accurately approach, separate, or even recognise the given linear SL rise at sites A and B. In other words, you may ask yourself: what if ...

- the water-level relationship of an indicator is based on insufficient or erroneous observations?
- the water-level relationship of an indicator has changed or is variable over time?
- an indicator has not been identified correctly?
- the variance of the reference water level itself is ignored?
- it is not recognised that the indicator is not in place? or that
- it has been subject to secondary local vertical displacement?
- it is unknown whether the tidal amplitude or symmetry of the tidal curve has remained constant, or if it is probable that tidal range has changed?
- it is ignored that a reference TL must also have been influenced by other factors than those controlling local tidal amplitude?
- it remains unnoticed that samples for radiometric dating are contaminated or otherwise influenced?
- the age of an indicator not consisting of dateable material is derived from samples with doubtful age relationship to the indicator?

This list of rhetorical questions can be expanded, but the above refer to the main sources of error that have to be taken into account in assessing reliability or accuracy of indicative meaning and range, height and age of SL indicators. Clearly, failure to do so seriously reduces the reliability, hence comparability, of results. Since most of the following chapters and many other publications are directly or indirectly concerned with these various error sources, they need not be dealt with here. A few general remarks may suffice.

As stated earlier, careful collection and evaluation of data is in itself no guarantee for comparability of results in the sense that too large error margins, especially vertical, may render it impossible to establish significant within- or between-area differences between sets of SL data or relative MTL graphs. Although the effect of error margins is smaller the larger the temporal scale of investigation, an effort should be made to obtain highest possible accuracy for future use in more detailed studies.

Apart from errors of height determination, which can be considerable, the three main sources of vertical error arise from determining the indicative range, conversion from a reference TL or GWL to MTL, and correction for compaction. The compaction problem should be avoided as much as possible when using SL indicators to study SL change; only compaction-free or -poor data yield workable results.

The problem is assimilated when SL change is expressed in directional sense only. One should, however, not overlook the possibility that indications for (continued) local SL rise resulted from compaction and consolidation.

The water-level relationship of an indicator can be a major source of error. With regard to the choice of reference level it is important that, if the position of the indicator relative to MSL is a function of tidal amplitude, the reference water level must be some high or low TL to allow for spatial and temporal variation in local tidal range (see, for example, Sollid *et al.*, 1973).

Ideally, a water-level relationship should be based on long-term and widespread measurements and observations to account for temporal variability of the reference water level and to know and understand the height variability of the indicator as a function of environmental and other factors. In practice, however, this ideal may not be (have been) feasible. References and/or the observational data base should therefore always be mentioned and taken into account when determining or evaluating the indicative range of a fossil SL indicator. However, even if a water-level relationship of a modern indicator has been established satisfactorily, it may not be fully applicable to its fossil counterpart if one or more of the factors determining the indicator range have changed with time. For example, the depth distribution of burrowing organisms is, in certain cases, a function of slope of the sediment surface. Therefore, if presentday slopes in a given area are, due to a change in sediment input, steeper or flatter than in the past, actual observations on the water-level relationship of these organisms may lead to smaller or larger errors when applied to fossil burrows of the same species. In water-level relationships it is important to understand the function of each variable. Finally, another case is that where no modern counterparts of a fossil indicator are present in or near the study area or no published information on comparable situations elsewhere is available. One may then proceed by assuming a water-level relationship, provided the assumption can be tested. In most, but in these cases in particular, systematically and carefully collected suites of time-overlapping data representing different conditions and/or indicators provide the best basis for evaluation and interpretation. This also applies to the problem of conversion from a derived TL(-related GWL) to derived MTL.

Due to SL topography and the possibility of topographic SL change, between-area study of relative SL change has to proceed in terms of the least irregular SL (surface), which is (X-) annual MSL. Since the indicative meaning of many SL indicators pertain to a TL(-related GWL) as reference water level, convertion to derived MSL is required. This, however, assumes knowledge of the local paleo-mean tidal range, paleo-tide curve symmetry and all other factors that controlled the local height of the reference water level.

Of these, the symmetry of the former tide curve is practically impossible to reconstruct. Therefore, derived MTL is the best possible approximation of derived MSL one might achieve. The influence of former within-area MSL topography, if present, remains practically indeterminable. Usually, the local paleo-mean tidal range can only be assumed or estimated with (relatively) large error margins, thus further increasing uncertainty or inaccuracy. The influence of other factors on the local reference water-level height is often equally hard to quantify. These problems are especially hard to deal with in the case that single sites yield no more than one or two samples (SL index points), though a complete suite of SL index points from one site, while ideal in many respects, may also be difficult or impossible to interpret in terms of local tidal range and contribution by other factors. In suitable study areas, however, smaller or larger series of strategically and systematically collected, time-overlapping dates from different localities, may aid interpretation (time-altitude-space analysis; Van de Plassche, 1982). The choice of sampling sites should then be based on a testable idea or hypothesis regarding the (changing) paleo-SL topography(ies) within (parts of) the study area (direction of increase or decrease of tidal range, GWL height, etc.). Model calculations can be quite useful too (Tooley, 1985).

Finally, radiocarbon dates often are or have to be taken for what they are for the reason that no means of evaluation exists *or has been created.* It is, however, most important that possibilities to evaluate the 'comparability' of radiometric datings are used to the full (cross-check with pollen-analytical dating, other dateable materials, twin samples, series of dates, etc.).

TEMPORAL ANALYSIS OF SEA-LEVEL CHANGE

The discussion so far has centered primarily on the quantitative vertical aspect of derived SLs and SL change. The difficulties and errors of establishing the indicative range and (primary) height of indicators as well as of convertion-to-MTL values may be considerable. Not surprisingly, therefore, the more convincing and useful conclusions in many a SL paper concern, or are based on a comparison of, the *time* at which changes in rate or sign of SL occurred. Neither is it surprising that researchers developed concepts and terminology for temporal analysis of SL data, pointing out possibilities and limits for exploiting otherwise useless or difficult to use field evidence of local, hence potentially regional, SL change (eg. Geyh, 1969, 1971; Roeleveld, 1974; Tooley, 1978; though especially Streif, 1979 and Shennan, 1980, 1982a,b, 1983). Or that, as a result, SL students have seen or will see the necessity to recall or revise earlier conclusions.

That all of the forementioned authors live and work in NW Europe largely stems from the fact that the 'vertical-error problem' is typical for alluvial coastal plains, which have a wide occurrence in that region. Built up of interfingering, compaction-susceptible organic and clastic deposits, their water-level relationships are often problematic and the space-time distribution is certainly not a function of (positive and negative) SL change alone. According to Streif (1979), the 'areas of pure clastic sediments or pure sedentary peat offer only limited possibilities for reconstructing the tendencies and velocities of sea-level fluctuations'. However, apart from the fact that these limited possibilities may be extremely important, he is careful to point out that although areas with interfingering organic and brackish to marine clastic deposits 'are often regarded as zones in which facies changes directly mirror the vertical changes of sea level and related groundwater level, ... the vertical changes of sea level are only one component among a great variety of factors which influence the paleogeographic tendencies of a retreating or prograding coastline'. Hence, it is 'misleading to interpret each regressive overlap as an indication of a sinking sea level'. Clearly, temporal analysis of SL change requires large and varied data bases, both in numbers and areal coverage, to reliably determine time-space distributions of significant phenomena such as regressive and transgressive overlap (see below) of clastic and organic deposits and the evidence of primary and secondary changes within these deposits (eg. species succession, see Menke, 1969; soil formation in subaquatic deposits, peat decomposition layers, etc.; see also Behre & Streif, 1980). The analysis increases in strength the better it is supported by geological mapping and knowledge of paleogeographic conditions.

Shennan (1980, 1982a,b, 1983) has extended Streif's stepping stones into an operationally defined route of temporal SL analysis, though with connecting lanes to the time-altitude road. He has argued repeatedly that, with adherence to its rules and signs of definition, scale and error, this road has a definite tendency to become (too) many possibilities wide, but without such adherence, narrows down to a way-at-will leading nowhere (see also Tooley, 1982; Shennan *et al,* 1983 and Shennan, in press). 'While it is unwise to interpret a single sea-level index point as representing a regionally significant process, each sea-level index point will reflect a tendency of sea-level movement. Cumulative tendencies of radiocarbon-dated sea-level index points showing the same direction of tendency of sea-level movement ... can be used to identify periods dominated by positive ... and negative tendencies of sea-level movement' (Shennan, 1982b). Statistical analysis (eg. time series, frequency histograms) of sufficiently large numbers of dated samples, with or without indicative meaning in terms of signs of local SL change, gives chronologies of tendencies of SL movement or of trans- and regressive overlap (*descriptive* terms of lithostratigraphic change irrespective of process), which can be correlated with similar data from other areas or considered together with time-altitude information from the same study locations. For details a reader should refer to the above-mentioned papers.

Temporal analysis of SL change is an indispensable (additional) approach, especially in vertical-error-sensitive study areas. It hinges rather heavily on the adequacy of the data base as regards numbers (Shennan mentions 40 dates/1000 years) and indicative strength of the evidence. Not every SL index point may be given a tendency of SL movement interpretation, and to do so forcefully would be to inter-change ends and means. Moreover, just as SL indicators differ widely in indicative value (accuracy), SL change indicators (often a complex of two or more SL indicators) vary in 'weight' from incontrovertible to weakly tendential; qualitative differences which have to be taken into account in the (statistical) analysis. Again, it is important to consider SL change indicators within the context of the geological development of the area. For example, a clear indicator of negative local SL change, such as a decomposition horizon within a peat bed, may be related to drainage of the peat by tidal gullies that penetrated the area as a result of SL rise.

SUMMARY AND CONCLUSION

'Sea-level studies have lacked an accepted methodology and raw data ... purporting to record marine events lack homogeneity and interfere with subsequent analysis and interpretation' (Tooley, 1978, 1985). As a result 'Meaningful comparisons of regional sea-level bands are not yet feasible' (Shennan, 1982a). Indeed, within- and between-area comparability of SL data and graphs is a prime *sine qua non* for the study of SL change. Comparability of SL data (sets) is principally a function of definition of terms, representativity of data with respect to time and space, evaluation of errors and uncertainties in determining indicative meaning, age and altitude of SL indicators (or only the first two for SL change indicators), and, depending on the temporal scale of study, degree of detail and accuracy.

The continuously changing height of the sea surface (instantaneous SL) is 'tamed' by means of tide-gauge measurements (average SL). For the study of SL change in the more distant past, the height range of present-day, natural or man-made, 'coastal' phenomena is related to average SLs (water-level relationship of SL indicators), from which, in fossil condition, former average SLs or SL-related GWLs can be determined (indicative meaning, derived SLs).

Because the height of an average SL depends on the period of observation, a change in average or derived SL can only be established correctly by comparing time-representative data. Each average SL (surface) has topography that changes or may change with time (topographic SL change). Consequently, a change in relative SL based on data from different localities can only be established correctly if the data are mutually site-representative and, because rates of land-level change also vary spatially, have not been subject to differential crustal movements (contamination with unrepresentative time-altitude information). This should be kept in mind when deciding on study area location and size and/or choice of sampling sites. The area for which a SL index point or a SL graph is considered representative should always be indicated (place dependency).

In time-altitude analysis of SL change, vertical error parameters are: determination of altitude; indicative meaning; range and primary height of indicator (compaction); and conversion to derived MTL. Problems of indicative meaning or range can develop from incorrect recognition or identification of the indicator, incomplete or incorrect establishment of the water-level relationship of an indicator, change of indicator range with time, or absence of modern counterparts for fossil indicator. Conversion to derived MTL presupposes information on the local paleo-mean tidal range, tide-curve symmetry and all other factors that controlled the local height of the reference water level. Where possible, time-altitude analysis should be expanded to or complemented by time-altitude-space or temporal analyses. In some cases, the latter approach may be the only one feasible.

To conclude, the demand for more, more detailed, better and better global distribution of SL data and graphs, together with the variety and complexity of SL change causes, make the primary occupation and objective of SL students (to obtain and explain differences among SL records) a hard, but increasingly rewarding excercise in postponed satisfaction, provided they produce pieces for the same puzzle.

ACKNOWLEDGEMENTS

This chapter has benefited from a draft paper by Rhodes W. Fairbridge on 'Spectra of sea level'. Drs. Chriet Titulaer is thanked for permission to reproduce Figure 1. Full-grant support was obtained from the Netherlands organisation for the advancement of pure research (H75-262). Lines at chapter heading are taken and partly adapted from the song 'Once in a lifetime' by Talking Heads.

REFERENCES

Barnett, T.P., 1983. Recent changes in sea level and their possible causes. *Climatic Change*, **5**, 15-38

Barth, M.C. & Titus, J.G. (eds.), 1984. *Greenhouse effect and sea level rise. A challenge for this generation.* (Van Nostrand Reinhold, New York)

Behre, K.-E. & Streif, H., 1980. Kriterien zu Meeresspiegel- und darauf bezogene Grundwasserabsenkungen. *Eiszeitalter und Gegenwart*, **30**, 153-160

Blaha, J.P., 1984. Fluctuations of monthly sea level as related to the intensity of the Gulf Stream from Key West to Norfolk. *Journal of Geophysical Research*, **89**, **C5**, 8033-8042

Bray, J.A., 1977. Preliminary differences in mean water level between tide gauges along the South American Pacific coast. *Marine Geodesy,* **1(2)**, 177-197

Cazenave, A. & Dominh, K., 1985. Le géoide marin: altimétrie par satellite et structure des océans. *Géochronique,* **14**, 14-17

Chappell, J., 1974. Late Quaternary glacio- and hydro-isostasy on a layered earth. *Quaternary Research,* **4**, 429-440

Clark, J.A., Farrell, W.E. & Peltier, W.R., 1978. Global changes in postglacial sea level: a numerical calculation. *Quaternary Research,* **9**, 265-287

Devoy, R.J.N., 1982. Analysis of the geological evidence for Holocene sea-level movements in southeast England. *Proceedings of the Geologists' Association,* **93**, 65-90

Fairbridge, R.W., 1961. Eustatic changes in sea level. *Physics and Chemistry of the Earth,* **5**, 99-185

Fairbridge, R.W., 1966. Mean sea level changes, long-term-eustatic and other. In: *Encyclopedia of Oceanography,* ed. R.W. Fairbridge, (Reinhold Publishing Company, New York), 479-485

Fairbridge, R.W. & Richards, H.G., 1967. The INQUA Shorelines Commission. *Zeitschrift für Geomorphologie,* N.F., **11**, 205-215

Figueiredo Jr., A.G., 1980. Response of water column to strong wind forcing, southern Brazilian inner shelf: implications for sand ridge formation. *Marine Geology,* **35**, 367-376

Fu, L.-L. & Chelton, D.B., 1984. Temporal variability of the Antarctic circumpolar current observed from satellite altimetry. *Science,* **226**, 343-346

Ganeko, Y., 1983. A 10' x 10' detailed gravimetric geoid around Japan. *Marine Geodesy,* **7(1-4)**, 291-314

Getijtafels voor Nederland, 1985. Rijkswaterstaat, Direktie Waterhuishouding en Waterbeweging, (Staatsuitgeverij, 's-Gravenhage).

Geyh, M.A., 1969. Versuch einer chronologischen Gliederung des marinen Holozans an der Nordseekuste mit Hilfe der statistischen Auswertung von ^{14}C-Daten. *Zeitschrift der deutschen geologischen Gesellschaft,* **118**, 351-360

Geyh, M.A. 1971. Middle and young Holocene sea-level changes as global contemporary events. *Geologiska Föreningens Stockholm Förhandlingar,* **93**, 679-690

Gill, E.D., 1968. Multiple factors in the description of Quaternary shorelines. *Proceedings of the VII INQUA Congress,* **8**, 375-379

Gill, E.D., 1971. The Paris symposium on world sea levels of the past 11 000 years. *Quaternaria,* **14**, 1-6

Glopper, R.J. de, 1973. Subsidence after drainage of the deposits in the former Zuyder Zee and in the brackish and marine forelands in The Netherlands. *Van Zee tot Land,* **50**, 205pp

Gornitz, V., Lebedeff, S. & Hansen, J., 1982. Global sea level trend in the past century. *Science,* **215**, 1611-1614

Hamon, B.V., 1970. A review of 'current effect' on sea level. In: *Report on the symposium on coastal geodesy,* Munich 1970, ed. R.S. Sigl, (Institute for Astronomical and Physical Geodesy, Munich), 395-410

Hamon, B.V. & Godfrey, J.S., 1980. Mean sea level and its interpretation. *Marine Geodesy,* **4(4)**, 315-329

Heyworth, A. & Kidson, C., 1982. Sea-level changes in southwest England and Wales. *Proceedings of the Geologists' Association,* **93**, 91-111

Jardine, W.G., 1975. The determination of former sea levels in areas of large tidal range. In: *Quaternary Studies,* eds. R.P. Suggate & M.M. Cresswell, (Royal Society of New Zealand, Wellington), 163-168

Jardine, W.G., 1981. Holocene shorelines in Britain: recent studies. *Geologie en Mÿnbouw,* **60**, 297-304

Jardine, W.G., (this volume) Determination of altitude

Johnson, D.S. & York, H.H., 1915. The relation of plants to tide levels. A study of factors affecting the distribution of marine plants. *Carnegie Institute, Washington Publication,* **206**, 162pp

Kaizuka, S., Matsuda, T., Nogami, M. & Yonekura, N., 1973. Quaternary tectonic and recent seismic crustal movements in the Arauco Peninsula and its environs, Central Chile. *Geographical Reports, Tokyo Metropolitan University,* **8**, 1-49

Kaye, C.A., 1964. The upper limit of barnacles as an index of sea-level change on the New England coast during the past 100 years. *Journal of Ecology,* **72**, 580-600

Kidson, C., (this volume) Sea level changes in the Holocene

Kidson, C. & Heyworth, A., 1979. Sea "level". *Proceedings of the 1978 International Symposium on Coastal Evolution in the Quaternary, Sao Paolo, Brazil,* 1-27

Kjerve, B., Greer, J.E. & Crout, R.L., 1978. Low-frequency response of estuarine sea level to non-local forcing. In: *Estuarine Interactions,* ed. M.L. Wiley, (Academic Press, New York), 497-514

Lennon, G.W., 1975. Coastal geodesy and the relative movements of land and sea levels. In: *Geodynamics today,* 97-104

Lisitzin, E., 1974. *Sea level changes.* (Elsevier, Amsterdam)

Louwe Kooijmans, L.P., 1976. Prähistorische Besiedlung im Rhein-Maas-Deltagebiet und die Bestimmung ehemaliger Wasserhöhen. *Probleme der Küstenforschung im Südlichen Nordseegebiet,* **11**, 119-143

Mather, R.S., 1974. Quasi-stationary sea-surface topography and variations of mean sea level with time. *Unisurv,* **G21**, University of New South Wales, Sydney, 18-72

Mather, R.S., 1975. Mean sea level and the definition of the geoid. *Unisurv G., Australian Journal of Geodesy and Photo Survey,* **23**, 68-79

Maul, G.A., Chew, F., Bushnell, M. & Mayer, D.A., 1985. Sea level variation as an indicator of Florida Current volume transport: comparisons with direct measurement. *Science,* **227**, 304-307

Meade, R.H. & Emery, K.O., 1971. Sea level as affected by river runoff, Eastern United States. *Science,* **173**, 425-427

Menke, B., 1969. Vegetationsgeschichtliche Untersuchungen und Radiocarbon-Datierungen zur holozänen Entwicklung der schleswig-holsteinischen Westküste. *Eiszeitalter und Gegenwart,* **20**, 35-45

Merry, C.L. & Vanicek, P., 1983. Investigation of local variations of sea-surface topography. *Marine Geodesy,* **7(1-4)**, 101-125

Molinari, R.L., 1980. Current variability and its relation to sea-surface topography in the Carribbean Sea and Gulf of Mexico. *Marine Geodesy,* **3**, 409-436

Montgommery, R.B., 1937. Fluctuations in monthly sea-level on eastern U.S. coast as related to dynamics of western North Atlantic. *Journal of Marine Research,* **1**, 165-185

Mörner, N.-A., 1976. Eustasy and geoid changes. *Journal of Geology,* **84**, 123-151

Pattullo, J.G., 1966. Mean sea level. In: *Encyclopedia of Oceanography,* ed. R.W. Fairbridge, (Reinhold Publishing Company, New York), 475-479

Pattullo, J.G., Munk, W., Revelle, R. & Strong, E., 1955. The seasonal oscillation in sea level. *Journal of Marine Research,* **14**, 88-156

Pinot, J.P., 1979. Méthodes et techniques permettant de determiner l'altitude relative des niveaux marins anciens en fonction de l'altitude absolue actuelle des divers indicateurs. In: *Les Indicateurs de Niveaux Marins,* ed. Groupe Nivmer, Oceanis, **5**, 335-355

Pirazzoli, P.A., 1979. Encoches de corrosion marine dans l'arc Hellenique. In: *Les Indicateurs de Niveaux Marins,* ed. Groupe Nivmer, Oceanis, **5**, 327-334

Plassche, O. van de, 1982. Sea-level change and water-level movements in The Netherlands during the Holocene. *Mededelingen Rijks Geologische Dienst,* **36(1)**, 1-93

Plassche, O. van de & Preuss, H., 1979. Explanatory guidelines for completion of the computer form for sample documentation (IGCP Project 61). *NIVMER Information*, **4**, 103-108

Pugh, D.T. & Alcock, G.A., 1976. Gradients on the mean and instantaneous sea surface level in a shallow water region. *International Hydrographic Review*, **53(2)**, 29-39

Roeleveld, W., 1974. *The Groningen coastal area. A study in Holocene geology and lowland physical geography*. Ph.D. thesis, Free University, Amsterdam. Also in: *Berichten van de Rijksdienst voor het Oudheidkundig Bodemonderzoek*. **20-21**, 1970-71, 7-25 and **24**, 1974, Supplement

Ronde, J.G. de, 1983. Changes of relative mean sea-level and of mean tidal amplitude along the Dutch coast. In: *Seismic risk in the offshore North Sea area*, eds. A.R. Ritsema & A. Gürpinar, (Reidel Publishing Company, Dordrecht), 131-141

Rossiter, J.R., 1962. Longterm variations in sea level. In: *The Sea*, **1**, (Wiley), 590-610

Rossiter, J.R., 1967. An analysis of annual sea level variations in European waters. *Geophysical Journal of the Royal Astronomical Society*, **12**, 259-299

Schott, F. & Zantopp, R., 1985. Florida Current: seasonal and interannual variability. *Science*, **227**, 308-311

Shennan, I., 1980. *Flandrian sea-level changes in the Fenland*. Unpublished Ph.D. thesis, University of Durham

Shennan, I., 1982a. Interpretation of Flandrian sea-level data from the Fenland. *Proceedings of the Geologists' Association*, **93**, 53-63

Shennan, I., 1982b. Problems of correlating Flandrian sea-level changes and climate. In: *Climatic Change in Water Prehistory*, ed. A.F. Harding, (Edinburgh University Press), 52-67

Shennan, I., 1983. Flandrian and Late Devensian sea-level changes and crustal movements in England. In: *Shorelines and Isostasy*, eds. D.E. Smith & A.G. Dawson, (Academic Press, London), 255-283

Shennan, I., in press. Global analysis and correlation of sea-level data. In: *Sea-surface Studies: A Global View*, ed. R.J.N. Devoy, (Croomhelm Ltd. Publishers, Beckenham)

Shennan, I., Tooley, M.J., Davis, M.J. & Haggart, B.A., 1983. Analysis and interpretation of Holocene sea-level data. *Nature*, **302**, 404-406

Sollid, J.L., Andersen, S., Hamre, N., Kjeldsen, O., Salvigsen, O., Sturød, S., Tveita, T. & Wilhelmsen, A., 1973. Deglaciation of Finnmark, North Norway. *Norsk Geografisk Tidsskrift*, **27**, 233-325

Streif, H., 1979. Cyclic formation of coastal deposits and their indications of vertical sea-level changes. In: *Les Indicateurs de Niveaux Marins,* ed. Groupe Nivmer, Oceanis, **5**, 303-306

Tooley, M.J., 1978. *Sea-level changes in North-West England during the Flandrian Stage.* (Clarendon Press, Oxford)

Tooley, M.J., 1982. Sea-level changes in northern England. *Proceedings of the Geologists' Association,* **93**, 43-51

Tooley, M.J., 1985. Sea levels. *Progress in Physical Geography,* **9(1)**, 113-120

Torge, W., 1980. *Geodesy.* (Walter de Gruyter, Berlin)

Vanicek, P., Castle, R.O. & Balazs, E.I., 1980. Geodetic levelling and its applications. *Reviews of Geophysics and Space Physics,* **18(2)**, 505-524

Walcott, R.I., 1972. Past sea levels, eustasy and deformation of the earth. *Quaternary Research,* **2**, 1-14

Wemelsfelder, P.J., 1970. Sea level observations as a fact and as an illusion. In: *Report on the Symposium on Coastal Geodesy,* Munich, 1970, ed. R.S. Sigl, (Institute for Astronomical and Physical Geodesy), 65-80

Wilson, B.W., 1966. Seiche. In: *Encyclopedia of Oceanography,* ed. R.W. Fairbridge, (Reinhold Publishing Company, New York), 804-817

Yoshikawa, T., 1970. On the relations between Quaternary tectonic movement and seismic crustal deformation in Japan. *Bulletin of the Department of Geography, University of Tokyo,* **2**, 1-24

Zeuner, F.E., 1958. Criteria for the determination of mean sea-level for Pleistocene shore-line features. *Quaternaria,* **5**, 143-147

Zonneveld, I.S., 1960. De Brabantse Biesbosch. A study of soil and vegetation of a fresh water tidal delta. *Verslagen Landbouwkundige Ouderzoekingen,* (Landbouw Hogeschool, Wageningen), 65.20, 3 volumes

2

Sea-level changes in the Holocene*

C. KIDSON

ABSTRACT

Over the last 30 years the emphasis in studies of the recovery from low last glaciation sea levels has changed significantly. The search for a eustatic sea-level curve having global relevance has ended. Studies into the rheology of the earth's crust, and recognition that the geoid has not remained stable over time, have resulted in the recognition that there must have been regional differences in eustatic response to deglaciation. As a part of this re-appraisal there has been a growing appreciation that crustal isostatic response to the removal of the weight of ice sheets has been accompanied by a consequential *hydro-isostatic* response, particularly in the areas of the shelf seas. In the later part of the post-war period attention has additionally been focussed on the much greater potential for error over the whole field of palaeoenvironmental reconstruction in the Holocene, including not only errors in dating, but also a large number of possible sources of errors in heighting. As a result of this, an increasing number of scientists are withholding judgement on the nature of sea-level rise and, more particularly, on the problem of higher than present late Holocene eustatic sea levels. The problems outstanding in the early 1960s have not yet been resolved but the bases of uncertainty have changed.

INTRODUCTION

The present or 'post-Daly' (Daly, 1920; 1934) phase of the study of the marine transgression which followed the melting of the ice of the last (Weichselian, Würmian, Wisconsinan, Devensian) glaciation began in earnest with the work of

---

*Reproduced with permission of Pergamon Press.

Fairbridge (1948, 1950) and Teichert (1950) in Western Australia. In the following decade, partly as a result of the stimulus given by the discovery of the ^{14}C dating method, a number of significant papers (Emery *et al.*, 1954; Shepard & Suess, 1956; Godwin *et al.*, 1958; Curray, 1960, 1961; Jelgersma & Pannekoek, 1960; Shepard, 1960; Schofield, 1960) showed, not only a world-wide interest in the phenomenon, but also an assumption that, since the oceans are interconnected, the change of ocean level would follow a universal pattern. Scientists around the world recognised that tectonic and isostatic adjustments would produce different relative (SL) curves in less stable regions. They assumed, however, that, if allowance could be made for crustal movements, these areas, together with the 'stable' regions, would conform to a common *eustatic* curve of SL change.

Two bench-mark review papers followed in the early 1960s. Fairbridge (1961) argued that the post-glacial rise of the sea was spasmodic and included regressive as well as transgressive phases. Additionally he argued that there were a number of stands of the sea in the late Holocene at heights up to 3.7 m (12 feet) above the present level at about 5700, 4900, 3700, 2400, 2200 and 1000 BP. Shepard (1963) indicated that there was general agreement on the nature of SL change in the late-glacial but that, in the later post-glacial, this agreement broke down. He argued that in this later period the rise in SL was a continuous one, at a rate diminishing in time but going on to the present day. He did not entirely rule out the possibility of a slightly higher than present late Holocene level but considered it to be unproven and drew attention to the fact that all Fairbridge's evidence for such higher levels came from Australia which remained, in his words, an 'enigma'.

There followed a sequence of papers from around the world which showed a broad division into either Fairbridge or Shepard schools of thought. Gill (1961), Mörner (1969a), Hopley (1971), Gill & Hopley (1972), Fedorov (1972), Ters (1973) and Tooley (1974) all produced evidence from their respective regions supporting a SL sequence as envisaged by Fairbridge. Conversely Jelgersma (1961), Bloom & Stuiver (1963), Bloom (1967), Thom *et al.*, (1969), Hails & Hoyt (1971) and Kidson & Heyworth (1973) among others, claimed to demonstrate the smooth exponential nature of SL rise and denied that the late Holocene sea had ever exceeded that of the present. Jelgersma (1966), in a review of SL studies, discerned a third group favouring a steadily rising sea during the Holocene reaching its present level between 5000 and 3600 BP and subscribing to the *standing SL after 3600 BP theory* (Fisk, 1951; Godwin *et al.*, 1958; McFarlan, 1961). Guilcher (1969) drew attention both to the equivocal nature of much of the evidence and to the fact that even some of the leading disputants in the controversy did not regard the problems as solved.

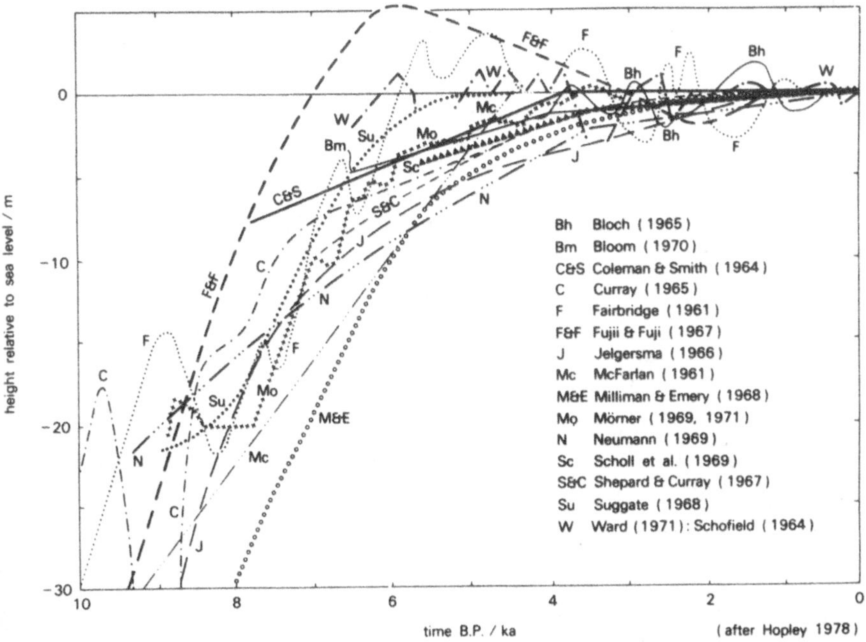

Figure 1. Different interpretations of sea-level history in the last 10 000 years.

In the 1970s a large number of time depth diagrams, or *SL curves*, were constructed from around the world, all reflecting, in the context of eustasy as a world-wide phenomenon, these wide differences of approach. Nowhere was the debate fiercer than in Australia. Thom *et al*. (1969) argued against SLs higher than present in Eastern Australia. Gill & Hopley (1972) replied with claims of definite evidence for such higher levels in Queensland and Victoria but Thom *et al*., (1972) argued that they had used equivocal evidence and mis-interpreted some of it without appreciating all the difficulties of interpretation involved. The debate continues with Belperio (1979), for example, giving additional evidence against higher Holocene SLs and Hopley (1980) replying.

EUSTASY AS A REGIONAL PHENOMENON

One of the earliest activities of Project No.61 of the International Geological Correlation Programme (IGCP 61) *Sea-level Movements during the Last Deglacial Hemicycle (About 15 000 years)*, scheduled to run from 1974 to 1982, was the production of an Atlas of Sea-Level Curves (Bloom 1977). This served to show that attempts to compare curves from different parts of the world emphasised their differences rather than the similarities which might have been expected from a world-wide eustatic event. One such comparison (Hopley, 1978) (Fig. 1) illustrates the near futility of such exercises in relation to the low amplitude but critical fluctuations around which the debate has been centred. Even an attempt to make the comparison meaningful, by separating groups of curves on a regional basis (Moriwaki, 1978) (Fig. 2) serves only to re-emphasise the differences. The difficulties of such comparisons are, of course, enhanced by the fact that some curves (eg. Mörner, 1969a) have been corrected for isostatic rebound, some (eg. Fujii & Fuji, 1967) for tectonic displacement, and others (eg. Kidson & Heyworth, 1973) for gravitational compaction.

During the course of the debate in Australia, on SLs higher than present in the late Holocene, it became evident that 'Holocene SL change relative to Australia differed substantially from the North Atlantic pattern' (Thom & Chappell, 1975). They showed that SL was close to, if not at, its present level, around the entire Australian continent 6000 years BP. They contrasted this to the position in North America and Europe, where the present level, according to Coleman & Smith (1964), Mörner (1969a), Tooley (1974), was reached not earlier than 3700 years ago or even as late as the present (Scholl *et al.*, 1969; Kidson & Heyworth, 1973). They questioned the supposed tectonic stability of Australia, advanced by earlier writers, while accepting the difficulty of tracing Pleistocene movements into the Holocene. They gave as the most important single factor causing divergencies between Holocene SL records as *glacio-* and *hydro-isostasy*. This accelerated a major revolution in thinking on the nature of eustasy and of the Holocene rise of SL signalled by Walcott (1972). The recognition that eustasy need not be globally uniform and that there were indeed significant regional differences began to dominate the literature. Adey working on coral reefs in the Atlantic and Pacific oceans concluded that:

> it seems clear that... the two predominant reef-building regions of the earth, the Indo Pacific and the Caribbean, have different relations to sea level resulting from deglaciation (Adey, 1978).

He produced *Atlantic* and *Pacific* SL curves over the last 10 000 years and extrapolated them to give predictions for the next 10 000 years, (Fig. 3).

The recognition that eustasy could involve differential SL changes owes much to three studies into the rheology of the earth (Walcott, 1972; Chappell, 1974; Clark *et al.*, 1978). To some degree these advanced the work of Bloom (1967) who recognised that, if the ocean floor deforms under the load of water from post-glacial rise of SL, submergence of a coast should be in part a function of the proximity of deep water. Walcott also revived Daly's (1925) argument that changes in surface load over the oceans because of eustatic changes in SL would deform the earth and that this deformation would have to be considered before proceeding with the *inverse problem* of determining eustatic change from observations of past SLs. He pointed out that evidence for relative SL rise in the later post-glacial (6000-0 BP) comes largely from areas of land subsidence related to adjacent upward post-glacial rebound movements in deglaciated areas,

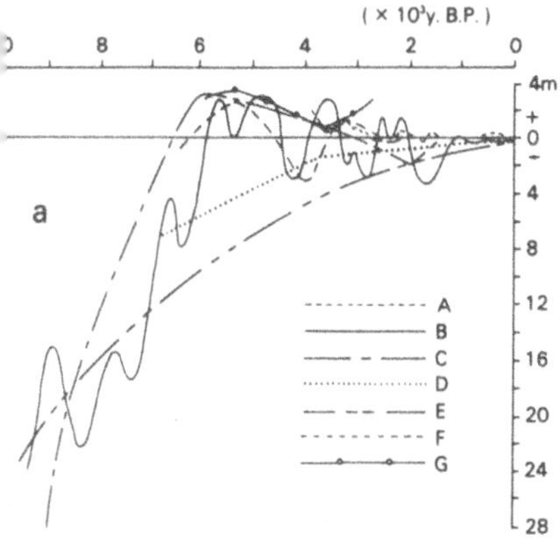

World sea - level curves except for Europe and North America

A : New Zealand (Schofield, 1960),
B : World (Fairbridge, 1961),
C : Japan (Fujii & Fuji, 1967),
D : Caroline and Marshall Islands (Bloom, 1970b),
E : Bermuda (Neumann, 1971),
F : Mauritania (Einsele et al. 1974),
G : Sunda Land (Tjia et al., 1975).

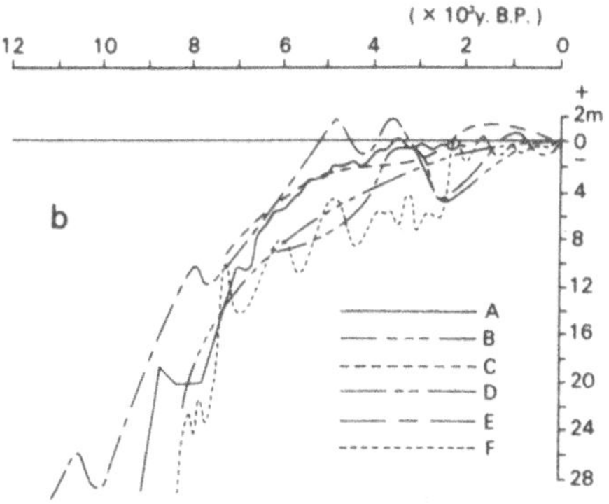

European sea - level curves except for Great Britain

A : Kattegatt Sea and Swedish West Coast (Mörner, 1969, 1971),
B : Netherlands (Jelgersma, 1971),
C : France, North Sea and Atlantic Coast (Delibrias and Guillier, 1971),
D : France, Bretagne (Delibrias et al., 1971),
E : Black Sea Coast (Fedorov, 1972),
F : France, Atlantic Coast (Ters, 1973).

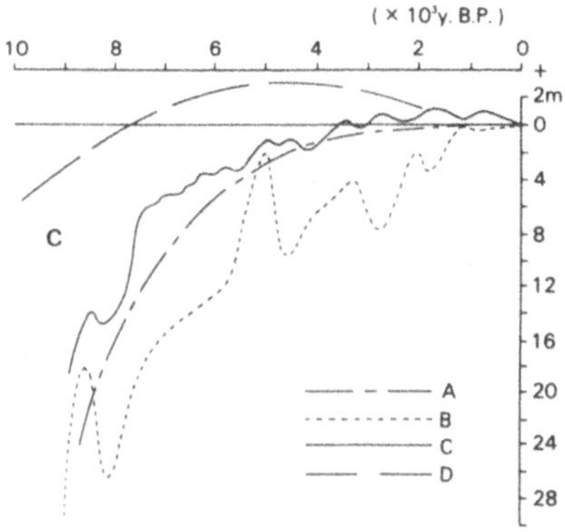

Sea - level curves in Great Britain

A : Bristol Channel (Kidson and Heyworth, 1973),
B : Thames Estuary (Greensmith and Tucker, 1973),
C : Northwest England (Tooley, 1974),
D : Southwest Scotland (Jardine, 1975).

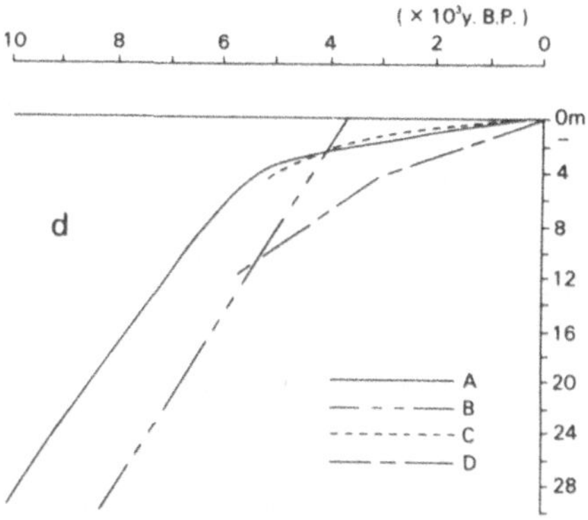

Sea - level curves in North America

A : the United States, Netherlands (Shepard, 1964),
B : Southwest Louisiana (Coleman and Smith, 1964),
C : South Florida (Scholl and Stuiver, 1967,; Scholl et al., 1969),
D : Delaware (Kraft, 1971).

(after Moriwaki 1978)

Figure 2. Sea-level curves by regions

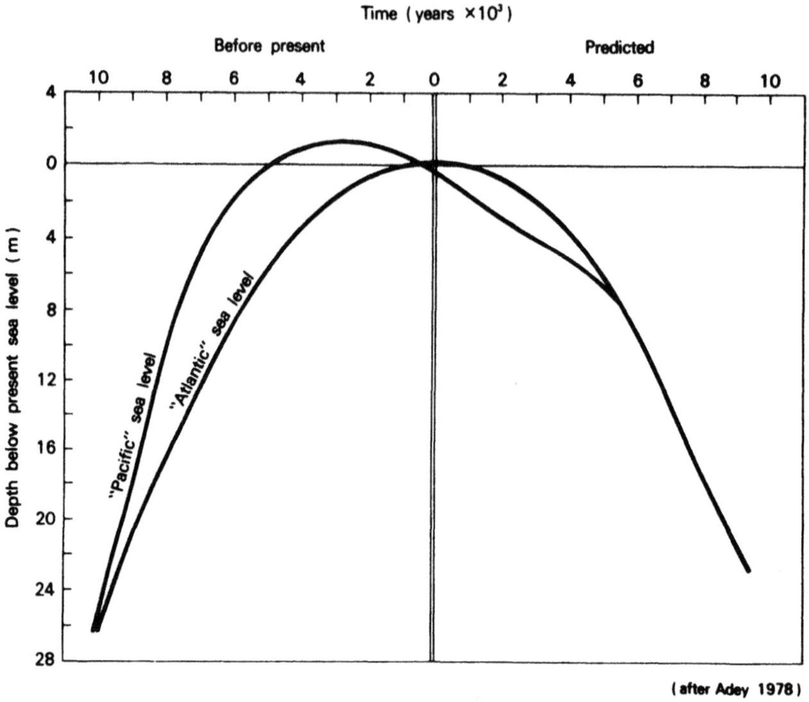

Figure 3. Holocene sea-level curves derived from coral-reef growth in the Pacific and the Caribbean

the areas of the so-called *forebulge*. Conversely, evidence for a relative fall in SL comes from coastlines well removed from areas of rebound and which have been affected by a rise of the continental areas through compensation for eustatic load. He concluded that no substantial change of SL in the past 6000 years is required to explain post-glacial SLs. Chappell elaborated the concept of hydro-isostasy and suggested that:

> broad scale changes in the figure of the earth cause the magnitude of glacio-eustatic transgression to vary around the globe (Chappell, 1974).

He argued that deformation of the ocean basins and their margins, arising from the fluctuating SLs of the Quaternary, can explain some of the regional differences in SL response in the post-glacial period. Figure 4, modified from Chappell (1974), summarises his interpretation of Holocene SLs in terms of hydro-isostatic deformation.

Clark *et al.* (1978) proposed a number of numerical models based on a spherical viscoelastic earth with varying layered structures and on different assumptions of the rate of northern hemisphere ice melt. They incorporated into their models some of the known 'facts' of relative SL change. They were careful to point out that so little is known of the elastic and viscous behaviour of the earth, that the

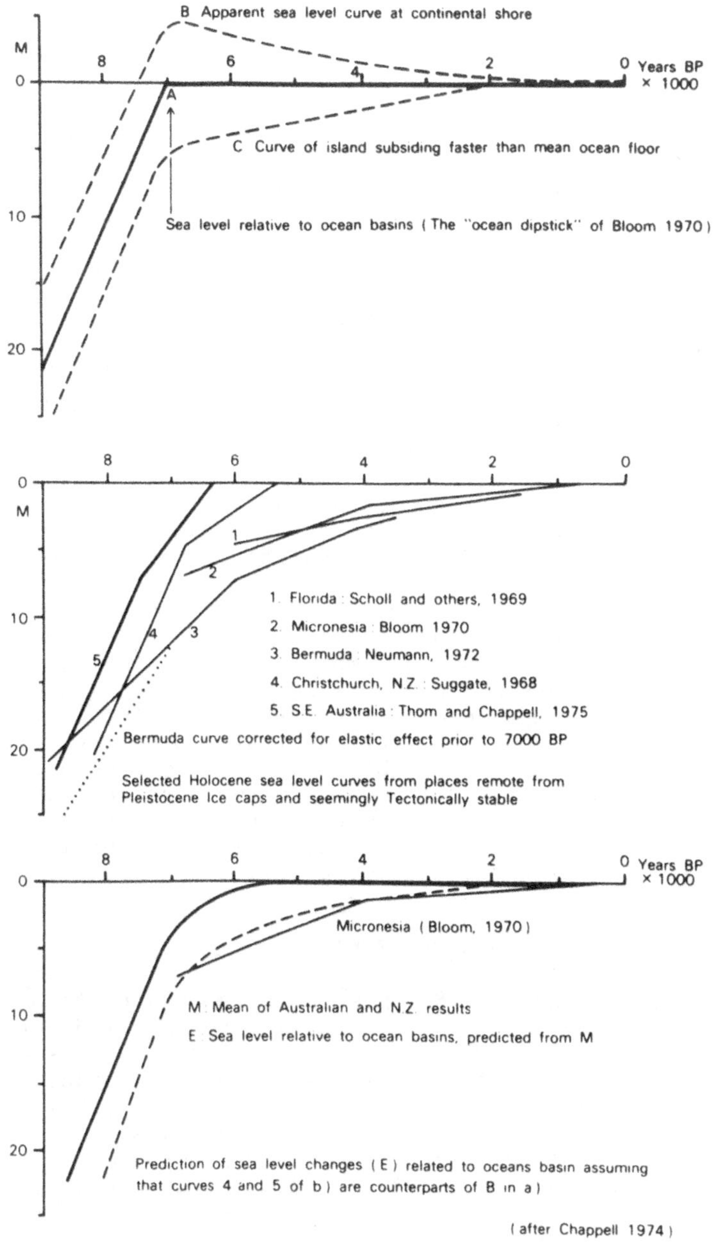

Figure 4. Interpretation of Holocene sea-levels in terms of hydro-isostasy

rates of ice melt are also little known particularly in a regional context; and that the effects of Antarctic ice movements had been ignored, so that their predictive models owed much to these 'known facts'. Despite the in-built drawbacks which this circular reasoning involved, they were able to extend Walcott's arguments on regional responses to deglaciation. Figure 5 illustrates their six SL zones for the 'realistic melting case' together with examples of predicted curves in the Pacific for zones where submerged (Zone IV) and emerged beaches (Zones III and V) are predicted. Differences between prediction and observation indicated that assumptions on ice melt and on the rheology of the lithosphere needed refinement. They suggested that as much is to be learned of glacial history and the earth's rheology from observations on SL change as of the history of SL change from studies of the earth's rheology.

Despite the limitations of the models, to which the authors themselves were careful to draw attention, these studies have had three significant consequences. They have underlined the growing belief that no part of the earth's crust can be considered stable. They have shown that the relationship between eustatic change and isostatic adjustment is much more complex than is frequently assumed. They have also confirmed the demise of the concept of world-wide eustasy. Even if further work to refine the models and to improve basic assumptions is required, the principle of regional differences in eustatic response has been widely accepted. However, it must be emphasised that the inadequacy of the assumptions on which these models are based make them incapable of contributing significantly, at present, to the debate on the reality or otherwise of the low amplitude fluctuations of SL which have been such an important element in the differences between Fairbridge and Shepard schools of thought. The coarseness of the data on which they are based also limits their present contribution to discussion of higher than present Holocene SLs.

GEOIDAL EUSTASY

In considering changes of SL, most authors have plotted their data in relation to a horizontal line labelled *zero* or *present SL*. Cartographically this is reasonable, but as a concept such a presentation of the *still water* surface of the oceans is unacceptable to geodesists. Defant (1941) demonstrated that, at the present time, the 'absolute dynamic topography' of the Atlantic Ocean varies by up to 1.5 m. Lisitzin (1965) presented a similar world-wide picture of the varying heights of mean sea level (MSL). Kidson & Heyworth (1979) drew attention to the importance of these irregularities of the ocean surface when comparing SL data from different parts of the world.

Mörner extended the discussion of the irregularity of the ocean surface and of its importance for studies of Quaternary SLs in a series of penetrating and seminal papers (1976, 1977, 1978, 1979a, b). He showed that, in

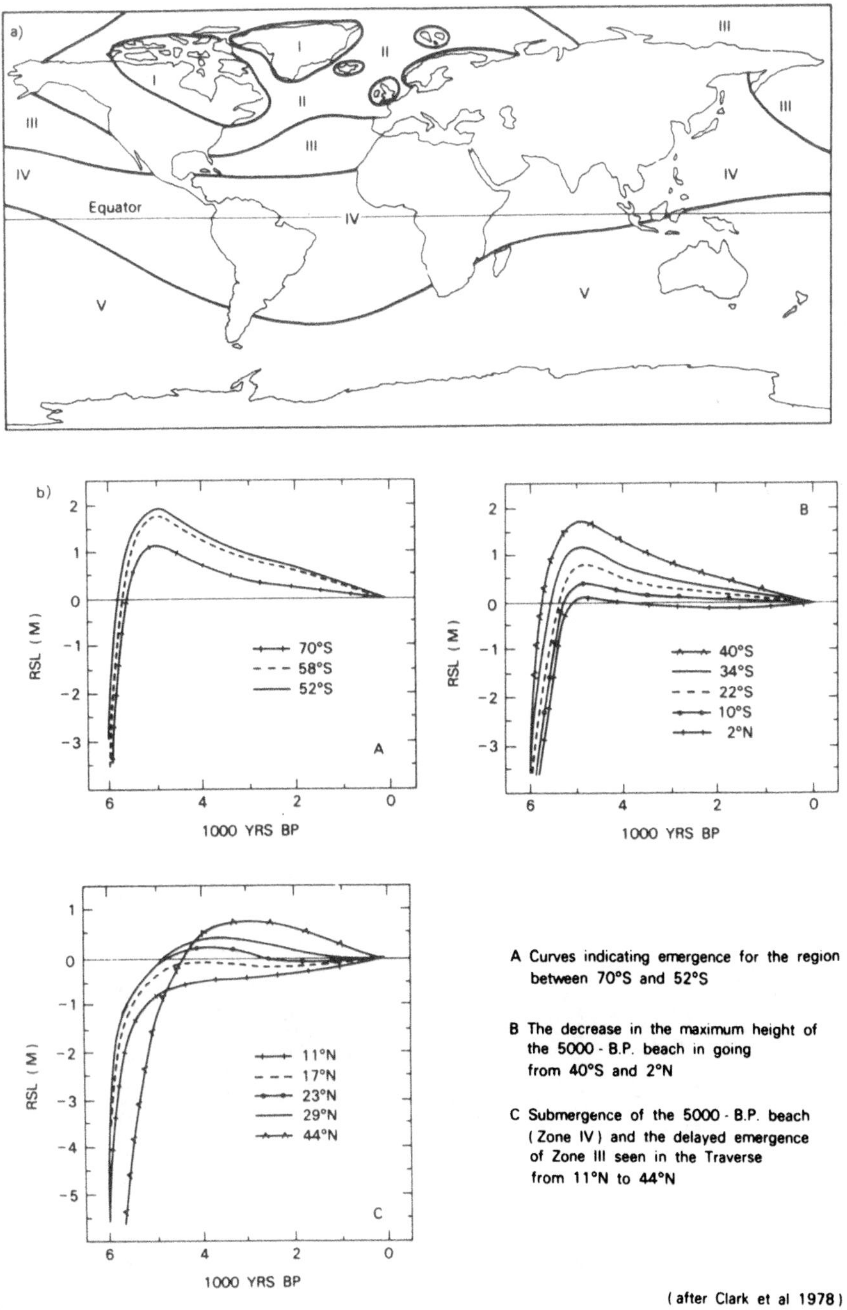

Figure 5. (a) Illustration of the six sea-level zones for the realistic melting case. (b) Predicted relative sea-level curves for a transect from north to south of the Pacific along the meridian 165°W

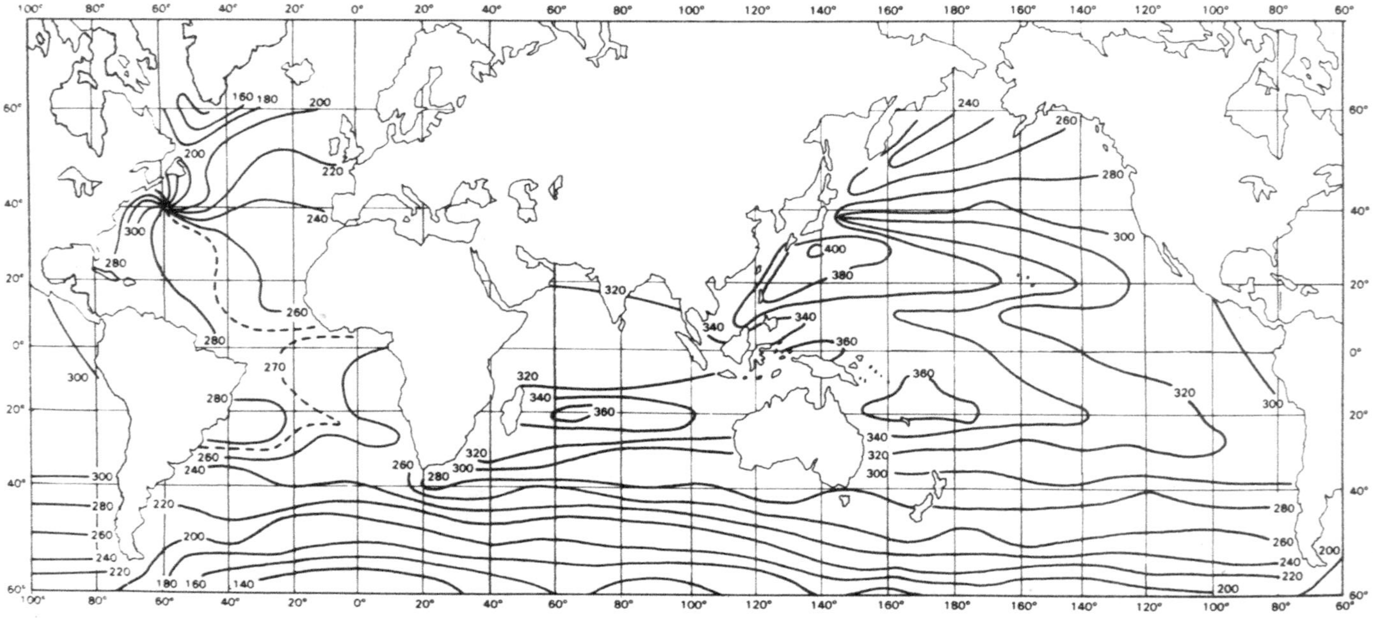

(after Lisitzin 1965)

Figure 6. The distribution of different heights of mean sea level (in dyn. cm) in the world's oceans

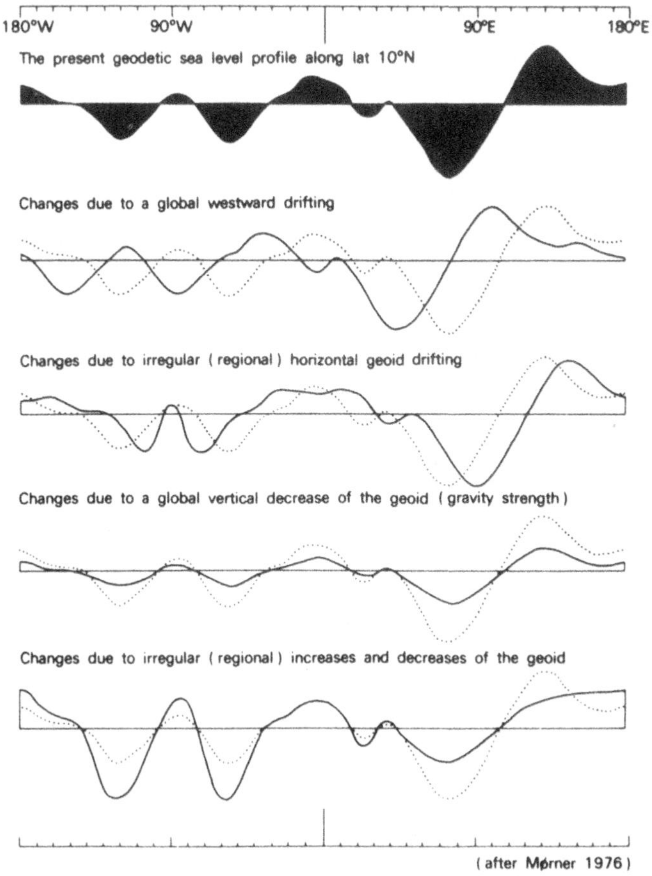

Figure 7. The effect of horizontal and vertical geoid changes on the present geodetic sea level

what he called 'the real ocean surface, or 'the geodetic sea level' or 'the equipotential surface of the geoid', there are 'bumps and depressions' showing SL differences, with respect to the earth's centre, of as much as 180 m. He argued that the present geoid configuration is not stable and must have changed with variations in gravity and the factors controlling it. Instability in the geoid is distinct from though not unrelated to crustal instability of tectonic or isostatic origin. This arises because the geoid is a function of the earth's gravity field. It can thus be only indirectly related to crustal instability. Instability of the geoid configuration, as far as the oceans are concerned, is instability of the ocean surface as distinct from instability of the ocean bed. Mörner coined the term *geoidal eustasy* to distinguish changes in SLs, resulting from changes in the geoid, from those of glacio-eustatic origin. He argued that the recognition of geoidal variation meant that SL changes could no longer by taken as *a priori* evidence of glacial volume changes. His ideas on the effects of horizontal and vertical geoid changes on present geodetic SL are exemplified in Figure 7.

The acceptance of the concept of *geoidal eustasy* means that eustasy as a world-wide phenomenon, as understood up to and beyond the beginning of IGCP 61, can no longer be regarded as valid. There must be significant regional differences. The importance of these ideas appears not to have been fully comprehended by all students of SL change. It is quite clear, for example, that SL curves which used data from all over the world (eg. Fairbridge, 1961) and even those which covered large regions (eg. Ters, 1973) and did not take geoidal variations into account cannot any longer be regarded as having more than historical value. Yet such curves are still frequently used as a basis for comparison even for curves based on detailed local studies (eg. Hopley, 1978). Increasingly, however, the concept of geoidal change over both time and space is having an impact. Already greater stress is laid on detailed local studies as distinct from those using data from large regions.

ERRORS

Almost from the inception of the modern phase of SL studies it has been recognised that the drawing of lines joining data points gives a false sense of accuracy to the resulting *SL curve*. Scholl *et al.* (1969), Grant (1970), Thom & Chappell (1975), Easton & Olson (1976) and Chappell & Polach (1976), among others, have argued that all that can be achieved realistically is the containment of SL rise within an envelope delimiting potential errors. More recently the realisation has grown that error terms built into SL graphs have been wholly inadequate. This, together with the recognition that many *SL indicators* frequently have only a crude relationship to *SL*, and that problems of datums have been largely neglected, has led to what is, in effect, yet another revolution in thinking on Holocene SL changes. Kidson & Heyworth, (1979) and Pinot (1979) have surveyed the potential for error and misunderstanding and Heyworth & Kidson (1982) (in a review of SL changes in south-west Britain) have made an approximation of the allowance necessary to cover a wide range of such errors. However, curves have been and continue to be published claiming to show oscillations of SL, with amplitudes of as little as a metre or less, and demonstrating close correlation with a wide variety of palaeo environmental indicators (Ters, 1973; Tooley, 1974; Schofield, 1975; 1977; Fairbridge, 1976; Mörner, 1980). Kidson & Heyworth argued that the wide range of potential sources of error, and the possible cumulative significance of these, reduces the value of a good deal of published data. It is suggested that the differences between authorities are smaller than the potential errors and that until technology is improved we should remain agnostic both on the nature of SL rise in the late Holocene and on the question of higher than present Holocene SLs.

Even before attempts are made to derive eustatic from relative SL curves, involving errors of correction for isostatic and tectonic deformation, a wide range of other potential errors have to be taken into account. These can be grouped into those resulting from choice of datum and definition of SL, those of tidal origin, those arising from choice of SL indicators, those relating to the altitude of sample collection points and finally those stemming from age determinations. In all of these, changes over time may be of critical importance.

Definition of datum

As Pinot observed (1979), it is adequate for Pleistocene studies to know SLs to within a metre or so but for the Holocene an accuracy of a few decimetres is essential 'surtout depuis le niveau de la mer est proche de l'actuel'. Since this is so, it is vital to begin with precision in the datum to which SL data are related. This is particularly important when SL curves are to be compared. Much of the literature ignores the problem and refers simply to *present SL*. For this reason alone, much of the published information is of limited value. Most investigators do not determine local SL, however defined. They refer their height information to their national geodetic datum, whether it be N.G.F. in France, the Sea Level Datum of 1926 in the U.S.A., O.D. in the United Kingdom or whatever. While these confer the highest standards of accuracy available, it is important to understand their limitations. Pinot noted:

> Sur les côtes atlantiques de la France, l'écart est de plusieurs décimètres entre le zéro N.G.F. et le niveau moyen de la mer, mais il est rarement connue avec précision (Pinot, 1979).

His statement that many heights given in the literature are very approximate is manifestly true. Shalowitz wrote:

> The Sea Level Datum of 1929 must not be confused with local mean sea-level. Whilst both are of tidal definition, the first is an adjusted datum arrived at by holding sea level fixed as observed at a selected number of tide stations whereas the second is derived entirely from observations made at the local tide station...hence where only a geodetic bench mark (one that is part of the national level net) is available in an area where a tidal boundary is to be established *a correction must be applied* to the geodetic elevation to bring it into harmony with the local datum plane (Shalowitz, 1964).

In Seymour (1980) it is recorded that:

> The outcome of the third Geodetic Levelling of England and Wales was something of a disappointment. The strong possibility that it was appreciably affected by a systematic error made it impossible to base firm conclusions about land and sea levels upon it or to confirm or controvert the northward rise in sea-level which the 2nd Geodetic Levelling had appeared to establish.

Some, at least, of the differences between SL curves revealed in Figure 1 may simply reflect differences and errors in datums! Quite clearly, therefore, each SL study begins with built in error, often of unknown dimensions. It is salutary to remember that, even in advanced countries, different determinations of SL produce significantly different answers. MSL based on tidal observations at Liverpool in 1844 provided the datum to which the Second Geodetic Levelling of England and Wales was referred. The Newlyn datum, based on tidal readings at Newlyn in Cornwall, between 1915 and 1921, replaced it for the Third Geodetic Levelling. Even on the coast between the two tide gauges, levels from the two surveys varied by more than 30 cm in places while agreeing in others (from data supplied by the Ordnance Survey of Great Britain and Ireland).

Definition of sea level

In his early work in Australia, Fairbridge used Mean Low Water of Spring Tides as the datum to which he referred his Older Peron, Younger Peron, Abrolhos and Rottnest higher levels. This was unexceptionable in the context of his major evidence; erosion forms, where solution of limestone rock is a dominant process. Other authors have adopted different *SLs*, including MSL, mean tide level, high water of neap tides, high water of spring tides and so on. Such authors have frequently compared their SL curve, without discrimination of datum, to the Fairbridge and other curves. Quite clearly, even in a non-tidal situation data from different areas cannot be compared until *SLs* or *zeros* have been reconciled. Where tidal ranges are significant the vertical spread over which marine processes operate becomes very wide and potential errors multiply. In extreme situations this vertical spread is several orders of magnitude greater than postulated differences in SL over time spans in millenia. In the Bristol Channel, United Kingdom, for example, predicted spring tidal ranges attain values of upwards of 14 m. Predicted heights of high water can be regularly exceeded (*not* in storm surges) by as much as 2 m (as in January 1982). When wave heights, reaching up to 7 m at the coast, are added, *SL*, depending on definition, could fall anywhere within a 23 m band. Since evidence of former SLs ranges from the *marine platforms* or *reefs*, formed at low water of spring tides at Point Peron in Western Australia (Fairbridge, 1950), to the storm beach gravels of the Portland, England *raised beach* quoted by Zeuner (1958), the need for definition seems obvious. The problem is, however, consistently ignored in many papers.

Tidal levels

Large variations in tidal range can occur over quite small distances. This is particularly so when, as in Western Europe, continental shelves are wide and coasts are crenulate. SL curves from such areas clearly have to be adjusted before they can be compared, more especially when they are plotted against an absolute datum such as O.D. or N.G.F.

Those from the Bristol Channel (Kidson & Heyworth, 1973) (maximum tidal range i.e. Highest Astronomical Tide (HAT) minus Lowest Astronomical Tide (LAT)) (14.4 m), Cardigan Bay, Wales (Wilks, 1977) (6.0 m), Morecambe Bay, England, (Tooley, 1974) (10.4 m) and Thames Estuary, England, (Devoy, 1979) (7.6 m), all show significantly smaller disagreement when adjusted even for differences in mean high water levels in relation to O.D. (Kidson & Heyworth, 1978). If maximum compensation for extreme tidal range were to be applied, differences between the curves, which might be attributed to other causes, would be reduced further. Even where the need for adjustment for tidal variation is as obvious as this, it has sometimes been overlooked. Where tidal ranges are lower, the contribution to the sum of potential errors is smaller but still important. It has frequently been ignored.

Many studies of SL change are conducted in estuaries and embayments, like the British examples cited above, because it is in such situations that sediments, both clastic and biogenic, survive. Long sedimentary sequences are unusual on open coasts. Frictional forces in such shallowing and narrowing arms of the sea result in the enhanced tidal ranges to be found there. It is important to note that this means that the height of high water above the geodetic datum is also increased. In the Bristol Channel the height of High Water of Spring Tides above O.D. increases by more than 3 m from the mouth of the Channel to the head of the estuary. Values for the other British examples are: Thames Estuary 1.6 m, Solway Firth 1.2 m, Humber Estuary 1.0 m and Morecambe Bay 0.7 m (Kidson & Heyworth, 1979). SL data even within individual such inlets reflects this progressive elevation of the interface between the marine and the terrestrial. Their use in the construction of curves of Holocene SL change should, but rarely does, reflect it. This is a strong argument for detailed very local studies such as that by Delibrias & Morzadec-Kerfourn (1975) or Devoy (1979) as distinct from studies of large regions such as that by Ters (1973).

Other problems of tidal levels, which need to be understood before interpretation of data relating to former SLs can properly be attempted, are mentioned in the literature so rarely that it must be assumed that many authors are not aware of them. All tend to indicate a higher than actual SL. Tidal heights contained in Tide Tables such as those issued by the British Admiralty are based on harmonic analysis of planetary relationships, usually adjusted to take into account average meteorological conditions. Where such conditions differ from the average, corresponding differences between predicted and actual tides can occur and are sometimes very large. Variations in atmospheric pressure and of wind direction are often significant and occasionally result in storm surges which can raise water levels alarmingly. The 1953 storm surge (Steers, 1953), resulting from the passage of an intense depression from north to south in the North Sea, led to high water levels which exceeded those predicted by nearly 2.4 m on the coast between the rivers Humber and Thames in eastern England and by more than 3 m on the Netherlands coasts (Admirality, 1982). Gottschalk (1971, 1975, 1977) has provided evidence, from a study of floods in the Netherlands, that the return period of surges of a similar order of magnitude to that of 1953 may be as low as 50 years. Predicted levels can also be exceeded where winds blow continuously onshore for sustained periods. The event of January 1982, in the Bristol Channel, already referred to, when tide levels 2 m above those predicted resulted in the over-topping of sea defences, was such an event. In other parts of the world, seismic activity may produce long wave forms similar to tides (hence *tidal waves*). These are also defined as surges and have similar effects (Admiralty, 1969). Tooley has argued (1979) that the 1953 storm surge in Eastern England has left virtually no trace. This is not surprising since hundreds of millions of dollars were expended in returning the coastal situation to the status quo ante-surge. By implication Tooley argues that 'rare events' have no place in the interpretation of SL data. Yet such events must have left an accumulation of evidence of the sea's presence at levels well *above* that which is now defined as *present SL* in the period before major civil engineering works interfered with the natural development of the coast. Two other sources of potential error must be added. Tidal predictions based on harmonic analysis are usually only made for ports scores of kilometres apart. Where sites for SL studies are a significant distance away from *primary* or *secondary* ports errors of upwards of 1 m may result (Kidson & Heyworth, 1979). Finally it must also be remembered that in estuarine situations, *fresh water effects* lead to uncertainties in tidal measurement and prediction. Admiralty Tide Tables are modified from time to time, often for this reason, to give revised relationships between tidal levels and the geodetic datum (Heyworth & Kidson, 1982).

Tidal problems are so complex that not to take them fully into account can only lead to loss of precision in palaeoenvironmental reconstruction. Since many of them cannot fully be taken into account real precision is probably unattainable. Most of the uncertainties, including the effects of *rare events*, tend to suggest levels of the sea higher than actual. The effects of high tidal or storm levels tend to remain in the sedimentary or erosional record until eliminated by still higher ones. One facet of the problem which has so far not been seriously tackled is the change of tidal regimes over time. Amphidromic systems at lower glacial SLs cannot but have been significantly different from those at higher levels including that of the present. Where coastal configuration changed significantly during the Holocene SL rise, tidal relationships must also have changed significantly. Jardine (1975b) has illustrated some of the problems involved. Heyworth & Kidson (1982) drew attention to the special case of the changing status and morphology of the Straits of Dover.

There must have been many similar cases around the world where the deepening or reoccupation by the sea of straits and channels resulted in major changes in the relationships between different amphidromic systems. Yet, of necessity, all SL studies have proceeded on the assumption that tidal relationships have remained similar to those now obtaining. This author now believes that uncertainties stemming from tidal factors alone make it necessary to cease presenting Holocene SL data in the form of *single line* curves.

Errors arising from choice of sea-level indicators

Pinot (1979) has rightly argued that the use of ambiguous indicators of former SLs has led to an artificial multiplication of oscillations on already published Holocene SL curves. Erosional features such as shore platforms and notches can only rarely be used by themselves as precise indicators of former SLs*. Even marine deposits cover such a wide range of heights, from the seaward margin of the nearshore sand wedge to the crests of storm beaches, that they must be interpreted with great care. The use of transported debris, such as marine shells to demonstrate Holocene SLs above that of the present, as in Schofield (1960), can scarcely be justified when such debris can be thrown to considerable heights by waves and when any ^{14}C dates on the shell do not relate to the age of erection of the shell ridges, only to the death of the organism. In such cases it is critical to show that shoreline features are attributable to SL change and not to changes of energy levels resulting from, say, shallowing by deposition or the development of protection behind a barrier.

The only wholly reliable indicators are organic remains in growth position where relationships to SL or to the water table can be determined within acceptable limits. Laborel & Delibrias (1976) and Laborel (1979a,b) have shown the value of fixed marine organisms, especially Vermetids, as indicators of former shorelines. Scott & Medioli (1978) have claimed that *Higher High Water* (HHW) of former SLs can be deduced to an accuracy of ± 5 cm from the dramatic reduction in the number of species of foraminifera which occurs in the upper levels of salt marshes. Kidson & Heyworth (1979) have argued that, around the British coasts, oak stumps in submerged forests can give accurate relationships to former tidal levels, when compared to the lower limits of oaks in the present succession from high salt marsh to oak woodland. Here the lower limit of growth is set by the levels at which acorns can germinate, and is found to be mid-way between MHWST and HAT. Geyh *et al.*, (1979) have used mangrove peats to give what are claimed to be accurate determinations of former SLs in the Malacca Strait. In many studies salt marsh peats have been taken to give the most accurate indications of former SLs. However, Kidson & Heyworth (1979) have argued that in many ways these are unsuitable and Heyworth & Kidson (1982) have suggested that the interface between saline and freshwater forms, which at present also occurs between MHWST and HAT, gives the best starting point for determining

*see, however, Pirazzoli, this volume (ed.)

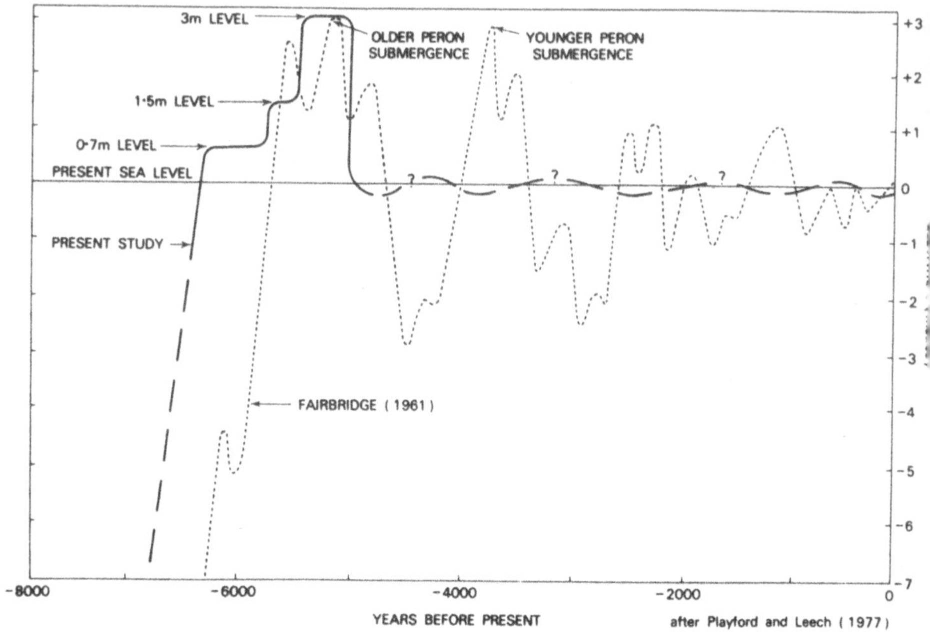

Figure 8. Diagram illustrating approximate changes of sea level relative to land at Rottnest Island from middle Holocene times to the present compared with the eustatic sea-level curve of Fairbridge (1961)

former SLs. They suggested that this could be done to an accuracy varying from ±15 cm to ±35 cm, depending on tidal range, by taking samples from the lowest levels of fresh-water peats.

However desirable the use of such precise biogenic indicators, many SL curves have been constructed using much more equivocal evidence both alone and in combination with more reliable information. In the period before IGCP 61 was initiated this was, in part, due either to a failure to recognise the limitations of erosional forms and transported materials or to a lack of understanding of the very wide vertical range covered by the sea's action. It is now widely recognised that even where biogenic material giving precise relationships to SL is not available, this other evidence must only be used with full recognition of its limitations. Frequently the same evidence can be interpreted in widely differing ways. Thus Playford & Leech's view (1977) of SL changes at Rottnest Island, Western Australia, differs significantly from Fairbridge's (1950, 1961) interpretation based partly on the same evidence (Fig. 8) Kidson (1981) provided yet another possible interpretation! It is salutary to remember that this type of evidence from Western Australia was originally the sole support, according to Shepard (1963), of higher than present Holocene SLs, and the basis of what he called the Australian 'enigma'.

It is frequently claimed that many of the pitfalls of interpretation can be avoided by the simple expedient of comparing present shoreline forms with those of the past. However, equally simple environmental changes have to be allowed for in such exercises if large errors are to be avoided. Delibrias & Morzadec-Kerfourn (1975), for example, in presenting evidence for the evolution of Dol-de-Bretagne in the Holocene, argued that differential compaction of sediments made it impossible to present a SL curve. Such caution is highly desirable. The disposition to ignore such factors must have led to significant errors in published data. Pinot (1979) has suggested that uncertainties of this kind introduce errors of several decimetres in the best cases and of several metres in the worst.

The present trend is to attempt to use only those indicators where a precise relationship to SL can be established. Where this is not possible, other more equivocal evidence is treated with much greater caution than has sometimes been found in the past, and larger allowances are made for the possible errors involved.

Sample height

Much attention has been focussed in the literature on error terms in ^{14}C dates of samples related to SL. Much less attention has been given to the equally important problem of errors in sample heights to which most other errors contribute. Those which have been connected by precise levelling to geodetic bench marks may be theoretically accurate to within a few centimetres. However, such accuracy may be more apparent than *real*. A further quotation from Seymour illustrates the point:

> In addition to the problem of interpreting the results of the 3rd Geodetic Levelling, there arose the very practical problem of how, if at all, to apply these results to the O.S.'s (Ordnance Survey's) published B.M. heights....it was clearly inadvisable to adopt the new height values *in toto*. This would have introduced changes of about half a foot in places in England and even more in Scotland and would certainly have created difficulties for engineers, especially during the transition period. On the other hand, the more recent levelling was certainly more consistent with the present height of B.M.s relative to each other locally. A compromise was, therefore, adopted for England and Wales. The odd heights of the Fundamental B.M.s were maintained and the new levelling was *adjusted* to these for the purpose of controlling lower order levelling (Seymour, 1980).

Bench marks, therefore, even when newly heighted have their problems. Many may not be at their original heights due to human interference or ground movement. Since many bench marks used for the control of coastal work rest on unconsolidated sediments, this is more of a problem than appears at first sight. Where samples are obtained from boreholes the accuracy derived from lines of levels ceases at the ground surface. Measurements in boreholes may be accurate to ±10 cm but this can only be achieved by rigid adherence to a strict procedure. Where samples are obtained from commercial boreholes put down for other purposes it is doubtful if anything like this accuracy can be assumed. Probably the greatest source of heighting error applies to samples obtained by ship from offshore coring. Problems of accurate measurement of tidal and sea state at considerable distance from coastal recording sites, and of drilling and depth measurements at sea, probably lead to uncertainties of up to ±2 m (Heyworth & Kidson, 1982). Many of the older Holocene dates available to us were obtained in this way.

Further aspects of the heighting problem concern the identification of dated horizons. For example, samples taken from submerged forest trees, the dates from which have been used in many SL curves may, if taken from the top of a stump, be up to 1 m above the SL controlled water table close to the roots responsible for the death of the tree. Reed swamp species, dates from which have been included in many European SL studies, are particularly tolerant of water-level changes and height uncertainties of ±50 cm may have been introduced from this source.

Dating errors

There is a significant literature on errors associated with ^{14}C dating, including recent commentaries by Pardi & Marcus (1977), Ogden (1977), and Stuckenrath (1977) which include references to most of the significant papers. Despite the caveats in the literature, however, many authors still appear to believe that the age of a sample probably lies within the one standard deviation given by the dating laboratory. This is, in fact, merely an indication of the precision of the measurement of the sample's radioactivity and of the likelihood that this represents a particular age. No allowance is made for non-counting errors. Many potential inaccuracies widen the possibility that the true date lies outside the one sigma limits. Ogden lists some six factors concerned, including a variety of forms of contamination, both 'natural' and of an anthropomorphic origin, but the greatest single query concerns past variations in atmospheric radiocarbon (see also Mook & Van de Plassche, this volume). Tree ring calibration studies such as those by Suess (1970) and more recently by Ralph *et al.* (1973) and Suess (1980) have clealy shown that such variations are of major significance. A new dimension has been introduced by similar calibration studies, involving some hundreds of ^{14}C dates including a number of groups of 20-30 dates on individual cross sections of pine and oak, cut from submerged forest trunks on the west coast of Britain (Campbell & Baxter, 1979; Heyworth & Kidson, 1982). All these studies tend to show that the precision sometimes attributed to ^{14}C ages in some, particularly earlier, SL studies is not justified. The significance of the submerged forest work is that it indicates clearly that many dates on wood, almost the best of all material for dating, incorporated in published SL curves, can be in error by several hundred radiocarbon years. Indeed the discrepancies revealed in the study were so great that it became essential to rule out laboratory errors before considering their significance. Further samples of submerged forest wood from the sites discussed in Campbell & Baxter (1979) and Heyworth & Kidson (1982) were submitted for replicated re-dating by 28 radiocarbon laboratories around the world. Preliminary reports of their conclusions (International Radiocarbon Study Group, 1982) show that: 'results reveal the existence of systematic bias and unexplained variability. Quoted errors should be multiplied by a factor of between 2 and 3'. In recent years several SL studies have adopted two sigma as the error term for plotting SL related ^{14}C dates on SL curves. The submerged forest work shows that even this may not cover the full range of possible dating errors.

PLOTTING SEA-LEVEL DATA

The growing realisation of the potential for errors has meant that, particularly since the inception of IGCP 61, a major attempt has been made to ensure that these are reduced to the minimum. The presentation of new information to a standard, computer based, pattern, as described in Tooley (1978a), Preuss (1979) and emanating from the German working group of IGCP 61, was designed for this end. This ensures that only *high quality* data is used and ensures also that it is treated, as far as possible, on standardised lines (see, for example, Shennan, 1979; Geyh, 1980). However, when every care has been taken to eliminate unsuitable data and to reduce error to a minimum it is still impossible to regard any SL information as a data *point* on a graph. Even carefully screened data can only realistically be plotted in an *error box* of varying dimensions. Usually the horizontal axis is related to errors of age and the vertical axis to errors in heighting. A recent example from a rigorous and carefully controlled study (Devoy, 1982) is given in Figure 9. A single line curve drawn through such an assemblage of data could only be arrived at by *eyeballing* or by using simple regression techniques. What is clearly not possible is the recognition and identification of minor oscillations in SL rise. The scale of potential error is manifestly greater than that of the possible fluctuations. Many authors, including the present writer, who have drawn single line *SL* curves in the past, have come to recognise that such a presentation can be so misleading that it should only be done with careful explanation of the limitations and pitfalls, even if it is done at all. Warner (1975) has drawn attention to the generally unsatisfactory nature of past graphical presentation of radiocarbon dates and recommended the use of equal area curves using, significantly, two standard deviations. While his method is unsuitable for plotting SL information, it does illustrate the fact that the use of *error boxes* makes allowance only for errors in age and some of those in height

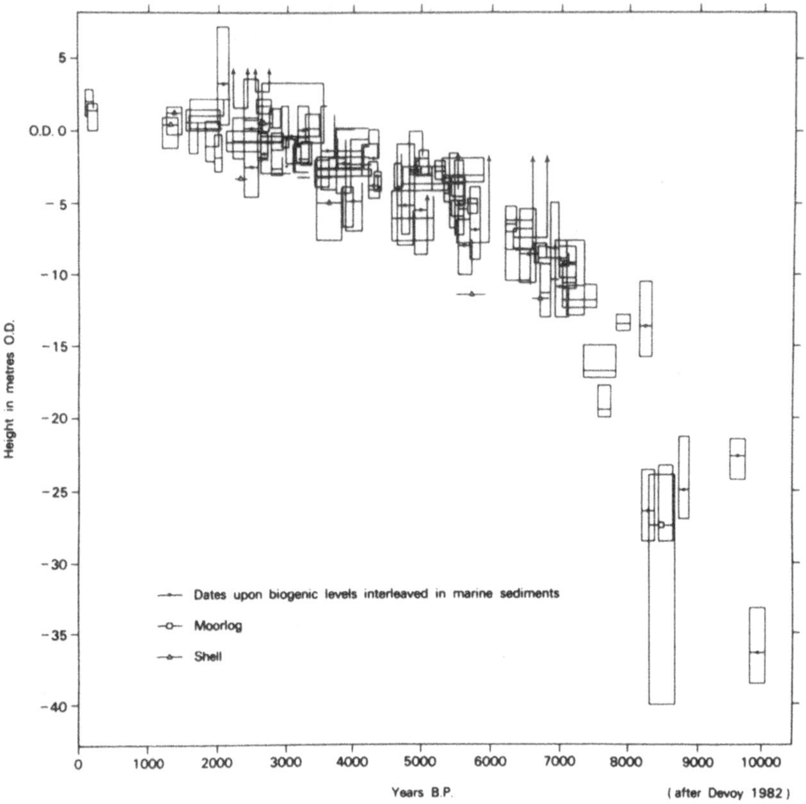

Figure 9. Time depth graph of sea-level points from southeast England

and does not incorporate all the sources of potential error outlined above. For this reason Heyworth & Kidson (1982) constructed *error ellipses*, as a modification of the Warner method, assuming a range of two standard deviations for the vertical axis. While they recognised that the standard deviations for altitudinal measurements have no statistical basis as have those for the radiocarbon dates, on the horizontal axis, they argued that they are as good a general approximation of all the errors involved in relating the sample to SL, as it is possible to find. Figure 10 shows the application of the method to data from southwest Britain in which each point is plotted at its altitude relative to that of the present MHWST at the site.

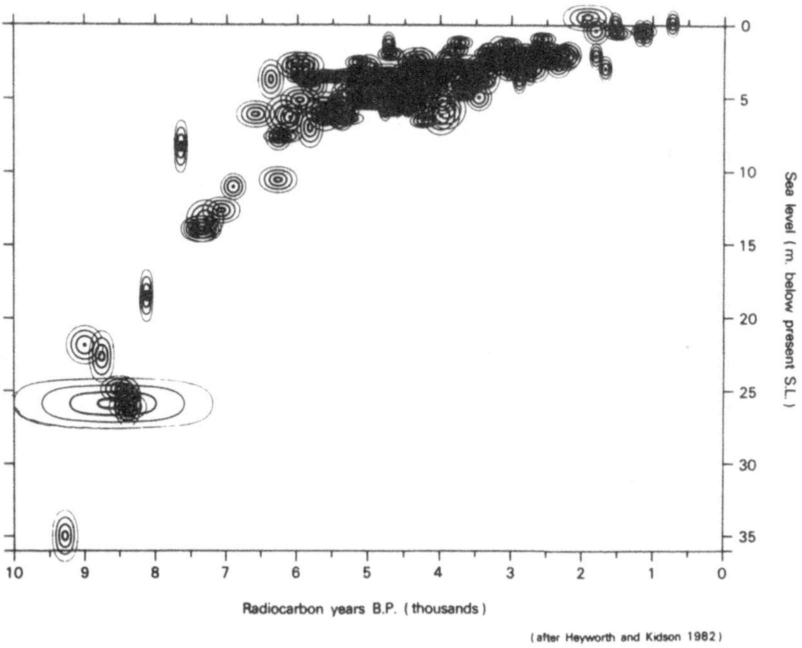

Figure 10. Graph of radio-carbon dates related to sea level from southwest England and Wales plotted as ellipses denoting potential errors. For explanation see text.

> The four concentric circles for each date present the probability distribution of the true point lying within each of four zones each representing half a standard deviation on each axis. The *total* line thickness is the same for each sample, so that if a sample has small associated uncertainties it will appear as a small dense circle or ellipse, while a sample with large uncertainties will be large and faint. The object of this method is to overcome the problem of imprecise results appearing much more important than precise ones, because they occupy a greater area, and also to allow a more realistic graphical display of the reinforcing effects of overlapping results, so that dense and faint areas are produced (Heyworth & Kidson, 1972).

There are a number of theoretical objections to the method used but it does seem to have the overriding advantage of giving a more objective picture of the true precision involved. It is immediately apparent that to draw a single line SL curve through these data, whether it be smooth and exponential or of an irregular character, could only be a subjective exercise. A recognition of the degree of uncertainty even in what appear to be precise SL data has caused the authors, who used much of the Bristol Channel data incorporated in Figure 10, to draw a smooth exponential curve of SL rise in that area (Kidson & Heyworth, 1973), to appreciate that such an interpretation is no longer acceptable. A range of *single line* curves could be drawn through the data but it would be extremely difficult to identify that which most nearly represents actual Holocene eustatic SL change.

PROBLEMS OF INTERPRETATION

In reviewing the work and literature of more than thirty years it is appropriate to attempt an assessment of how far the problems which exercised researchers at the beginning of the period have been resolved by its end. In the 1950s, three basic differences of interpretation of the world-wide evidence of the rise of the sea from the low levels of the last glaciation were apparent. Firstly, the nature of the Holocene rise was variously seen as smooth and almost exponential or alternatively as spasmodic with temporary falls superimposed on the general rise. Secondly, there was disagreement on the date at which the sea attained its present level and, finally, there was no agreement on possible late Holocene high stands of the sea above the present level.

Smooth or spasmodic rise

Almost by definition, much of the evidence of Holocene SL change is below present SL, either beneath the waters of the present shelf seas or within the sediments of the peripheral, particularly the estuarine, regions. Elevated Holocene strand lines are, of course, common wherever isostatic rebound has occurred as in Scandinavia (Mörner, 1969b, 1979b) and the Canadian Shield (Hillaire-Marcel & Fairbridge, 1978; Hillaire-Marcel, 1979) or where tectonic activity has caused vertical movements as in the Pacific Island arc (Pirazzoli, 1978; Ota *et al.*, 1981) or in Sunda shelf region (Bloom *et al.*, 1974). Frequently beach ridge and terrace staircases are spectacular and relative SL movements over time can easily be plotted from them. However, when eustatic SL curves are derived from such data, it is on the basis of estimates of crustal movements which can only be approximate. It is frequently forgotten that any inaccuracy in the estimates must, of necessity, appear as irregularities in the *eustatic* resultant. By the same token, evidence from sediments, which may also have been subjected to crustal movements, can easily be misinterpreted. Indeed the additional problems of gravitational compaction and difficulties in precise heighting make error more probable as suggested by Delibrias & Morzadec-Kerfourn (1975).

The evidence from Holocene sediments, infilling today's coastal lowlands and the deep channels cut by Pleistocene rivers both beneath these lowlands and the present sea bed, has given rise to considerable disagreement and argument. Frequently these deposits include, in their upper sections, intercalating marine and terrestrial sediments. Quite clearly such alternating beds indicate oscillations of the land and sea margins with the sea variously advancing and retreating over low-lying coastal flats. Some authors (eg. Tooley, 1974, 1978b, 1979) have interpreted these transgressions and regressions as resulting from rises and falls of SL. Others (eg. Kidson & Heyworth, 1973, 1976) argued that such an interpretation is too simplistic. They pointed out that SL rise has been accompanied by sedimentation on a comparable scale. They suggested that as the rate of SL rise slowed in the later Holocene, it would variously fall below or exceed the rate of sedimentation. Alternations of marine and terrestrial sediments, particularly when the former are clastic and the latter biogenic, could be emplaced without any falls of SL. In the absence of more unequivocal evidence of falls, they preferred this explanation. Streif (1978) suggested that, when fen peat layers occur, as they often do 'in regressive overlap' on top of marine or brackish deposits, it is 'misleading to interpret each regressive overlap as an indicator of a sinking SL'. He also concluded that phases with a relatively low rate of SL rise seem to offer the most favourable conditions for the formation of intercalated peat layers. Recognising the possibility of such interpretive difficulties, Tooley (1982) now accepts the terms transgressive and regressive overlap as alternatives to presumed rises and falls of SL but Shennan (1982), echoing Menke (1969) quoted in Streif (1978), complements them with 'positive and negative tendencies' of SL movement (see editor's note page 55) All now appear to agree that the limitations of present techniques and the possibility of in-built error mean that the drawing of SL curves, whether smooth or irregular, has been attempted, in the past, with much too facile confidence.

It is clear from the literature, however, that the smooth and the spasmodic schools of thought still remain. It may be significant that there is a tendency for a degree of agnosticism on this subject to creep in, wherever detailed consideration has been given to potential errors and data have been plotted with this clearly in mind. The confident assumptions frequently made by proponents of both schools of thought are becoming more muted with the acceptance of the fact that no part of the crust can be regarded as stable but merely as less unstable than some other parts, and that sediments can be interpreted in a wide variety of ways.

Date of attainment of present sea level

It can be safely asserted that few students of the Holocene now base their work on the assumption, common at the beginning of the present phase of study, that the present ocean levels were achieved synchronously around the world. Recognition of mobility in the crust and of deformation of the geoid over time involves also the perception that Holocene SL change differed from one region to another. While accepting that there is now something of a consensus on this point, it may be as well to remember that the scale of disagreement in the past can be exaggerated. Thom & Chappell (1975) claimed that SL was close to or at its present level 6000 years ago in Australia. They contrasted that continent to Western Europe and North America, outside areas of major isostatic rebound. Yet even here SLs were only a very few metres below present (see, for example, Shepard & Curray, 1967; Milliman & Emery, 1968; Krog, 1965; Delibrias & Guillier, 1971; Kidson & Heyworth, 1973). Indeed, Godwin *et al.* (1958) argued that present SL was reached 5500 years ago in the Somerset Levels of South-west England and Coleman & Smith (1964) gave a date of 3650 for the Louisiana coast. Godwin *et al.*'s argument was based on a ^{14}C date from Tealham Moor (Q 120) of 5412 ± 130 BP, for a peat at O.D. (MSL). However, when the large spring tidal range (in excess of 12 m) in the Bristol Channel is taken into account, it is clear that such an interpretation is not justified. Actual SL must be taken as close to *high water*. In this area the two differ by 4.6 m to >6 m. SL here has continued to rise since 5500 BP by upwards of 4 m (Kidson & Heyworth, 1973). In this case misinterpretation of the data appeared largely to eliminate the differences between the North Atlantic and Australian regions. Coleman & Smith (1964) argued that the apparent rise of the sea in the later Holocene can be explained in terms of massive and long continued subsidence in the Mississippi delta. Their date for the attainment of the present level by the rising Holocene sea (3650 BP), arrived at after allowing for a rate of subsidence of 0.24 feet (7.3 cm) per century, would also appear largely to cancel out the regional differences which are now widely found to be convincing. Studies such as these characterised the earlier assessments of SL change. At the present time it is generally accepted that there are variable but significant regional differences in the date of attainment of present SL. Whether these differences follow the pattern suggested by the rheological models, discussed earlier in this paper, only future detailed work will demonstrate.

Late Holocene high stands

Many parts of the world provide evidence of higher than present strand lines dating from the late Holocene. Hundreds of relative SL curves have been published showing oscillations above present levels. The problem in the 1980s, as it was in the 1950s, is to derive *eustatic* changes from relative data. In the earlier period two basic assumptions were made in this process. Firstly, it was assumed that there are stable parts of the crust where neither isostatic nor tectonic deformation complicates interpretation. The most persuasive arguments for higher than present late Holocene SLs came from the Australian work of Fairbridge (1948, 1950). The influence of Fairbridge's early presentations, including his 1961 paper, has been enormous in influencing later interpretations. His basic assumption that Western Australia is 'such a highly stable region during the recent epoch' so that 'we would expect old sea levels to be truly eustatic without danger of distortion by isostatic or other local causes', was scarcely questioned at the time. However, Playford & Leech (1977) have since argued that movement along the Darling Fault in the late Quaternary cannot be ruled out. They pointed to the 1968 Meckering earthquake centred on the

Darling plateau, only 100 km east of Point Peron, Rottnest Island and the Abrolhos, as evidence of regional instability. Many other authors, including the present writer (Kidson & Heyworth, 1973) have used arguments similar to Fairbridge's with as little justification. There is now general agreement that the so-called stable parts of the crust are only less stable than some others.

The second assumption was that ice-volume changes must be and have been reflected in SL changes. However, Mörner (eg. 1981) has repeatedly argued that eustasy is not a direct measure of glacial volume changes because of palaeo-geoidal variation. His argument now commands general acceptance.

Because these two assumptions have been largely discarded by many, if not most, workers, there is now much greater caution in attempting to derive eustatic information from the available data. By the same token, there is also much less certainty in interpreting higher relative land/SLs as evidence of higher eustatic SLs. Indeed, some authors (eg. Newman *et al.*, 1980) argue 'the near futility of searching for a purely eustatic curve'. Most authors share Black's (1980) view, from his work in the Aleutian Islands, that separation of isostatic, tectonic and eustatic movements 'remains difficult'. Nevertheless, claims continue to be made that crustal stability allows interpretation of evidence of high strand lines as of eustatic definition. Tjia *et al.* (1976) and Tjia (1977) in presenting a eustatic curve for Peninsula Malaysia claim 'tectonic stability of the region during at least late Quaternary time'. In consequence they interpret their data as indicating four stands 'above present high tide level' since 6000 BP which 'represent actual sea level change and are not the result of crustal movement'. Geyh *et al.* (1979) to some extent support this eustatic argument from studies of Mangrove peats.

Even in areas which have clearly been subjected to crustal deformation attempts to derive eustatic from relative SL curves continue. Perhaps the best known of these is Mörner's (1980) curve from Scandinavia which he claims represents a true eustatic curve for the north-east Atlantic region and which he relates to a very wide variety of palaeo-climatic and crustal parameters. The significant difference between this curve, and some earlier ones, which also show higher than present late Holocene stands of the sea, is that it is limited in areal relevance and makes no claim to world-wide validity. Whether the adjustments to the relative SL data for Scandinavia, applied by Mörner, will be generally accepted only time will show. In some other parts of the world, for example in Brazil (Martin *et al.*, 1979-80), high SLs are interpreted as due to variations in the geoid surface during the Holocene and no attempt is made to derive a eustatic curve.

The problem remains a difficult one. Even without the regions defined by the geophysicists (eg. those of Clark *et al.*, 1978 illustrated in Fig. 5) there remain differences of view. Thus McLean *et al.* (1978) interpret evidence from the Great Barrier reef off Queensland as indicating a high Holocene eustatic SL while Thom & Chappell (1978) seek an explanation of the same data in terms of hydro-isostatic theory. Similarly while Heyworth & Kidson (1982) find no evidence in Cardigan Bay, Wales, of higher than present levels, Mitchell & Stephens (1979) report a eustatic transgression to 3.5 m above modern SL about 5000 BP in Dublin Bay in the same latitude on the opposite side of the Irish Sea! Badyukov & Kaplin (1979) report similar differences of view within the regions of the Soviet Union and comparable discussion can be reported from many other regions of the world. These are not opposing views which can be explained in the context of *regional eustasy* and changes in the geoid. They arise from different interpretations of the eustatic component of *relative* data or from different compensation for potential errors. No single interpretation of world data for the whole of the Holocene has yet been presented which would permit us to make an assessment of how far the available evidence of high levels conforms to the pattern suggested by the rheological models of Clark *et al.* (1978) and others. Pirazzoli (1977) has attempted this for the last two thousand years and even within this shorter period, while noting regional differences of submergence or emergence, concludes that it would be hazardous to attempt any precise determination of *eustatic* sea level. He does, however, venture the opinion that possible eustatic fluctuations 'seem not to have exceeded decimetric order'.

No clear view on higher eustatic SLs in the Holocene has emerged. The differences of view in the 1980s seem as wide as they were in the 1950s and 1960s. A much more cautious approach can, however, be detected and many who have formerly written in support of one or other view now seem to prefer to suspend judgement.

CONCLUSIONS

IGCP 61 has run its course. National working groups have published or are preparing their final reports and the International Group is considering what form, if any, the next phase of co-ordinated study of SL change should take. (JGCP 200, ed.). This review has indicated that the problems to be resolved in such future work, while different, remain as considerable as at any time in the last thirty years. Some firm conclusions can, however, form the basis of this future work:

1) The search for a universal eustatic curve must be regarded as over.
2) Regional differences in response to changes in the geoid mean that eustatic SL curves can have only regional validity.
3) No part of the earth's crust can be regarded as wholly stable. Tectonic activity in the late Quaternary was significantly more widespread than assumed when the

present phase of SL studies began after the second World War. The visco-elastic nature of the earth's crust means that the concept of isostatic adjustment has to be widened to include that of *hydro-isostasy*.

It is now clear that progress in resolving the considerable difficulties which remain in understanding the nature of late and post-glacial SL change must depend on greater rigour in acquiring and interpreting data than has frequently been shown in the past. One significant outcome of recent work has been the recognition that the potential for error from a wide variety of causes is much greater than has been formerly accepted. This means that *screening* of SL data in the future, along the lines suggested by the working groups of International Geological Correlation Programme No. 61, will become increasingly important. Even if greater rigour is universally achieved it is by no means certain that some of the problems, outstanding from past studies, will be solved. The difficulties of separating the eustatic, tectonic and isostatic components of SL change are so complex that subjective analysis will always be possible. For the moment the only possible course is to keep an open mind about outstanding issues such as smooth or spasmodic change of level and higher than present stands of the sea. Only detailed and rigorous studies in individual localities may provide an adequate accumulation of data which may make an eventual solution of these outstanding problems possible.

ACKNOWLEDGEMENTS

I wish to thank my colleague Alan Heyworth who read the manuscript and whose collaboration over the last 18 years has had a significant influence on my thinking. The support of the Natural Environment Research Council (GR3/3683) is gratefully acknowledged.

EDITOR'S NOTE

Shennan (1983), in a note on 'A problem of definition in sea-level research methods', explains that the terms *transgressive overlap* and *regressive overlap* should be used as lithostratigraphic descriptive terms in which no process, such as SL rise or fall, is implied. These terms describe only a change in sediment type and should not be used in connection with the interpretation as to the cause(s) of this change.

Furthermore, Shennan defines a *positive (negative) tendency of SL movement* as an apparent increase (decrease) of the marine influence. This interpretative term refers to the process(es) operating on a wider scale as identified from a comparison of all available lines of evidence. Tendencies, it is stressed, are not synonyms for altitudinal changes. Whether a positive or negative tendency of SL movement can be shown to actually indicate a rise or fall of a particular SL, or tide level, is a further step in the analytical process.

REFERENCES

Adey, W.H., 1978. Coral reef morphogenesis: a multi-dimensional Model. *Science,* 202, 831-837

Admiralty, 1969. *Admiralty Manual of Hydrographic Surveying. Vol.II Tides and Tidal Streams,* Hydrographer of the Navy, 119 pp

Admiralty, 1982. *Tide Tables. Vol.I European Waters.* Hydrographer of the Navy, 382 pp

Badyukov, D.D. & Kaplin, P.A., 1979. Sea-level changes on the coasts of the U.S.S.R. during the last 15 000 years. *Proceedings of 1978 International Symposium on Coastal Evolution in the Quaternary.* Sao Paulo, Brazil, September 1978, 135-169

Belperio, A.P., 1979. Negative evidence for a mid-Holocene high sea level along the coastal plain of the Great Barrier Reef Province. *Marine Geology,* 32, M1-M9

Black, R.F., 1980. Isostatic, tectonic and eustatic movements of sea level in the Aleutian Islands, Alaska. In: *Earth Rheology, Isostasy and Eustasy,* ed. N.A. Mörner, (Proceedings of Symposium, Stockholm, 1977) John Wiley, New York, 231-248

Bloch, M.R., 1965. A hypothesis for the change of ocean levels depending on the albedo of the polar ice caps. *Palaeogeography, Palaeoclimatology, Palaeoecology,* 1(2), 127-142

Bloom, A.L. & Stuiver, M., 1963. Submergence of the Connecticut coast. *Science,* 139, 332-334

Bloom, A.L., 1967. Pleistocene shorelines: a new test of isostasy. *Geological Society of America Bulletin,* 78, 1477-1493

Bloom, A.L., 1970. Paludal stratigraphy of Truk, Ponape and Kusare, Eastern Caroline Islands. *Geological Society of America Bulletin,* 81, 1895-1904

Bloom, A.L., 1970b. Holocene submergence in Micronesia as the standard for eustatic sea-level changes. *Quaternaria,* 12, 145-154

Bloom, A.L., Brocker, W.S., Chappell, J.M.A., Mathews, R.K. & Mesolella, K.J., 1974. Quaternary sea-level fluctuations on a tectonic coast: new ^{230}Th/^{234}U dates from the Huon Peninsula, New Guinea. *Quaternary Research,* 4, 185-205

Bloom, A.L., 1977. *Atlas of Sea-Level Curves.* Cornell University, New York, (International Geological Correlation Programme 61)

Campbell, J.A. & Baxter, M.S., 1979. Radiocarbon measurements on submerged forest floating chronologies. *Nature,* 278, 409-413

Chappell, J., 1974. Late Quaternary glacio- and hydro-isostasy, on a layered earth. *Quaternary Research,* 4, 429-440

Chappell, J. & Polach, H.A., 1976. Holocene sea-level change and coral-reef growth at Huon Peninsula, Papua New Guinea. *Geological Society of America Bulletin*, 87, 235-239

Clark, J.A., Farrell, W.E. & Peltier, W.R., 1978. Global changes in Post-glacial sea level: a numerical calculation. *Quaternary Research*, 9, 265-287

Coleman, J.M. & Smith, W.G., 1964. Late recent rise of sea level. *Geological Society of America Bulletin*, 75, 833-840

Curray, J.R., 1960. Sediments and history of Holocene transgression, continental shelf, north-west Gulf of Mexico. In: *Recent sediments, Northwest Gulf of Mexico*, eds. F.P. Shepard, F.B. Phleger & Tj.H. van Andel, (American Association of Petroleum Geologists), Tulsa, 221-266

Curray, J.R., 1961. Late Quaterary sea level: a discussion. *Geological Society of America Bulletin*, 72, 1707-1712

Curray, J.R., 1965. Late Quaternary history, continental shelves of the United States. In: *The Quaternary of the United States*, eds. H.E. Wright & D.G. Frey, Princeton, New Jersey, 723-735

Daly, R.A., 1920. A general sinking of the sea in recent times. *Proceedings of National Academy of Science*, 6(5), 246-250

Daly, R.A., 1925. Pleistocene changes of level. *American Journal of Science*, (5th Series) 10, 281-313

Daly, R.A., 1934. *The changing world of the Ice Age*. Yale University Press, New Haven, Connecticut, 272 pp

Defant, A., 1941. Untersuchungen zue Statik und Dynamik des Atlantischen Ozeans, 5. *Wissensch. Ergebn. Dtsch. Atlant. Exped. 'Meteor'*, 6, 191-260

Delibrias, G. & Guillier, M.T., 1971. The sea level on the Atlantic coast and the Channel for the last 10,000 years by the ^{14}C method. *Quaternaria*, 14, 131-135

Delibrias, G., Giot, P.R., Gouletuer, P.L. & Morzadec-Kerfourn, M.-T., 1971. Evolution de la ligne de rivage le long du littoral armoricain depuis le Néolithique. *Quaternaria*, 14, 175-179

Delibrias, G. & Morzadec-Kerfourn, M.-T., 1975. Evolution du marais de Dol-de-Bretagne au Flandrien (Ille-et-Vilaine, France). *Ass. Francais pour l'Etude du Quaternaire*, Bulletin 11(43), 59-70

Devoy, R.J.N., 1979. Flandrian sea-level changes and vegetational history of the lower Thames Estuary. *Philosophical Transactions, Royal Society of London*, B, 285, 355-410

Devoy, R.J.N., 1982. Analysis of the geological evidence for Holocene sea-level movements in southeast England. *Proceedings of the Geologists' Association*, 93, 65-90

Easton, W.H. & Olson, E.A., 1976. Radiocarbon profile of Hanauma reef, Oahu, Hawaii. *Geological Society of America Bulletin*, 87, 711-719

Einsele, G., Herm, D. & Schwartz, H.U., 1974. Sea-level fluctuation during the past 6000 yr at the coast of Mauritania. *Quaternary Research*, 4, 282-289

Emery, K.O., Tracey, J.L. & Lada, H.S., 1954. Geology of Bikini and nearby atolls. Part I. Geology. *U.S. Geological Survey Professional Paper 260A*, 265 pp

Fairbridge, R.W., 1948. Notes on the geomorphology of the Pelsart group of the Houtman's Abrolhos Islands. *Royal Society of Western Australia Journal*, 33, 1-43.

Fairbridge, R.W., 1950. The geology and geomorphology of Point Peron, Western Australia. *Royal Society of Western Australia Journal*, 34, 35-72

Fairbridge, R.W., 1961. Eustatic changes in sea level. *Physics and Chemistry of the Earth*, 5, 99-185

Fairbridge, R.W., 1976. Shellfish-eating preceramic Indians in coastal Brazil. *Science*, 191, 353-359

Fedorov, P.V., 1972. Postglacial transgression of the Black Sea. *International Geological Review*, 14, 160-164

Fisk, H.N., 1951. Loess and Quaternary geology of the Lower Mississippi Valley. *Journal of Geology*, 59, 333-356

Fujii, S. & Fuji, N., 1967. Postglacial sea level in the Japanese Islands. *Journal of Geoscience*, (Osako City University), 10, 43-50

Geyh, M.A., 1980. Holocene sea-level history: case study of the evaluation of ^{14}C dates. *Radiocarbon*, 22, 695-704

Geyh, M.A., Kudrass, H.-R. & Streif, H., 1979. Sea-level changes during the late Pleistocene and Holocene in the Strait of Malacca. *Nature*, 278, 441-443

Gill, E.D., 1961. Change in the level of the sea relative to the land in Australia during the Quaternary era. *Zeitschrift für Geomorphologie*, Supplement III, 73-79

Gill, E.D. & Hopley, D., 1972. Holocene sea levels in eastern Australia - a discussion. *Marine Geology*, 12, 223-233

Godwin, H., Suggate, R.P. & Willis, E.H., 1958. Radiocarbon dating of the eustatic rise in ocean-level. *Nature*, 181, 1518-1519

Gottschalk, M.K.E., 1971, 1975, 1977. *Stormiloeden en Rivieroverstromingen in Nederland, 600-1700 A.D.* 3 Vols. Van Gorcum, Essen

Grant, D.R., 1970. Recent coastal submergence of the Maritime Province, Canada. *Canadian Journal of Earth Sciences*, 7, 676-689

Greensmith, J.R. & Tucker, E.V., 1973. Holocene transgressions and regressions on the Essex coast, outer Thames estuary. *Geologie en Mijnbouw*, 52, 193-202

Guilcher, A., 1969. Pleistocene and Holocene sea level changes. *Earth Science Reviews*, 5, 69-97

Hails, J.R. & Hoyt, J.H., 1971. The question of the late Quaternary changes of sea level in New South Wales, Australia. *Quaternaria*, 14, 255-264

Heyworth, A. & Kidson, C., 1982. Sea-level changes in south-west England and Wales. *Proceedings of the Geologists' Association*, 93, 91-111

Hillaire-Marcel, C., 1979. *Les mers post-glaciaires du Quebec: quelques aspects*. Thèse de Doctorat de l'Etat Univ. Pierre et Marie Curie, Paris, 293 pp

Hillaire-Marcel, C. & Fairbridge, R.W., 1978. Isostasy and eustasy of Hudson Bay. *Geology*, 6, 117-122

Hopley, D., 1971. Sea-level and environment changes in the late Pleistocene and Holocene in North Queensland, Australia. *Quaternaria*, 14, 267-274

Hopley, D., 1978. Sea-level change on the Great Barrier reef: an introduction. *Philosophical Transactions Royal Society of London*, A, 291, 159-166

Hopley, D., 1980. Mid-Holocene sea levels along the Coastal plain of the Great Barrier Reef Province: a discussion. *Marine Geology*, 35, M1-M9

International Radiocarbon Study Group, 1982. An inter-laboratory comparison of radiocarbon measurements in tree-rings. *Nature*, 298, 619-623

Jardine, W.G., 1975a. Chronology of Holocene marine transgression and regression in South-western Scotland. *Boreas*, 4, 173-196

Jardine, W.G., 1975b. The determination of former sea levels in areas of large tidal range. In: *Quaternary Studies*, eds. R.P. Suggate & M.M. Cresswell, (Royal Society of New Zealand), Wellington, 163-168

Jelgersma, S., 1961. Holocene sea-level changes in the Netherlands. *Meded. Geologische Stichting*, Series C-V1, 101 pp

Jelgersma, S., 1966. Sea-level changes during the last 10,000 years. *Royal Meteorological Society Proceedings of Symposium on World Climate 8000-0 B.C.*, 54-71

Jelgersma, S., 1971. Sea-level changes during the last 10,000 years. In: *Introduction to Coastline Development*, ed. J.A. Steers, Macmillan, 25-48

Jelgersma, S. & Pannekoek, A.J., 1960. Post-glacial rise of sea level in the Netherlands (a preliminary report). *Geologie en Mijnbouw*, 39, 201-207

Kidson, C., 1981. Criteria for the evaluation of 'evidence' for higher Holocene sea levels. *Oceanis*, 7, 381-388

Kidson, C. & Heyworth, A., 1973. The Flandrian sea-level rise in the Bristol Channel. *Proceedings of the Ussher Society*, 2, 565-584

Kidson, C. & Heyworth, A., 1976. The Quaternary deposits of the Somerset Levels. *Quarterly Journal of Engineering Geology*, 9, 217-235

Kidson, & Heyworth, A., 1978. Holocene eustatic sea-level change. *Nature*, 273, 748-750

Kidson, C. & Heyworth, A., 1979. Sea 'Level'. *Proceedings 1978 International Symposium on Coastal Evolution in the Quaternary*. Sao Paulo, Brazil, September 1978, 1-27

Kraft, J.C., 1971. Sedimentary facies patterns and geologic history of a Holocene marine transgression. *Geological Society of America Bulletin*, 82, 2131-2158

Krog, H., 1965. On the Post Glacial development of the Great Belt. *Baltica*, 2, 47-60

Laborel, J., 1979a. Fixed marine organisms as biological indicators for the study of recent sea level and climatic variations along the Brazilian tropical coast. *Proceedings 1978 International Symposium on Coastal Evolution in the Quaternary*. Sao Paulo, September 1978, 193-211

Laborel, J., 1979b. Les Gasteropodes Vermetidés: leur utilisation comme marqeurs biologiques des rivages fossiles. *Oceanis*, 5, 221-238

Laborel, J. & Delibrias, G., 1976. Niveaux marins récents à Vermetidae du littoral ouest africain. *Bulletin liaison. Assoc. Sénégale. Etudes Quaternaire Afrique*, 47, 97-110

Lisitzin, E., 1965. The mean sea level of the world oceans. *Comment. Phys. Math. Helsingf*, 30, 35 pp

Martin, L., Suguio, K., Flexor, J.-M., Bittencourt, A. & Vilas-Boas, G., 1979-80. Le quaternaire marine brésilien (littoral pauliste, sud Fluminense et Bahianais). *Cahiers ORSTOM. Serie Géologie*, 11(1), 95-124

McFarlan, E., 1961. Radiocarbon dating of Late Quaternary deposits, South Louisiana. *Geological Society of America Bulletin*, 72, 129-158

McLean, R.F., Stoddart, D.R., Hopley, D. & Polach, H., 1978. Sea-level change in the Holocene on the Northern Great Barrier Reef. *Philosophical Transactions Royal Society of London*, A, 291, 187-194

Menke, B., 1969. Vegetationgeschitliche Untersuchungen und Radiocarbon-Datierungen zur holozänen Entwicklung der schleswig - holsteinischen Westkuste. *Eiszeitalter und Gegenwart*, 20, 35-45

Milliman, J.D. & Emery, K.O., 1968. Sea levels during the Past 35,000 years. *Science*, 162, 1121-1123

Mitchell, G.F. & Stephens, N., 1979. Is there evidence for a Holocene sea level higher than that of today on the coasts of Ireland? *Colloques Internationaux du CRNS*, 219, 115-125

Moriwaki, H., 1978. Problems concerning Holocene sea-level changes. *Geographical Reports*, Tokyo Metropolitan University, 13, 49-64

Mörner, N.-A., 1969a. The late Quaternary history of the Kattegatt Sea and the Swedish west coast: deglaciation, shore-level displacement, chronology, isostasy and eustasy. *Sveriges Geol. Undersökning,* Series C, no.640, 487 pp

Mörner, N.-A., 1969b. The Fennoscandian uplift and late Cenozoic geodynamics: geological evidence. *Journal of Geology,* 3.3, 287 318

Mörner, N.-A., 1971. Eustatic changes during the last 20,000 years and a method of separating the isostatic and eustatic factors in an uplifted area. *Palaeogeography, Palaeoclimatology, Palaeoecology,* 19, 63-65

Mörner, N.-A., 1976. Eustasy and geoid changes. *Journal of Geology,* 84, 123-151

Mörner, N.-A., 1977. Eustasy and instability of the geoid configuration. *Geologiska Föreningens i Stockholm Förhandlingar,* 99, 369-376

Mörner, N.-A., 1978. Palaeoclimatic, palaeomagnetic and palaeogeoidal changes: complexity and interaction. *Proceedings CNES Colloque International 'Evolution of planetary atmospheres and climatology of the earth',* Nice, 1978, 221-232

Mörner, N.-A., 1979a. Paleogeoid changes. *Abstracts 6th E.G.S. Meeting,* Vienna, 1979, 563-564

Mörner, N.-A., 1979b. South Scandinavian sea-level records; a test of regional eustasy, regional paleoenvironmental changes and global paleogeoid changes. *Proceedings 1978 International Symposium on Coastal Evolution in the Quaternary,* Sao Paulo, Brazil, September 1978, 77-103

Mörner, N.-A., 1980. The Northwest European 'sea-level laboratory' and regional Holocene eustasy. *Palaeogeography, Palaeoclimatology, Palaeoecology,* 29, 281-300

Mörner, N.-A., 1981. Eustasy, palaeoglaciation and palaeoclimatology. *Geologische Rundschau,* 70(2), 691-702

Neumann, A.C., 1969. *Abstracts VIII Inqua Congress Paris,* 1969, 228-229

Neumann, A.C., 1971. Quaternary sea-level data from Bermuda. *Quaternaria,* 14, 41-43

Neumann, A.C., 1972. Quaternary sea-level history of Bermuda and the Bahamas. *American Quaternary Association (AMQUA) 2nd National Conference,* Miami, Florida, (Abstracts) 41-44

Newmann, W.S., Marcus, L.F., Pardi, R.R., Paccione, J.A. & Tomecek, S.M., 1980. Eustasy and deformation of the geoid: 1000-6000 radiocarbon years BP. In: *Earth Rheology, Isostasy and Eustasy,* ed. N.-A. Mörner, (Proceedings of Symposium, Stockholm 1977) John Wiley, New York, 555-567

Ogden, J.G. III., 1977. The use and abuse of radiocarbon dating. *Annals New York Academy of Science,* 288, 167-173

Ota, T., Matsushima, Y. & Moriwaki, H., 1981. *Atlas of Holocene Sea-Level Records in Japan*. Presented at the Meeting of the Quaternary Shoreline Holocene, and Neotectonic Commissions of Inqua and of the International Working Group of the International Geological Correlation Programme 61, Columbia, S. Carolina, April 1981

Pardi, R. & Marcus, L., 1977. Non-counting errors in ^{14}C Dating. *Annals New York Academy of Science*, 288, 174-180

Pinot, J.-P., 1079. Methodes et techniques permettant de determiner l'altitude relative des niveaux marins anciens en fonction de l'altitude absolue actuelle des divers indicateurs. *Oceanis*, 5, 335-355

Pirazzoli, P.A., 1977. Sea-level relative variations in the world during the last 2000 years. *Zeitschrift für Geomorphologie*, N.F. 21(3), 284-296

Pirazzoli, P.A., 1978. High stands of Holocene sea levels in the Northwest Pacific. *Quaternary Research*, 10, 1-29

Playford, P.E. & Leech, R.E.J., 1977. Geology and Hydrology of Rottnest Island. *Geological Survey of Western Australia. Report, 6*. Perth, 98 pp

Preuss, H., 1979. Progress in computer evaluation of Sea-level data within the IGCP Project NO.61. *Proceedings International Symposium on Coastal Evolution in the Quaternary*, Sao Paulo, Brazil, September 1978, 104-134

Ralph, E.K., Michael, H.N. & Han, M.C., 1973. Radiocarbon dates and reality. *MASCA Newsletter*, 9, 1

Schofield, J.C., 1960. Sea-level fluctuations during the last 4000 years as Recorded by a Chenier Plain, Firth of Thames, New Zealand. *New Zealand Journal of Geology and Geophysics*, 3, 467-485

Schofield, J.C., 1964. Post-glacial sea-levels and isostatic uplift. *New Zealand Journal of Geology and Geophysics*, 7, 359-370

Schofield, J.C., 1975. Sea level fluctuations cause periodic postglacial progradation, South Kaipara barrier, North Island, New Zealand. *New Zealand Journal of Geology and Geophysics*, 18, 295-316

Schofield, J.C., 1977. Late Holocene sea level, Gilbert and Ellice Islands, west central Pacific Ocean. *New Zealand Journal of Geology and Geophysics*, 20, 503-529

Scholl, D.W. & Stuiver, M., 1967. Recent submergence of Southern Florida: a comparison with adjacent coasts of other eustatic data. *Geological Society of America Bulletin*, 78, 437-454

Scholl, D.W., Craighead, F.C.Sr. & Stuiver, M., 1969. Florida submergence curve revised: its relation to coastal sedimentation rates. *Science*, 163, 562-564

Scott, D.S. & Medioli, F.S., 1978. Vertical zonations of marsh foraminifera as accurate indicators of former sea levels. *Nature*, 272, 528-531

Seymour, W.A. (ed.), 1980. *The History of the Ordnance Survey*. Wm. Dawson & Sons, Folkestone, England, 394 pp

Shalowitz, A., 1964. *Shore and Sea Boundaries*. U.S.Dept. of Commerce, Coast and Geodetic Survey. Vol.2, p.49. Washington U.S. Govt. Printing Office.

Shennan, I., 1979. Statistical evaluation of sea-level data. *Information Bulletin of International Geological Correlation Programme Project No.61*, 1, 6-11

Shennan, I., 1982. Interpretation of Flandrian sea-level data from the Fenland, England. *Proceedings of the Geologists' Association*, 93, 53-63

Shennan, I., 1983. A problem of definition in sea-level research methods. *Quaternary Newsletter*, 39, 17-19

Shepard, F.P., 1960. Rise of sea level along north-west Gulf of Mexico. Recent sediments. In: *Recent Sediments, Northwest Gulf of Mexico*, eds. F.P. Shepard, F.B. Phleger & Tj.H. van Andel, American Association of Petroleum Geologists, Tulsa, 338-344

Shepard, F.P., 1963. Thirty-five thousand years of sea level. In: *Essays in Honor of K.O. Emery*, University of Southern California Press, Los Angeles, pp 1-10

Shepard, F.P., 1964. Sea-level changes in the past 6000 years: possible archaeological significance. *Science*, 163, 574-576

Shepard, F.P. & Curray, J.R., 1967. Carbon-14 determination of sea-level changes in stable areas. *Progress in Oceanography*, 4, 283-291

Shepard, F.P. & Suess, H.E., 1956. Rate of postglacial rise of sea-level. *Science*, 123, 1082-1083

Steers, J.A., 1953. The east coast floods. *Geographical Journal*, 119, 280-298

Streif, H., 1978. Cyclic formation of coastal deposits and their indications of vertical sea-level changes. *Oceanis*, 5, 303-308

Stuckenrath, R., 1977. Radiocarbon: Some notes from Merlin's diary. *Annals New York Academy of Science*, 288, 181-186

Suess, H.E., 1970. Bristlecone fine calibration of the radiocarbon time scale 5200 B.C. to the present. In: *Radiocarbon Variations and Absolute Chronology*, ed. I. Olsson, 12th Nobel Symposium, Uppsala, 303-312

Suess, H.E., 1980. The radiocarbon record in tree rings of the last 8000 years. *Radiocarbon*, 22, 200-209

Suggate, R.P., 1968. Post-Glacial sea levels in the Christchurch metropolitan area. *Geologie en Mijnbouw*, 47, 291-297

Teichert, C., 1950. Late Quaternary changes of sea level at Rottnest Island, Western Australia. *Proceedings Royal Society, Victoria*, 59, 63-79

Ters, M., 1973. Les variations du niveau marin depuis 10,000 ans, le long du littoral Atlantique français. In: *Le Quaternaire, Géodynamique, Stratigraphie et Environment*, Le Comité National francais de l'Inqua, 114-135

Thom, B.G., Hails, J.R. & Martin, A.R.H., 1969. Radiocarbon evidence against higher post-glacial sea levels in eastern Australia. *Marine Geology*, 7, 161-168

Thom, B.G., Hails, J.R., Martin, A.R.H. & Phipps, C.V.G., 1972. Post-glacial sea levels in eastern Australia - a reply. *Marine Geology*, 12, 233-242

Thom, B.G. & Chappell, J., 1975. Holocene sea levels relative to Australia. *Search*, 6, 90-93

Thom, B.G. & Chappell, J., 1978. Holocene sea-level change: an interpretation. *Philosophical Transactions Royal Society of London*, A, 291, 187-194

Tjia, H.D., Fujii, S., Kigoshi, K. & Sugimura, A., 1975. Additional dates on raised shorelines in Malaysia and Indonesia. *Sains Malaysiana*, 4, 69-84

Tjia, H.D., Fujii, S., Kigoshi, K. & Sugimura, A., 1976. Quaternary sea level in the southern South China Sea area. *ESCAP, CCOP Symposium Quaternary Geology*, Kuala Lumpur, November 1976

Tjia, H.D., 1977. Inqua Quaternary Shorelines Commission: Pacific and Indian Oceans Sub-commission Region: Indonesia and Malaysia. Status report 1975-1976. *Berita Direkt. Geological/Geosurvey Newsletter*, 9, 89-92

Tooley, M.J., 1974. Sea-level changes during the last 9000 years in North-West England. *Geographical Journal*, 140, 18-42

Tooley, M.J., 1978a. The UNESCO-IGCP Project on sea-level changes. *International Journal of Nautical Archaeology and Underwater Exploration*, 7, 75-77

Tooley, M.J., 1978b. Interpretation of Holocene sea-level changes. *Geologiska Föreningens i Stockholm Förhandlingar*, 100(2), 203-212

Tooley, M.J., 1979. Sea-level changes during the Flandrian stage and the implications for coastal development. *Proceedings 1978 International Symposium on Coastal Evolution in the Quaternary*, Sao Paulo, Brazil, September 1978, 502-533

Tooley, M.J., 1982. Sea-level changes in northern England. *Proceedings of the Geologists' Association*, 93, 43-51

Walcott, R.I., 1972. Past sea levels eustasy and deformation of the earth. *Quaternary Research*, 2, 1-14

Ward, W.T., 1971. Postglacial changes in level of land and sea. *Geologie en Mijnbouw*, 50, 703-718

Warner, R.B., 1975. Some tables for users of radiocarbon dates, including the construction of 'Equal-area bisymmetrical normal curves'. *Irish Archaeological Research Forum*, 11(2), 29-47

Wilks, P.J., 1977. *Flandrian sea-level change in the Cardigan Bay area*. Unpublished Ph.D. Thesis, University of Wales

Zeuner, F.E., 1958. *Dating the Past*, Methuen, London, 516 pp

SEA-LEVEL INDICATORS

3

Elevation and age relationships: raised marine deposits and landforms in glaciated areas:

Examples based on Canadian Arctic data

J. T. ANDREWS

ABSTRACT

This paper outlines approaches to four basic questions that arise in the study of raised marine sediments and shoreline forms in glaciated regions. These questions are: 1) what is the relationship of the deposit to glacial ice? 2) what is the present altitude of the deposit or landform and how does it specify a former sea level? 3) what are the actual or inferred relative ages of the sediment or features? and, finally, 4) what methods can be used to portray the patterns of relative sea-level change and shoreline deformation?

INTRODUCTION

Sediments of glacio-marine and marine origin extend from 0 to *c*.300 m above sea level (SL) within areas that were formerly covered by the Pleistocene ice sheets. In addition to raised marine sediments, these areas of glacio-isostatic recovery also record traces of former shorelines, such as marine-cut terraces, and the lower limit of till and/or perched boulders (Sim, 1960; Bird, 1954; Andrews, 1970). The study of raised marine features in areas of the former Laurentide and Fennoscandian ice sheets has provided significant information on the viscoelastic structure of the earth (Mörner (ed.), 1980; Peltier & Andrews, 1976; Cathles, 1975; Gutenberg, 1941) and in turn provides vital data on the geometry and chronology of the ice sheets (eg. Clark, 1980a; Peltier & Andrews, 1976; Walcott, 1972; Cathles, 1975; Prest, 1969; Bryson *et al.*, 1969) and on global changes of SL.

There are four basic questions that need to be asked when investigators are working in areas of glacio-isostatically raised marine sediments and are concerned

1a

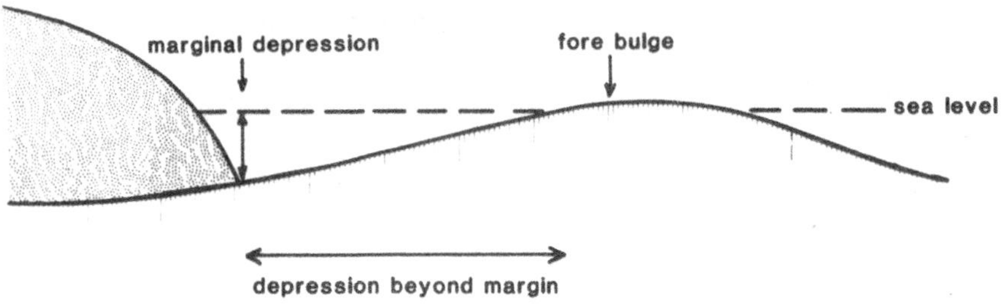

1b

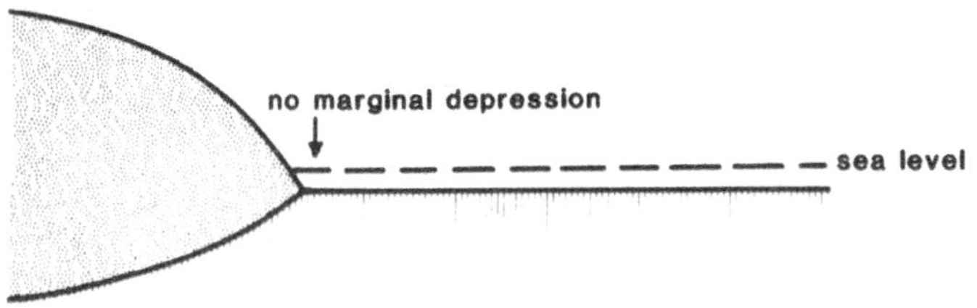

Figure 1a. Schematic model of a visco-elastic earth with a lithosphere showing the flexing of the lithosphere at the ice margin (after Walcott, 1970)
Figure 1b. Schematic diagram of a purely 'plastic' earth in which the effect of the ice load terminates at the margin of the load and does not extend seaward

with SL variations with time:

1) What was the relationship of the sediment or landform to glacial ice?
2) What is the altitude (above present SL) of the paleo-sea into which specific sediments were deposited or a landform produced?
3) What are the relative and absolute ages of the sediment or feature?

4) What methods can be used to show the amount and degree of crustal warping?

In the field these questions must be kept in mind and the relationships continually evaluated. In the remainder of this chapter each of these concerns will be examined. There is a very large literature on the topics I will discuss. In particular much research stems from the long-continued efforts of Fennoscandian researchers to understand the process and result of glacial isostatic and eustatic SL changes. However, in this chapter I draw most heavily upon my own experience and the published literature for North America. This bias in the in the literature citations is obvious and is acknowledged. A major source of literature on raised shorelines is contained in the *Annotated Bibliography of Quaternary Shorelines* (first published in 1965). Its volumes contain much additional information on other countries; nevertheless, the methods and problems presented here with a marked North American bias have been applied universally within formerly glaciated areas (eg. Richards & Fairbridge, 1965; Richards & Shapiro 1969)

RELATIONSHIP OF SEDIMENT OR LANDFORM TO GLACIAL ICE

This question has important ramifications in two distinct fields of enquiry. First, it relates to the problem of what is an appropriate model for the earth (eg. Brotchie & Silvester, 1969; Walcott, 1970); *whereas* secondly it is tied closely to the problems of evaluating the extent of Pleistocene ice sheets around the coasts of the world, especially in arctic areas (Andrews, 1978; Boulton, 1979; Grosswald, 1980; Denton & Hughes, 1981).

Figures 1a & b show two simple earth models. In the first (Fig. 1a) the earth is considered to have an elastic lithosphere which bends under the applications of an ice load. A simple but crude analogue is a sheet of thin but strong plywood to which a load has been applied, floating in a bath tub. Toward the edge of the load the strength of the lithosphere extends the depression to and beyond the physical ice margin. Figure 1a thus implies that raised marine deposits may be found *beyond* the maximum limits of glaciation. Conversely, Figure 1b shows an earth model with no lithosphere and depression is zero at the ice margin. In this latter model marine deposits clearly may be interpreted as representing rebound from beneath an overlying ice mass. This model is favoured by Denton & Hughes (1981) in their attempts to reconstruct Late Pleistocene ice sheets, whereas the model implicit in Figure 1a is favoured by most geophysicists and indeed is supported by field studies from arctic regions (Clark, 1980b; Boulton, 1979; Miller *et al.*, 1977; Dyke, 1979). What is now required of field workers is a detailed evaluation of the association between raised marine sediments or landforms, and the presence or absence of adjacent glacial ice (eg. Fig. 1). This question introduces the need for adequate sedimentation models and specifically requires distinctions to be made between sedimentary sequences that are, and are

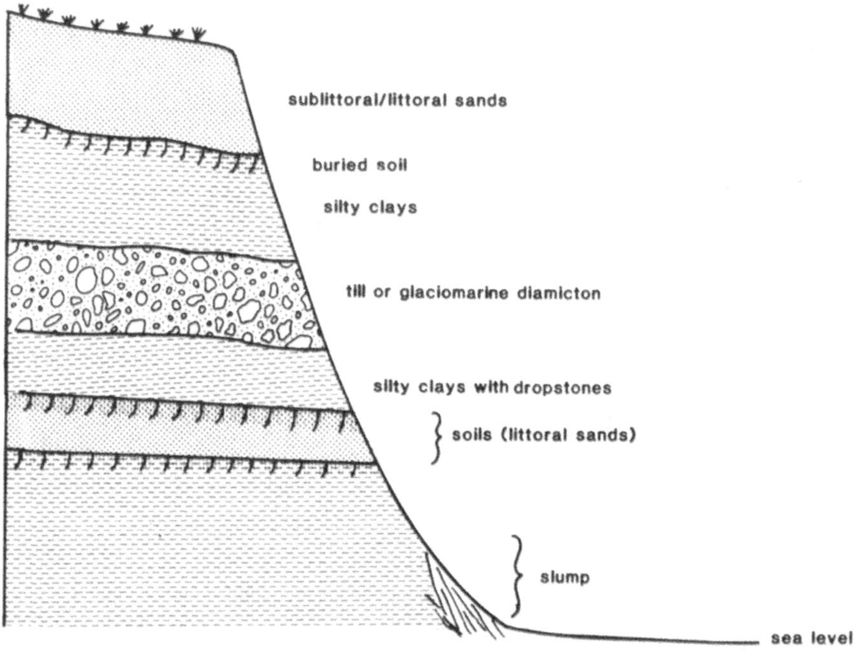

Figure 2. Diagram of a typical sequence of sediments from raised cliff sections along the coast of eastern Baffin Island or Spitsbergen (eg. Miller *et al.*, 1977; Nelson, 1981; Boulton, 1979b) in which alternating sequences of littoral, offshore, and glacial sediments alternate and reflect variations in the proximity of glacial ice and variations in relative sea level (*cf.* Andrews, 1978). Buried soils are commonly found intercalated and reflect subaerial exposure and pedogenesis.

not, directly associated with a glacial sediment source. In addition, great attention must be paid to the search for, and recognition of, erosional unconformities or disconformities caused by cycles of erosion and deposition during glacio-isostatically induced transgressions and regressions. Thus the traditional interpretation of postglacial raised marine sediments as offlap deposits only applies to areas within the late glacial margins. Beyond, or at the ice margin itself, the sedimentary records from some areas (eg. Miller *et al.*, 1977; England *et al.*, 1978; Boulton, 1979) indicate repetitive sedimentary sequences (Nelson, 1978, 1981; Mode, 1980) with raised marine strata separated by buried soils, peaty mucks, or erosional unconformities (Fig. 2).

Figure 3a & b represents an idealized situation but portrays the sequence of deposits and landforms that typify ice-marginal depositional sites and sediments deposited at sites distal from an ice margin. Of critical importance is the common misfit in sediment volume between deltas formed in drainage basins occupied by large ice lobes and those deposited at the seaward terminus of unglacierized drainage basins. Church (1972) has documented the obvious difference in run-off and sediment transport between basins with and without an ice cover. The most explicit evidence for the ice-contact nature of a raised marine unit is: 1) that the ice marginal features (lateral moraines/kame terraces) can be traced downvalley until a contact is made with one or more littoral marine facies; and 2) *large*, abandoned deltas occur in positions where they could not have received sediment except from a former ice front. Figure 4 is a sketch of such a situation from the outermost southeast coast of Baffin Island. The sketch shows a delta with its ice-contact slope on the seaward side of the small bay. This relationship indicates that deposition must have taken place during an episode reconstructed as in Figure 4b.

One of the major points illustrated by Figure 3 is that the highest raised marine landform/deposit, which is usually referred to as the 'marine limit', need not necessarily be associated directly with glacial retreat and the subsequent rebound and sequence of off-lap marine deposits. In places, the highest marine unit of a particular age may represent the upper limit of a marine transgression reflecting glacial events tens of kilometers distance. It is also important to note that the marine limit may be formed during deglaciation, or long after depending on the interplay between isostatic and eustatic SL rates of change.

DETERMINATION OF PALEO-SEA LEVELS FROM RAISED MARINE SEDIMENTS AND FEATURES

Raised marine sediments and landforms require that their mode of origin be known before the present elevation of the features can be adjusted, as far as possible, so that paleo-SLs can be determined. Such studies require answers to the following: 1) what can be used as a measure of present SL? 2) how can the elevations of raised marine deposits be calculated? and 3) what aspects (structural, sedimentological, biological) of a deposit can be used to infer the surface of a former water plane?

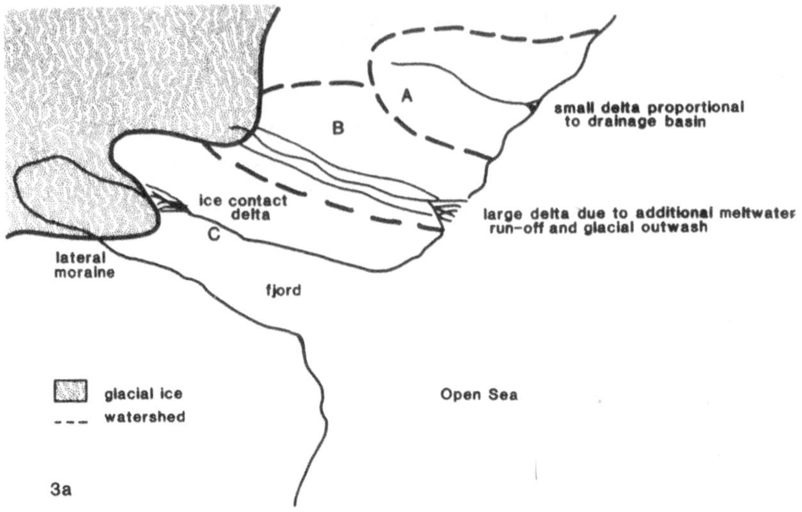

3b

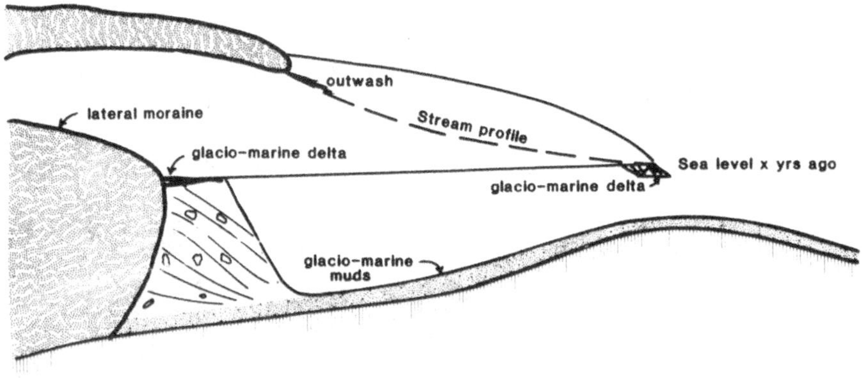

Figure 3a. Plan view of ice margin during a glacial readvance and stillstand. In the far watershed (A) no glacial ice exists within the watershed and sedimentation into a raised sea-level results in a small delta; however, in watershed (B) the delta is *over*-sized for the watershed because of the influx of glacial sediment; in area (C) the actual ice/margin contact can be traced by the continuation of the lateral moraine into a raised marine delta.

Figure 3b. Sketch and projection of 3a onto the plane X-X' showing the situation at sites (B) and (A) of Figure 3a.

4a

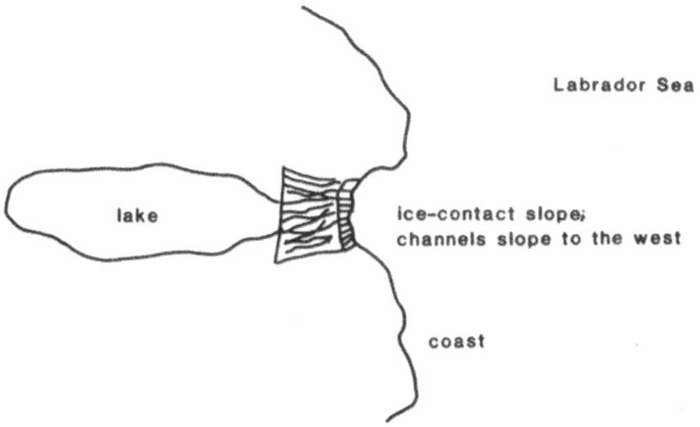

4b

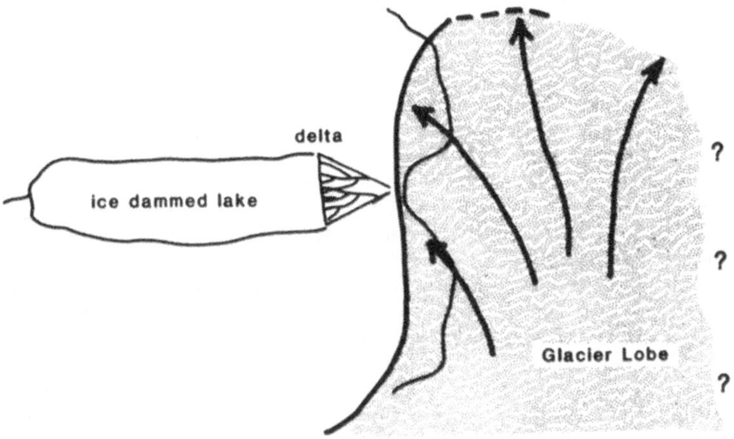

Figure 4a. Sketch of the field relationships at the southernmost tip of SE Baffin Island where a large raised delta, containing *in situ* marine molluscs has an ice-contact slope facing *seaward*.

Figure 4b. Sketch of the necessary glacial situation to result in the sequence illustrated as Figure 4a, with an ice lobe flowing northward on the Baffin Island shelf.

1. Present sea level

In remote areas of the world, or in areas not covered by detailed surveys, the first fundamental problem is what can be used to determine the present SL and how do we define it? This is not a trivial problem because of the natural variations along coasts of tidal amplitudes, fetch, and sediment availability. SL datum indicators must also apply to a specific point on the tidal curve. In areas that are settled, where surveys exist, and where tide tables can be used, most investigators emphasize elevations with respect to some survey datum or convert this to mean tide level (MTL). In remote areas, especially where large stretches of coast are being investigated, the researcher has to select some observable feature within the present tidal range. In certain areas this level might be the elevation of a certain organism, such as *Balanus balanoides* (Linné) which lives within the intertidal zone but with an upper limit close to the MTL (Fig. 5). In areas where sea ice covers the intertidal zone for many months of the year, the intertidal fauna and flora are scarce or absent. In these areas the upper limit of 'fresh' marine debris has to be used. This level approximates to 'high tide' but it is certainly affected by exposure and meteorological conditions.

In remote areas it is critical to try to determine the tidal range as accurately as possible by installing short-term tide-gauge stations, or by measuring the tidal range throughout a field season. In addition, it is important to note the present-day elevational range of shore sediments and features so that these 'present' observations may be used as guides to reconstruct former SLs.

The accuracy of the datum that is selected is partly a function of the tidal range. In areas of small tidal range (<3 m) only areas affected by extreme fetch or by tidal surges are likely to result in significant error. But in some coastal areas, for example, the tidal range today is 9 to 14 m and in such areas determinations of a present datum could vary by ±7 m along a complex coastline. In general, the littoral biological indicators of tidal range are less variable than such physical features as gravel ridges and highest seaweed/debris limit. Figure 5 shows a cross-profile of a littoral zone and some of the indicators that are used for datums in paleo-SL surveys. The basic need in such studies is the application of a *consistent* datum. Jardine (1975, 1976) discussed these and other problems at length (see also Jardine, this volume).

2. Determination of paleo-sea levels

Analysis of raised marine deposits with a view to reconstructing former SLs is carried on through two phases. In the *first*, the physical characteristics of the sediment or landform are used to infer the depth of water during deposition or formation; whereas the *second* approach is to examine the sediment or landform for biological indicators of former SLs. However, in both instances the necessary

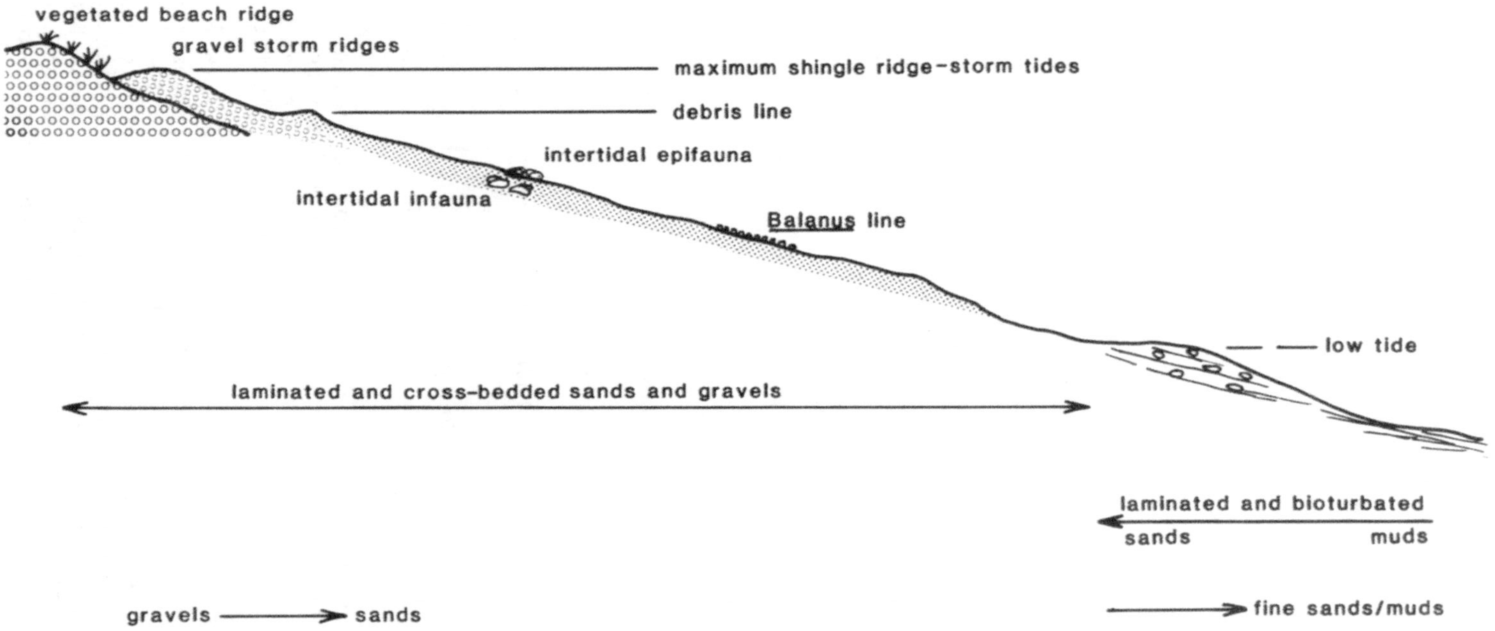

Figure 5. Illustration of some of the characteristic faunal, floral, and sedimentological conditions that occur between the high- and low-tide limits for an average coast. In areas of long fetch, erosion probably dominates and only the upper limit of wave action will be recorded

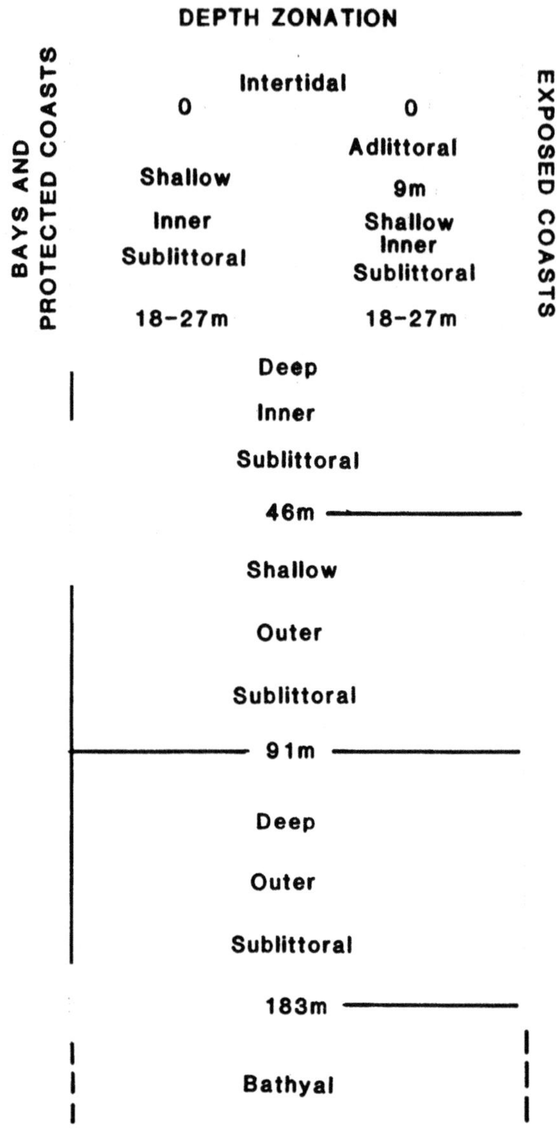

Figure 6. General scheme of depth zonation for molluscs on exposed and protected coasts. In raised marine deposits, former depths of *in situ* molluscan faunas vary between the intertidal to deep outer sublittoral

models for reconstruction of former SLs require either a *lithofacies or biofacies model* from which inferences about past conditions can be derived. In areas of glacio-isostatic movement (as compared with active tectonic coasts) these facies represent marine and glacial marine transgressive and regressive sequences deposited in up to 300 m of water, but most typically in 50 to 200 m of water and now raised above SL. Figure 6 illustrates one depth zonation

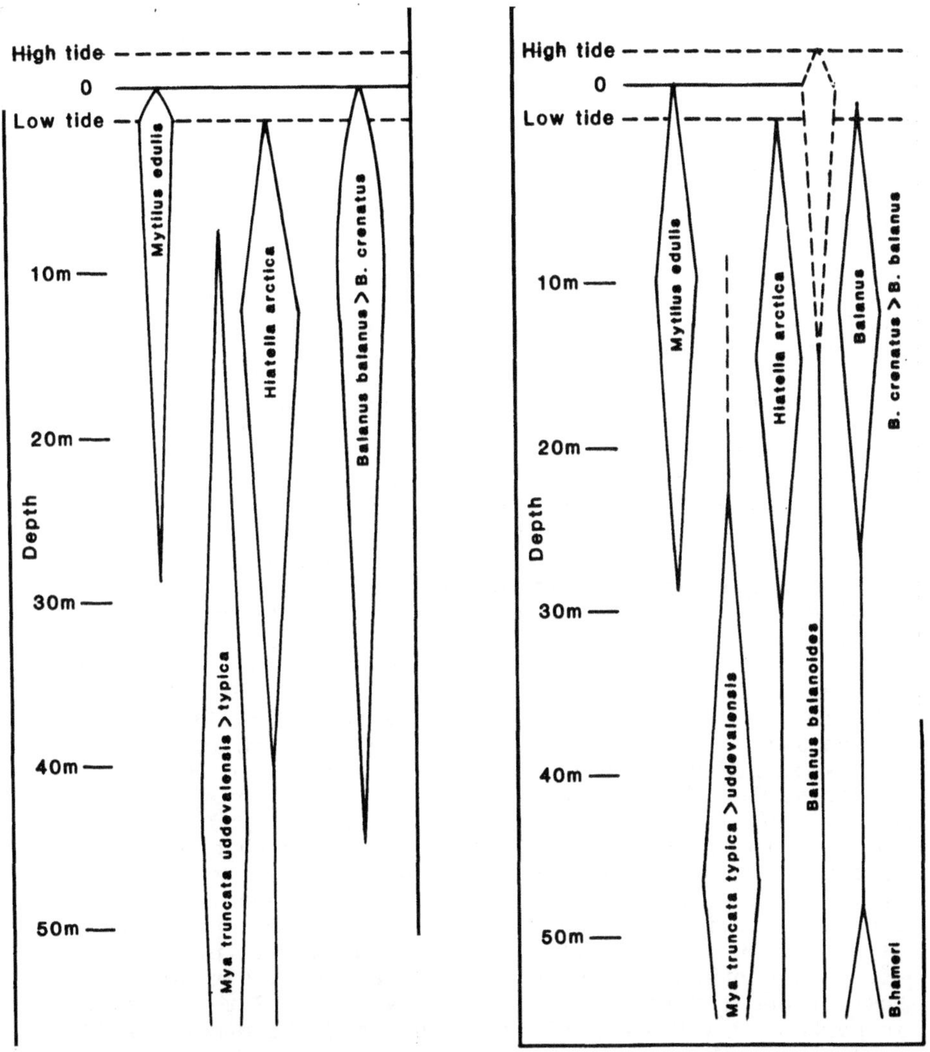

Figure 7. From Hillaire-Marcel (1980): Depth preferences of certain epifauna and infauna molluscs for what the author (*op.cit.*) terms 'northern basins' and 'closed basins'

used in mollusc studies (Valentine, 1961; Kennedy, 1978) whereas Figure 7 illustrates the various proportions of various sessile epifauna taxa at different depths in eastern Canada (Hillaire-Marcel, 1980).

The shallow-water deposits are dominated by sands and/or gravels whereas in deeper water and in more glacial distal environments (Fig. 5) clayey silts and silty clays are more likely to be laid down. Although there are several lithofacies models for marine transgression/regression sequences based on Holocene shelf sequences or on older (pre-Quaternary rocks), these do not always adequately portray the situation in areas of raised marine sediments because: 1) postglacial[1] rebound is a geologically rapid process involving relative falls in SL of 1 to 10 m/100 yr during the early stages of rebound (Andrews, 1970; Walcott, 1972); and 2) the changes in sediment supply and texture are influenced by *both* water depth and nearness of the glacial margin.

a) Marine limits

The first and most important determination to be made in an area is the elevation of the *marine limit*. There can be only *one* marine limit along a stretch of coast and theoretically both younger and older sediments and landforms can extend elevationally below this limit. In some countries the term 'highest shoreline' is used rather than 'marine limit'. This elevation is the highest trace of marine action preserved on a coast. In glaciated areas this limit is most frequently determined by:

- the *lower* limit of perched glacial boulders
- the *lower* limit of till
- the *lower* limit of continuous terrestrial deposition
- the *upper* limit of wave-washed bedrock
- the *upper* limit of shore deposits - beach ridges
- the *upper* limit of *in situ* marine fossils

These criteria frequently apply to exposed coasts or where sedimentation rates during deglaciation were very low (Figure 8). In an ice-contact situation (Figs. 3a, 8b) the maximum recorded elevation of the paleo-SL is commonly expressed by the flat and channelled surfaces of massive glacio-marine deltas and extensive littoral sediments associated with longshore drift. The former position of low tide is taken as the elevation at which channels disappear from the surface of the delta (Fig. 8b, inset) Away from the ice margin, however, the time-equivalent shoreline facies will be either erosional or depositional depending on the rate of supply of sediment to protected sites, steepness of the coast, and degree of exposure. Most frequently the upper level of a particular transgression is marked by well-formed beach ridge(s), which may overlie soils or weathered sediments.

There will be differences in the elevation of marine limits depending on which of the criteria above are mappable within each geographic area. Some of the markers definitely relate to the maximum height achieved by storm action and hence might be 1 to 10 m (in extremes) above normal high-tide level. Other features are associated with minimum estimates of SL, such as the highest elevations of marine fossils in non-specific sediments (non-specific in the sense of paleo-SL definitions).

---

[1] 'postglacial' as used here has no chronostratigraphic implications. It is used in the literal sense, eg. on eastern Baffin Island 'postglacial' deposits at 80 m above SL are probably 80 000 to 120 000 years old (cf. Miller *et al.*, 1977; Szabo *et al.*, 1981).

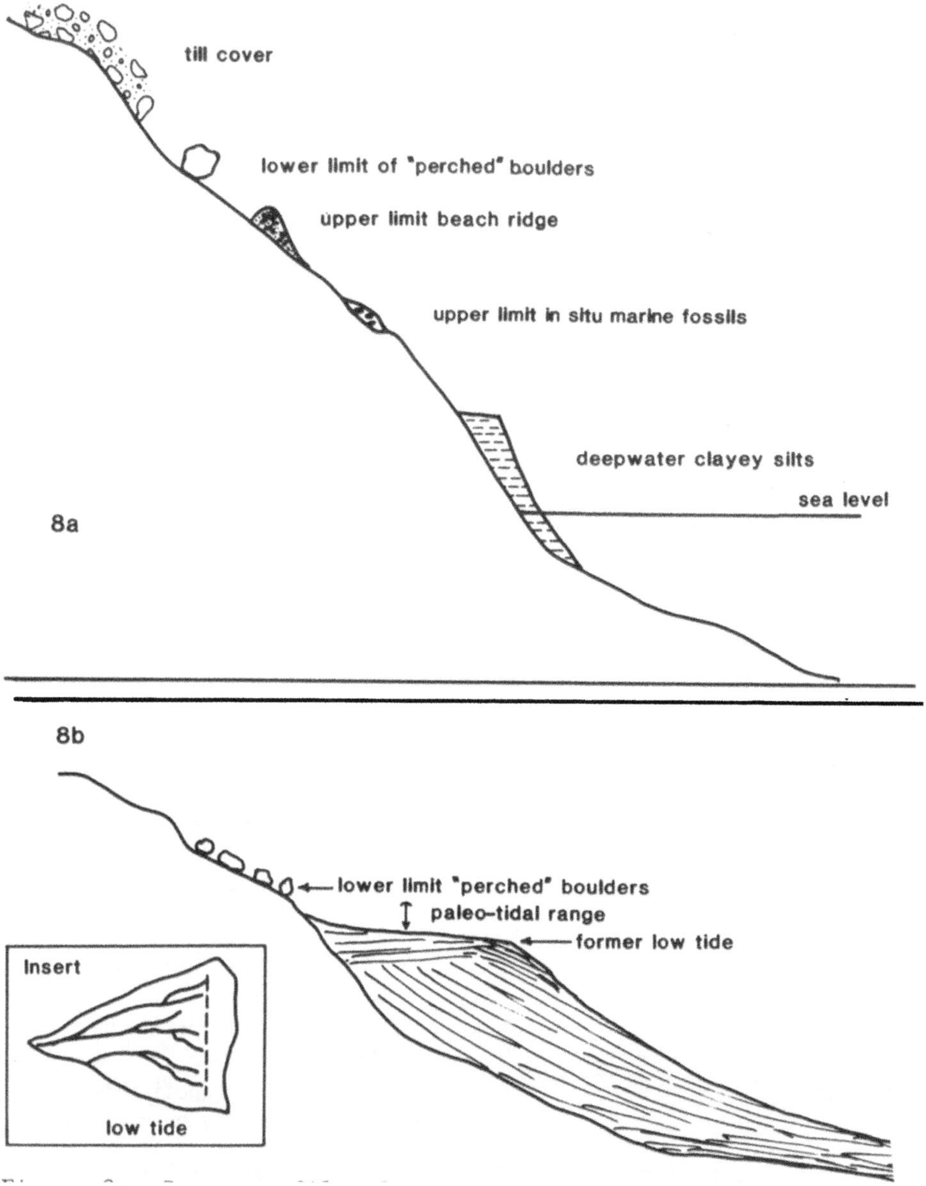

Figure 8a. Cross-profile along a valley wall of an open coast, or fjord coast, not influenced by massive proximal glacial deposition. The marine limit and lower marine sediments and features provide evidence for the magnitude of the submergence although precise determinations of former sea levels may be difficult.

Figure 8b. In areas where large glacio-marine deltas were constructed at the ice front, the lower limit of distributary channels is a close approximation to the low tide. The difference in elevation between 'high-tide limits' (eg. perched boulders) and the delta lip may be used as a first approximation of the paleo-tidal range.

Table 1. Suggested errors (maximum) on different features that are used to delimit the elevation of paleo-sea levels. These estimates are influenced by the tidal range and fetch

Marine limits	Error (±)m	Problems
lower limit perched boulders	10	boulders may roll down slope
lower limit till	10	solifluction lobes can move deposits well below marine limit but can be recognized
uppermost gravel beach ridge	5	may reflect an extreme storm
upper *in situ* molluscs	1-100	depends on nature of enclosing sediments and whether shells have a specific life habitat
surface ice-contact delta	2	by far the least ambiguous feature for an elevation estimate
Lower deposits/features		
gravel ridges	5	can be influenced by extreme storm conditions
beach sediments	5	need to interpret sediment structures
littoral marine fossils	1-?	depends on whether they are found in living position or have been regimented (moved up or down by wave action)
sublittoral marine fossils	5-60	depends on how specific a particular mollusc's life system is, very few have very restricted sublittoral depth range
deltas	1	good sea-level estimates possible
terrace	5	erosional or depositional? Origin possibly non-marine?

Table 1 lists common criteria used for the determination of marine limits and some assessment of the errors.

b) Elevation measurements

Once one or more paleo-SL point(s) have been determined, there remains the important question of what method(s) of elevation determination are sufficiently accurate for time/elevation studies. In remote areas and during reconnaissance studies, surveying aneroid altimeters are frequently used. The writer's experience with a Wallace & Tiernan aneroid is that, if the traverse to the marine limit and back to the present datum is closed, comparisons between altitudes derived by this method and those measured with an automatic level are within ±1.5 m of each other. However, this result is probably an upper level of precision and the accuracy depends on a number of meteorological factors. It is recommended that in most cases elevations be determined by profiling the raised beach sequences with an automatic level and stadia rod. A theodolite and stadia rod can also be used. If traverses are closed, both these methods should give elevations above datum with errors of less than ±0.25 m, which is usually well within the ability of most methods to determine the relationships between paleo-SLs and specific depositional/erosional forms. Sissons' work in Scotland is an example of the benefits gained by detailed elevation determinations of raised shorelines (eg. Sissons, 1966).

The use of Abney level and stadia rod, or Brunton compass and stadia rod might have some merits in reconnaissance studies, especially if traverses are closed. However, this approach is not recommended as there is no real saving of time between use of an Abney level and use of an automatic level.

c) Deposits and landforms below the marine limit

Once ice has uncovered a site isostatic rebound will uplift the site at rates of <1 to 10 m/100 yr. The actual change of SL will also be influenced by the change in global SL and even by the change in gravitational attraction of the sea surfaces toward the ice mass (Clark, 1976; Clark *et al.*, 1978). Particularly in areas where large sediment accumulations occurred during deglaciation, these sediments may be reworked during the marine regression phase. In such areas deposits below the marine limit frequently consist of lower delta surfaces and/or beach ridges, whereas if marine erosion is more dominant, delta fronts can be eroded into one or more broad marine terraces. In some regions, for example northeast England, the glacial isostatic depression at deglaciation was less than global SL lowering, hence deglaciation and the formation of the marine limit were not coeval.

One of the important aspects of studies below the marine limit is to determine if there were specific periods when shore processes produced deposits or landforms - these are frequently called *strandlines* and are isochronous surfaces. Considerable research in Fennoscandia and North America has been concerned with the recognition and mapping of these traces of paleo-SLs (eg. Jardine, 1975; Blake, 1970). Many of the better-formed strandlines appear to be associated with short intervals when relative SL *rose* producing a minor marine transgression. In Arctic Canada, however, the origin of some of the pronounced strandlines is still being debated. They could be due to variations in the rate of change of RSL or they could be associated with variations in such parameters as sediment supply, or the duration of a period of reduced sea ice (cf. Andrews *et al.*, 1972; Blake, 1975). Beautiful sequences of beach ridges in Hudson Bay

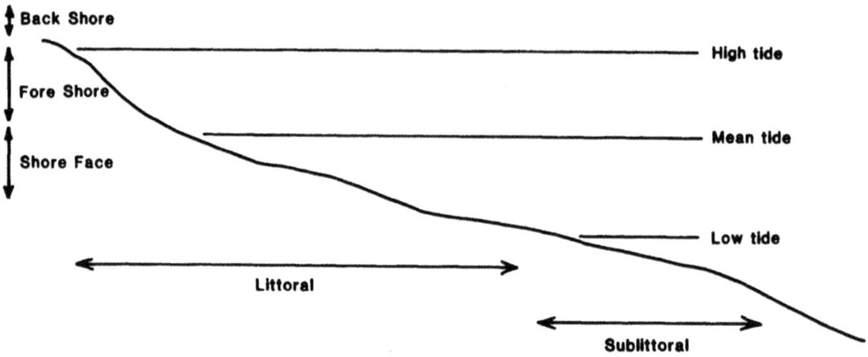

Figure 9. Terminology applied to the areas from above high tide to below low tide

(Hillaire-Marcel & Fairbridge, 1978) may represent evidence for pronounced periodicities of astronomical origin during the last 7000-8000 years.

The determination of paleo-SLs from deposits and features below the marine limit is controlled by many of the problems discussed above. The elevations of constructional forms such as beach ridges, or erosional forms such as marine terraces, should be accurately determined and, ideally, profiles should be surveyed. In these cases, determinations of paleo-SLs involves judgments on the extent to which storm tides may have exceeded normal high tides. In addition it is useful to attempt to estimate the former tidal range. What are much more difficult are attempts to reconstruct paleo-SLs from sedimentary strata that have been eroded and dissected, and which therefore cannot be traced easily to a specific backshore zone. The most difficult sediments to associate with an elevation of a paleo-SL are glacial marine silts and clays that were deposited below the shoreface (Fig. 9). These medium-depth deposits are often overlain by prograding deltaic beds that build out over the bottomset beds as SL falls. However, if silt bluffs outcrop, the maximum elevation that the silts can be traced represents the youngest minimum estimate of SL. However, in practice, these bottomset sediments frequently accumulate during deglaciation and dates from them are often close to the age of the local marine limit (Andrew *et al.*, 1970).

The structures within isolated sand bodies can give clues as to their depositional position at the time of formation (Roep, this volume). Cross-laminated sands occur within the back- and foreshore zones due to the effects of tides and wave swash. Below the limit of low tides the stratification is pseudo-horizontal and in slightly deeper water the sands can be massive, especially in areas of strong benthic bioturbation below the zone of frequent wave erosion (Figure 9). In areas where off-lap deposits are scattered, examination of the sediments and structures should enable minimum estimates of paleo-SL to be established.

d) Offshore sediments

Some areas affected by glaciation have been areas of submergence rather than of emergence during the last several thousand years. Examples would be the coasts of non-glaciated parts of the North Sea and the glaciated coastal areas of the Canadian Maritimes (eg. Grant, 1970, 1980) and the Canadian Pacific Coast (Clague *et al.*, 1982). In these areas information on the offshore stratigraphy has advanced enormously over the last 10 years with the application of various acoustic/seismic profilers operated from research vessels. 'Seismic stratigraphy' provides a broad stratigraphic reference for sediments on the shelves and enables piston-core and grab samples to be taken in specific sedimentary units. Thus radiocarbon dates from cores can now be more accurately associated with specific sedimentary strata.

e) Biofacies

Inference on paleo-SLs can be drawn from the nature of the biological content of sediments. Precise indicators are certain marsh foraminifers that have limited environmental tolerances (Scott & Medioli, this volume) as well as diatoms (Palmer & Abbott, this volume). However, in most cases raised marine sediments are relatively coarse and therefore have been examined mainly for their molluscan assemblages (Petersen, this volume). Specific depth assemblages can be determined from modern analogues as long as the assemblage is *in situ*. Examination of fossil locations requires the following questions to be answered:

1) Do the biota represent an infauna, an epifaunal, or a mixed assemblage?

2) Are the molluscs in living or growth position?

3) What species are present and do these provide any indications of limits on water depth?

4) If the answer to 2) (above) is 'no', can estimates be made of the degree of reworking of the shells and the extent to which their depth relationships might have been disturbed?

Thus the essential question is whether modern molluscan assemblages (biocoenoses) adequately reflect death assemblages (thanatocoenoses). For example, in raised marine sediments in northern Canada the *dominant* fossil is frequently *Hiatella arctica* Linné - a small (1 to 4 cm) bivalve that lives in either a sessile or a boring mode (Fig. 7). This mollusc is usually *not* the dominant species in the biocoenoses. It is difficult in this volume to extend the present discussion other than through the use of examples because each region has its own biocoenoses. Changes of inshore marine conditions during the last 18 000 years will have affected the mollusc assemblages and indicator species frequently reveal changes from 'warmer' to 'cooler' conditions, and *vice versa,* at specific periods (ecostratigraphy). Nevertheless, in many parts of the world, especially in the polar latitudes, it is radiocarbon dates on molluscs that give the ages of the enclosing sediments and thus provide time/depth curves (eg. Salvigsen, 1977; Blake, 1975; Andrews, 1970). For this reason questions 1) and 3) mentioned above must be dealt with. Of special interest is the recognition of littoral indicators. These species can, of course, be transported into the sub-littoral zone upon death of the animal and it is for this reason that examination of the mollusc assemblages has to proceed hand-in-hand with analysis of the sediments themselves. Deposition of shells by currents can be inferred from orientation of valves and presence of cut-and-fill structures. In most cases the elevation of a sample of marine molluscs in raised marine sediments provides *minimum* estimates on paleo-SLs and dates on such shell samples provide *maximum* age estimates for SL at that elevation. However, in cases where shell fragments, or even whole valves are collected from raised beaches, these samples have been moved into the backshore zone. Since the sea is receded, the zone of wave erosion is displaced into surfaces of older and deeper-water sediments. Hence shells in beach ridges may consist of a mixture of shells of different habitats and different ages. However, members of the epifauna, such as *Mytilus edulis,* certain gastropods and barnacles, can be thrown onto storm ridges during high seas; if beach ridges are to be dated it is *possible* that determinations on whole valves of this epifauna might provide reliable indices for age and elevation (Walcott, 1980).

Depth zonal schemes, such as Figures 6 and 10, are important models for paleoenvironmental studies (*cf.* Hillaire-Marcel, 1980) but, by themselves, shell assemblages of the sublittoral zone cannot be tied to specific paleo-SLs. For this, the total lithofacies and biofacies of the raised marine strata must be considered.

DATING METHODS

Within the scope of IGCP Projects 61 and 200 most raised marine features in formerly glaciated areas can be successfully dated by the radiocarbon technique. Raised beaches frequently contain a variety of materials suitable for radiocarbon dating. These include: driftwood (Blake, 1975, 1972); whalebone (Blake, 1975; Birkenmaier & Olsson, 1971); and marine shells (Washburn & Stuiver, 1962; Walcott, 1980; Andrews *et al.,* 1970). Some problems associated with shells will be discussed. At times, marine regression followed by a transgression caused littoral facies to migrate over and bury peaty soils and the organic materials can also be dated to provide age estimates for the period

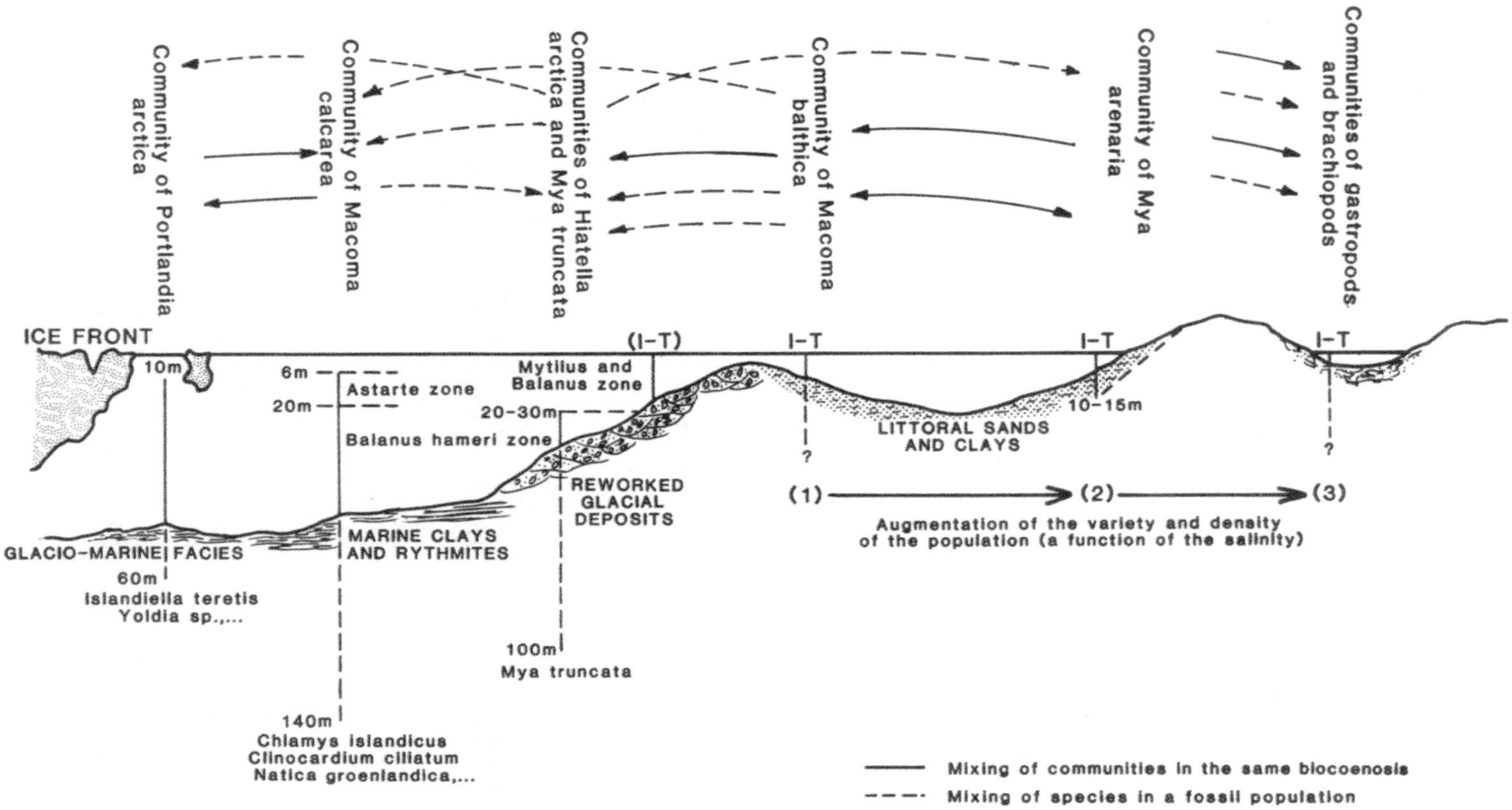

Figure 10. Diagram of the major marine communities in the postglacial seas of eastern and northeastern North America (from Hillaire-Marcel, 1980)

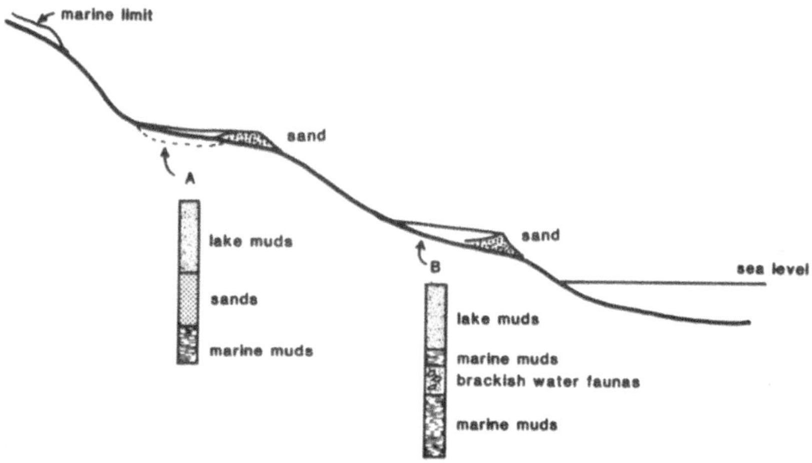

Figure 11. Sketch of the stratigraphy of two small lakes dammed by beach ridges on a coast affected by glacial isostatic rebound. In Lake A, a core shows a sequence from marine offshore muds, to littoral sands, to lake muds. However, in Lake B, two marine muds are encountered which would indicate an increase in water depth and a marine transgression. Small rock basins are also frequently cored for such studies.

prior to marine transgression (Andrews, 1980; Hjort, 1973). In addition the change in diatom assemblages may be used (the so-called 'isolation method') to define in a lake core the change from a marine to a terrestrial environment (Fig. 11) (Mőrner, 1969; Scott & Medioli, this volume).

Sublittoral sediments also contain a variety of materials that may be dated. The dominant materials in such situations are marine shells, but wood and terrestrial organics can be washed into the marine environment and deposited offshore, especially along delta fronts. Sublittoral sediments may also contain bones of marine mammals. Finally, if SL reconstructions are based on lake or marine cores, dates are usually obtained on total sediment and disseminated organic matter concentrated from the sediment. Dates from such situations are usually reliable in inshore areas except for those obtained on the lowermost sediments, which mark deglaciation. However, ^{14}C dates from sediments on the continental shelf may be consistently *too old* (Fillon & Harmes, 1982). In northern Canada an increasing number of apparently spurious 'old' dates are being reported from basal core samples (Short, 1981). It is suggested that these organically poor sediments contain finely disseminated organic matter that has been reworked by the ice caps and glaciers. Ideally, dates can be corrected by examination of the ^{13}C value (Mangerud, 1972).

Specific problems associated with radiocarbon dating of materials in raised marine sediments include the following:

1. Mangerud (1972) has discussed at length the problems of reporting radiocarbon dates on marine shells. Different laboratories use different approaches. Thus questions on marine shell dates include: are the dates based on a ^{13}C value of 0‰ or the NBS standard? Are the data corrected for the apparent age of modern sea water from that area? Krog & Tauber (1973) believe that fractionation and the old sea-water effect are esentially self-cancelling (see Blake, 1975).

2. A slightly different problem concerns the equivalence of shell dates and terrestrial dates (eg. Jardine, 1978). Because of the reservoir effect of the oceans, and because radiocarbon years are not calendar years it may be asked if an 8000 BP date on marine shells is the same age as an 8000 BP date on peat, or other terrestrial materials? Blake (1979) has presented data to indicate that such dates are directly comparable but further studies are urgently needed.

3. Dates on driftwood are an important element of the chronologies of arctic raised marine deposits. Driftwood is abundant in Spitsbergen and is not uncommon in the Canadian Arctic islands (Salvigsen, 1979; Blake, 1972, 1975). The source of the driftwood appears to be primarily Siberia. The residence time for driftwood in the Arctic Ocean may well be short (10 to 100 years?) because it can be expected that the wood would soon become waterlogged and would sink. However, driftwood requires open water to move and studies (eg. Blake, 1972) suggest that the rate of supply of such wood has varied during the Holocene.

Although radiocarbon dates provide the bulk of the information on time/elevation fluctuations of SL from raised marine sequences, other dating methods have been employed in specific situations. Generally, radiocarbon dates are the most accurate methods for deposits <20 000 BP. Between 20 000 to 40 000 BP, radiocarbon dates on shells are frequently difficult to interpret (Olsson & Blake, 1961), but experience suggests that they generally provide *minimum* age estimates in the range between 30 000 to 260 000 BP. Recently, considerable reliance has been placed on amino-acid chronology as a means of elucidating the SL history of Arctic areas (eg. Miller *et al.*, 1977; Nelson, 1980, 1981; Andrews, 1980). The method works well to differentiate suites of strata >40 000 years old, and when combined with U-series dates it *may* offer the prospect for 'absolute' dates in the range 40 000 to 300 000 BP (Szabo *et al.*, 1981). Amino-acid geochronology is temperature dependent and hence can be used over a range of 2 million years in the U.K. (Miller *et al.*, 1979), whereas in Arctic areas it can distinguish broad pulses of SL variations back to the Miocene.

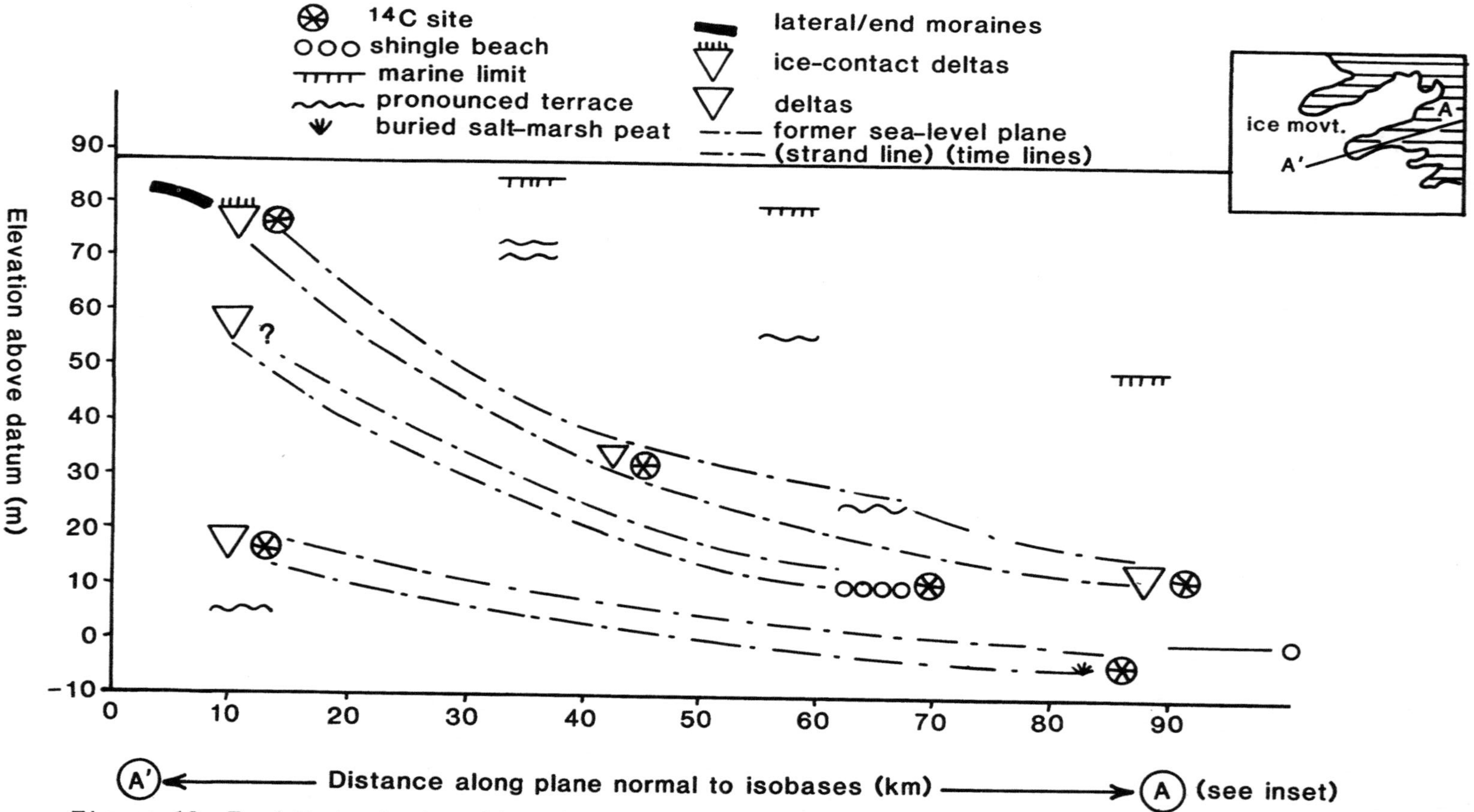

Figure 12. Equidistant shoreline diagram drawn in the plane normal to local isobases. Such graphs should indicate the nature of each data point (delta, terrace, etc.), the location of radiocarbon control points, and envelopes which contain paleo-sea levels for any specific time

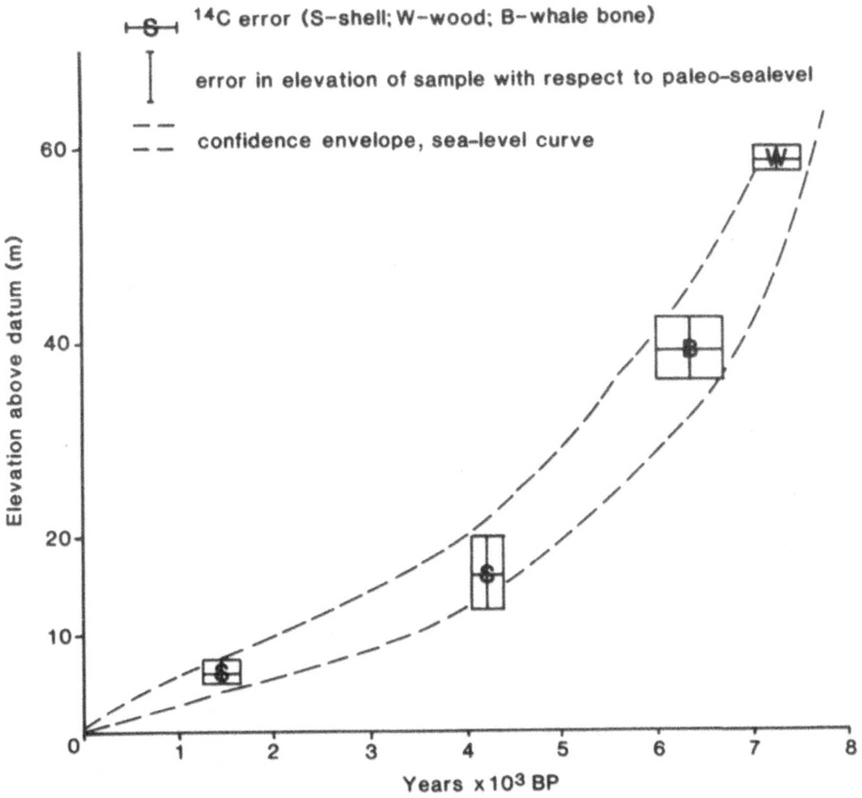

Figure 13. Graph showing how errors in radiocarbon dates and former sea-level estimates can be drawn as a series of rectangles. The 'true' time/elevation curve for sea level at the point (then) lies between the lines of envelope which connects the corners of the error boxes.

CONSTRUCTION OF TIME/ELEVATION CURVES

In studies on raised marine sediments from glaciated areas two basic time/elevation curves are usually presented for interpretation. One has a long history of use in Fennoscandia and is commonly called a 'shoreline equidistant diagram' (*cf.* Andrews, 1970) (Fig. 12), whereas the other shows the change of RSL at a single point (Fig. 13) and is called a SL curve. The information in it is also implicit in the shoreline diagram (Fig. 12).

Although it is common to draw shoreline diagrams and SL curves as lines, or a line through the data points, it is clear that there are errors associated both with the determination of the elevation of the site and with inferences in paleo-SL. The value of this latter error will vary depending upon the nature of the raised marine features or sediment and how closely it can be associated, in fact or in theory, with a former SL (Table 1). Nevertheless, it is important that the size of the uncertainty be expressed in some useful fashion (Fig. 13). Radiocarbon dates are supplied by laboratories with an error that reflects the uncertainties in the counting procedure. Most of the laboratories report dates ±1α, but some (Geological Survey of Canada) report all their dates with ±2α where α is the standard deviation. These errors bear no relation to geological uncertainty! The errors can be reduced by longer counting but the relationship between the age and counting time is non-linear and thus increased precision is costly (Stuckenrath, 1977). The geological uncertainty is unknown, but dates should be plotted on the diagrams with a 95% confidence interval based on counting statistics. For all practical purposes this simply means doubling (precisely x 1.96) the quoted standard deviation.

As each data point is commonly represented on a graph by a rectangle of varying size, time/elevation graphs (eg. Washburn & Stuiver, 1966) should be considered as *probability envelopes* rather than as a line or surface in time and space. This procedure would certainly assist researchers concerned with fitting glacial rebound curves to the responses of different ice and/or earth rheological models (eg. Peltier & Andrews, 1976; Clark *et al.*, 1978; Clark, 1980a).

Time/elevation information is also available from shoreline diagrams, especially where several dates are scattered along the axis of the diagram and through a range of elevations.

The question of whether SL histories for a particular region record marine transgression within a general period of emergence has to be answered on the basis of well-founded stratigraphic studies. Peats underlying beach gravels may indicate a change in RSL, but even here in polar areas, this situation might reflect changes in the extent of fast and pack ice so that storms become more effective and cause higher tide levels. Detailed studies of small SL variations ($\leq \pm$ 3 m) can be made by carefully coring lakes, bogs and marshes (eg. Mörner, 1969; Tooley, 1974; Van de Plassche, 1982) or by many radiocarbon dates (Blake, 1975). But even in these cases it is difficult to separate out short-term processes (eg. storm surges, or an interval of storminess) from our main interest - regional changes in SL related to glacio-eustatic and glacio-isostatic processes.

ACKNOWLEDGEMENTS

This manuscript has been rendered more useful by the comments of several colleagues including Claude Hillaire-Marcel, Jan Mangerud, and Doug Grant. My work on these problems has been supported by the Geology program, National Science Foundation.

REFERENCES

Andrews, J.T., 1970. A geomorphological study of post-glacial uplift, with particular reference to arctic Canada. *Institute of British Geographers Special Publication*, No. 2, 156 pp

Andrews, J.T., 1980. Progress in relative sea level and ice sheets reconstructions, Baffin Island, NWT, for the last 125,000 years. In: *Earth heology, sostasy and ustasy*, ed. N.-A. Mörner (John Wiley & Sons, New York), 157-200

Andrews, J.T., Buckley, J.T. & England, J.H., 1970. Late-glacial chronology and glacio-isostatic recovery, Home Bay, east Baffin Island. *Bulletin of the Geological Society of America*, 81, 1123-1148

Andrews, J.T., Mears, A., Miller, G.H. & Pheasant, D.A., 1972. Holocene late-glacial maximum and marine transgression on the eastern Canadian Arctic. *Nature*, 239, 147-149

Bird, J.B., 1954 Post-glacial marine submergence in central Arctic Canada. *Bulletin of the Geological Society of America*, 65, 457-467

Birkenmaier, K. & Olsson, I.V., 1971. Radiocarbon dating of raised marine terraces at Hornsand, Spitsbergen, and the problem of land uplift. *Norsk Polarinstitutt Arbok*, 1969

Blake, W., Jr., 1970. Studies of glacial history in Arctic Canada, I. Pumice, radiocarbon dates, and differential postglacial uplift in the eastern Queen Elizabeth Islands. *Canadian Journal of Earth Sciences*, 7, 634-664

Blake, W., Jr., 1972. Climatic implications of radiocarbon dated driftwood in the Queen Elizabeth Islands, Arctic Canada. In: Climate changes in arctic areas during the last 10,000 years. *Acta Universitatis Ouluensis, Series A., Geol. No. 1*, 77-104

Blake, W., Jr., 1975. Radiocarbon age determinations and postglacial emergence at Cape Storm, southern Ellesmere Island, Arctic Canada. *Geografiska Annaler*, 62A, 1-71

Blake, W., Jr., 1979. Age determinations on marine and terrestrial materials of Holocene age, southern Ellesmere Island, Arctic Archipelago. *Geological Survey of Canada Paper*, 79-1C, 105-109

Boulton, G.S., 1979. Glacial history of the Spitsbergen archipelago and the problem of a Barents Shelf ice sheet. *Boreas*, 8, 31-58

Brotchie, J.F. & Silvester, R., 1969. On crustal flexure. *Journal of Geophysical Research*, 74, 5240-5252

Bryson, R.A., Wendland, W.M., Ives, J.D. & Andrews, J.T., 1969. Radiocarbon isochrones on the disintegration of the Laurentide Ice Sheet. *Arctic and Alpine Research*, 1, 1-14

Cathles, L.M.,III, 1975. *The viscosity of the earth's mantle.* (Princeton University Press, Princeton)

Church, M., 1972. Baffin Island sandurs: a study of arctic fluvial processes. *Geological Survey of Canada Bulletin,* 216, 208 pp

Clague, J., Harper, J.R., Hebda, R.J. & Howe, D.E., 1982. Late Quaternary sea levels and crustal movements, coastal British Columbia. *Canadian Journal of Earth Sciences,* 19, 597-618

Clark, J.A., 1976. Greenland's rapid postglacial emergence: a result of ice-water gravitational attraction. *Geology,* 4, 310-312

Clark, J.A., 1980a. The reconstruction of the Laurentide Ice Sheet of North America from sea level data: method and preliminary results. *Journal of Geophysical Research,* 85, 4307-4323

Clark, J.A., 1980b. A numerical model of worldwide sea level changes on a visco-elastic earth. In: *Earth rheology, isostasy and eustasy,* ed. N.-A. Mörner, (John Wiley & Sons, New York), 525-534

Clark, J.A., Farrell, W.E. & Peltier, W.R., 1978. Global changes in post-glacial sea level: a numerical calculation. *Quaternary Research,* 9, 265-287

Denton, G.H. & Hughes, T.J. (eds), 1981. *The Last Great Ice Sheets.* (John Wiley & Sons, New York)

Dyke, A.S., 1979. Glacial and sea level history of the south-western Cumberland Peninsula, Baffin Island, NWT, Canada. *Arctic and Alpine Research,* 11, 179-202

England, J., Bredley, R.S. & Miller, G.H., 1978. Former ice shelves in the Canadian High Arctic. *Journal of Glaciology,* 20, 393-404

Fillon, R.A. & Harmes, R.H., 1982. Northern Labrador shelf glacial chronology and depositional environments. *Canadian Journal of Earth Sciences,* 19, 162-192

Grant, D.R., 1970. Recent coastal submergence of the Maritime Provinces. *Canadian Journal of Earth Sciences,* 7, 679-689

Grant, D.R., 1980. Quaternary sea level change in Atlantic Canada as an indication of crustal delevelling. In: *Earth Rheology, Isostasy and Eustasy,* ed. N.-A. Mörner, (John Wiley & Sons, New York), 201-214

Grosswald, M.G., 1980. Late Weichselian Ice Sheet of northern Eurasia. *Quaternary Research,* 13, 1-32

Gutenberg, B., 1941. Changes in sea level, post-glacial uplift, and mobility of the earth's interior. *Bulletin of the Geological Society of America,* 52, 721-772

Hillaire-Marcel, C., 1980. Les faunes des mers post-glacaires du Quebec: quelques considerations paléo-écologiques. *Géographie Physique et Quaternaire,* 34, 3-60

Hillaire-Marcel, C. & Fairbridge, R.W., 1978. Isostasy and eustasy of Hudson Bay. *Geology*, 6, 117-122

Hjort, C., 1973. The Vega Transgression: A hypsithermal event in central East Greenland. *Bulletin of the Geological Society of Denmark*, 22, 25-38

Jardine, W.G., 1975. The determination of former sea levels in areas of large tidal range. In: *Quaternary Studies*, Royal Society of New Zealand, Wellington, 163-168

Jardine, W.G., 1976. Some problems in plotting the mean surface level of the North Sea and the Irish Sea during the last 15,000 years. *Geologiska Föreningens i Stockholm Förhandlingar*, 98, 78-82

Jardine, W.G., 1978. Radiocarbon ages of raised-beach shells from Oronsay, Inner Hebrides, Scotland: a lesson in interpretation and deduction. *Boreas*, 7, 183-196

Jardine, W.G., (this volume) Determination of altitude

Kennedy, E.G., 1978. *Pleistocene paleoecology, zoogeography and geochronology of marine invertebrate faunas of the Pacific Northwest Coast (San Francisco Bay and Puget Sound)*. Ph.D. thesis, University of California

Krog, H. & Tauber, H., 1973. Radiocarbon chronology of late and post-glacial marine deposits in North Jutland. *Dansk Geologiske Undersøgelse Arbok*, 93-105

Mangerud, J., 1972. Radiocarbon dating of marine shells, including a discussion of apparent age of recent shells from Norway. *Boreas*, 1, 143-172

Miller, G.H., Andrews, J.T. & Short, S.K., 1977. The last interglacial/glacial cycle, Clyde Foreland, Baffin Island, NWT: stratigraphy, biostratigraphy and chronology. *Canadian Journal of Earth Sciences*, 14, 2824-2857

Miller, G.H., Hollin, J.T. & Andrews, J.T., 1979. Aminostratigraphy of UK Pleistocene deposits. *Nature*, 281, 539-543

Mode, W.N., 1980. *Quaternary stratigraphy and palynology of the Clyde Foreland, Baffin Island, NWT, Canada*. Ph.D. thesis, University of Colorado, Boulder

Mörner, N.-A., 1969. The late Quaternary history of the Kattegatt Sea and the Swedish West Coast. *Sveriges Geologiska Undersökning*, Series C, Nr. 640, 487 pp

Mörner, N.-A., (ed.), 1980. *Earth rheology, isostasy and eustasy*. (John Wiley & Sons, New York)

Nelson, A.R., 1978. *Quaternary glacial and marine stratigraphy of the Qivitu Peninsula, northern Cumberland Peninsula, Baffin Island, Canada*. Ph.D. thesis, University of Colorado, Boulder

Nelson, A.R., 1980. Chronology of Quaternary land forms, Qivitu Peninsula, northern Cumberland Peninsula, Baffin Island, NWT, Canada. *Arctic and Alpine Research*, 12, 265-286

Nelson, A.R., 1981. Quaternary glacial and marine stratigraphy of the Quivitu Peninsula, northern Cumberland Peninsula, Baffin Island: Summary. *Bulletin of the Geological Society of America*, 92, Part I, 512-518

Olsson, I.V. & Blake, W., Jr., 1961. Problems of radiocarbon dating of raised beaches based on experience in Spitsbergen. *Norsk Geografisk Tidsskrift*, 18, 1-18

Palmer, A.J.M. & Abbott, W.H., Diatoms as indicators of sea level. (this volume)

Peltier, W.R. & Andrews, J.T., 1976. Glacial isostatic adjustment - I: The forward problem. *Geophysical Journal of the Royal Astronomical Society*, 46, 605-646

Prest, V.K., 1969. *Retreat of Wisconsin and Recent ice in North America*. Geological Survey of Canada, Map 1257A

Richards, H.G. & Fairbridge, R.W., 1965. *Annotated bibliography of Quaternary shorelines (1945-1964)*. Academy of Natural Sciences, Philadelphia

Richards, H.G. & Shapiro, E.A., 1979. Annotated bibliography of Quaternary shorelines. Third Supplement 1974-1977. *Geo Abstracts*, Norwich, U.K. Bibliography No.5

Roep, Th.B., (this volume). Sea-level markers in coastal barrier sands; examples from the North Sea coast

Salvigsen, O., 1977. Radiocarbon datings and the extension of the Weichselian ice sheet in Spitsbergen. *Norsk Polarinstitutt Arbok*, 1976, Oslo (1977), 209-224

Salvigsen, O., 1979. Holocene emergence and finds of pumice, whalebones, and driftwood at Svartknausflya, Nordaustlandet. *Norsk Polarinstitutt Arbok*, 1977, 217-228

Scott, D.B. & Medioli, F.S., (this volume). Foraminifera as sea-level indicators

Short, S.K., 1981. Radiocarbon date list I Labrador and Northern Quebec, Canada. *INSTAAR Occasional Paper*, No.36, University of Colorado, Boulder, 33 pp

Sim, V.W., 1960. Maximum post-glacial marine submergence in northern Melville Peninsula. *Arctic*, 13, 178-193

Sissons, J.B., 1966. Relative sea level changes between 10300 and 8300 BP in part of the Corse of Stirling. *Institute of British Geographers Transactions*, No.39, 1-19

Stuckenrath, R., 1977. Radiocarbon: some notes from Merlin's Diary. *Annals of the New York Academy of Sciences*, 288, 181-188

Szabo, B.J., Miller, G.H., Andrews, J.T. & Stuiver, M., 1981. Comparison of uranium-series, radiocarbon and amino acid data from marine molluscs, Baffin Island, Arctic Canada. *Geology*, 9, 451-457

Tooley, M.J., 1974. Sea level changes during the last 9000 years in Northwest England. *Geographical Journal*, 140, 18-42

Valentine, J.W., 1961. Paleoecologic molluscan geography of the Californian Pleistocene. *University of California Publication Geological Sciences*, 34, 304-442

Van de Plassche, O., 1982. Sea-level change and water-level movements in The Netherlands during the Holocene. *Mededelingen Rijks Geologische Dienst*, 36(1), 1-93

Walcott, R.I., 1970. Isostatic response to loading of the crust in Canada. *Canadian Journal of Earth Sciences*, 7, 716-726

Walcott, R.I., 1972. Past sea levels, eustasy, and deformation of the earth. *Quaternary Research*, 2, 1-14

Walcott, R.I., 1980. Rheological models and observational data of glacio-isostatic rebound. In: *Earth Rheology, Isostasy and Eustasy*, ed. N.-A. Mörner, (John Wiley & Sons, New York), 3-10

Washburn, A.L. & Stuiver, M., 1962. Radiocarbon-dated postglacial delevelling in northeast Greenland and its implications. *Arctic*, 15, 66-73

4

Sea-level markers in coastal barrier sands: examples from the North Sea coast

Th. B. ROEP

ABSTRACT

From a review of recent literature on coastal barrier environments, the methods and techniques used to derive former sea-level positions from barrier deposits are discussed. These comprise: 1) coastal morphology, 2) lacquer peels of cores from recent coastal sediments and fossil barrier deposits, 3) depth distribution of sedimentary structures and shells along the present coast, and 4) sequences observed in deep, open pits in barrier sands. The discussion is largely based on work carried out in the coastal area of the western Netherlands. Examples are given to infer former high and low tide levels and from these the amount of sea-level rise over a given period of time.

INTRODUCTION

In the literature on modern coastal barrier deposits (eg. Reinson, 1979; Hubbard *et al.*, 1979; Elliott, 1979; Kraft & John, 1979; Hayes, 1980; Davies *et al.*, 1971; King, 1972; Reineck & Singh, 1973, 1980) emphasis has been on general environment, sedimentary processes or specific sedimentary structures. Few studies have been undertaken with a view to sea-level (SL) research. As a consequence, morphological cross-sections are frequently presented without reference to tidal levels or to mean sea level (MSL). However, waves, tides, plants and animals in a modern coastal barrier environment (Fig. 1) are seen to produce sedimentary structures that can be related to high and low tide levels. To the extent that these structures are preserved during coastal progradation, and provided compaction is negligible, they can be used to derive paleo-tide levels from fossil barrier deposits.

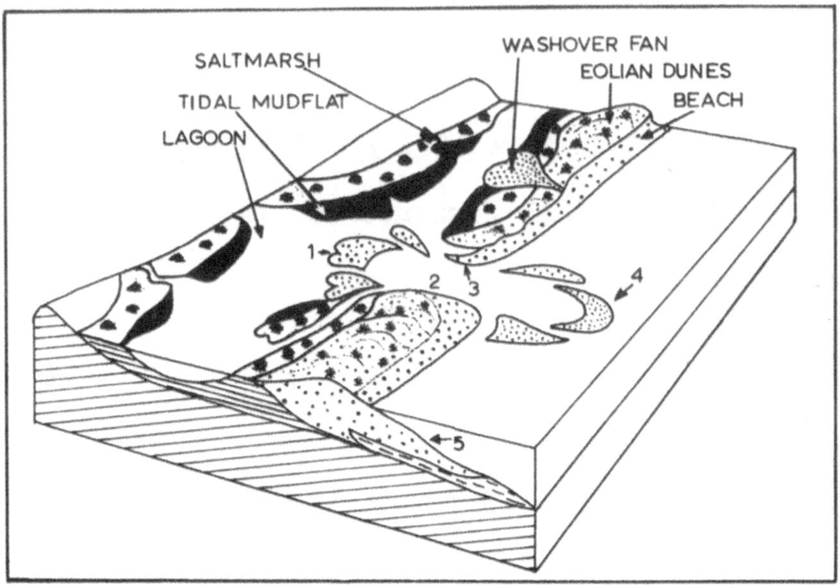

Figure 1. Coastal barrier and associated environments: 1. flood tidal delta, 2. tidal inlet, 3. spit, 4. ebb tidal delta, 5. shoreface (mainly after Reinson, 1979)

Analysis of sedimentary structures in barrier deposits in terms of paleo-tide levels is, of course, mainly based on knowledge of modern analogues. This chapter therefore begins with a brief survey of recent literature on sandy coastal deposits, followed by a summary of work on coastal morphology, sedimentary structures, and shell distribution that has been carried out lately on present-day beaches along the western Netherlands and northern Germany. In the second part, examples are given of how sediment characteristics observed in pits in fossil barrier deposits in the western Netherlands can be interpreted in terms of paleo-SLs. Finally, some practical recommendations for field study are given.

MODERN ANALOGUES: SOME GENERAL REMARKS

Studies on modern sandy coastal deposits with emphasis on sedimentary structures, have been made by Clifton *et al.* (1971; non-barred, high-energy coast of the Pacific Ocean), Davidson-Arnott & Greenwood (1976), Greenwood & Davidson-Arnott (1979), and Hubbard *et al.* (1979; barred coasts of the Atlantic Ocean), Reineck & Singh (1973, 1980; barred, non-tidal Mediterranean coasts), and Howard & Reineck (1972; low-energy coast of Sapelo Island, U.S.A.) among others. In profiles surveyed from the beach (dunes) to subtidal depths, surface morphology, sedimentary structures, grain-size distributions, etc. were studied. From this a hypothetical vertical sequence during and after beach progradation may be inferred, thus providing a model for analyzing fossil sequences.

From the afore mentioned papers it appears that the most suitable sub-environments for SL study are the down-drift side of a laterally migrating inlet and the seaward side of a prograding barrier ridge (Fig. 1). Here, sequences are dominantly sandy (little compaction) and, provided enough sand is supplied, accumulation can be relatively rapid.

Inlet sequences are characterized by an erosive base with lag deposits of shells, clay pebbles or gravel, followed by cross-bedded ebb- and flood-tidal delta sediments. These are covered by tidal mudflat and salt-marsh deposits and/or beach and eolian sands. This type of sequence has been described by Kumar & Sanders (1974), Nio *et al.* (1980) and Van den Berg (1982), among others.

The other type of sequence shows a more or less gradually coarsening upwards going from shelf mud to coastal sands covered by beach sands and somewhat finer-grained eolian sands (see Van Straaten, 1965 and Davies *et al.*, 1971 for descriptions of actual cases).

Usually, both types of sequences are deposited under the influence of tides and waves. The inlet sequences, however, mainly show tidal influences, whereas the coastal sequences mainly show wave influences. Inlet sequences are characterized by frequent clay intercalations deposited during slack water and neap tides, and are therefore somewhat less suitable (compaction) for paleo-SL studies than the cleaner coastal sequences with wave-beaten sands on top. Furthermore, inlet sequences are more affected by local tidal amplification or reduction. An example of both types of sequences from the North Sea, based on assumed progradation is given in Figure 10. In this chapter, however, emphasis will be on the barrier-front type of sequence.

A problem with hypothetical vertical sequences as a model for interpreting fossil barrier deposits is that it is unknown what will be preserved in an actual case of beach progradation. Usually, surveys as described above are made under quiet summer-weather conditions. Storms may result in quite different sedimentological aspects and exceptionally heavy storms in still others. And there is reason to believe that storms leave a disproportionally larger record than normal or quiescent conditions. Structures produced by storms, with significant sediment reworking below the pre-storm sediment surface, are more likely to be preserved. On the other hand, preservation also depends on factors such as the intensity of burrowing organisms and the rate of coastal progradation. With respect to the latter, when looking at actual vertical sequences, the question should be considered whether all

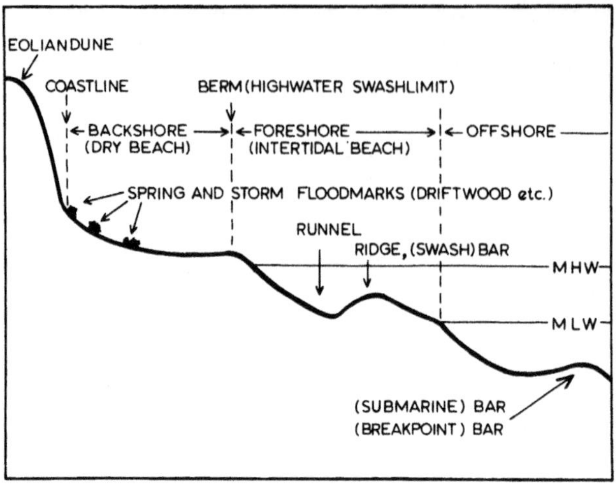

Figure 2. Beach nomenclature (mainly after King, 1972). Apart from the terms for wave-built bars given here, the general term (longshore) bar and trough will be used.

parts of a sequence belong to one shifting, prograding depth profile of about the same age, or whether the preserved section resulted from environmental changes that occurred over a longer time period.

Figure 2 summarizes the nomenclature of shoreline features used in this chapter.

SEA-LEVEL MARKERS ON THE MODERN COAST OF THE WESTERN NETHERLANDS AND NORTHWEST GERMANY

The following sections summarize what is presently known of coastal morphology and depth distribution of sedimentary features along the present-day coast of the western Netherlands and Northwest Germany.

In terms of hydrographic conditions the North Sea coast is characterized by storm-wave, mesotidal environments; tides are semidiurnal (De Jong, 1977). Tidal current velocities at a distance of a few kilometers off-shore vary commonly from 1 to 2 m/sec. West winds are dominant and most waves (height < 2 m, period ⩽ 7 sec) come from WNW, followed in frequency by waves (height ⩾ 2 m, period > 7 sec) from WNW; waves (height < 2 m, period > 7 sec) from NW, and waves (height ⩾ 2 m, period ⩽ 7 sec) from WSW (Van Straaten, 1961) (Fig. 3).

Median sand-grain sizes along the coast of the province of North-Holland at the foot of the dunes have the range 170-310µ; at the dry beach 175-340µ, and at the wet beach 180-400µ (Glim, 1980).

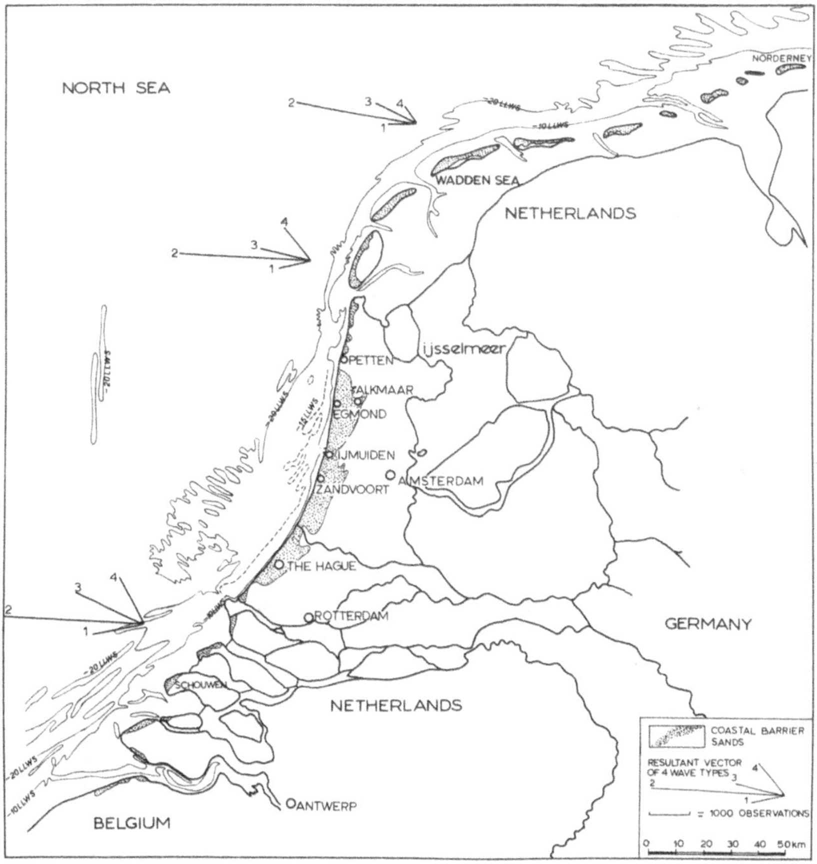

Figure 3. Map showing the coastal barrier complex of the Netherlands and location of sites mentioned in the text. Depth contours in meters below mean lowest low water springs. (LLWS is the hydrographic datum; it varies along the open coast between about -1 and -2 m NAP.) After maps from the Geological Survey (1975) and Rijkswaterstaat, Directie Noord Zee (1979). For resultant vectors of four main wave types, see text.

1. Coastal morphology

The present-day depth distribution of morphological features of the foreshore and shoreface may help the SL researcher in analyzing and interpreting fossil barrier deposits.

The coastline of The Netherlands has been surveyed by Rijkswaterstaat since 1860 (Edelman, 1967, cited in King, 1972). It appears that areas of retreat show steepest shoreface gradients (eg. near Petten: 1:60 between MSL and -3 m NAP[1], and *c.*1:120 between -3 and -7 NAP) and are

[1]NAP: Dutch ordnance datum, very near to present MSL

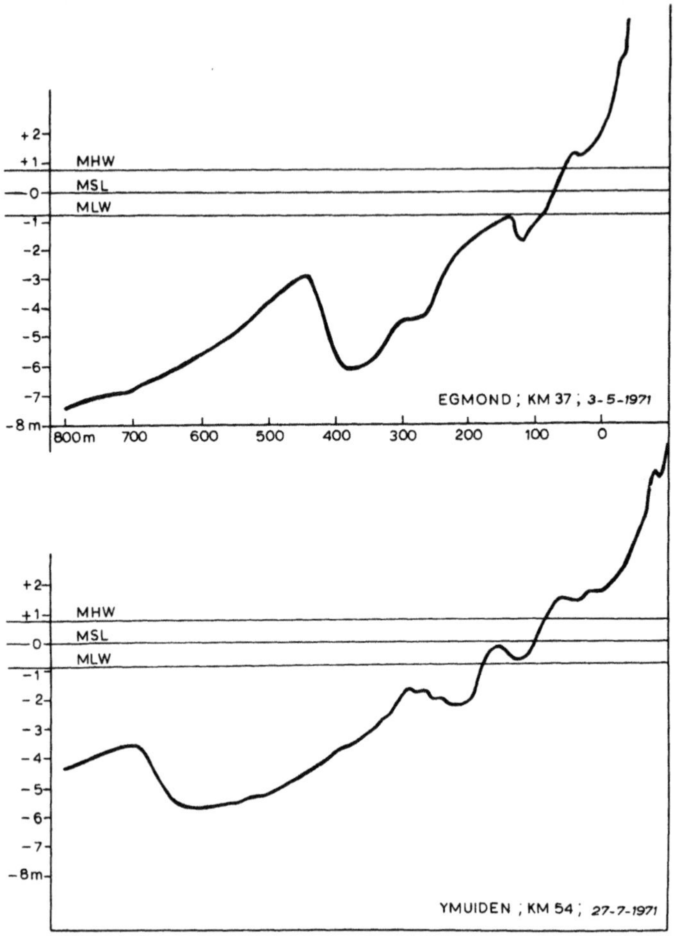

Figure 4. Two examples of beach profiles, sounded by Rijkswaterstaat in the summer of 1971. The heights of these bars, among others, have been used in the bar-depth distribution given in Figure 5.

characterized by the presence of only one large, more or less stable subaquatic bar. Oppositely, areas of coastal progradation have lower shoreface gradients (eg. near IJmuiden: 1:75 between MSL and -3 m NAP; 1:250 between -3 and - 6 m NAP) and have 2 or 3 mobile subaquatic bars. Stable parts of the coastline show intermediate characteristics. Figure 4 shows two examples of summer profiles recorded by Rijkswaterstaat.

Figure 5 shows the depth distribution of all bars near IJmuiden and Egmond measured by Rijkswaterstaat each summer for more than ten years. The Figure was constructed by combining all bar heights, taken as the vertical distance from the crest of a bar to the base of the next trough to landward. Between +2 and -7 m NAP the number of bars was counted at each level in 5 cm steps downward. Two strong

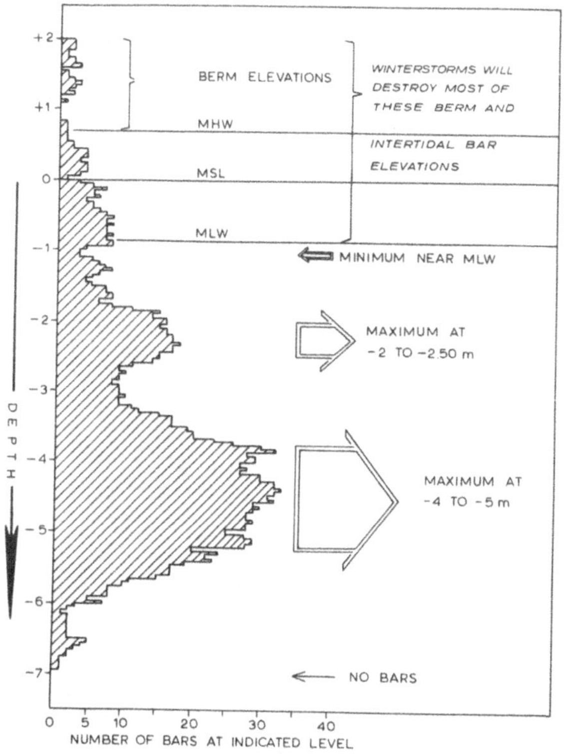

Figure 5.

Depth distribution of bar elevations. Construction of the graph is explained in the text. The data are derived from soundings made by Rijkswaterstaat during the summer, at km pole 54 near IJmuiden (1961-1973), at km pole 40 near Egmond (1960-1966) and at km pole 37 near Egmond (1960-1973).

maxima of bar numbers were found (Fig. 5), one between -2 and NAP and the other between -4 and -5 NAP. Minima in bar numbers occur just below MLW and just above MHW. Berm elevations occur just above MHW, between +1 and +2 m NAP. Finally, the number of bars decreases markedly below -5.5 m NAP.

From the complete series of soundings of Rijkswaterstaat from IJmuiden and Egmond (not given here) it appears that the crests of 0.75 m high bars are most frequently situated between MSL and -2 m NAP, whereas 2-3 m high bars mostly occur between -2 and -6 m NAP. Furthermore there is a concentration of bar crests and troughs between -3 and -4 m NAP and between -1 and -2 m NAP. It is conceivable that after progradation of such summer profiles erosion surfaces (or set boundaries) will be observed at these levels in the sequence. Below -7 m no bars are present in the profiles at our disposal. This implies that wave-built mega-crossbedding below that depth will be absent in the sequence after the assumed progradation.

Below (page 122) these Rijkswaterstaat data will be compared with the vertical distribution of combined sedimentological, structural data of fossil barrier deposits near Alkmaar. In general, however, direct comparison of the depth distribution of modern morphological features

Figure 6 North Sea beach at low tide near Wijk aan Zee (4 km north of IJmuiden) October 1971. In the foreground a cross-sectioned intertidal bar (see also Figure 9b). Landwards of the bar the runnel floor with wave ripples made at high tide. Towards the background the limit of wet sand (grey on photo) marks the highest swash limit of the last tide against dry sand (white). The swash limit is marked by lines of algae. Dunes, planted with bent-grass, form the skyline. Mean tidal range is *c*. 1.60 m (photograph by J. Smit).

with a distribution of older sedimentary features is difficult. For one reason, the form taken by a modern bar usually can not be observed in a fossil barrier sequence, because the form will have been modified by later erosion and sedimentation during progradation. Also a recent barrier is usually composed of more than one crossbedded set. What is needed is to combine long-term studies of beach-profile changes with sedimentological analysis of cores from the same spot.

2. Depth-distribution of sedimentary features on modern beaches of the western Netherlands

This section mainly deals with results of studies on the relation between beach profile and internal sedimentary structures, depth distribution of sedimentary features, and their preservation potential in the intertidal zone (foreshore). These studies were carried out on the modern North Sea beach at Zandvoort, Schouwen and Egmond (see Fig. 3). Figure 6 shows the Egmond beach at low tide; the cross-sectioned bar in front is pictured in detail in Figure 9b.

The investigations at Zandvoort and Schouwen (Van den Berg, 1977a) were focussed on the question of preservation of sedimentary structures at a site more exposed to waves (Zandvoort; mean tidal range: 1.6 m, max. spring tide range: 2.2 m) and at a more sheltered one (Schouwen; mean tide range: 2.4 m, max. spring tide range: 3.2 m). Van den Berg, using selected profiles from Rijkswaterstaat, measured the distribution of preserved coastal sands over an interval of several years. He also used the profiles to determine the vertical distribution of low-angle seaward-dipping zones and of bar crests and troughs. The bar crests are characterized by similarly oriented swash lamination, the forests by landward-dipping mega-crossbedding with gradational contacts covering trough-floor deposits (see also Doeglas, 1954). Subsequent coring (method described in Van den Berg, 1977b) was done to verify the analysis from the profiles.

a) Zandvoort survey

From the work at Zandvoort beach it appears that the highest preserved trough facies occurs at +1.6 m NAP and should normally occur at +1.2 m NAP. The deepest preserved back-shore facies comprises low-angle seaward-dipping swash lamina associated with eolian deflation horizons and eolian sands. This facies occurs at +0.9 m NAP and more normally at +1.2 m NAP. All these features occur just above MHW (+0.9 m NAP). On the other hand, around low tide level the most significant sedimentary structure is trough-shaped mega-ripple crossbedding caused by longshore currents at the lower part of the foreshore (in the runnels or on low bars).

b) Schouwen beach

Figure 7 summarizes the results of the investigation at the more sheltered beach on the island of Schouwen. The profile at the top of the Figure is that measured on 14 October 1975. The irregular line near the bottom connects the deepest eroded parts on the shore due to trough migration and swash activity between 1962-67 and 1968-75. The structures at Schouwen beach are of interest because they represent:

1) a phase of coastal retreat (1955-1962), followed by
2) a phase of slow progradation (1962-1967) with steeper gradients (up to 1:50), characterized by seaward-inclined swash laminations (occurring from MSL up to +3.80 m NAP), and
3) a phase of rapid progradation (1968-1975) with smaller gradients (up to 1:90), characterized by an increased number of intertidal bars and a much higher preservation potential of these bars.

In the cores several distinctive features were noted: clay drapes covering small-scale ripples were found up to the level of MHW; landward-inclined bar lamination up to

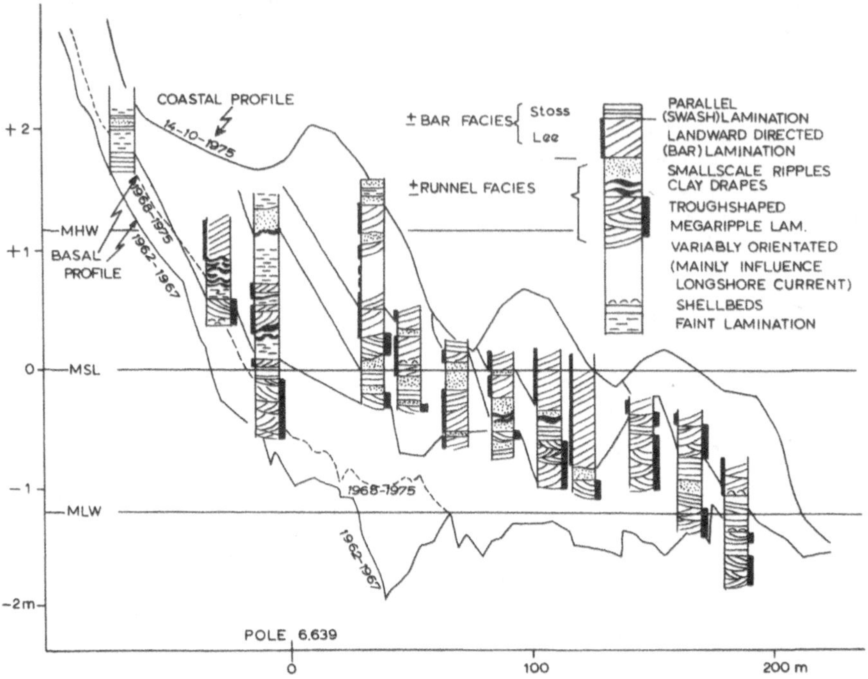

Figure 7. Beach profiles at Schouwen (slightly modified after Van den Berg, 1977a). The basal profiles are lines connecting the deepest erosion points observed during the periods 1962-67 and 1968-75. Note that the preserved part deposited between 1962 and 1967 originated during a period of *slow* progradation. This resulted in a steeper beach gradient of *c*. 1:50, less preserved bar structures, and more and deeper preserved swash lamination. The preserved part deposited between 1968 and 1975 originated during a period of *rapid* progradation, hence a less steep beach gradient of *c*.1:90 and more preserved bar structures. The sedimentary structures of the core were studied from lacquer peels.

+1.40 m NAP (the crest of the highest longshore bar in 1975 reached up to +2 m NAP); small-scale ripple laminations were found up to +2.10 m NAP; trough-shaped mega-crossbedding (dm-scale) due to longshore currents in the troughs usually up to MSL (in one case up to +0.6 m NAP); (sub)parallel swash laminations were found from MLW (-1.20 m NAP) up to +3.80 m NAP, but were almost lacking around MSL.

Another interesting feature stressed by Van den Berg is that the trough bases landward of bars remain at the same depth, or even become slightly deeper, during the landward migration of the bars up to a level around MSL (see the lines connecting individual bars and troughs between the cores; Fig. 7). Above this level, around MSL, successive runnel bottoms follow the average beach gradient in landward direction. This important feature may probably be recognized in ancient prograded sequences as truncation planes which are either landward dipping, horizontal or seaward dipping, respectively.

In conclusion, the Schouwen beach survey indicates that if a barrier progrades during longer periods of varying wave intensity, one may expect beach sequences in which seaward-inclined low-angle (swash) lamination is predominant, alternating laterally with sequences with a predominance of landward-inclined crossbedding associated with small-scale ripples and clay drapes (preserved parts of bars and troughs). Further, in these more wave-protected sequences thin clay drapes occur up to MHW; landward oriented mega-crossbedding reaches up to spring tide level; major set boundaries of the latter dip seaward above MSL; and longshore oriented trough-shaped mega-crossbedding occurs up to MSL.

c) Egmond survey

In order to achieve a better understanding of the depth distribution of sedimentary structures on the beach in relation to tide levels, the author examined the same beach profile near Egmond at low water once a week between 27 January and 25 May 1980. All surveyed beach profiles are indicated on the left of Figure 8. Profiles typical for fair and stormy weather are marked by heavy lines. The depth distribution of those sedimentary characteristics that may also be recognized in subrecent coastal sequences, is given on the right of Figure 8: eolian features, HW swash limit, shell concentrations, shell orientation, longshore bars, small-scale wave ripples, current ripples, and thin clay drapes. It should be born in mind that the preservation potential of these features is generally difficult to evaluate, as this depends on exposure to waves and rate of progradation (*cf.* Roep *et al.*, 1975; van den Berg, 1977a; Beets *et al.*, 1981; also compare Figure 8 with Figures 13 and 15). Median sand-grain sizes at the Egmond locality in 1967 were 240μ, 250μ, and 290μ for dune foot, dry beach, and wet beach respectively (Glim, 1980).

A general conclusion from the Egmond survey is that the best lithologic marker in relation to SL is the HW swash limit separating the backshore from the foreshore. Backshore features include isolated large shells (often with a hole in the apex, Mr.A. Schermer, pers. comm.), driftwood, dry eolian sand layers and, on wetter parts, deflated surfaces with sand shadows behind shells or other objects, and adhesian ripples. The upper foreshore beneath the swash limit is characterized by low-angle, seaward-dipping sand beds, with primary-current lineation and often with small-scale rhomboid ripples, very low transverse ripples, sand shadows and current crescents around shells.

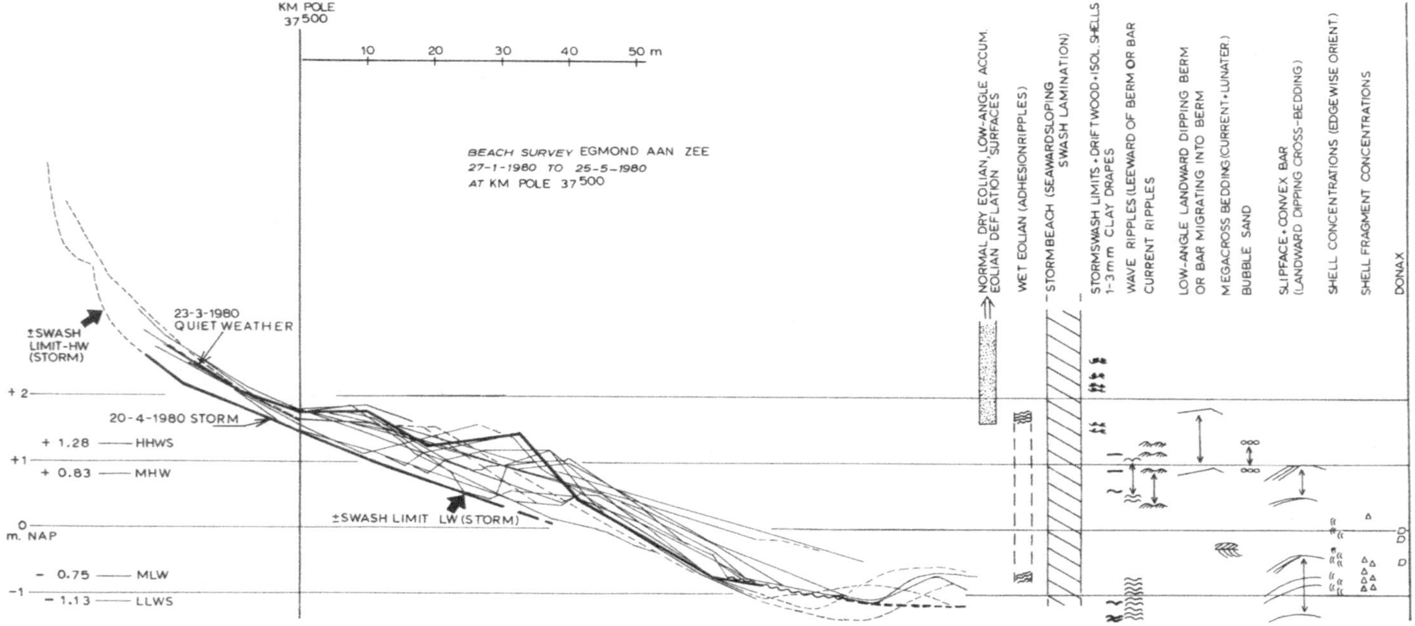

Figure 8. Beach survey near Egmond. The left part of the Figure shows all measured beach profiles during the period indicated. The typical quiet-weather profile and the storm profile are emphasized by heavy lines. The right part of the Figure gives the depth distribution of those features which also may be recognized in fossil barrier sediments.

With respect to individual sedimentary characteristics mentioned above (Fig. 8), it is noted that low-angle seaward-dipping swash lamination reaches down to -1 m NAP. One may expect this type of lamination to extend somewhat deeper into the subtidal zone, down perhaps to -1.5 m NAP or even deeper. Dry eolian activity starts above +1.6 m NAP. Hence, the total range of swash lamination on progradation will be in the order of 3 m. The best high tide indicators on Figure 8 are highest wave- and current-ripple laminations, and thin clay drapes (1-3 mm) in combination with very low-angle, landward-dipping berm laminations and air bubble sand Good berm descriptions are given by Reineck & Singh (1973, 1980), Hine (1979), Augustinus (1980), and others. At Egmond, landward-dipping slip faces of longshore bars reach up to +1 m NAP. Often the slip faces of intertidal longshore bars are reshaped by the waves during falling or rising tide or deeply truncated during storms (Wunderlich, 1972; Van den Berg, 1977a, Beets *et al.*, 1981). Two drawings from lacquer peels are given in Figure 9. Above +0.5 m NAP landward-displacing trough bottoms only move upward (see also lefthand part of Figure 8). The lower foreshore is characterized by packed shell concentrations and shell fragments (*cf*. Van Straaten, 1965; Roep *et al*., 1975) and the occurrence of *Donax* (see next section). During the survey, megaripples due to longshore currents draining the runnel were rarely found. As Van den Berg observed at Zandvoort (page 105), they occurred on the lower part of the foreshore. On one particular day these current megaripples were associated with wave-built lunate megaripples, their steeper sides facing towards the land and situated on a low part of the longshore bar.

To conclude this section some observations are made on the height variations of the swash limit. The swash limit naturally differs from high tide levels as predicted by tide tables, because run-up depends on wave spectrum, beach slope, grain size, and position of bars. The HW and LW swash limits during the storm of 20 April 1980 (Fig. 8) were at +3 m and + 0.5 m NAP respectively. Predicted HW and LW for that day were at +0.97 m and -0.93 m NAP. During fair weather (23 March 1980, Fig. 8) the measured swash limit was at +1.75 m NAP, whereas the predicted HW level was at +0.74 m NAP. On another fair day (9 March 1980) the measured swash limit variation was between +1.35 m and +1.60 m NAP; predicted HW was at +0.75 m NAP. The variation on that day depended on the position of bars and rip-channel depressions cutting through the bars, causing a difference in exposure to the incident breaking waves at high tide. Even on calm days, therefore, the swash limits may reach 0.5 to 1 m above predicted HW.

The Egmond survey should be regarded as an example. As yet, the depth variation over longer tracts of the beach is unknown. Beaches elsewhere may show different structures and depth distributions. Future work of this kind could include an analysis of wave spectra and the employment of small lacquer peels of surface structures, in combination with their depth distribution.

a

b

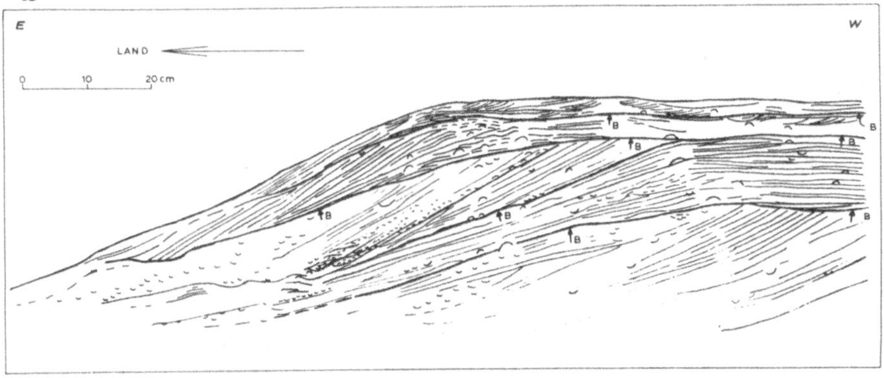

Figure 9a. Intertidal longshore bar and runnel deposit (redrawn from photograph of lacquer peel; Wijk aan Zee, 6-10-1972). Top of the bar is at about MSL. Bubble sand levels, indicating the preservation of at least 4 high-tide submersions (last stage included); T: undulating layer, thinning landwards due to overwash; S: slipface lamination with most shells lying concave side up. The slip-face lamination shows lateral flattening, probably a falling tide effect. The uppermost sedimentary unit contains convex lamination, with most shells lying convex side up, overlain by wave ripples formed at high tide. Subsequent convex truncation occurred during lowering of the water table. Two bar sets occur on top of each other.

b. Intertidal longshore bar (redrawn from photograph of lacquer peel; for position see Figure 6). The top of the bar is at *c.*-0.2 m NAP. B: bubble sand levels (like in Figure 9a, at least 4 high tide submersions recorded). The lateral steepening and flattening of the slip-face lamination is interpreted as a result of breaking waves over the bar during respectively rising and falling tidal stages. On the top of the bar truncated, wave-induced current ripples may be seen. The uppermost units of this bar were deposited during calm weather.

3. Shell distribution on beaches

Van der Baan (1977-1978) studied the distribution of dead shells and their size distribution along the beach of the province of North-Holland by counting 100 specimens at 6 to 10 sites, lying along a profile from dune foot to the low-water spring tide line at each kilometer pole. She found that the average size of all species decreased from the dune foot seaward. From the dune foot towards the first trough floor (from *c.*+3 m to *c.*0 to -1 m NAP) she observed a decrease in the percentage of *Cerastoderma (Cardium)* and an increase of the percentage of *Spisula*. Both species are most numerous along the present beach; in storm-swash lines *Cerastoderma* may comprise 80-90% of the sample. A similar decrease in *Cerastoderma* percentage was found on the seaward side of each longshore bar occurring further down in a seaward direction. Smaller species like *Donax, Macoma, Tellina* and *Abra* were only found in or near the troughs. *Venus* is found also in the troughs, but more frequently on the bar near low tide level. All these shell differentiations are attributed to size and shape sorting by wave action (see Lever *et al.*, 1964, cited in Van der Baan, 1977-78).

In conclusion, the study of the depth distribution of loose shells on the beach seems very worthwhile for the study of former SLs, if done in combination with studies of the kind mentioned above.

4. Sedimentary structures from the modern shoreface

Reineck and coworkers took short, undisturbed cores along profiles perpendicular to the coast in order to study the depth distribution of sedimentary structures of the *subtidal* part of the beach (Reineck & Singh, 1973, 1980; Chowduri & Reineck, 1978). Here a summary is given of such distributions for an inlet sequence and a barrier-front sequence in Northwest Germany.

From lacquer peels of the cores the depth distribution and percentage of the following structures were noted: mega-crossbedding, micro-crossbedding, parallel laminations, and bioturbated sand and mud layers (Fig. 10). The coastal sequence near Norderney rests with a gradational contact on shelf muds of the North Sea and contains abundant burrows in its basal part. The sequence near Wangerooge is an inlet sequence and begins with a channel-lag deposit followed by abundant tidal mega-ripple crossbedding up to a level of *c.*6 m below MLWS level. Both sequences may be subdivided into 4 units (see Fig. 10).

This type of study leads to objective results which are easily compared with sediment-structure distributions of other modern coasts. However, for comparison with subrecent sequences the classification of sedimentary structures is somewhat too schematic. Furthermore, major divisions are based on morphological criteria such as sawtooth bars. While these may be easily recognized in a

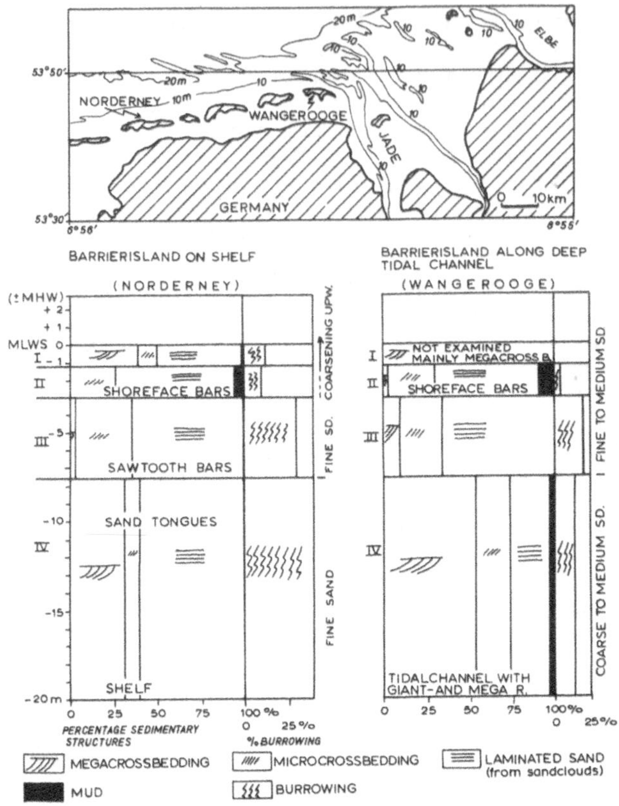

Figure 10. Depth distribution of sedimentary features in percentages (data from Reineck, 1976 and Chowduri & Reineck, 1978). Location map at the top shows the position of Norderney along the North Sea shelf and Wangerooge along a deep tidal channel. The data were obtained from profiles, cored at right angles to the coast and presented as if progradation had taken place The Norderney sequence shows abundant burrowing in its basal part, whereas the Wangerooge sequence shows tidal mega-ripple lamination. Sand tongues and sawtooth bars are named after their morphological appearance on the hydrographic maps.

modern environment, this may be very difficult in fossil sequences (page 103). Emphasis on actual depth distribution of structures as well as on trends in crossbedding set heights etc. would make them more useful for paleo-SL studies.

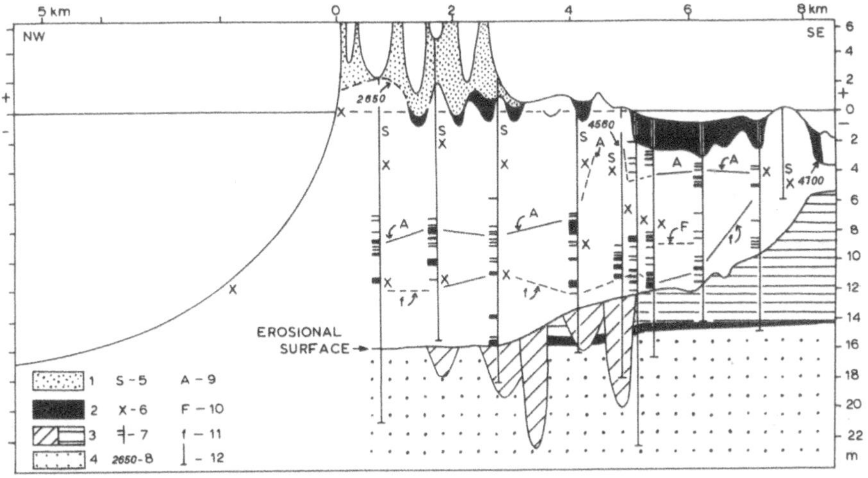

Figure 11. Cross section through Subboreal coastal barrier complex near The Hague (slightly modified after Van Straaten, 1965; pollen zones from Zagwijn, 1965). 1: Medieval eolian sand (Younger Dunes); 2: Holocene peat deposits; 3: Holocene lacustrine, lagoonal, tidal mudflat and channel deposits older than the barrier complex; 4: Pleistocene sands; 5: highest shell layers; 6: maximum median sand diameters in higher and lower zone; 7: clay layers, simplified; 8: age in C14 yr BP, after Van Straaten, 1965; Jelgersma *et al.*, 1970 and the Geological Survey; 9: pollen-curve maximum for *Alnus*, low value *Quercus*; higher value *Corylus*; 10: pollen-curve maximum for *Fraxinus*; 11: regular occurrence of *Fagus* pollen; 12: location of boring by the Geological Survey.

EXAMPLES OF PALEO-TIDE LEVEL DERIVATION FROM SEDIMENTARY STRUCTURES IN FOSSIL BARRIER DEPOSITS OF THE NETHERLANDS

Van Straaten (1965) and Zagwijn (1965), in a classic study of the West-Netherlands barrier complex, analyzed coastal progradation and SL change (summary in King, 1972). Their work is based on a series of corings through the entire barrier system (Fig. 11). Van Straaten, using lacquer peels of the cores, was able to infer approximate paleo-tide levels. These were defined on the basis of 3 zones occurring above each other. From top to bottom:

zone 1: horizontally laminated and crossbedded sands with intercalated, homogenized humic levels (soils) and root traces;

zone 2: mainly horizontally laminated sands with incidental intercalations of cm-thick crossbedded sands;

zone 3: 'wildly' cross-laminated sands.

Zone 3 was interpreted as the lower part of the beach and the upper part of the subtidal zone with strong influence of waves and currents. The top of this zone was taken to roughly coincide with MLW. Zone 2 was interpreted as beach sands with the highest shell layer lying at or above MHW. Finally, zone 1 sediments were recognized to be of eolian origin. The base of this zone must have occurred well above the contemporaneous MHW level.

Below zone 3 Van Straaten distinguished a zone with decreasing crossbedding and increasing numbers of clay layers. Still deeper occurred a zone characterized by maximum median sand diameters (Figure 11), which was interpreted to belong to a large subtidal longshore bar, which moved landward while shallower parts of the coastal zone prograded.

Van Straaten and Zagwijn also came to the conclusion that the former coastal gradient, between MSL and about 5 m water depth, was about 1:200, compared to the present-day gradient (1:100).

For further information about the development of the coastal barrier system a reader should refer to Van Straaten (1965), Jelgersma *et al.* (1970), Raven (1979), Roep *et al.* (1983), Roep (1984) and Zagwijn (1984). Postglacial SL rise and development of the North Sea during the Holocene are discussed by Zagwijn (1974), Jalgersma (1979) and Van de Plassche (1982).

During the last ten years several paleo-tide level studies have been carried out. Two examples are discussed below. The first concerns a barrier sequence near IJmuiden, which was formed around 2300 BP; the sequence can best be interpreted in terms of the modern coastal environment at IJmuiden. The second example concerns a barrier segment near Alkmaar. It originated between 4200 and 3650 BP. The coastline was more protected from waves than at IJmuiden. However, two types of coastal sequences are recognized (at Alkmaar) which indicate greater and lesser wave influence during the period of progradation.

1. Barrier sands near IJmuiden

The top 6 m of a prograded barrier sequence near the sluices of IJmuiden could be studied in a deep excavation (Roep *et al.*, 1975). Figure 12a shows a section at right angles to the former coastline. Even on this photograph three major units are immediately apparent below a peat bed. Figure 13 depicts the sequence of sedimentary structures in a slighly schematic way. It proved possible to define approximate paleo-MHW and -MLW positions with the help of the sedimentary structures.

The highest unit (unit 1) is characterized by crossbedded sand without a preferred orientation of the crossbeds. Large troughs can be seen (upper part Fig. 12b) often with undulating boundaries, smoothly undulating

a

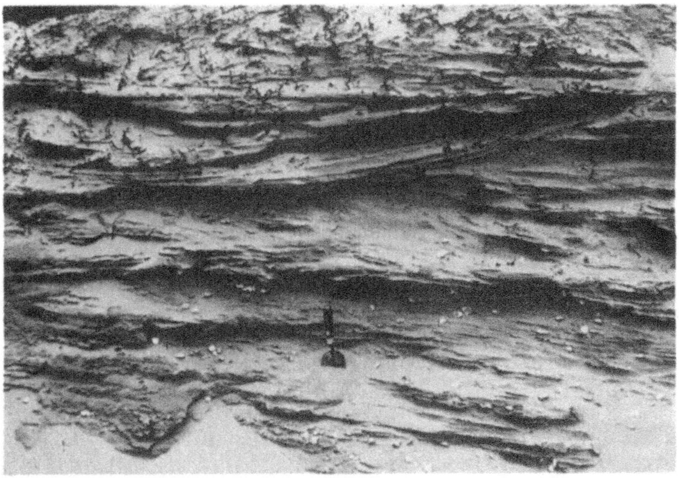

b

Figure 12a. Section in coastal barrier sands exposed in pit just N of IJmuiden sluices in 1971. The height of the section on the left is about 6m (see also Fig. 13). Sea is and was to the right. In the lower part a unit with high-angle bars may be seen. This unit is largely of subtidal origin (see text). Above this unit a shell-rich zone occurs with a temporary road on top (halfway the section). The top of the shell-rich zone approximately corresponds with paleo-MSL (see text). Higher up in the section low-angle seaward-dipping swash lamination can be seen, which slightly steepens upward. Note that in the inferred intertidal part only flat storm-beach lamination is preserved. Broken parts from a peat layer overlie eolian sand at the top with trough-shaped lamination. b. Contact between eolian sand (no shells, trough-shaped crossbedding, reed roots) and seaward-sloping swash lamination with shells, pieces of wood, etc. Shell doublets from this zone dated 2310 ± 35 BP. Shaft of spade is *c*. 0.1 m long. For location see Figures 13 and 12a.

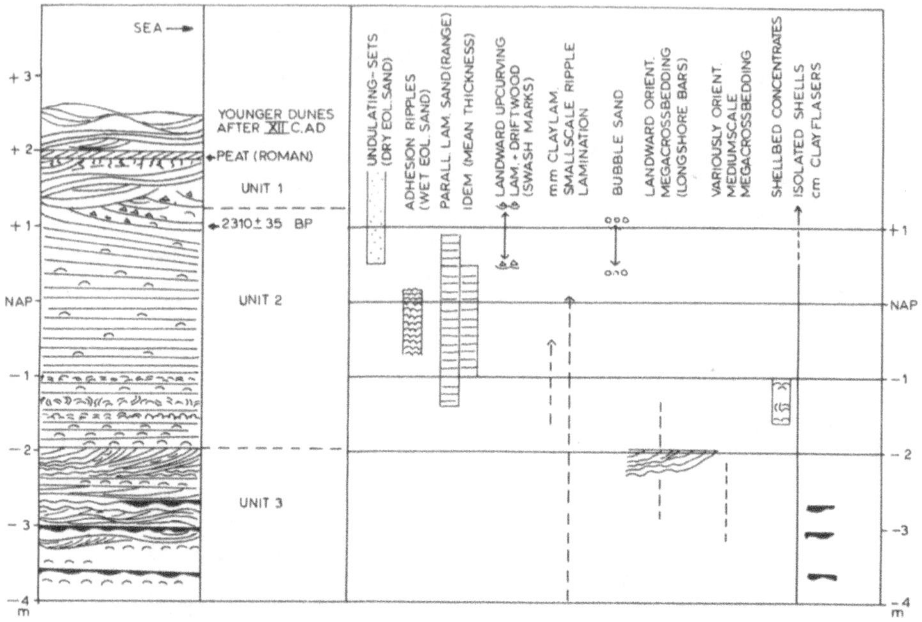

Figure 13. Schematic section in coastal barrier sands (see also Figure 12a) just N of the IJmuiden sluices, *c*. 1.8 km E of the modern coast (after Roep *et al.*, 1975 and Ruegg, 1975). Depth distribution of sedimentary features are indicated on the right. Compare this distribution with that of the modern beach near Egmond (Fig. 8). Inferred paleo-MSL is near -1 m NAP; paleo-MHW near 0 m NAP and paleo-MLW near -2 m NAP.

and tapering crossbeds and strata, penetrated by *Phragmites* roots from the overlying peat. Unit 1 is devoid of shells. The crossbeds of unit 1 interfinger with shell-bearing, very low-angle seaward-dipping strata which show landward-upcurving tips with pieces of driftwood, brittle stars and also bubble sand. The zone of interfingering is about 0.75 m thick (Fig. 13, right hand part) and clearly represents the HW swash zone at its contact with a low dune foot (unit 1).

Unit 2, the intermediate unit, is mainly characterized by very low-angle seaward-dipping strata with abundant shells and shell beds especially in its lower part (see Figure 13, right part). Towards the top of this unit the strata dip at an angle of 3 to 3½° (about 1:15), becoming flatter in the down-dip direction to around 2° (about 1:30). Such gradients also are found at the modern beach of Egmond (see Fig. 8).

The deepest unit (unit 3) is characterized by complex,

landward-dipping mega-crossbedding associated with wave ripples (Figs. 12a and 13). Set boundaries have a seaward gradient of 1:30 (at a level of -3 m NAP: (Ruegg, 1975). Up to the limit of exposure, the mega-crossbedded sets decrease downwards in height, and are replaced by variously oriented dm-scale mega-crossbedding. Still deeper, cm-thick clay flasers are intercalated and increase downward in number and thickness (Fig. 13). A similar situation was described by Van Straaten (1965; Fig. 14). The mega-crossbedded sets with landward dip are interpreted to be longshore bars on account of their orientation, association with wave ripples, complexity, and setting in the sequence. Some crossbeds are inclined at a low angle, others are slip-face lamina at high angle. On lacquer peels, the high-angle slip-face lamina contain shells lying concave side up, whereas most other shells lie convex side up.

What are the SL indicators in such a sequence? The largest lithologic breaks in this excavation occurred at *c*.+1 m and *c*.-2 m NAP (see the boundaries of unit 2 in Figure 13). It was reassuring to find that the vertical distance between these breaks is not very different from the present tidal range of 1.5 to 2 m, allowing for the swash run-up. More details are needed to decide if these boundaries coincide with MHW and MLW. Clearly the sand of unit 1, with dry eolian structures, must lie above HW and the cm-thick clay laser at -2.70 m NAP below LW. An indication for the latter is that cm-thick clay layers along the present North Sea beach away from the tidal inlets are only found subtidally (Glim, Rijkswaterstaat, pers.comm.).

The dry eolian processes, especially deflation, must take place above the water table. Ground-water level of recent beaches with low dunes is near HW level, though it may be higher in areas with higher dunes. Deepest eolian scour, here at +0.50 m NAP, is therefore at or above HW level. Roep *et al*. (1975) concluded that paleo-MHW was at +0.50 m NAP. After the Egmond survey (Fig. 8), and a more careful analysis of lacquer peels of the IJmuiden pit, it now seems that 0 m NAP is a better estimate for MHW around 2310 BP (Fig. 13).

Highest adhesion ripples (dry eolian sand blown over the wet upper part of the intertidal zone), highest small-scale ripples (formed in the highest aquatic, wave-protected parts of the beach), the berm, and deepest bubble sand all occur around 0 m NAP. At the recent beach of Egmond similar features occur around +1 m NAP (or about modern MHW). In both localities the zone with swash marks and pieces of driftwood occur regularly up to a level lying about 1.5 m higher. It is presumed that still higher swash marks, formed during storms in eolian sand, are destroyed by dry eolian activity normally occurring at these higher levels.

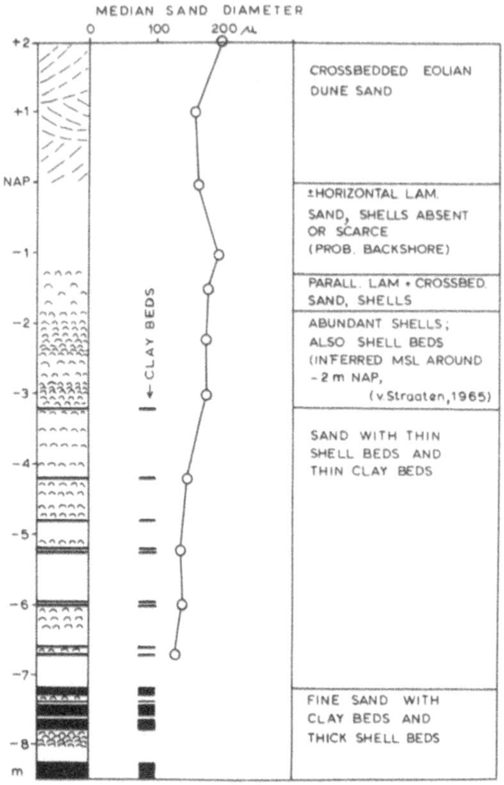

Figure 14.

Schematic section in coastal barrier sands, Fishing Harbour IJmuiden, *c.*1km E of the present-day coast (slightly modified after Van Straaten, 1965). Note the downward increase in number and thickness of clay layers and upward increase of median sand-grain diameters up to -1 m NAP. According to Van Straaten paleo-MSL occurred near -2 m NAP.

At IJmuiden paleo-MLW is inferred to lie close to -2 m NAP (Fig. 14) at the major lithologic break between unit 3 and 2. At the modern Egmond beach, storm profiles often reach down to the MLW zone (here at about -1 m NAP). Shell beds at IJmuiden regularly occur up to -1 m NAP and at Egmond up to 0 m NAP (MSL). In conclusion, paleo-MHW at IJmuiden is inferred to be at a level around 0 m NAP and paleo-MLW at -2 m NAP with MSL in between at -1 m NAP. This means that in the intertidal parts at IJmuiden, no bar lamination has been preserved, but only sub-parallel, low-angle seaward-inclined (swash) lamination. In other words, only the storm profile with swash laminations has been recorded.

2. Barrier sands near Alkmaar

The internal structures of barrier deposits in the vicinity of Alkmaar have been studied in a large number of pits and trenches (Roep *et al.*, 1975, 1979; Beets *et al.*, 1981). Two types of coastal sequences have been recognized: 1) with more parallel laminations in the top, below a relatively thick eolian cover, and 2) with low-angle longshore bars at the top, below a thinner eolian veneer, slightly more clayey and burrowed (Beets *et al.*, 1981). The second type occurs more frequently.

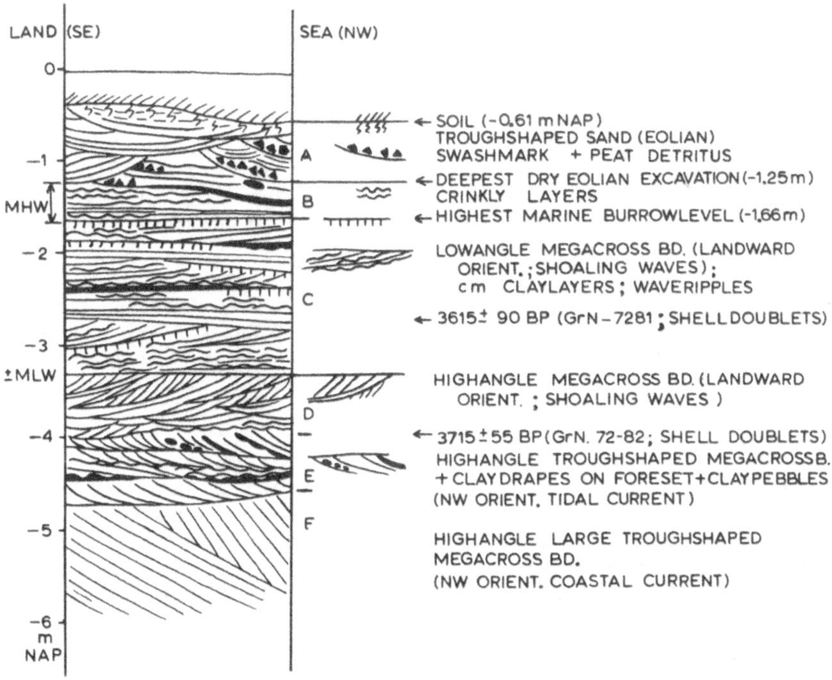

Figure 15. Coastal barrier sand exposed in a pit just S of Alkmaar (slightly schematized). The level of inferred MHW is between -1.25 and -1.66 m NAP, and of MLW near -3.33 m NAP. In contrast to the IJmuiden sequence (see Figs. 12a and 13), the intertidal part is composed of low-angle bars, flattening upward. This indicates that the coast at Alkmaar was more protected from waves than at IJmuiden. Below -4 m NAP megaripple lamination formed by tidal (?) currents replaces bar laminations formed by shoaling waves.

Type 1 sequences resemble in many respects those studied at IJmuiden, though parallel laminations are less sharp, more undulating and contain fewer shells. Another point of difference is that both the Alkmaar sequences originated in a more wave-protected environment than the IJmuiden sequence. This is apparent from the fact that the Alkmaar sequences are finer grained, and contain thin clay layers and frequent burrow levels.

Type 2 sequences are characterized by the presence of low-angle bars (described in detail by Beets *et al.*, 1981; see also Figure 15). The low-angle bars are landward oriented sets 0.15 - 0.75 m high, with internal lamination up to several meters long which dip from 15° to almost 0°. Foreset dip usually becomes flatter in successively higher low-angle crossbedded sets. Internally, the bars show frequent alternations of wave-rippled units and cm-thick massive or slightly graded sandbeds, as well as erosion levels and shell layers. The wave-rippled units were interpreted as HW features and the massive or graded units as LW features resulting from waves breaking over the bar crest in shallow water. Invariably, low-angle bars occur in the top of the coastal sequence, below horizontal burrow levels and above a unit with high-angle landward-dipping cross-bedded sets. In some aspects low-angle bars resemble modern berm deposits on the North Sea coast. The latter, however, are lacking in burrow levels and thin clay drapes, and composite units are much thinner.

Figure 15 shows a typical type 2 sequence near Alkmaar. Unit C is the unit with low-angle bar laminations. The lowermost units, F and E, probably constitute the top of an inlet sequence. The sedimentary structures are indicative of currents.

The difference between the two types of sequences is explained by a difference in coastal gradient and rate of progradation. As in the case of the modern Schouwen beach (page 105), slow progradation with steeper gradients results in preservation of parallel swash lamination, whereas rapid progradation with flatter gradients results in preservation of long-shore bar lamination. Furthermore, a steep beach face promotes dune building as it implies relatively high wave energy on the upper part of the beach and hence deposition of sand high on the backshore (Beets *et al.*, 1981). A difference between the modern Schouwen beach and the paleo-beach at Alkmaar is the type of longshore bars. The Schouwen beach (Fig. 7) mainly shows convex bars and high-angle bars with slipface lamination (*cf.* Figs. 8 and 9a, whereas the Alkmaar sequence shows low-angle bars.

In the pits around Alkmaar it was easier to establish paleo-high tide levels than in the IJmuiden pit. Here MHW levels could be found with the help of dry eolian erosion features *and* marine burrow levels. Wind scour can only take place in dry sand above the groundwater (or high water) table. The burrow levels on the other hand were made by marine worms making straight vertical tubes with a coating of clay, mostly resembling those of *Pygospio elegans*. The burrow levels occurred throughout the sections up to a certain level. Being suspension feeders the worms must have been flooded daily. MHW therefore had to be sought between the deepest eolian scour and the highest burrow level (Roep *et al.*, 1975). Figure 15 is based on the sequence exposed in a tunnel pit under the railway just S of Alkmaar (for details see Beets *et al.*, 1981). The highest burrow level in the tunnel pit was found at -1.66 m, the deepest eolian scour at -1.25 m NAP; thus MHW occurred between these levels. Articulated shells at -2.76 m NAP dated 3615 ± 90 BP (GrN-7281).

Swashmarks with pieces of driftwood were found between -0.75 m NAP and -1.25 m NAP. Wave ripples occurred regularly up to a level of -1.30 NAP, and bubble sand

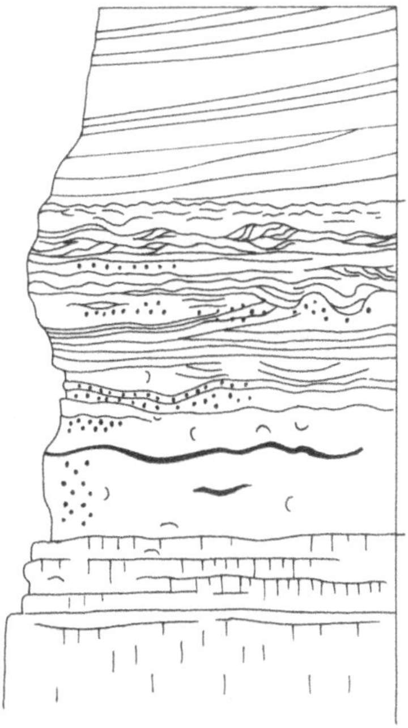

Figure 16. Transition unit between the highest marine burrow layer and the deepest eolian sand exposed in a pit *c*.500 m SE from the pit discussed in Figure 15. The transition unit consists partly of wet eolian adhesion ripples and possible washover-fan sediments (with shells on edge). Note that the highest burrow level and the base of eolian sand occur somewhat deeper than in the younger sequence of Figure 15. This difference is attributed to a rise of sea level.

between -1.50 and -1.20 m NAP. In our previous paper we therefore chose -1.25 m NAP as the paleo-MHW level, but -1.66 m NAP possibly is a better estimate, because living *P. elegans* are found up to HW levels. In any case, former MHW is between these levels.

A detailed drawing of a transition zone between highest burrow level and eolian sand, in another pit 500 m SE from the tunnel pit, is given in Figure 16. Some layers with isolated shells on edge in a massive sandy matrix suggest massflow sedimentation, which may be the result of episodic washover deposition. In other pits this transition zone is often characterized by sandy beds with crinkly lamination and thin clay beds. The transition unit is not unlike washover-fan sections pictured by Dickinson, *et al*. (1972).

In some cases the crinkly lamination can be attributed to adhesion ripples, in other cases to plastic deformation comparable to convolute lamination.

In the pits around Alkmaar the greatest structural change usually occurs between predominantly low-angle crossbedding and predominantly high-angle crossbedding. Both types have crossbeds dipping towards the land and are, therefore, interpreted as wave-built longshore bars. In the tunnel pit (Fig. 15) this structural change, between units C and D, occurs at -3.33 m NAP. In 7 pits around Alkmaar the thickness of the unit between highest burrow level and the level of greatest structural change, has a mean value of $\bar{x}$ (7) = 1.79 m and a standard deviation of s = 0.36 m. The average thickness between greatest structural change to base of eolian sand is $\bar{x}$(8) = 2.24 m with s = 0.47 m. The interval thicknesses of 1.79 m and 2.24 m are so near the value of present-day mean tidal range along the coasts of the provinces of North- and South-Holland (1.75 m), that we assumed that this interval corresponds largely to the palaeo-intertidal zone (its top being around paleo-MHW). For this reason we interpret the low-angle bars as intertidal and believe that their base corresponds to MLW or just below it (Roep *et al.*, 1975).

Still deeper parts of the coastal sequence near Alkmaar have only been seen in 2 or 3 pits. These deeper parts usually contain thicker clay layers, but otherwise show a great diversity of mega-crossbedding and parallel laminated sand. Probably, large scale longshore bars, such as occur along the present coast, are concealed within these complex structures of the deeper sequence.

Interpretation of the deeper parts of coastal sequences, such as near Alkmaar are also difficult because their modern analogue is still poorly known. Still it is tempting to compare the vertical bar distribution of the modern coast (Fig. 5) with a vertical distribution of the combined sedimentological, structural data from 13 pits around Alkmaar (Fig. 17). Such a comparison has to be done with the greatest caution on account of the following problems:

- During deposition of the Alkmaar barrier system a SL rise of *c*.0.8 m occurred between about 4200 and 3600 BP (Jelgersma, 1979; Van de Plassche, 1982).
- The vertical distribution given in Figure 17 is biased to an unknown extent by the few exposures of deeper parts of the sequence (11 pits could be studied down to -4 m NAP; 5 down to -4.5 m; 3 to -5.5 m; 2 to -6.5 m and 1 down to -8 m NAP).
- Modern bar heights are not the same as fossil set heights, as bars often are composed of more than one truncated crossbedded set.
- The individual sequences around Alkmaar have sharper structural breaks around the inferred low tide level than is apparent from the combined data of Figure 17.

With the above considerations in mind, the following tentative conclusions may be drawn from Figure 17:

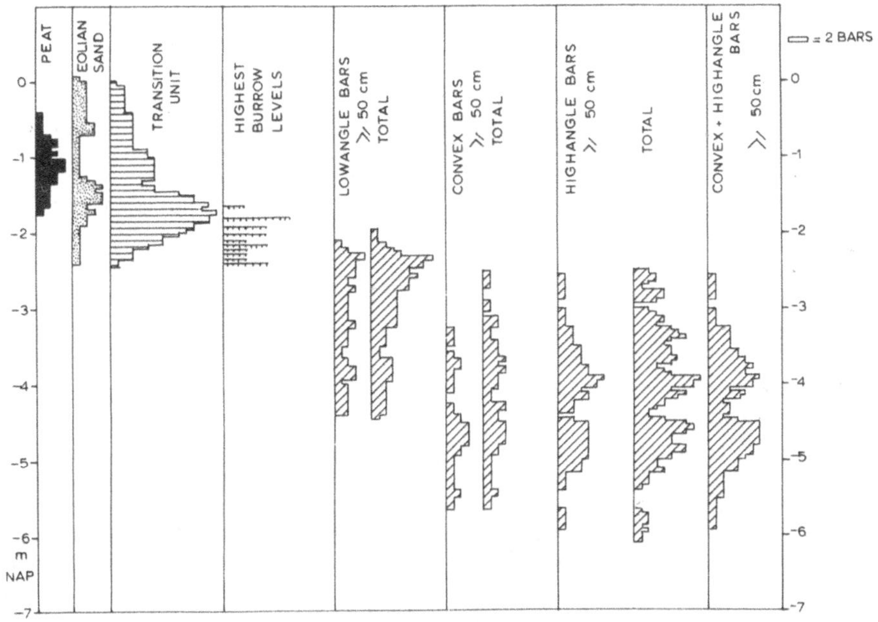

Figure 17. Vertical distribution of combined data of 13 pits in the barrier complex around Alkmaar. The ages may vary by a few hundred years. This distribution suggests that the coastal gradient was less than at present (*cf*. Fig. 5). See text for details and discussion.

(1) Deepest dry eolian sand levels (second column in Figure 17) occur between *c*.-2.5 and 0 m NAP. Highest burrow levels occur between *c*.-2.5 and -1.75 m NAP, most frequently around -2 m NAP. Hence eolian sand levels and highest burrow levels show an overlap of 0.75 m, which is not far from the inferred SL rise of 0.8 m that took place during deposition of these sands.

(2) The level of -2 m NAP separates a mode in the transition unit, interpreted as washover-fan sediments in part (Fig. 16), and a slight mode in the distribution of low-angle bars, interpreted as largely intertidal sediments.

(3) Weak modes in the high-angle bar distribution occur at -4 m and -4.75 m NAP, whereas the bars disappear around -5.5 m NAP. If one assumes that paleo-MSL for the combined data roughly lies around -2.5 m NAP (Jelgersma, 1979; Van de Plassche, 1982), then these modes originally occurred around -1.50 and -2.25 m below paleo-MSL, and bars disappeared below *c*.3 m paleo-MSL. Along the modern coast (Fig. 5), modes in vertical bar distribution occur at

-2.25 and -4.5 m NAP and bars disappear below -7 m NAP. The difference in depth between recent and fossil modes suggest a former flatter gradient for the Alkmaar coast than for the present North Sea. This seems to be supported by a gradient estimate for the Alkmaar coast based on the vertical distribution of C14 ages in Figure 15 and the mean progradation rate of 238 m/100 yr from the horizontal distribution of C14 ages (Beets *et al.*, 1981). The gradient found is *c.*1:200 (or less if progradation was temporarily more rapid). Van Straaten (1965) previously found a coastal gradient of 1:200 for the The Hague barrier complex (as contrasted with 1:100 for the modern coast). These results therefore, agree rather well with the suggestions inferred from Figure 17.

SOME RECOMMENDATIONS

1) Whenever possible survey the modern beach and offshore morphology.

2) Study the modern sedimentary structures by means of *oriented* lacquer peels from pits and undisturbed cores. Establish depth ranges of individual structures, dead and living shells, and burrows. Be aware of the problem of preservation potential of modern sedimentary features.

3) Do not ignore the problem of compaction. Select (the most) sandy sequences. Coastal barriers often overlie clayey and peaty marsh deposits, implying possible lowering of the beach sequence due to compaction.

4) Try to obtain cores (and sections) at right angles to the contemporaneous coastline. This facilitates interpretation of crossbedding resulting from longshore currents or of landward-oriented wave-built longshore bars.

5) Concentrate on height and nature of major divisions within the vertical sequence. Pay close attention to individual structures such as deepest trough in eolian sand, indicating dry eolian action above the water table; to adhesion ripples, indicating wet eolian action on the wet beach surface at low tide; and to those features indicating daily flooding (highest marine burrows or shells in living position).

6) 'Never' use single shell-halves to date the barrier deposit; these may be reworked from older deposits. Use articulated shell pairs only. The bivalves need not necessarily be in living position.

7) Be aware of the fact that the age of one radiocarbon dated sample from, say the top of a coastal sequence, need not be representative of the entire sequence. Thus, dated articulated shell pairs from an ancient swash zone gives the age of the HW zone. The age of deeper levels in the sequence such as the LW zone and still deeper levels depend on rate of progradation, paleo-coastal gradient and preservation of high or low bar structures. Recent gradients of the intertidal zone of *c.*1:50 to 1:90

(Schouwen beach; see Fig. 5) and 1:25 to 1:45 (Egmond beach, Figure 8) indicate that 2 m of section will be formed after 50 to 180 m of progradation. With a progradation rate of 200 m/100 yr (*cf.* example of Alkmaar coast) an intertidal zone of 2 m in a vertical section will be recorded in about 25 to 100 years. In a C14-dated example (Fig 15) 1.30 m of section around inferred MLW was deposited in 100 years. If possible, therefore, take more than one sample above each other in a sequence.

ACKNOWLEDGEMENTS

I thank Dr. G.H.J. Ruegg of the Netherlands Geological Survey for providing unpublished data from the IJmuiden excavation and a photograph (Figure 12a); Dr. J. Smit for providing a photograph (Figure 6) and Mr. G.W. Glim for providing depth charts and grainsize data of the Dutch coast recorded by Rijkswaterstaat. Thanks are due to Prof. Dr. L.M.J.U. van Straaten; Dr. O. van de Plassche; Dr. R.W.G. Carter and an unknown referee for criticially reviewing earlier drafts of the manuscript. The correction by Dr. D.R. Grant and Dr. J. Grocott of the English is greatly appreciated. Illustrations were done by F. Kievits and J. Wiersma; the typewriting by Mrs. L.Gonggryp. Jens and Elisa Roep assisted during fieldwork on the Egmond beach. Most of all I want to thank Dr. D.J. Beets for years of pleasant and fruitful cooperation.

REFERENCES

Augustinus, P.G.E.F., 1980. Actual development of the chenier coast of Suriname (South America). *Sedimentary Geology*, 26(1/3), 91-114

Baan, G. van der, 1977-78. Een onderzoek naar het aanspoelen van schelpen. *Het Zeepaard*, 37, 94-96; 38, 5-9; 38, 34-39

Beets, D.J., Roep, Th.B. & Jong, J. de, 1981. Sedimentary sequences of the subrecent North Sea coast of the western Netherlands near Alkmaar. *International Association of Sedimentologists, Special Publication*, No 5, 133-145

Berg, J.H. van den, 1977a. Morphodynamic development and preservation of physical sedimentary structures in two prograding recent ridge and runnel beaches along the Dutch coast. *Geologie en Mijnbouw*, 56(3), 185-202

Berg, J.H. van den, 1977b. A fast procedure for sampling sedimentary structures down to 1.1 m in unconsolidated wet sands. *Geologie en Mijnbouw*, 56(3), 259-262

Berg, J.H. van den, 1982. Migration of large-scale bedforms and preservation of crossbedded sets in highly accretional parts of tidal channels in the Oosterscheide, S.W. Netherlands. *Geologie en Mijnbouw*, 61(3), 253-264

Chowduri, K.R. & Reineck, H.E., 1978. Primary sedimentary structures and their sequence in the shoreface of barrier island Wangerooge (North Sea). *Senckenbergiana Maritima*, 10(1/3), 15-19

Clifton, H.E., Hunter, R.E. & Phillips, R.L., 1971. Depositional structures and processes in the non-barred high energy nearshore. *Journal of Sedimentary Petrology*, 41, 651-670

Davidson-Arnott, R.G.D. & Greenwood, B., 1976. Facies relationships on a barred coast, Kouchibouguac Bay, New Brunswick, Canada. In: *Beach and Nearshore Sedimentation*, eds. R.A. Davis & R.L. Ethington, *Society of Economic Paleontologists and Mineralogists, Special Publication*, 24, 149-168

Davies, D.K., Ethridge, F.G. & Berg, R.R., 1971. Recognition of barrier environments. *Bulletin of the American Association of Petroleum Geologists*, 55, 550-565

Dickinson, K.A., Berryhill, H.L. & Holmes, C.W., 1972. Criteria for recognizing ancient barrier coastlines. In: *Recognition of Ancient Sedimentary Environments*, eds. J.K. Rigby & W.K. Hamblin, *Society of Economic Paleontologists and Mineralogists, Special Publication*, 16, 192-274

Doeglas, D.J., 1954. The origin and destruction of beach ridges. *Leidse Geologische Mededelingen*, 20, 34-47

Edelman, T., 1967. Systematic measurements along the Dutch coast. *Proceedings of the 10th Conference on Coastal Engineering*, 489-501

Elliott, T., 1979. Clastic shore lines. In: *Sedimentary Environments and Facies*, ed. H.G. Reading, (Blackwell), 2nd edition, 143-177

Glim, G.W., 1980. Korrelgrootte strandmonsters Noord-Holland en waddeneilanden. *Internal Report Rijkswaterstaat*, notitie W.W.K.72-80, H254, 1-34

Greenwood, B. & Davidson-Arnott, R.G.D., 1979. Sedimentation and equilibrium in wave-formed bars: a review and case study. *Canadian Journal of Earth Sciences*, 16(2), 312-332

Hayes, M.O., 1980. General morphology and sediment patterns in tidal inlets. *Sedimentary Geology*, 26(1/3), 139-156

Hine, A.C., 1979. Mechanisms of berm development and resulting beach growth along a barrier spit complex. *Sedimentology*, 26, 333-351

Howard, J.D. & Reineck, H.E., 1972. Georgia coastal region, Sapelo Island, U.S.A.: Sedimentology and biology, IV Physical and biogenic sedimentary structures of the nearshore shelf. *Senckenbergiana Maritima*, 4, 81-123

Hubbard, D.K., Oertel, G. & Nummedal, D., 1979. The role of waves and tidal currents in the development of tidal-inlet sedimentary structures and sand body geometry: examples from North Carolina, South Carolina and Georgia. *Journal of Sedimentary Petrology*, 49(4), 1073-1092

Jelgersma, S., 1979. Sea-level changes in the North Sea basin. In: *The Quaternary History of the North Sea*. eds. E. Oele, T.E. Schüttenhelm & A.J. Wiggers, Acta Universitatis Uppsalauensis Symposium University Uppsala, Annum Quingentesimum Celebrantis, 2, 233-248

Jelgersma, S., Jong, J. de, Zagwijn, W.H. & Regteren Altena, J.F. van, 1970. The coastal dunes of the western Netherlands, geology, vegetational history and archeology. *Mededelingen Rijks Geologische Dienst*, N.S. 21, 93-167

Jong, J.D. de, 1977. Dutch tidal flats. *Sedimentary Geology*, 18, 13-23

King, C.A.M., 1972. *Beaches and coasts*. (Arnold, London), 2nd edition

Kraft, J.C. & John, C.J., 1979. Lateral and vertical facies relation of transgressive barriers. *Bulletin of the American Association of Petroleum Geologists*, 63(12), 2145-2163

Kumar, N. & Sanders, J.E., 1974. Inlet sequence: a vertical succession of sedimentary structures and textures, created by the lateral migration of tidal inlets. *Geology*, 21, 491-532

Lever, J., Bosch, M. van den, Cook, H., Dijk, T. van, Thiadens, A.J.K. & Thijssen, R., 1964. Quantitative beach research III. An experiment with artificial valves of *Donax vittatus*. *Netherlands Journal of Sea Research*, 2(3), 458-492

Nio, S.D., Berg, J.H. van den, Goesten, M. & Smulders, F., 1980. Dynamics and sequential analysis of a mesotidal shoal and inter-shoal channel complex in the eastern Scheldt (South-western Netherlands). *Sedimentary Geology*, 26(1/3), 263-279

Plassche, O. van de, 1982. Sea-level change and water-level movements in The Netherlands during the Holocene. *Mededelingen Rijks Geologische Dienst*, 36(1), 1-93

Raven, J.G.M., 1979. The subboreal coastal barriers at Leidschendam, with a description of the faunas (province of Zuid-Holland, The Netherlands). *Mededelingen Werkgroep Tertiair en Kwartair Geologie*, 16(1), 17-54

Reineck, H.E. & Singh, I.B., 1973 and 1980 (2nd revised ed.). *Depositional sedimentary environments*. (Springer Verlag, Berlin, etc.)

Reinson, G.E., 1979. Barrier Island Systems. In: *Facies Models*, ed. R.G. Walker, Geoscience Canada, Reprint Series 1, 6, 57-64

Roep, Th.B., 1984. Progradation, erosion and changing coastal gradient in the coastal barrier deposits of the western Netherlands. In: *Geological Changes in the Western Netherlands During the Period 1000-1300 A.D.* eds. H.J.A. Berendsen & W.H. Zagwijn, *Geolgie en Mijnbouw*, special issue, 63(3), 249-258

Roep, Th.B., Beets, D.J. & Ruegg, G.H.J., 1975. Wave-built structures in sub-recent barriers of the Netherlands. *Proceedings of the 9th International Congress of Sedimentology,* (I.A.S., Nice, 1975), 6, 141-146

Roep, Th.B., Beets, D.J. & Jong, J. de, 1979. Het zeegat tussen Alkmaar en Bergen van ca. 1900 tot 1300 voor Chr., *Alkmaarse Historische Reeks,* III, (De Walburg Pers), 9-36

Roep, Th.B., Van de Plassche, O., Van der Valle, L. & Ruegg, G.H.J., 1983. Sedimentologie van de strandwal-afzettingen onder 's-Gravenhage en Rijswijk. In: *De bodem van 's-Gravenhage,* ed. H.W.J. van Amerom, Mededelingen Rijks Geologische Dienst, 37-1, 63-95

Ruegg, G.H.J., 1975. Sedimentologisch onderzoek van vroeg-subatlantische strandafzettingen bij IJmuiden. *Geological Survey of the Netherlands, Internal Report,* 10, 1-7

Straaten, L.M.J.U.van, 1961. Directional effects of winds, waves and currents along the Dutch North Sea Coast (Part 2). *Geologie en Mijnbouw,* N.S. 23, 363-391

Straaten, L.M.J.U.van, 1965. Coastal barrier deposits in South- and North-Holland, in particular in the areas around Scheveningen and IJmuiden. *Mededelingen Geologische Stichting,* N.S. 17, 41-76

Wunderlich, F., 1972. Georgia coastal region, Sapelo Island, U.S.A., sedimentology and biology, III. Beach dynamics and beach development. *Senckenbergiana Maritima,* 4, 47-79

Zagwijn, W.H., 1965. Pollen - analytic correlations in the coastal - barrier deposits near The Hague (The Netherlands). *Mededelingen Geologische Stichting,* N.S., 17, 83-88

Zagwijn, W.H., 1974. The palaeo-geographic evolution of The Netherlands during the Quaternary. *Geologie en Mijnbouw,* 53, 369-385

Zagwijn, W.H., 1984. The formation of the Younger Dunes on the west coast of The Netherlands (AD 1000-1600). In: *Geological Changes in the Western Netherlands During the Period 1000-1300 AD,* eds. H.J.A. Berendsen & W.H. Zagwijn, Special issue *Geologie en Mijnbouw,* 63(3), 259-268

5

Marine molluscs as indicators of former sea-level stands

K. S. PETERSEN

ABSTRACT

Molluscs are a well-known geoscientific tool. Within the large group of molluscan species, those indicating shallow-water environments are outlined taxonomically. Subfossil molluscan assemblages are described in the light of modern sea-floor communities. Examples of the use of marine molluscs for recognizing earlier sea-level stands are given. The environmental implications of *post-mortem* phenomena are briefly discussed, and recommendations for sample selection are given. In short, there are few definitive reliable sea-level indicators in the mollusc group and of those most are exposed to *post-mortem* transport. However, on the basis of quantitative studies it is possible to differentiate between littoral, shallow- and deep-water environments.

INTRODUCTION

'The molluscs have the demonic character in which Nature has been pleased to write all her most curious documents' (Rudwick, 1972), and from mankind's earliest thoughts about the history of the earth, they have certainly been of great importance. This is primarily due to their good preservation and large numbers. To some extent it is related to their aesthetically pleasing shape. Their history covers a lengthy time span; their main origins are traced to the beginning of the fossil record in the Paleozoic Era (Moore, *et al.*, 1952). They have had the time to spread out and fill in nearly all biotopes, the shell often reflecting their way of living (Stanley, 1970). However, this contribution is limited to subrecent deposits and is based on knowledge of modern communities. This enables us to survey the enormous field of 6 classes comprising about 70 000 species, most of which live on the shelf and possess hard parts which make them useful for evaluating former sea levels (SLs). The huge number of species renders it necessary to limit the discussion to those which are useful indicators of SL. However, this study does not disregard all other living species, because they make it possible to gain information on the paleoenvironment indicated by the association of organisms. Applying this concept to bottom communities, less definitive indicators can also be used. The aims of this paper are therefore firstly, to outline the basic concept of bottom communities as applied to determination of former SL stands and secondly, to identify the most reliable SL indicators. However, given the large number of mulluscan species, a survey on high taxonomic rank is added in condensed form as an appendix.

WORKING PROCEDURES

A major problem with even the best mollusc SL indicators is *post-mortem* reworking and transportation, which raises the question of preservation potentials. When looking for criteria of whether shell material has been transported, one can consider whether the shells represent a community, and one can determine whether single species are in life position. In the latter case the infauna species of bivalves certainly deserve most attention.

As this chapter is not centered on taphonomy, the reader cannot expect details about environmental implications of the *post-mortem* phenomena, viz. fragmentation, accumulation and sorting, all factors linked to environment and presenting part of the question raised when dealing with bioturbation.

During the last decades much effort has been given to studies on trace fossils. The now classical book by Abel (1935) on 'Vorzeitliche Lebensspuren' is worth mentioning, also because it gives a fine example of using biological knowledge on fossil data, which is the main task in actuopaleontology as presented by Schäfer (1962). However, the use of traces as facies indicator was highly stimulated during the first Meeting on Trace Fossils held in Liverpool in 1970 (Crimes & Harper, 1970). The work by Seilacher (1967) emphasizes the use of trace fossils as bathymetric indicators. Furthermore, reference is made to the revised edition of the text book: *Depositional Sedimentary Environments* by Reineck & Singh (1980), especially the chapter on biological parameters.

Differences in working procedures in obtaining material from sections, cores and surface deposits need comments. It is clear that sections offer the best means of evaluating sedimentary structures. Moreover, the sampling of molluscs is facilitated by the fact that the sedimentary structures are the key to individual beds and their macrofossil content. However, even if the sample is intended mainly for radiocarbon age determination, the nature of the material should be considered as well. For instance, if the shell material is voluminous, a large sample of individual specimens should be supplemented by a bulk sample. The bulk sample allows the characterization of the community later on in the laboratory, when it is treated quantitatively for all the species found.

The size of each sample may vary both according to the nature of the sediment and the sampling situation, but it is the author's experience that 1-2 kg bulk samples suffices for further investigations. However, the number of specimens in 1000 g samples of sediment may vary a great deal (Petersen, K. Strand, 1982, Fig. 2). If it is impossible to transport heavy samples, washing at the site on a mesh of 0.5 mm can be done. It is always wise, however, to take a small portion of sediment, enough for granulometric and microfossil analysis - this requires about 100-200 g.

For core samples, in which the study of structures is impossible or reduced, granulometric data are even more necessary. It is also important to retain all the shell material >0.5 mm in each core level. If the sediment is silty the material must be caught by some sort of net as shown on Figure 1.

A third sampling situation is often encountered in the arctic where molluscs have to be collected from the surface at different levels. In this case representative samples are out of the question as the surface sediment has been stirred by cryoturbation and permafrost prohibits deeper sampling. Here, attention must be given to the minor fragments of the fragile, thin specimens as well as to robust specimens which are preferred for ^{14}C dating. The associated faunal elements have high indicative value when the water depth of the assemblage found is being considered.

Other sampling situations might be discussed here, but 'what is most required is a study of and comparison between the fossil faunas and the molluscan faunas now living'. This statement by C.G.J. Petersen (1910) has been the basis of my own approach in the study of molluscs, which is based on the community concept.

THE COMMUNITY CONCEPT

The general notion of bottom communities was developed from studies in Danish waters by C.G.J. Petersen and published in a series of papers from the beginning of this century (1911, 1913, 1914, 1915), *cf.* Thorson (1957, p 467): 'In this classical basis for marine level-bottom ecology, the most numerous, most conspicuous, and for the area most characteristic species of animals were used to designate the community'. In the paper from 1914, Petersen called attention to the possible zoogeographical significance of his investigations. Subsequently, his studies were followed up by investigations in other latitudes all over the world, as compiled by Thorson (1957). The worldwide studies based on the concept of parallel bottom communities within the shallow shelf are of great importance, as they allow marine

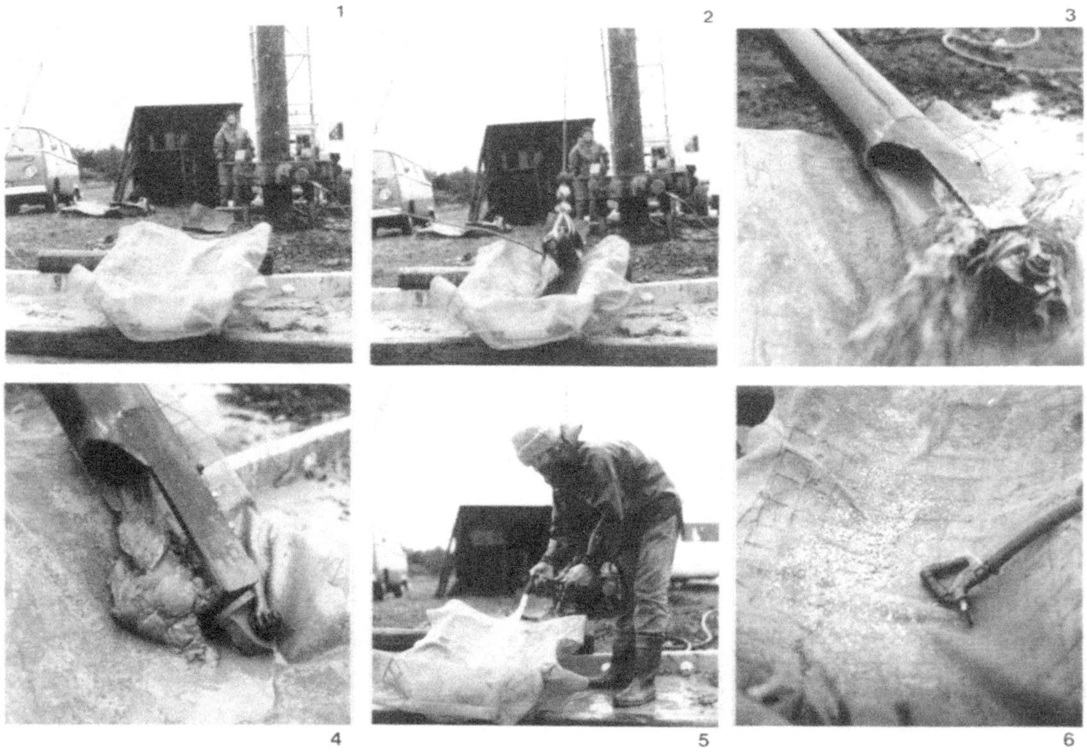

Figure 1. Working procedure (1-6) at the Vust well-site, northern Jutland, Denmark. Appr. 0.04 m^3 material was sieved at the site from each 0.5 m core interval, when silty beds were penetrated.

biologists in different areas to cooperate (Thorson, 1961, p 74) - and to demonstrate some general rules regarding the behaviour of shallow-water marine molluscs found in subfossil deposits.

Another very important result of the quantitative method introduced by Petersen, is the recognition of the difference between the epifauna and the infauna. On shallow bottoms where plants are present, most of the animals are more closely associated with the plants than with the bottom itself (Elton, 1927). It is clear, therefore, that an analysis of a subfossil assemblage must be worked out in such details as to allow an evaluation of the ecology. Furthermore, within modern zoological work it is stressed that the characterization of a level-bottom community must be based upon more than one species, ie. on a combination of species of different groups of animals (Thorson, 1957, p 476). But for the geologist, working with subfossil material, different groups of animals do not have the same preservation potential. As an example one could mention the *Astarte borealis-Pectinaria* shallow-water zone within the *Macoma calcaria* community. The *Pectinaria* are rarely preserved while the bivalves remain. However, there seems to be reason to believe that especially the shallow-water infauna community is adequately characterized by its molluscan components, while the communities of intermediate deep waters have a higher degree of important elements from other phyla than the molluscs, for example *Amphiura-*, *Amphiodia-*, *Amphioplus-* and the *Macoma sarsi-Ophiura sarsi* communities at depths of 20-300 m (Thorson, 1957).

The infauna on level sea bottoms from shallow waters and intertidal zones are depicted in Table 1. Parallel communities at other latitudes are given by Thorson (1957, pp 504-521) where Figure 10 illustrates the parallelism between Arctic, Boreal and Northeast Pacific *Macoma* communities.

By studying the interrelationship between sediment and infauna species in samples of autochthonous material one may determine whether or not the burrowing bivalve species are in life position (Stanley, 1970). Furthermore, the bivalve burrowing patterns can be used as bathymetric indicators (*cf.* McAlester & Rhoads, 1967); intertidal and shallow subtidal environments, for instance, are demonstrated by species adapted for deep vertical burrowing.

As illustrated by Thorson (1952, Fig. 14), the number of prosobranchs on level bottoms are rather small and constant at the different latitudes compared to the epifaunal elements.

Epifaunal communities from rocks and stones are well known all around the world, but in addition to these we have the epifauna attached to the vegetation. Thus within the artic there is the *Margarita groenlandica* community corresponding to the *Macoma calcaria* community, and in temperate seas the *Rissoa, Lacuna* community which

Table 1. Characteristic infauna of level-sea bottom communities

Community		Characteristic infauna	Depth in metres
ARCTIC	shallow water	*Macoma calcaria* *Astarte* spp.	(0)4 to 50-60
	sand bottom	*Venus fluctuosa*	3-7 to 20-30
	soft bottom	*Astarte crenata* *Arca glacialis* *Portlandia* spp. *Dacrydium vitreum*	30-40 to 550
BOREAL	shallow water (E Atlantic)	*Macoma baltica* *Mya arenaria* *Cerastoderma edule*	tidal to 8-10
	shallow water (E Pacific)	*Macoma nasuta* *Cardium corbis*	tidal to 8-10
	shallow water (N Atlantic)	*Tellina tenuis*	tidal to 10
	sand bottom	*Donax vittatus*	tidal to 10
	soft bottom	*Abra alba* *Corbula gibba* *Cultellus pellucidus*	8-10 to 20
BOREAL-MEDITERRANEAN	sand bottom	*Venus gallina* *Spisula* spp.	8-10 to 40-50
LUSITANIAN-MEDITERRANEAN	shallow water	*Tapes aureus* *Loripes lacteus* *Ensis siliqua* *Scrobicularia plana*	?
	sand bottom	*Venus verrucosa*	?

corresponds to the *Macoma baltica* community have about the same depth range. As pointed out earlier, the subfossil assemblage will contain both of these faunas; hence, zoological principles must be employed to establish the original composition of the environment. This is realized, for instance, when considering the attack by predatory gastropods in an interglacial marine molluscan fauna from East Greenland (Petersen, 1981). The drill-holes observed on the epifauna elements are made by muricacean gastropods, whereas drill-holes on the infauna species are attributable to naticean gastropods - all within the *Astarte borealis-Pectinaria* zone belonging to the upper part of the arctic *Macoma calcaria* community in shallow water.

As stated by Thorson (1961), the epifauna comprises roughly 80% of all species of benthic invertebrates. In the polar region the number of epifauna elements equals the number of infauna species, but in the tropics the number of epifauna species is much greater. The coral reefs are prolific, especially with regard to gastropods. According to Stephenson & Stephenson (1972), tropical shores differ from other shores in two respects, viz.: mangroves and coral reefs, which intermingle to produce the greatest complexity.

The epifauna on rocks and stones in the intertidal zone of temperate seas also has a regular representation of certain genera of molluscs, treated by Stephenson & Stephenson (1972), ie.: *Acmaea, Callistoma, Cerithium, Crepidula, Fissurella, Gibbula, Haliotis, Helicon, Littorina, Margarita, Nerita, Patella, Planaxis, Strombus, Thais, Urosalpinx, Arca, Brachidontes, Chama, Crassostrea, Hiatella, Mytilus, Ostrea* and *Tridacna*.

Highly dominant within this fauna in number of species are various gastropods feeding on algae, namely *Patella, Acmaea, Littorina* and *Neritina* or predators such as *Purpura*, although the predominant genus varies regionally.

Among the few filter-feeding gastropods, the Vermetids from the tropical shores deserve mention. While the Vermetidae, Spondylidae, Ostreidae and Chamidae have a high preservation potential due to their cemented attachment to the substrate, most of the other species characterizing the intertidal rocky shore are easily transported after death. Bioerosion may further reduce them if they are not subsequently buried. The *post-mortem* transportation involves the 'belt of movement' treated by Johansen (1901, 1902) and Fischer (1943).

In the light of recent possibilities of radiocarbon age determination of individual species, Cullen (1970) has demonstrated that a mixture of molluscan population may date from the earliest Holocene to the present day in shelf deposits near New Zealand. In this case the anomalous depths of certain Mollusca are attributed by Cullen to a subsequent rapid rise of SL and little sedimentation during the early Holocene.

To sum up, the concept of epifauna and infauna communities of recent marine environments from the shallow and intertidal regions of the sea provides us with some very good information on assemblages of molluscs with a high indicative value on depth zonations. The preservation potential of such faunas, however, is often very low due to their susceptibility to redeposition. The evaluation of SL data thus requires a thorough study of the subfossil assemblage based on bulk samples to determine the interrelationship between the species found and the habitus of the individual specimens.

As stated by Sorgenfrei (1958) in his work on molluscan assemblages from the marine Middle Miocene, students of fossils have since the earliest paleontological research endeavoured to apply the principles of modern environmental relationships. Faced with the entire world of shallow-water marine Mollusca, it would be helpful to use the community concept as pointed out in the previous chapter embracing the huge quantities of species from within our topic in that way. However, it should be mentioned that the community concept is not applicable universally. For instance, the Bahamian communities, pointed out by Newell *et al.* (1959), comprise eleven major habitats, some of which are characterized by their molluscan components, viz.: *Littorina* community on rocky shores; *Strombus samba* community on sand bottoms of the outer platform, generally deeper than 3.5 m; *Tivela* community on unstable oolite sand bottoms; *Strombus costatus* community in shelf lagoons and marginal lagoons, on stable sand bottoms, generally at depths less than 2.5 m. Furthermore, many molluscs are also found on mangrove roots, for instance *Saccostrea cuccullata* (Born) at mean tide level. In such cases the specific area must be treated and the molluscan fauna evaluated according to the local environment.

Regarding the distribution of benthic marine Mollusca on the Atlantic shelf from Gibraltar to Murmansk (Thorson, 1963), this is said to be the most thoroughly investigated of our globe (references in Thorson, 1963). No less than 1245 species of benthic Mollusca are found in these waters to depths of 300 m. It is seen (Thorson, 1963, Table 1) that gastropods are the dominating element (65.2%) and, together with lamellibranchs, constitute more than 90% of the total number of molluscan species.

Of the common species occurring along the European west coast from all areas mentioned by Thorson (1963, pp 7-8, nos. I-XV), the following can be mentioned, along with their depth affiliation:

Loricate:	*Tonicella subra* (L.)	0-270 m
Scaphopoda:	*Dentalium entale* L.	20-400 m (incidently also on tidal flats)
Prosobranchia:	*Patina pellucida* (L.)	tidal
	Patelloida virginea (Müller)	tidal-55 m
	Gibbula tumida (Montagu)	tidal-174 m
	Lacuna vincta (Montagu)	tidal-38 m
	Lacuna pallidula (da Costa)	tidal-73 m
	Littorina littorea (L.)	tidal-54 m
	Littorina saxatilis (Olivi)	tidal
	Littorina littoralis (L.)	tidal-73 m
	Cingula semicostata (Montagu)	tidal-46 m
	Rissoa parva (da Costa)	tidal-21 m

	Skeneopsis planorbis (Fabr.)	tidal-29 m
	Eulima trifasciata (Adams)	tidal-146 m
Prosobranchia: (continued)	*Velutina velutina* (Müller)	tidal-81 m
	Nucella lapillus (L.)	tidal-5 m
	Buccinum undatum L.	tidal-2200 m
	Nassarius incrassatus (Strom)	tidal-92 m
Ophistobranchia:	*Chrysallida spiralis* (Montagu)	tidal-?
	Odostomia unidentata (Montagu)	tidal-28 m
Lamellibranchia:	*Monia patelliformia* (L.)	shallow-considerable depth
	Heteranomia squamula (L.)	shallow-?
	Mytilus edulis L.	tidal-5 m
	Modiolus phaseolinus (Philippi)	tidal-considerable depth
	Lima sulcata Brown	18 m-considerable depth
	Lucinoma borealis (L.)	tidal-100 m
	Trutonia minuta (Fabr.)	tidal
	Parvicardium ovale (Sowerby)	4-110 m
	Venus ovata Pennant	4-183 m
	Venus striatula (da Costa)	tidal-55 m
	Macoma baltica (L.)	tidal-5 m
	Spisula subtruncata (da Costa)	tidal-37 m
	Mya truncata L.	tidal-73 m
	Hiatella arctica	shallow water-considerable depth
	Hiatella striata (Fleuriau)	in low region of shore as a borer in soft rock

Only a few of these common species are unique to the tidal zone (high indicative value), while most others may also be found in deeper water. For this latter category the indicative range can be somewhat reduced by considering where the particular species is most abundant. Such calculations should be based on quantitative investigations since they must be considered in the light of the other species found. Different species occur both in the warm temperate zone and in the arctic zone in the north, not at all stations between but locally abundant. For instance, *Hydrobia ventrosa* (Montagu), *Hydrobia ulvae* (Pennant) and others usually occur above midtide level, but may extend to depths of 100 m. Furthermore, a species such as *Cyprina islandica* (L.) occurs in the northern part as a littoral form due to the warmer surface water, while to the south its range is from shallow water to considerable depths (Jensen, 1902). Some other faunal elements may change their indicative value/depth range according to the well-known tropical 'submergence', viz.: species which inhabit quite shallow water in arctic regions live at increasingly greater depths the further south they occur (Thorson, 1951, p 98).

Considering only the region between Gibraltar and up to West Finmark, where the northward transport of pelagic larvae by the Gulf Stream occurs, 62 additional species are found both to the south and in the north. The following common species are recorded from all over this region (Thorson, 1963, p 14):

Prosobranchia:	*Emarginula reticulata* Sowerby	tidal-165 m
	Patella vulgata L.	tidal
	Rissoa inconspicua Alder	tidal-42 m
	Balcis lubrica (Monterosato)	37-134 m
	Capulus ungaricus L.	13-156 m
	Aporrhais pespelicani (L.)	7-128 m
	Natica alderi Forbes	tidal-165 m
	Natica montagui Forbes	15-183 m
	Lamellaria latens (Müller)	tidal-165 m
	Philbertia linearis (Montagu)	28-146 m
Ophistobranchia:	*Menestho divisa* (Adams)	tidal-165 m
	Odostomia turrita Hanley	tidal-73 m
Lamellibranchia:	*Musculus marmoratus* (Forbes)	tidal-?
	Chlamys opercularis (L.)	subtidal-183 m
	Similipecten similis (Laskey)	4-55 m
	Kellia suborbicularis (Montagu)	tidal-120 m
	Montacuta substriata (Montagu)	subtidal-85 m
	Mysella bidentata (Montagu)	tidal-120 m
	Acanthocardia echinata L.	4 m-considerable depth
	Parvicardium scabrum (Philippi)	9 m-considerable depth
	Cerastoderma edule (L.)	tidal
	Dosinia lupinus (L.)	tidal-128 m
	Venerupis pullastra (Montagu)	tidal-37 m
	Tellina tenuis da Costa	tidal-4 m
	Tellina fabula Gmelin	tidal-55 m
	Tellina pygmaea Lovén	tidal-146 m
	Gari fervensis (Gmelin)	tidal-considerable depth
	Cultellus pellucidus (Pennant)	4-110 m
	Ensis ensis (L.)	tidal-4 m
	Mya arenaria L.	tidal-73 m
	Thracia distorta Montagu	tidal-medium depth

Among the common species mentioned above, some live in shallow water but have no representation in the tidal zone, so having a low indicative value. However, in order to define a certain fauna according to depth, knowledge about deeper living species is an additional prerequisite in quantitative studies.

EXAMPLES OF QUANTITATIVE INVESTIGATION

A quantitative method of investigation is well illustrated by Norton (1967) in his work on marine molluscan assemblages in the Early Pleistocene of Norfolk. Furthermore, the relation of fauna to sediment is considered in an attempt to interpret the ecological and depositional environment in which the defined series of molluscan assemblage zones accumulated. This makes possible a configuration as in Figure 2 in Norton (1967, p 188) where the grouping of the subfossil species follows their relation to depth (littoral, just below or normally sublittoral) and habitat (shores with rocks, sandy shores, mud, coarse muddy bottoms, clean coarse bottoms). Also, the epifauna elements are considered as well as commensals, predators, etc.

When working with the molluscan assemblages in the Tjörnes sequence in Iceland, Norton (1975) stated that the arrangement of communities found today was not necessarily the same in the past. He proposed that a more unsorted community had existed originally, which later diverged into climatically specialized groups. However, the paleoecological groupings (Norton, 1975, Table 1) are an approach to differentiate littoral and sublittoral environments by analysing the epifauna and infauna elements.

Parker, in his work on macro-invertebrate assemblages in the East Mississippi Delta region, on the other hand, tries to answer the question of the application of recent macro-invertebrate distributions to ancient marine environments by tracing the geological range of species and genera from this delta region (Parker, 1956; Table VII). Parker concludes that 'since many of the macro-invertebrates serving to define the environments [in his study] have also been found as fossils as far back as the Miocene it should be possible to reconstruct some ancient environments on certain characteristics'. To sum up some of these environments close to the present topic regarding their molluscan assemblages (number from Parker):

1) River deltas and mainland shores ... *Neritina reclivata* and *Littorina irrorata*.

2) The shorelines of deltas and river mouths ... *Rangia cuneata, R. flexnosa, Macoma mitchelli, M. tageliformis* and *Littoridina sphinctostoma*.

8) The gulf side of barrier islands ... *Donax, Terebra cinerea, Olivella*.

9) The shallow continental shelf out to about 20 m should be characterized by a large number of molluscs ... *Dinocardium, Atrina, Aequipecten, Callocardia, Cyrtopleura, Labiosa, Macoma, Spisula, Dosinia, Phalium, Olivia, Tonna, Murex* and *Strombus*.

However, Parker points out that 'the shells of many of the shallow-shelf species of molluscs often occur in large drifts on the beaches after storms, and their presence should characterize Gulf beach deposits in older formations'. This is the 'belt of movement' (Johansen, 1901) quoted earlier and taking place within a vertical zone of 15 m.

While the faunas in most of the above-mentioned studies are far too old to be dated by radiocarbon, the molluscan fauna from the Disko Bugt area in West Greenland, treated by Donner & Jungner (1974), is from marine Holocene deposits and is thoroughly dated by ^{14}C. It is shown that the distribution of shells in the sediments is largely dependent on habitats and water depth. Thus, it is demonstrated that clays with *Portlandia arctica,* which have been described as Late-glacial, are in fact early Holocene in age and were deposited in deep water at about the same time as marine sands and gravels with a warmer fauna were deposited in more shallow water.

On the basis of quantitative analyses of molluscan assemblages from raised marine deposits in East Greenland, the author (Petersen, K. Strand, 1978) has demonstrated how the subfossil communities parallel the recent arctic bottom communities within their known depth range.

Fifty samples of 1-2 kg were taken from 4 sections (Suselv S, Suselv N, Loch Fyne Hytten 1 km N and Loch Fyne Hytten 100 m N; Fig. 2), ranging in elevation between 1 and 14 m. Samples at the top and bottom have been ^{14}C dated to the interval 9000-6000 BP and are related to the emergence curve of Funder & Hjort (pers.comm.). The biogene components of the samples are conditional on a water depth of about 45 m for the older part of the sections. It is possible to distinguish the shallow - water *Astarte borealis-Pectinaria* zone and the deeper lying *Ophiocten* zone within the arctic *Macoma* bottom community (Fig. 3). Besides the Ophiuroidea, the *Ophiocten* zone is characterized by *Nucula tenuis* (Montagu), *Leda pernula* (Müller), *Modiolaria nigra* (Gray) and *Pecten groenlandicus* Sowerby. Furthermore, *Macoma calcaria* (Chemnitz), *Hiatella arctica* (L.) and *Mya truncata* L. are found, species with a wider depth range, but quantitatively not dominant as in the *Astarte borealis* zone. From the *Astarte borealis* zone, *Mytilus edulis* is a characteristic species of the subtidal zone, while the others only give information within a range of 10 m. The grain-size distribution of the samples seen on Figure 2 shows a general coarsening upwards in the three profiles covering the *Astarte borealis* zone, thus indicating the shallowing of the sea. This emphasizes that marine molluscs in the arctic, if not sampled in bulk, are only useful when their relationship to a known SL position can be established (Blake, 1975).

On the basis of a quantitative study of a Holocene marine fauna in the North Sea region, comprising 111 species of molluscs (Petersen, K. Strand, 1981), there seems to be reason to conclude that rather few species have a high indicative value. The area of investigation is situated between the North Sea and Limfjord which functioned as a strait during part of the Holocene. The highest marine limit has been demonstrated already in the beginning of this century by Mertz (1924). Later studies have described the molluscan fauna and the SL changes based on ^{14}C datings (Petersen, K. Strand, 1976). Recently, the fauna from a well by Vust within the same area has been dated to give information on the sedimentation rate in the strait connecting the North Sea with Limfjord.

Considering the fauna at different water depth through time, it is possible to follow the shifting faunas as a function of water depth and sedimentation rate (Petersen, K. Strand, 1981). Five samples have been chosen from the well at different core levels, dated by ^{14}C and shown on Table 2 with all the molluscan species. The present depth range is given for each species to compare with the known paleo-water depth when the individuals were living. In the lowermost sample (No. 2869, Table 2) a mixture of species with different depth ranges is found. Carbon-14 dating of different species from within this sample indicates that deposition and burial have taken place during a 500 years interval and that the ages of certain species reflect the deepening of the water, viz.: *Mytilus edulis* older than *Ostrea edulis* which again is older than *Arctica (Cyprina) islandica*. These results are interpreted like the situation described by Cullen from the Pacific Ocean, referred to earlier in this text. During a period with rapid SL rise and low sedimentation rate, a deposit was formed containing a highly mixed fauna from various depths and ages.

In the other examples on Table 2, the molluscan species from samples derived from paleo-depths of *c*.25 m, *c*.18 m, and *c*.6 m and from the former tidal zone are given in number of individuals. The dispersal of the specimens found in relation to the depth zonation given from the paleo-environment is, with few exceptions, in accordance with the present depth range. This range, however, is very broad for most of the species found in the sediment from the sandy shore. Few are limited to the tidal zone or just below, eg. among the bivalves *Mytilus edulis, Telina tenuis, Ensis ensis* and the *Tapes* species - although the *Tapes* is reported from depths exceeding 50 m - and among the gastropods the *Rissoa* species, *Bittium reticulatum* and the *Littorina* species - in this place dominated by the *Littorina littorea*. The indication of recent depth range in brackets for the *Littorina* species refer to few dredged specimens. However, the littoral position, hence the high indicative value, can be considered the characteristic feature for these species.

CONCLUSION

The general conclusion of this chapter is that most molluscan species can not be assigned a high indicative value. Considering all the fauna elements on the basis of quantitative investigations, it is possible to differentiate only between littoral, shallow- and deep- water environments. Furthermore, the few species that do occur within a (relatively) narrow range are often subject to *post-mortem* transport

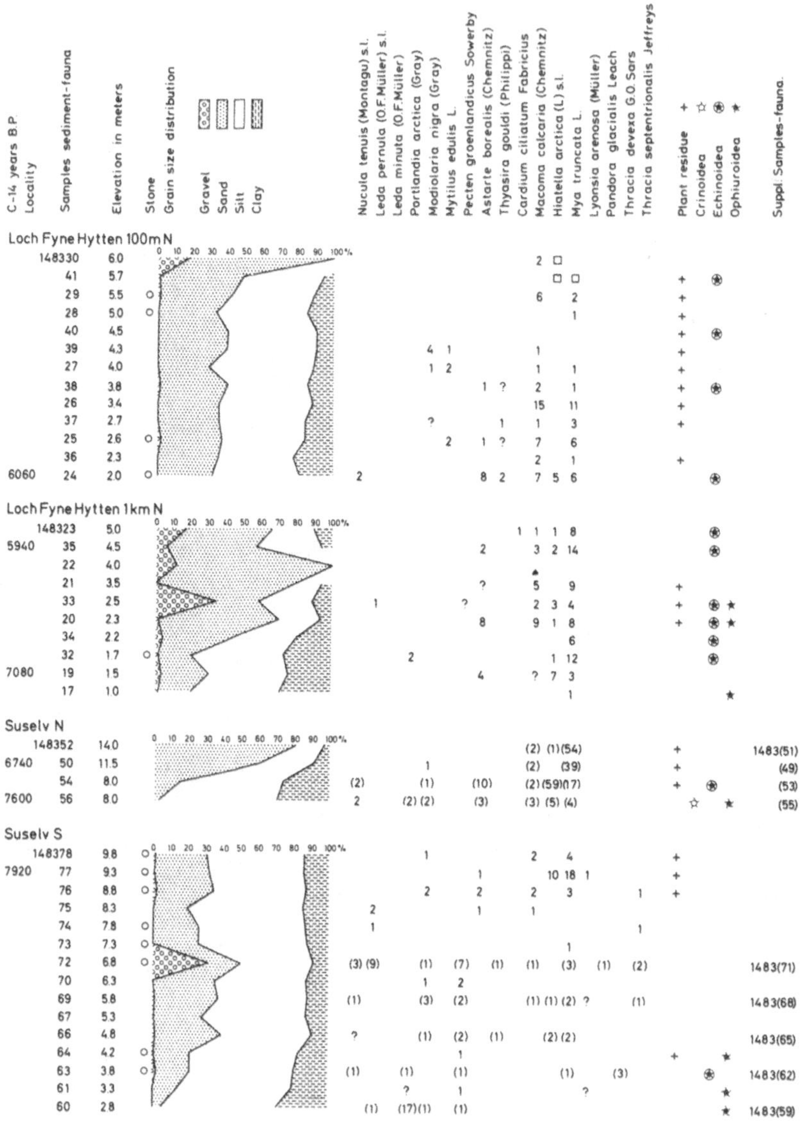

Figure 2. Summary of sedimentological and faunistic data from the sections in raised marine deposits at Loch Fyne, East Greenland. The numbers of individuals in brackets refer to supplementary samples listed at the right in the diagram.

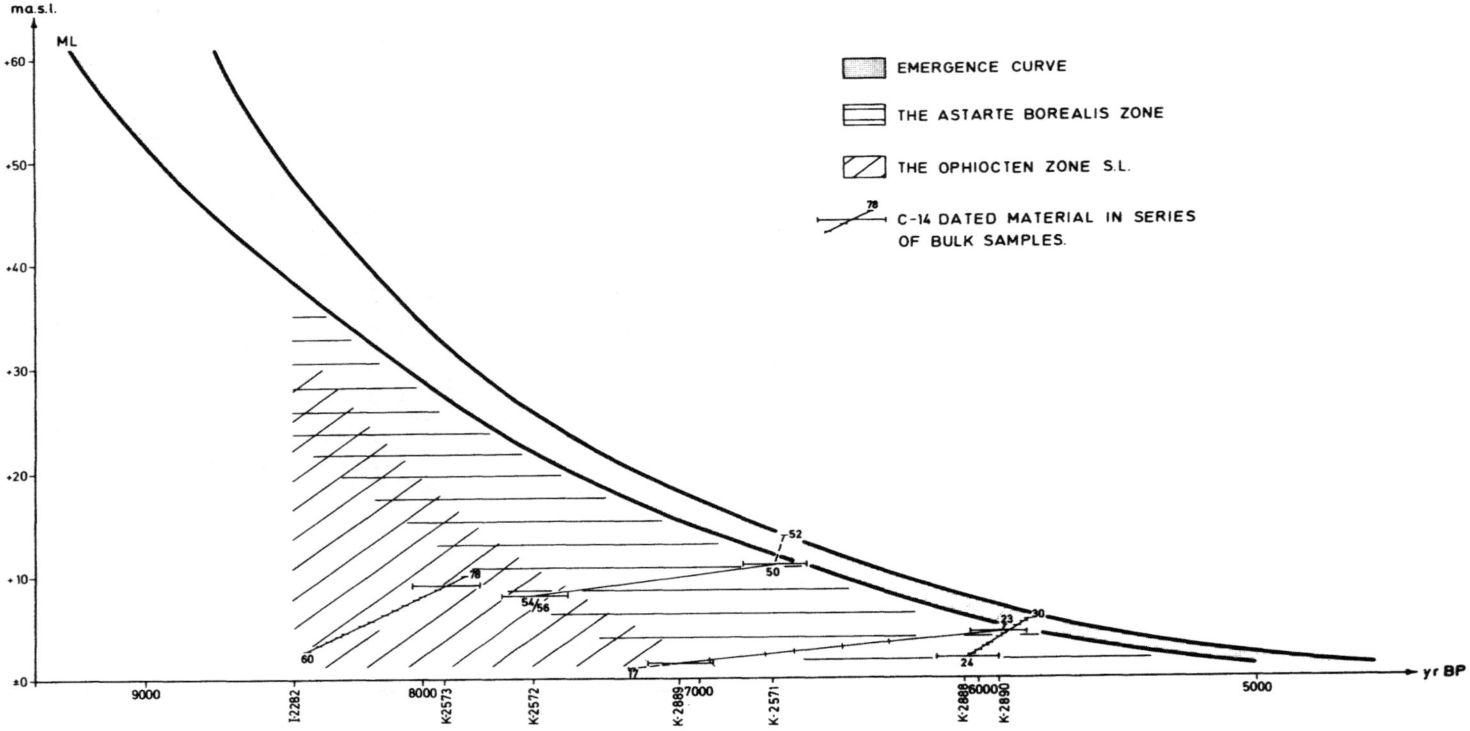

Figure 3. The relation between the emergence curve (Funder & Hjort) and the demonstrated faunistic zones through time in the arctic *Macoma* community, East Greenland.

Table 2. The paleo-depths at different core levels of molluscan species from the Holocene in the Skagerrak-Limfjord region, compared to their modern depth ranges. Zero m means below the tidal zone.

Elev. m	-25	-21	-14	-2	+1		
Sample	2869	2861	2848	2823	2816		
Paleo-depth	tidal-*c.* 15 m	*c.* 25 m	*c.* 18 m	*c.* 6 m	tidal	Molluscan species Bivalvia	Modern depth range (m)
	3		2	1		*Nucula nitida* Sowerby	6-45
	90		1	6	3	*Ostrea edulis* Linné	0-82
				1		*Anomia squamula* Linné	0-150+
	2					*Pecten varius* Linné	0-82
	26	1		6	28	*Mytilus edulis* Linné	0-5
	1					*Modiola modiolus* (Linné)	5-150
					1	*Modiolaria discors* (Linné)	0-?
		8	4	1		*Thyasira flexuosa* (Montagu)	11-180
				1		*Lucina borealis* (Linné)	0-100
	1		3		2	*Montacuta ferruginosa* (Montagu)	?
	9	12		1	2	*Montacuta bidentata* (Montagu)	0-120
	1		2	1		*Cardium echinatum* Linné	4-150
	1			4	23	*Cardium exiguum* Gmelin	0-55
	13					*Cardium edule* Linné	tidal
	10					*Cyprina islandica* (Linné)	10-50
	?					*Tapes decussatus* (Linné)	0-?
	?				21	*Tapes aureus* (Gmelin)	0-86
	?			1	1	*Tapes pullastra* (Montagu)	0-36
	3			1	2	*Venus gallina* Linné	0-55
					1	*Venus ovata* Pennant	4-180
		1	3			*Venus fasciata* (Da Costa)	4-110
	3					*Macoma baltica* (Linné)	tidal
	2				14	*Tellina tenuis* Da Costa	0-5
	3			14	4	*Tellina fabula* Gronovius	0-55
	2		7			*Abra prismatica* (Montagu)	0-55
	9	40	4	2		*Abra alba* (Wood)	0-65
	1	1	1			*Cultellus pellucidus* (Pennant)	4-110
				1		*Ensis ensis* (Linné)	0-5
	19	70	17	19	14	*Spisula subtruncata* (Da Costa)	0-36
	8	5		1	3	*Hiatella arctica* (Linné)	0-1400
	6			1	1	*Mya truncata* (Linné)	0-73
	31	175	22		1	*Corbula gibba* (Olivi)	4-50
	3				1	*Zirphaea crispata* (Linné)	0-7
	3					*Barnea candida* (Linné)	0-5
	1	1	1	3	1	*Thracia papyracea* (Poli)	0-55

Table 2 continued

Elev. m	-25	-21	-14	-2	+1		
Sample	2869	2861	2848	2823	2816		
Paleo-depth	tidal- c. ÷15 m	c. 25 m	c. 18 m	c. 6 m	tidal		
						Molluscan species Gastropoda	Modern depth range
				3		*Acmaea testudinalis* (Müller)	0-36 m
	1	2		5	6	*Gibbula cineraria* (Linné)	0-126 m
	5	2		14	15	*Lacuna vincta* (Montagu)	0-38 m
					3	*Lacuna parva* Da Costa	0-?
				47	21	*Lacuna pallidula* (Da Costa)	0-54 m
	1				1	*Littorina saxatilis* (Olivi)	tidal
	13			1	17	*Littorina littorea* (Linné)	0-(63) m
					1	*Littorina obtusata* (Linné)	0-(72) m
	15				17	*Hydrobia ulvae* (Pennant)	0-(126) m
			1			*Cingula vitrea* (Montagu)	10-117 m
				1	2	*Cingula semicostata* (Montagu)	0-45 m
					91	*Rissoa albella* Lovén	0-15 m
				20	5	*Rissoa albella* var. sarci Lovén	tidal
				75		*Rissoa parva* (Da Costa)	0-22 m
	1			2	14	*Rissoa membranacea* (Adams)	tidal-?
				2	1	*Rissoa violacea* Desmarest	0-36 m
				5	5	*Skeneopsis planorbis* (Fabr.)	0-19 m
					1	*Caecum glabrum* (Montagu)	5-63 m
				38	244	*Bittium reticulatum* (Da Costa)	0-18 m
	9	15	11		1	*Natica nitida* Donovan	0-162 m
	5					*Nucella lapillus* (Linné)	0-5 m
	1					*Nucella lapillus* var. imbricata Linné	0-? m
	11					*Buccinum undatum* Linné	0-2000 m
	4	2	6			*Nassa reticulata* (Linné)	0-36 m
	2	2		2		*Nassa pygmaea* (Lamarck)	4-108 m
	4		3			*Lora turricula* (Montagu)	18-270 m
	1	1		1		*Chrysallida obtusa* (Brown)	0-80 m
				2		*Chrysallida spiralis* (Montagu)	0-5 m
	2					*Odostomia plicata* (Montagu)	0-13 m
				1		*Eulimella nitidessima* (Montagu)	0-80 m
	1	1				*Eulimella* sp.	
			1			*Turbonilla* sp.	
			1	1		*Actaeon tornatilis* (Linné 1758)	10-100 m
	4	6				*Cylichna alba* (Brown 1827)	10-300 m
				1		*Akera bullata* Müller 1776	5-10 m
	1				9	*Retusa obtusa* (Montagu 1803)	?
					13	*Retusa truncatula* (Bruguière 1792)	5-80 m

and are, therefore, only useful when it can be shown that they are autochthonous.

It is most important to collect and treat subfossil material such that information about the paleoenvironment can be obtained. The concept of parallel molluscan communities should be the tool which allows us to interpret any coastal situation and the huge quantity of species around the world.

APPENDIX: AN OUTLINE ON HIGH TAXONOMIC RANK OF SHALLOW-WATER AND INTERTIDAL MARINE MOLLUSCA

The phylum Mollusca comprises six classes, viz.: Monoplacophora, Amphineura, Gastropoda, Scaphopoda, Bivalvia and Cephalopoda - in all some 70 000 species. The Mollusca are very diverse in morphology as a consequence of differences in life style. The marine Mollusca are most varied in the tidal zone and to a depth of 100 m, due to the wide range of external conditions found there. The majority of the marine Mollusca are found as benthos, both epifauna and infauna. The six classes mentioned above are not of equal importance.

The Monoplacophora Wenz in Knight, 1952, well known from the Paleozoic deposits of epicontinental seas, are today only represented by the recently found *Neopilina* Lemche, 1957, (3 species) from the deep sea, having a limpet-like shell.

The Scaphopoda Bonn, 1862, with the families Dentaliidae Gray, 1840, and Siphonodentaliidae Simroth, 1895, have a cosmopolitan genera, eg. *Dentalium* L., 1758, and *Cadulus* Philippi, 1844. They are, with few exceptions, as for instance *Dentalium vulgare* Da Costa at Saint Malo, practically never found in the tidal zone, but known from the shallow waters to the deep sea. *Siphonodentalium galathea* Knudsen, 1964, was dredged from a depth of 7000-6900 m. About 350 species are known.

The Cephalopoda Cuvier (about 1000 species) are all predatory, swimming carnivores occupying the major marine habitats. Some attack their prey by drilling holes in the shell (Bromley, 1981), as known from the class Gastropoda. Only few, such as *Nautilus,* have true shells in which the animal occupies the last chamber of the coiled test. The shell of some is a calcareous plate. Both kinds of shells may be found on the beaches - though drifted ashore, often far from the place where the animal actually lived. As reported by Fischer (1943), *Nautilus pompilius* has been found drifted to Australia and a sudden multitudinous occurrence of *Sepia offinalis* L. shells at the Faroes Is. described by Spärck (1923).

The Amphineura von Ihering, 1876, are represented by the Polyplacophora de Blainville, 1816, which generally have 8 separate plates. They are found from temperate to tropic seas and they usually live on rocky surfaces close to shore. All orders within the Polyplacophora are represented in shallow water, ie. Lepidopleurina Thiele, 1909, Ischnochitonina Bergenhayn, 1930, and Acanthochitonina Bergenhayn, 1930. However, the genus Lepidopleurus is more common in depths exceeding 150 m. About 600 species are known, most of them from the Pacific Ocean. They belong to the epifauna.

The Gastropoda Cuvier, 1797, are by far the largest of the six classes and comprise 50 000 to 60 000 species. It is the most varied group within the Mollusca and is divided into three subclasses, viz.: Prosobranchia, Opisthobranchia and Pulmonata - all to be considered regarding the present topic.

Within the subclass Prosobranchia Milne Edwards, 1848, most of the species are marine, occurring from the tidal zone into the deep part of the sea. They are divided into three orders, viz.: Archaeogastropoda, Mesogastropoda and Neogastropoda.

Order Archaeogastropoda Thiele, 1925. The family Haliotidae Rafinesque, 1815, with only one genus *Haliotis* Linné, 1758, comprising 100 species, has many representatives in shallow waters of warmer seas on rocks close to the shore, eg. *Haliotis cracherodii* Leach from the west coast of Northern America.

Fissurellidae Fleming, 1822, with 500 species of 30 genera, are found mostly in the lower part of the tidal zone in warmer seas. *Emarginula, Scutus* and *Macrochisma* are all grazers on algae.

The Patelloidae Raf., 1815, comprising several hundred species, are typical on rocky coasts in all seas with the families Patellidae Raf., 1815, and Acmaeidae Phil., 1853, while the Lepetidae Dall, 1882, live in deeper waters in the northern seas and in the Antarctic.

The Trochacea Raf., 1815, number several thousand species which are widely distributed from the tidal zone and coral reefs to depths of 10 000 m. Some of the families as Trochidae Raf., 1815, are distributed in warmer seas within shallow water, while few members thrive in cold waters as *Margarita groenlandica* (Chemnitz) in the arctic shallow-water algal zones. *Margarita* has the same role there as *Littorina* and *Lacuna* have in the boreal algal assemblages (Thorson, 1944).

The Neritacea Raf., 1815, known from far back in geological history (Trias), have become accustomed to different environments such as the sea, fresh water and land. Of the several hundred species of Neritidae Raf., 1815, some typify the tidal zone, especially in tropical areas. Some genera, such as *Theodoxus,* are known from coasts with brackish water.

Among the order Mesogastropoda Thiele, 1925, the Littorinoidea Gray, 1840, are widespread inhabitants of the intertidal zone. The family Littorinidae Gray, 1840, comprises about 20 genera - *Littorina,* with nearly 100 species, is eulittoral, belonging to the grazing gastropods *Littorina obtusata* (L.), *L. littorea* (L.), *L. saxatilis* (Olivi) and *L. neritoides* (L.).

The family Lacunidae Gray, 1857, belongs to the shallow-water epifauna on algae, mainly in cooler waters of the northern hemisphere: *Lacuna parva* (da Costa), *L. pallidula* (da Costa) and *L. vincta* (Montagu).

Rissoidae Gray, 1847, are tiny gastropods that characterize the shallow- water environment as epifauna on the vegetation. Hydrobiidae Troschel, 1857, form communities on tidal flats. The occurrence of *Hydrobiae* in northern Europe is very useful for obtaining data on the relation to substratum and salinity. It is quantitatively the most important species on the flats in estuaries and shallow lagoons (Muus, 1967): *Hydrobia ulvae* (Pennant), *H. ventrosa* (Montagu) and *H. neglecta* Muus.

Within the Cerithioidea Fleming, 1822, the Vermetidae Raf., 1815, occur in great abundance in the tidal zone and on coral reefs cemented to the substratum, while Cerithiidae Fleming, 1822, are often found in the northern seas as epifauna on algae living on debris in shallow water. The genus *Bittium* is found in all seas, eg. *Bittium reticulatum* (da Costa) on both sides of the Atlantic Ocean. The Planaxidae Gray, 1850, resembling *Littorina,* are common in the warmer seas. *Planaxis sulcatus* occurs in the intertidal zone together with *Littorina* species. In the mangrove the Potamididae Adams, 1854, are found near high tide level.

Among the Calyptralidae Blainville, 1824, the genus *Crepidula* is found on coasts of warm and temperate seas from shallow water to moderate depths. For example, *C. fornicata* (L.) occurs gregariously in chains and solitary, on stony beaches at low water down to about 10 m (Fretter & Grahm, 1962).

The family Strombidae Raf., 1815, which lives on algae, is found in shallow water mainly within the Indo-west-Pacific region.

The family Cypraeidae Raf., 1815, having about 200 species, is restricted to the warmer seas, especially to the Indo-west-Pacific region, with the majority living on coral reefs, while some other species are found on rocky shores within the tidal zone and on sandy bottoms of shallow seas.

Within the order Neogastropoda Thiele, 1929, the family Muricidae Raf., 1815, contains many species of predatory gastropods, especially known on rocky shores and coral reefs in warmer seas. The Thaididae are common on rocky shores in the intertidal zone. The genus *Thais* Rødding, 1798, with different species, is known from all seas, feeding on barnacles and mussels: *T. neritoidea* (L.), *T. haemastoma* (L.) and *T. lapillus* (L.).

Among the Buccinoidea Raf., 1815, the Columbellidea Swainson, 1840 (400 species) are common on rocky shores, and Nassariidea Iredale, 1916, occurs in shallow water in all seas. The Melongenidea Gill, 1871, are mainly found in the tidal zone of tropical waters or in the mangrove swamps.

Many of the Marginellidea Fleming, 1818, and Olividea Latreille, 1825, comprising 600 and 60 species respectively, are living in the sublittoral and littoral zones on sand in the warmer seas.

The Mitridea Swainson, 1831, count 600 species, all living in the shallow water of warmer seas mainly in the Indo-Pacific region.

Among the Conoidea Raf., 1815, the family Turridae Adams, 1853, represents the largest family of marine gastropods with 600 genera, among which also species living in the intertidal zone.

The subclass Opisthobranchia contains mainly tiny animals, some without shell, living within a wide depth range. Some, however might be useful as depth indicators, eg. *Retusa obtusa* var. *pertenuis* (Mighels) which seems to be exclusively bound to the arctic *Macoma* community, behaving as a typical boreal-arctic-littoral species in the East Greenlandic waters (*cf.* Lemche, 1941). Among the Bullidae Lamarck, 1801, *Bulla occidentalis* Adams could be mentioned, being common on mud flats at low tide (Abbott, 1954).

Among the subclass Pulmonata Cuvier, 1817, a few families within the order Basommatophora Schmidt live in the sea, usually in the tidal zone: Siphonariidae Gray, 1840, on rocky shores in warmer seas, together with rock-encrusting oyster (Stenzel, 1971, Fig. 51). They resemble the limpets, eg. *Siphonaria pectinata* L. and *S. alternata* Say. The family Ellobiidae Adams, 1855, with several subfamilies, is known from warm seas, eg. Malampinae Stimpson, 1851, which includes the salt-marsh snail *Melampus olivaceus* Carp.

The Bivalvia, comprising about 10 000 species, are mainly found in areas close to the coast as infauna elements.

Some of the Arcidae Bronn, 1824, are represented in the deep sea, but most of the species occur in shallow waters at the lower region of the shore on sandy bottom, and on rocks and coral colonies all attached by byssus, eg. *Arca umbonata* Lamarck, while the Noetiinae Stewart, 1930, live at sandy places in shallow waters, eg. *Noetia ponderosa* Say.

The Mytiloidea Raf., 1815, with the family Mytilidae Raf., 1815, are known as eulittoral forms in most seas outside the high-arctic waters, eg. *Mytilus californianus* Conrad, while the genus *Lithophaga* drills the rocks in the warmer seas, eg. *Lithophaga nigra* Orbigny.

Families within the Eterioidea Broderip, 1839, like Pennidae Leach, 1819, Pteriidae Broderip, 1839, and Isognomonidae Woodring, 1925, are common in shallow tropical seas.

Ostreidae Raf., 1815, comprising, according to Ranson (1948, in Franc, 1960) 3 genera, *Pycnodonta, Crassostrea* and *Ostrea,* are, with the exception of *Pycnodonta,* widely distributed within shallow water, and often occur in the intertidal zone, except in the arctic. However, northern species of *Crassostrea* are known to survive freezing solid for several weeks in winter (Stenzel, 1971). The rock-encrusting oyster *Saccostrea cuccullata* (Born) and the mangrove oyster *Crassostrea virginica* (Gmelin) are, according to Stenzel (1971), characteristic of the intertidal zone.

Among the tiny Laptonidae Gray, 1847, some are found from the intertidal zone to depths of a few meters (eg. *Lepton squamosum*). Numerous species are living as commensals.

Among the Cardiidae Schweigger, 1820, many species are known from shallow water and the intertidal zone in most seas, some of them with high indicative value such as the common edible cockle from European waters. Furthermore, *Cerastoderma edule* L. may occur in large numbers just below the surface of the sand with densities exceeding 1000 specimens/m^2. However, other species, like *Acanthocardia echinata* (L.), are characteristic of deeper waters.

Among the Tridacnidae Bronn, 1824, the genus *Tridacna* occurs together with the corals, eg. *T. gigas* L., while the *Hippopus* lives in the sand of shallow seas, eg. *H. hippopus* (L.)

The Solenidae Menke, 1828, are commonly found on shores. Some, like *Ensis ensis* (L.), burrow from low in the intertidal zone to depths of a few meters.

Among the Tellinoidea Blainville, 1814, the Tellinidae Blainville, 1814, (eg. *Tellina tenuis* da Costa) are known from the intertidal zone. *Macoma* (eg. *Macoma baltica* L.) is living from the tidal zone to depths of 8-10 m; in the Baltic, however, as deep as 140 m (Thorson, 1957). *Scrobicularia plana* da Costa must be considered as intertidal, often found on the tidal flats, and so must *Abra tenuis* (Montagu), while other *Abra* species known from Europe occur in deeper waters. On wave-exposed beaches, where instability of the sand greatly limits the population, the species of Donacidae Fleming, 1828, with a worldwide distribution, are found in great quantities at the surface, eg. *Donax variabilis* Say described by Arnold (1901) from the

Florida beaches. Their method of migrating with the tide is figured by Hedgpeth (1957), and Mortensen (1922) described the habitus of *Donax vittatus* (da Costa) from the North Sea shores. However, *Donax vittatus* (da Costa) also occurs at depths of 30 m as demonstrated by Petersen (1977) from the Dogger Bank area of the North Sea, having almost the same distribution as *Tellina tenuis* da Costa.

The Veneridae Raf., 1815, are known from sandy bottoms in the shallow water of warmer seas. Some of these have a high indicative value, eg. *Mercenaria mercenaria* L. being most abundant a short distance above low water marks and at a depth range of only a few meters.

Among the *Venerupis*, which have a vast distribution, *V. saxatilis* (Fleuriau) lives between the tide marks usually attached by byssus threads and *V. decussata* (L.) in sand below mid-tide level (Tebble, 1966).

Among the Veneroidea Raf., 1815, the Petricolidae d'Orbigner, 1835, have species living in older borings from other bivales. Species such as *Petricola pholadiformis* Lamarck can bore into clay, chalk, etc. from mid-tide level; their depth range, however, is not known (Tebble, 1966).

Among the Hiatellidae Winckworth, 1932, and the Myoidae Bronn, 1824, the genera *Hiatella* and *Mya* are characteristic of prodeltaic deposits in the arctic, eg. *Hiatella arctica* Linné and *Mya truncata* (Linné), but elsewhere were recorded from deeper waters. They are the two most common bivalves in arctic raised marine deposits, as noticed by many authors (eg. Blake, 1975; Hillaire-Marcel, 1980).

The Pholadidae Raf., 1815, the great borers, are usually found from the lower region of the tidal zone to depths of a few meters, but they depend on a firm substrate such as peat, wood or rocks, eg. *Barnea costata* L. and *Zirphaea crispata* Linné.

ACKNOWLEDGEMENTS

The editor is thanked for his encouraging suggestions, and Dr. Høpner Petersen, Zoological Museum of Copenhagen, for reading and commenting on the manuscript. Douglas Grant, Geological Survey of Canada, kindly improved the English and Birgit Nielsen, at the Geological Survey of Denmark, typed the manuscript.

REFERENCES

Abel, O., 1935. *Vorzeitliche Lebensspuren.* (Fischer, Jena)

Abbott, R.T., 1954. *American seashells.* (D. van Nostrand Company Inc., Toronto)

Arnold, A.F., 1901. *The sea-beach at ebbtide.* (The Century Company, New York)

Blake, W., 1975. Radiocarbon age determinations and Post-glacial emergence at Cape Storm, Southern Ellesmere Island, Arctic Canada. *Geografiska Annaler,* 57 A

Bromley, R.G., 1981. Concepts in ichnotaxonomy illustrated by small round holes in shells. *Acta Geologica Hispànica,* 16(1-2), 55-64

Crimes, T.P.& Harper, J.C. (eds), 1970. *Trace fossils.* Proceedings of an International Conference held at Liverpool University (Seel House Press, Liverpool)

Cullen, D.J., 1970. Radiocarbon analyses of individual molluscan species in relation to Post-glacial eustatic changes. *Palaeogeography, Palaeoclimatology, Palaeo-ecology,* 7, 13-20

Donner, J. & Jungner, H., 1974. Radiocarbon dating of shells from marine Holocene deposits in the Disko Bugt area, West Greenland. *Boreas,* 4, 25-45

Elton, C., 1927. *Animal ecology.* (Sidgwick & Jackson, London)

Fischer, P.H., 1943. La répartition actuelle des mollusques marins envisagée commes base d'appréciation pour les géologues. *Bulletin Société Géologique de France,* Ser.5, 13, 441-458

Franc, A., 1960. Classe des Bivalves. *Traité de Zoologie,* 5, 2, 1845-2133

Fretter, V. & Graham, A., 1962. *British prosobranch molluscs.* (Ray Society, London)

Hedgpeth, J.W., 1957. Sandy beaches. *Geological Society of America,* Memoir 67, 1, 587-608

Hillaire-Marcel, Cl., 1980. Les faunes des mers Post-glaciaires du Québec, quelques considerations paléo-écologiques. *Géographie physique et Quaternaire,* 34, 3-59

Hjort, C. & Funder, S., 1973. The subfossil occurrence of *Mytilus edulis* L. in central East Greenland. *Boreas,* 3, 23-33

Jensen, A.S., 1902. Studier over nordiske mollusker, II. *Cyprina islandica. Videnskabelige Meddelelser fra Dansk naturhistorisk Forening i København,* 54, 33-42

Johansen, A.C., 1901. Om aflejringen af molluskernes skaller i indsøer og i havet. *Videnskabelige Meddelelser fra Dansk naturhistorisk Forening i København,* 53, 5-47

Johansen, A.C., 1902. On the hypotheses on the sinking of sea-beds. *Videnskabelige Meddelelser fra Dansk naturhistorisk Forening i København*, 54, 393-435

Lemche, H., 1941. Gastropoda Opisthobranchiata. *Meddelelser om Grønland*, 121, 7

McAlester, A.L. & Rhoads, D.C., 1967. Bivalves as bathymetric indicators. *Marine Geology*, 5, 383-388

Mertz, E.L., 1924. Oversigt over de sen- og postglaciale niveauforandringer i Danmark. *Danmarks Geologiske Undersøgelse* II. Række, 41

Moore, R.C., Lalicker, C.G. & Fischer, A.G., 1952. *Invertebrate fossils*, (McGraw-Hill, New York)

Mortensen, T., 1922. Biologiske studier over sandstrandsfaunaen. *Videnskabelige Meddelelser fra Dansk naturhistorisk Forening i København*, 74, 23-56

Muus, B.J., 1967. The fauna of Danish estuaries and lagoons. *Meddelelser fra Danmarks Fiskeri- og Havundersøgelser*, Ny Serie, 5, 1

Newell, N.D., Imbrie, J., Purdy, E.G. & Thurber, D.L., 1959. Organism communities and bottom facies, Great Bahama Bank. *Bulletin of the American Museum of Natural History*, 117, Article 4, 181-228

Norton, P.E.P., 1967. Marine molluscan assemblage in the early Pleistocene of Sidestrand, Bramerton and the Royal Society borehole at Ludham, Norfolk. *Philosophical Transactions of the Royal Society, London*, Series B, No.784, 253, 161-200

Norton, P.E., 1975. Paleoecology of the Mollusca of the Tjörnes sequence, Iceland. *Boreas*, 4, 97-110

Parker, R.H., 1956. Macro-invertebrate assemblages as indicators of sedimentary environments in East Mississippi Delta Region. *Bulletin of the American Association of Petroleum Geologists*, 40(2), 295-376

Petersen, C.G.J., 1910. Some considerations on the study of the post-Glacial climatic changes. In: *Die Veränderungen des Klimas seit dem Maximum der Letzten Eiszeit*, (Generalstabens Litografiska Anstalt, Stockholm), 329-331

Petersen, C.G.J., 1911. Valuation of the sea. I. Animal life of the sea bottom, its food and quantity. *Report Danish Biological Station*, 20

Petersen, C.G.J., 1913. Valuation of the sea. II. The animal communities of the sea bottom and their importance for marine zoogeography. *Report Danish Biological Station*, 21

Petersen, C.G.J., 1914. Appendix to report 21. On the distribution of the animal communities of the sea bottom. *Report Danish Biological Station*, 22

Petersen, C.G.J., 1915. On the animal communities of the sea bottom in the Skagerrak, the Christiania Fjord and the Danish waters. *Report Danish Biological Station*, 23, 3-28

Petersen, G. Höpner, 1977. The density, biomass and origin of the bivalves of the Central North Sea. *Meddelelser fra Danmarks Fiskeri- og Havundersøgelser*, Ny Series, 7, 221-273

Petersen, K. Strand, 1976. Om Limfjordens postglaciale marine udvikling og niveauforhold, belyst ved mollusk-faunaen og C-14 dateringer. *Danmarks Geologiske Undersøgelse*, Arbog 1975, 75-103

Petersen, K. Strand, 1978. Kvantitative analyser af mollusk-faunaen i hævede marine aflejringer ved Loch Fyne, Østgrønland. Abstract 13. *Nordiske Geologiske Vintermode*, January 1978, Copenhagen

Petersen, K. Strand, 1981. The Holocene marine transgression and its molluscan fauna in the Skagerrak-Limfjord region, Denmark. In: *Holocene Marine Sedimentation in the North Sea Region*, eds. S.-D. Nio, R.T.E. Schüttenhelm & Tj.C.E. van Weering, Special publication of the International Association of Sedimentologists, 5, 497-503

Petersen, K. Strand, 1982. Attack by predatory gastropods recognized in an interglacial marine molluscan fauna from Jameson Land, East Greenland. *Malacologia*, 22(1-2), 721-726

Reineck, H.-E. & Singh, J.B., 1980. *Depositional sedimentary environments*. (Springer Verlag, Berlin, Heidelberg, New York), 2nd revised and updated edition

Rudwick, M.J.S., 1972. *The meaning of fossils*. (MacDonald, America-Elsevier)

Schäfer, W., 1962. *Aktuo-Paläontologie nach Studien in der Nordsee*. (Verlag Waldemar Kramer, Frankfurt am Main)

Seilacher, A., 1967. Bathymetry of trace fossils. *Marine Geology*, 5, 413-428

Sorgenfrei, T., 1958. Molluscan assemblages from the marine Middle Miocene of South Jutland and their environments. *Danmarks Geologiske Undersøgelse* II, række, 79

Spärck, R., 1923. En pludselig masseforekomst af *Sepia*-skaller ved Færöerne i foråret 1923. *Videnskabelige Meddelelser fra Dansk naturhistorisk Forening i København*, 76, 141-146

Spärck, R., 1949. On the zoogeographical significance of quantitative investigations of the bottom fauna. *13. Congrès international de Zoologie*, (Paris), 415-417

Stanley, S.M., 1970. Relation of shell form to life habits of the Bivalvia (Mollusca). *The Geological Society of America, Memoir* 125

Stephenson, T.A. & Stephenson, A., 1972. *Life between tide-marks on rocky shores*. (W.H. Freeman, San Francisco)

Stenzel, H.B., 1971. Oysters. *Treatise on invertebrate Paleontology*, Part N, 3, Mollusca 6, Bivalvia, (The Geological Society of America Inc.)

Tebble, N., 1966. *British bivalve seashells*. (The British Museum, London)

Thorson, G., 1944. Marine Gastropoda Prosobranchiata. *Meddelelser om Grønland*, 121, 13

Thorson, G., 1951. Scaphopoda, Placophora, Solenogastres, Gastropoda Prosobranchiata, Lamellibranchiata. *Meddelelser om Grønland*, 81, 2

Thorson, G., 1952. Zur jetzigen Lage der marinen Bodentier-Ökologie. *Verhandlungen der Deutschen Zoologischen Gesellschaft in Wilhelmshaven 1951*, 276-327

Thorson, G., 1957. Bottom communities (sublittoral or shallow shelf). *Geological Society of America*, Memoir 67, 1, 461-534

Thorson, G., 1961. *Livet i havet*. (Berlingske Forlag, København)

Thorson, G., 1963. The distribution of benthic marine Mollusca along the N.E. Atlantic shelf from Gibraltar to Murmansk. *Proceedings of the First European Malacological Congress*, 5-23

6

Beachrock as a sea-level indicator

D. HOPLEY

ABSTRACT

Although it is generally agreed that beachrock forms within the intertidal zone, the exact upper limit to the cementation processes is uncertain, particularly in areas of high tidal range. Problems occur with dating materials from outcrops: a maximum age may be obtained from constituent organic materials, a minimum age from the cementing matrix, but difficulties may be encountered in the interpretation of such dates. Where possible beachrock as an indicator of former sea levels should be used in conjunction with other types of evidence. Care needs to be taken not to confuse beachrock *sensu stricto* with other cemented materials which may outcrop in the intertidal zone.

INTRODUCTION

The term 'beachrock' is a very loose one and has been used to describe a very wide range of lithified materials in the littoral zone. Whilst cements of iron and silica have been reported, strictly speaking, the term is applied to beach sediments found within the intertidal zone, and cemented by calcium carbonate (Fig. 1). The location of modern beachrocks between the upper and lower limits of tidal oscillation have tempted sea-level (SL) researchers to use them as a guide to past SLs, especially where cementation levels clearly extend beyond the modern intertidal extremes (e.g. Hopley, 1971, 1975). Drawing on experience, this paper aims to show that great caution is necessary for the proper interpretation of such cementation levels.

The beach materials lithified have a great range in size from fine sands to large boulders (forming beach conglomerates) but are generally identical to the materials of the adjacent or overlying beach. Mineralogically they

Figure 1. Typical intertidal exposure of beachrock showing gently sloping frontal ramp, Three Isles, Great Barrier Reef. Tidal range 2.5 m.

Figure 2. Close up of beachrock. Beach bedding is retained, the beach materials in this example being a mixture of lithic and coral shingle. After exposure a number of chemical, biological and physical processes produce erosion pits of various sizes.

Figure 3. A coarse boulder conglomerate, Curacoa Island, north Queensland. Although the materials are very coarse, the cementing matrix is mainly aragonite and similar to that seen in finer beachrocks.

vary from pure silica sands of terrigenous origin to the biogenic carbonate sands of coral cays (Fig. 2). Lithic fragments may be common and dominate the massive conglomerates (Fig. 3). Other constituents may be olivene sands or quartzose sands with major heavy mineral assemblages. Beachrock *sensu stricto,* with a calcium carbonate cement, is limited to lower latitudes approximately 30° north and south of the equator and extending just outside the latitudinal range of modern coral reefs (Russell & McIntyre, 1965; Davies, 1972; Hopley, 1982).

THE FORMATION OF BEACHROCK

A wide range of hypotheses has been put forward to account for interidal cementation of beach deposits (see Guilcher, 1961; Kaye, 1959; Bathurst, 1971; Milliman, 1974; Hopley, 1982 for reviews). The majority recognises that it forms within the beach or beach ridges behind the present beach, and has been exposed in its present position on the beach by coastal retreat. Most of the outcrops retain the sedimentary structures of beach deposition. Coarse coral shingles for example, may show a very steep dip, whilst finer sediments may show gentle seaward dips and bedding typical of rhythmic deposition. The shape of the cemented zone varies. In freshly exposed outcrops there may be a frontal

Figure 4. Cemented basset edges typically formed within coral-shingle ridges on coral reefs. Processes are again similar to those in beachrock formation but cements are frequently calcitic and the dip is away from the direction of deposition as the ridges were formed as washover deposits during cyclonic storms.

ramp or nose apparently formed beneath the sloping beach (Fig. 1). This rises onto a surface that is horizontal or slightly fretted, following the strike ridges of the cemented internal structures (Fig. 4). It gives the impression of a truncated surface and explains why many early reports regarded beachrock as erosional in origin (eg. Agassiz, 1898; Steers, 1929, 1937, 1938). The thickness of the cemented strata varies with tidal range. The upper level may commonly approximate Mean High Water Springs (MHWS) but, as discussed below, may go higher; the lower level of cementation is flat and reported (Kuenen, 1933) as being about Mean Low Water Springs (MLWS). In microtidal or tideless areas the beachrock may thus be very thin, reflecting the seasonal fluctuation of mean sea level (MSL), but in areas of higher tidal range, such as along the Great Barrier Reef of Australia, the cemented strata may reach over 3 m in thickness.

There is much debate over the form of the beachrock cement. Various claims have been made for calcite cements (eg. Russell, 1959, 1962; Emery & Cox, 1956; Kaye, 1959), aragonite (Daly, 1920; Ginsburg, 1953; Illing, 1954), or magnesium calcite (Taft & Harbaugh, 1964; Ebanks, 1967). Ranson (1955) suggested that the cement may change with time from aragonite to calcite to amorphous $CaCO_3$. Similarly Taylor & Illing (1971) have suggested that magnesium calcite might represent an alteration product of aragonite, whilst Stoddart & Cann (1965) have suggested a two stage cementation process involving initial bonding by aragonite and void filling with calcite. These latter authors emphasised the importance of identifying the cement, as calcite is the form normally precipitated from fresh water and aragonite from solutions of high ionic strength such as sea water, with direct implications for the cementation process.

Explanations for the formation fall into three broad categories: a) organic; b) inorganic from fresh water; or c) inorganic from sea water.

a) Microbial action (eg. Nesteroff, 1954; Guilcher, 1961; Puri & Collier, 1967), decay of organic matter (eg. Field, 1920; Daly, 1920, 1924) or the activity of algae (eg. Cloud, 1959; Maxwell, 1962) have been the most frequent organic processes quoted. Milliman (1974), however, has dismissed organic mechanisms, quoting examples of the low amount of organic matter normally found in sands and pointing out that whilst algae were common on the surfaces of beachrock, they are normally absent from the interior. Davies & Kinsey (1973) came to similar conclusions in their examination of the Heron Island beachrock.

b) Inorganic precipitation from groundwater has been proposed most extensively by Russell (1959, 1962, 1963, 1967) and his co-workers (Jones, 1961; Deboo, 1962). They suggested that cementation takes place along the water table, beach sands above that level being typically polished as percolating sea water is supersaturated with $CaCO_3$ and non-aggressive. In contrast, beneath the water table circulating fresh water dissolves beach sands which are pitted and unpolished. At the water table sand grains are coated with calcite, which coalesces to form the initial bonding. At a later stage the rocks are filled by a mottled brown cement of uncertain origin. Although it is notable that Russell's calcite cements were all reported from high islands with significant groundwater, his hypothesis has come under considerable criticism. Milliman (1974) questioned Russell's identification of cement and pointed out that many of the outcrops examined in the eastern United States are actually outcrops of the Pleistocene Anastasia and Pamlico Formations. Stoddart & Cann (1965) were critical of a groundwater hypothesis on climatic grounds, ie. beachrock occurs on coral cays in excessively arid areas without groundwater such as the Red Sea, and on physiographic grounds, ie. beachrock is found on islands too small to support a permanent water table.

c) With more detailed studies of the cement of beachrock, particularly the use of SEM techniques, an aragonite cement *precipitated from sea water* is receiving increasing acceptance as the primary cement. This is usually acicular or fibrous although cryptocrystalline and drusy aragonite has been recognised (Milliman, 1974). Magnesium calcite may be present as semi-opaque cryptocrystalline cement consisting of small (4 to 8 micron) rhombohedra. Most workers favour precipitation from sea water, lithification either occurring at depth after percolation or sea-water evaporation and heating at the surface. However, any reversal of the $CaCO_3$ - H_2O - CO_2 process system can lead to precipitation. For example, Hanor (1978) has proposed that much of the cement is precipitated as the result of loss of CO_2 from carbonate saturated beach groundwaters. His mass transport calculations support the proposal that vertical fluid dispersion in the phreatic zones, resulting from tidal oscillation of the water table, is sufficient to induce degassing of CO_2 from a seaward-flowing groundwater. According to Hanor, loss of CO_2 is further enhanced by tidal pumping of the gas phase in the vadose zone across the sediment-atmosphere interface. As sediment porosity is lowered by precipitation of cement, the ability of the groundwater system to degas and form new cements is reduced. Thus as Matthews (1971) suggested, for cementation to take place all that is required is that the waters have sufficient residence time in the landward areas to acquire a high PCO_2 and dissolved $CaCO_3$ and sufficient residence time in the fore-beach areas that degassing can take place before the water is discharged into the sea. Hanor (1978) considered this the normal situation in tropical climates where seasonal rainfall results in episodic variations in groundwater discharge. Precipitation is less likely in areas where high groundwater discharge results in low residence time and less likelihood of degassing fore-beach sediments. Hanor further suggested that greater oscillation of the water table in areas of higher tidal range could greatly increase the dispersive transport of CO_2, and result in degassing and supersaturation of groundwaters. Distribution of massive beachrocks along the Great Barrier Reef, particularly on offshore islands where the seasonal variations in groundwater flow will not be masked by a more general groundwater flow as on the mainland, would appear to fit the model proposed by Hanor. The model does not require a permanent fresh-water lens as the groundwater involved in the degassing and precipitation may be brackish or saline. The major requirement is an overall annual flow of groundwater seawards. In highly arid regions, where landward flow and deep refluxing of sea water occurs, the degassing hypothesis may not be applicable. However, indirect biological precipitation may be possible in anaerobic environments where bacterial processes can consume CO_2 (Berner, 1971a, b; Morita, 1976) and calcium carbonate can be precipitated out of sea water by evaporation. Mixing of water types appears to be incidental to precipitation, but may influence cement chemistry and morphology. Hanor's (1978) experiments produced a low-Mg calcite cement, but clearly in other circumstances aragonite or high-Mg calcite may be precipitated.

The variety of cements, and by implication processes or environments of precipitation, in intertidal and immediately supratidal deposits of coral reefs, is illustrated by the northern Great Barrier Reef (Scoffin & McLean, 1978). Beachrocks (Fig. 1) normally have a fringe

Figure 5. Modern basset edges in foreground and mid-Holocene emerged equivalents forming a platform rock, about 1 m higher, Three Isles, Great Barrier Reef.

of acicular aragonite crystals, although some examples from windward cays which have formed close to mangroves display a micritic cement or a mixture of micrite and aragonite. Similarly the rampart rocks (Fig. 4, Fig. 5) of the windward side of some reefs, formed from the cementation of shingle ridges, are cemented by a white to rusty brown micrite containing 68 to 95% high-Mg calcite (14% $MgCO_3$) and from 0 to 25% aragonite. Clay minerals make up 5 to 8% of the matrix (mainly kaolinite with some illite and montmorillonite). Cemented boulder tracts show the same variation in cements with micritic calcite cements to windward, fibrous aragonite to leeward. These results suggest that at least two environments of precipitation were involved. Beaches on leeward sand cays are kept free from fines, being washed by open shelf sea water throughout the tidal cycle. Ramparts and windward boulder tracts in contrast are not cleared of trapped fines, and are immersed in open shelf water only at high tide and ponded reef-flat water at low tide. This reef-flat water may have been concentrated by evaporation at low tide, diluted by fresh water after heavy rain or influenced by passage through mangroves. The occurrence of solution fabrics in rampart rocks and their absence from leeward beachrocks suggests contrasting compositions of interstitial waters. Beachrock waters may be permanently saturated or supersaturated although rampart-rock waters may be undersaturated at times. A similar contrast between aragonitic cements associated with open-ocean beachrocks and high-Mg calcite on beaches of leeward lagoons with mangroves has been reported from Belize by Ebanks (1975).

Figure 6. Emerged mid-Holocene beachrock on Herald Island near Townsville, Queensland. Tidal range is 3.0 m, with HAT at the level of the figure's feet. The cemented horizons rise a further 4.0 m.

PROBLEMS IN THE USE OF BEACHROCK AS A SEA-LEVEL INDICATOR

1. Problems of height relationships

The upper level of cementation is not always easy to define in emerged outcrops of beachrock. Whilst horizontal, firmly cemented levels may exist, frequently a zone of friable partially cemented beach materials may form the uppermost part of the outcrop, or the beachrock *sensu stricto* passes upwards into a phosphatic cay sandstone or dune calcarenite (see below). The major question to be answered, however is, What is the uppermost level of cementation? There is little doubt that it is at least MHWS, but in areas of high tidal range this may be a metre or more below the highest astronomical tidal level (HAT). Russell (1959) proved that beachrock could form in a matter of a few hundred years. Frankel (1968) brought this figure down to tens of years by observing World War II debris in outcrops. Exposures on Magnetic Island near Townsville, Australia formed within six months. Thus, given sufficient stability of the beach materials, tidal levels achieved only a few times a year may allow cementation to take place above MHWS. A conservative approach to emerged outcrops is to compare their height to that of HAT (Fig. 6). A further insurmountable problem may result from any change in tidal range which can result from local changes in coastal configuration.

Cementation above the level of HAT is possible subsequent to the formation of an extensive area of beachrock. Porosity in partially cemented beachrock is only 30% (Kinsey & Davies, 1979) and much lower in well cemented examples, especially if it has been subaerially exposed. Permeability may also be low, and the opportunity is present for temporary perched water tables caused by either (a) water being thrown up over the outcrop or its overlying beach sediments by high storm waves, or (b) percolating meteoric waters moving down to the cemented level from the surface. In the former case, the cement is likely to be aragonitic, in the latter, calcitic, but an extensive horizontal surface is unlikely to develop. The irregular friable cemented horizon found at the upper limits of some beachrocks may result from such processes.

The lowest level of cementation, corresponding to the lowest tidal levels has not been identified in exposed beachrocks. Given a stable sediment body, the zone of cementation will rise then fall through a SL rise-and-fall cycle and the lowest level of cementation will correspond only to the lowest level at any time during that sediment body's existence. Cementation of the lowest level of an exposed outcrop may thus have taken place at an early stage during the rising SL, or more recently since SL dropped down through the lower levels or during any intervening SL fluctuation. Only the uppermost level of the true beachrock is a reliable indicator of a paleo-tidal level.

2. Problems of dating

The precise dating of beachrock is a highly difficult task, although constituent biogenic materials such as shell or coral may be easily dated by radiocarbon, and, if chosen carefully to avoid any recrystallisation can give a reliable age; the date they yield is the time that the organism died. Thereafter it has been transported from its *in situ* location, deposited in a beach (often a storm beach) and only subsequently cemented into the beachrock. Whilst it is probable that in many examples this sequence takes place in a matter of only a few years, it is possible that the time difference between the death of the organism and the cementation of the beachrock is hundreds and even thousands of years. The age yielded by the dating of a constituent organism can be regarded only as the maximum possible for the beachrock cementation.

A minimal date for the process may be obtained by dating only the cementing matrix. However, difficulties arise in trying to obtain sufficient cement, without fine grained beach materials, to allow radiocarbon dating. Further, it is likely that even after exposure any passage of water through an outcrop can cause an exchange of carbonate, continuously rejuvenating the apparent age of cementation. For example at Herald Island near Townsville, Australia, where a cliff of emerged beachrock rises to 4.9 m above MHWS or 4.0 m above HAT, *Trochus* sp. from near the top of the cemented layer gave a radiocarbon age of 4280 ± 100 BP, whilst a sample of whole fine grained

Figure 7a. Beachrock made up of coarse coral debris, the larger corals being in a range of attitudes and branching corals clearly rolled and water-worn.

Figure 7b. Emerged reef rock of *in situ* coral reef. In contrast to the beachrock all head corals are upright; branching corals, although fragmented, are not water-worn. Barbados, West Indies.

Figure 8. Phosphatic cay sandstone, Raine Island, Great Barrier Reef. Although the upper surface may be horizontal, the vertical piping is typical providing an indistinct lower level of cementation. The phosphatic cement is distinctive.

beachrock including up to 50% cement from near the base produced a date of only 3540 ± 90 BP (Hopley, 1971).

3. Problems of confusion with other cemented materials in the intertidal zone.

Cemented materials other than beachrock may outcrop on low latitude beaches which may have no direct relationship to SL or have a relationship different to that of beachrock *sensu stricto*. They include:

a) Emerged coral reef (Fig. 7a, 7b). Emerged reef frequently occurs with emerged beachrock (eg. Hopley *et al.*, 1982) but is related to MLWS not MHWS (see Hopley, this volume). Superficially it may resemble a beach conglomerate of coral debris and exposures of Pleistocene reef rock have been misinterpreted as beachrock (see Milliman's (1974) comments on Russell's identification in Florida). However, close examination should reveal the *in situ* location of coral heads and branching corals.

b) Cay sandstone (Fig. 8). On islands with large populations of nesting birds the downward percolation of meteoric waters can carry guano to temporary or perched water tables to form cay sandstone (eg. Scoffin, 1977; Scoffin & McLean, 1978; Stoddart, Gibbs & Hopley, 1981). In some cases this may directly overlie true beachrocks. It can be

Figure 9. Pleistocene dune calcarenite, Cockermouth Island, Queensland. This outcrop was mistakenly identified as emerged beachrock by Steers (1937). It has distinctive dune bedding and its upper level of cementation is non-horizontal. However, two horizontal platforms have been cut into it, the lower is modern, the upper is about 2.5 m higher and probably mid-Holocene.

differentiated on morphologic grounds, it frequently showing distinctive vertical piping and in the nature of the phosphatic cement, probably hydroxyapatite.

c) *Dune calcarenite* (Fig. 9). Percolating waters through dune fabrics rich in carbonates may cause widespread cementation of the dune. The process appears to be slow, and most examples examined in Australia are Pleistocene in age. When outcropping in the intertidal or immediately supratidal zone, dune calcarenite may be mistaken for beachrock, as for example by Steers (1937) on Cockermouth Island, Queensland (see Hopley, 1982, pp 362-3). This example of Pleistocene calcarenite has been planed by the sea to produce a horizontal surface similar to beachrock, whilst normally calcarenite has a highly irregular upper surface of cementation and vertical piping is very common. Dune bedding and low-Mg calcite cements contrast with true beachrock.

d) *Humate coffee rock.* Processes of soil development in humid areas may result in the formation of a B-horizon rich in organic matter and often iron (see for example McGarity, 1956). Coffee rock is common in many Australian Pleistocene barrier systems and can outcrop on the beach after coastal recession. Especially when covered by algae the outcrop

Figure 10. The ideal combination of emerged beachrock overlying emerged coral reef, Rattlesnake Island, Queensland. The amount of emergence is 1 m and the age of the reef just over 5000 BP.

may resemble beachrock, but the friable organic cements are quite different to the carbonate cements of true beachrock.

e) Silcrete. Riverine silcretes (Grant & Aitchison, 1970) occur along the banks of many Australian rivers which have maintained their Pleistocene courses. Seasonal water-table fluctuations and silica rich groundwaters seem necessary for this cementation. As the streams extended out onto the continental shelf during periods of lowered SLs, outcrops of silcrete may now occur in the littoral zone. The silica cement gives, however, a clear identification; some of the examples described by Russell & McIntire (1965) as beachrock in north Queensland are in fact silcrete.

CONCLUSION AND GUIDE TO THE USE OF BEACHROCK AS A SEA-LEVEL INDICATOR

Many problems exist in the use of beachrock as a SL indicator and its use as a primary indicator is not advised particularly in areas of high tidal range. Used in parallel with other evidence, such as emerged reef, it may provide a useful indicator of a paleo-HAT (Fig. 10). It is thought to be more reliable in areas of low tidal range, in coarse deposits where upward migration of carbonate rich groundwaters by capillary action is limited, in more arid areas where fresh groundwater tables are non-existent, or small cays or continental island spits and beaches where no significant groundwater-table gradient can develop due to lack of size.

Where beachrock is used the following guidelines should be followed:

1. Care should be taken that a cemented outcrop is not older coral reef, phosphatic cay sandstone, dune calcarenite, coffee rock or silcrete;
2. the cement should be carefully examined. An aragonite cement is more likely to be associated with a cemented outcrop related to a paleo-SL;
3. the uppermost level of cementation should be horizontal, not sloping and possibly related to a regional water table. A friable upper level of cementation is dubious;
4. beach structures should be sought in the outcrop;
5. an extensive dating programme should be followed on both constituent biogenic materials (shells, coral etc.) and cementing matrix. The wider the spread of dates, the less reliable is the outcrop as an indicator of an old SL.

Where possible, parallel evidence in the form of emerged reef, mangrove peats or *in situ* molluscs, vermetids etc. with the beachrock should be sought.

REFERENCES

Agassiz, A., 1898. A visit to the Great Barrier Reef of Australia in the steamer 'Croydon' during April and May 1896. *Bulletin of the Museum of Comparative Zoology,* Harvard College, 28, 95-148

Bathurst, R.G.C., 1975. Carbonate sediments and their diagenesis. *Developments in Sedimentology,* 12, (Elsevier, Amsterdam), 2nd enlarged edition

Berner, R.A., 1971a. Bacterial processes affecting the precipitation of calcium carbonate in sediments. In: *Carbonate Cements,* ed. O.R. Bricker, (Johns Hopkins Press, Baltimore), 247-251

Berner, R.A., 1971b. *Principles of chemical sedimentology.* (McGraw-Hill, New York)

Cloud, Jr., P.E., 1959. Geology of Saipan, Mariana Islands, 4. Submarine topography and shoal water ecology. *United States Geological Survey Professional Papers,* 280-K, 361-445

Daly, R.A., 1920. Origin of beachrock. *Carnegie Institution of Washington Yearbooks,* (1919), 18, 192

Daly, R.A., 1924. The geology of American Samoa. *Carnegie Institution of Washington Papers of the Geophysical Laboratory,* 340, 93-143

Davies, J.L., 1972. *Geographical variation in coastal development.* (Oliver & Boyd, Edinburgh)

Davies, P.J. & Kinsey, D.W., 1973. Organic and inorganic factors in recent beachrock formation, Heron Island, Great Barrier Reef. *Journal of Sedimentary Petrology,* 43, 59-81

Deboo, P.B., 1962. A preliminary petrographic study of beach-rock. *Proceedings of the National Coast and Shallow Water Research Conference,* 1961, 456-458

Ebanks, Jr., W.J., 1967. *Recent carbonate sedimentation and diagenesis, Ambergris Cay, British Honduras.* Ph.D thesis, Rice University

Ebanks, Jr., W.J., 1975. Holocene carbonate sedimentation and diagenesis, Ambergris Cay, Belize. In: *Belize Shelf-Carbonate Sediments, Clastic Sediments and Ecology,* eds. K.F. Wantland & W.C. Pusey, *American Association of Petroleum Geologists, Studies in Geology,* 2, 234-296

Emery, K.O. & Cox, D.C., 1956. Beachrock in the Hawaiian islands. *Pacific Science,* 10, 382-402

Field, R.M., 1920. Investigations regarding the calcium carbonate oozes at Tortugas and the beachrock at Loggerhead Key. *Carnegie Institution of Washington Yearbook,* (1919), 18, 197-198

Frankel, E., 1968. Rate of formation of beachrock. *Earth and Planetary Science Letters,* 4, 439-440

Ginsburg, R.N., 1953. Beachrock in south Florida. *Journal of Sedimentary Petrology,* 23, 89-92

Grant, K. & Aitchison, G.D., 1970. The engineering significance of silcretes and ferricretes in Australia. *Engineering Geology,* 4, 93-120

Guilcher, A., 1961. Le 'beachrock' ou grès de plage. *Annales de Géographie,* 70, 113-125

Hanor, J.S., 1978. Precipitation of beachrock cements: mixing of marine and meteoric waters versus CO_2 degassing. *Journal of Sedimentary Petrology,* 48, 489-501

Hopley, D., 1971. The origin and significance of north Queensland island spits. *Zeitschrift für Geomorphologie, N.F.,* 15, 371-389

Hopley, D., 1975. Contrasting evidence for Holocene sea levels with special reference to the Bowen-Whitsunday area of Queensland. In: *Geographical Essays in Honour of Gilbert J. Butland,* eds. J.E. Hobbs & J.J. Pigram, (Department of Geography, University of New England, (Armidale), 51-84

Hopley, D., 1982. *Geomorphology of the Great Barrier Reef: Quaternary development of coral reefs.* (John Wiley-Interscience, New York)

Hopley, D., Slocombe, A.M., Muir, F. & Grant, C., 1982. Nearshore fringing reefs in north Queensland. *Coral Reefs*, 1, 151-160

Illing, L.V., 1954. Bahamian calcareous sands. *American Association of Petroleum Geologists Bulletin*, 38, 1-95

Jones, C.L., 1961. *Petrography of some Recent beachrock from the Caribbean*. Unpublished MS thesis, Louisiana State University, Baton Rouge

Kaye, C.A., 1959. Shoreline features and Quaternary shoreline changes, Puerto Rico. *United States Geological Survey Professional Papers*, 317-B, 49-140

Kinsey, D.W. & Davies, P.J., 1979. Inorganic carbon turnover, calcification and growth in coral reefs. In: *Biogeochemistry of Mineral Forming Elements*, eds. P. Trudingar & D. Swaine, 131-162

Kuenen, P.H., 1933. *Geology of coral reefs*. Snellius expedition east part of Netherlands-East-Indies, 5, 2, 1-126

McGarity, J.W., 1956. Coastal sandrock formation at Evans Head, NSW. *Proceedings of the Linnean Society of New South Wales*, 81, 52-58

Matthews, R.K., 1971. Diagenetic environments of possible importance to the explanation of cementation fabric in sub-aerially exposed carbonate sediments. In: *Carbonate Cements*, ed. O.P. Bricker, (Johns Hopkins Press, Baltimore), 127-132

Maxwell, W.G.H., 1962. Lithification of carbonate sediments in the Heron Island Reef, Great Barrier Reef. *Journal of the Geological Society of Australia*, 8, 217-238

Milliman, J.D., 1974. *Marine carbonates*, recent sedimentary carbonates, 1, (Springer-Verlag, Berlin)

Morita, R.Y., 1976. Anaerobic microbial calcite formation: Great Barrier Reef studies. *Abstracts of the Joint Oceanographic Assembly, Edinburgh 1976 FAO*, Rome, 117

Nesteroff, W.D., 1954. Sur la formation des grès de plage ou 'beachrock' en Mer Rouge. *Comptes Rendus de l'Académie des Sciences, Paris*, 238, 2547-2548

Puri, H.S. & Collier, A., 1967. Role of micro-organisms in formation of limestones. *Gulf Coast Geological Society Transactions*, 17, 355-367

Ranson, G., 1955. La consolidation des sédiments calcaires dans les regions tropicales. *Comptes Rendus de l'Académie des Sciences, Paris*, 240, 640-642

Russell, R.J., 1959. Caribbean beachrock observations. *Zeitschrift für Geomorphologie, N.F.*, 3, 227-236

Russell, R.J., 1962. Origin of beachrock. *Zeitschrift für Geomorphologie, N.F.*, 6, 1-16

Russell, R.J., 1963. Beachrock. *Journal of Tropical Geography*, 17, 24-27

Russell, R.J., 1967. *River plains and sea coasts.* (Univ. California Press, Berkeley and Los Angeles)

Russell, R.J. & MacIntyre, W.G., 1965. Southern hemisphere beachrock. *Geographical Review,* 55, 17-45

Scoffin, T.P., 1977. Sea-level features on reefs in the northern Great Barrier Reef. *Third International Coral Reef Symposium,* 2, 319-324

Scoffin, T.P. & McLean, R.F., 1978. Exposed limestones of the northern province of the Great Barrier Reef. *Philosophical Transactions of the Royal Society, London,* Series A, 291, 119-138

Steers, J.A., 1929. The Queensland coast and the Great Barrier Reef. *Geographical Journal,* 74, 232-357, 341-370

Steers, J.A., 1937. The coral islands and associated features of the Great Barrier Reefs. *Geographical Journal,* 89, 1-28, 119-146

Steers, J.A., 1938. Detailed notes on the islands surveyed and examined by the geographical expedition to the Great Barrier Reef in 1936. *Reports of the Great Barrier Reef Committee,* 4, part 3, 51-96

Stoddart, D.R. & Cann, J.R., 1965. Nature and origin of beachrock. *Journal of Sedimentary Petrology,* 35, 243-247

Stoddart, D.R., Gibbs, P.E. & Hopley, D., 1981. Natural history of Raine Island, Great Barrier Reef. *Atoll Research Bulletin,* 254, 1-44

Taft, W.H. & Harbaugh, J.W., 1964. Modern carbonate sediments of southern Florida, Bahamas and Espiritu Santo Island, Baja California: a comparison of their mineralogy and chemistry. *Stanford University Publications in Geological Sciences,* 8(2), 1-133

Taylor, J.C.M. & Illing, L.V., 1971. Development of Recent cemented layers within intertidal sand flats, Qatar, Persian Gulf. In: *Carbonate Cements,* ed. O.P. Bricker, (Johns Hopkins Press), 27-31

7

Ooids as sea-level indicators

L. R. KUMP and A. C. HINE

ABSTRACT

Ooids are subspherical carbonate sand grains, composed of concentric laminae which surround detrital nuclei. This paper is an evaluation of the usefulness of ooids as paleo-sea-level indicators.

Although ooids form in a variety of shallow-marine environments they do not form precisely *at* sea level. Proper assignment of depth limits to the process of oolitization must await a clearer understanding of the conditions under which ooids form. After they form, ooids can be transported hundreds of kilometers away from, and to depths not indicative of, their place of origin. They may be reworked from older deposits, altered in appearance by diagenetic processes, or confused with similar grains of disparate origin. Laboratory dating of ooids by radiocarbon methods is confounded by such things as the non-uniformity of grain sizes and cortical thicknesses, nuclei which are older than their cortices, microboring and infilling, and the span of time represented by a single ooid.

Ooids provide a wealth of information about the environments of origin, deposition and diagenesis of a particular sedimentary deposit. Unfortunately, the complicated story they present often makes them unreliable indicators of former sea levels.

INTRODUCTION

The correct interpretation of paleoenvironment and paleobathymetry from the presence and distribution of ooids in carbonate rocks is based on understanding their mode of formation and the influence of transport and diagenesis on

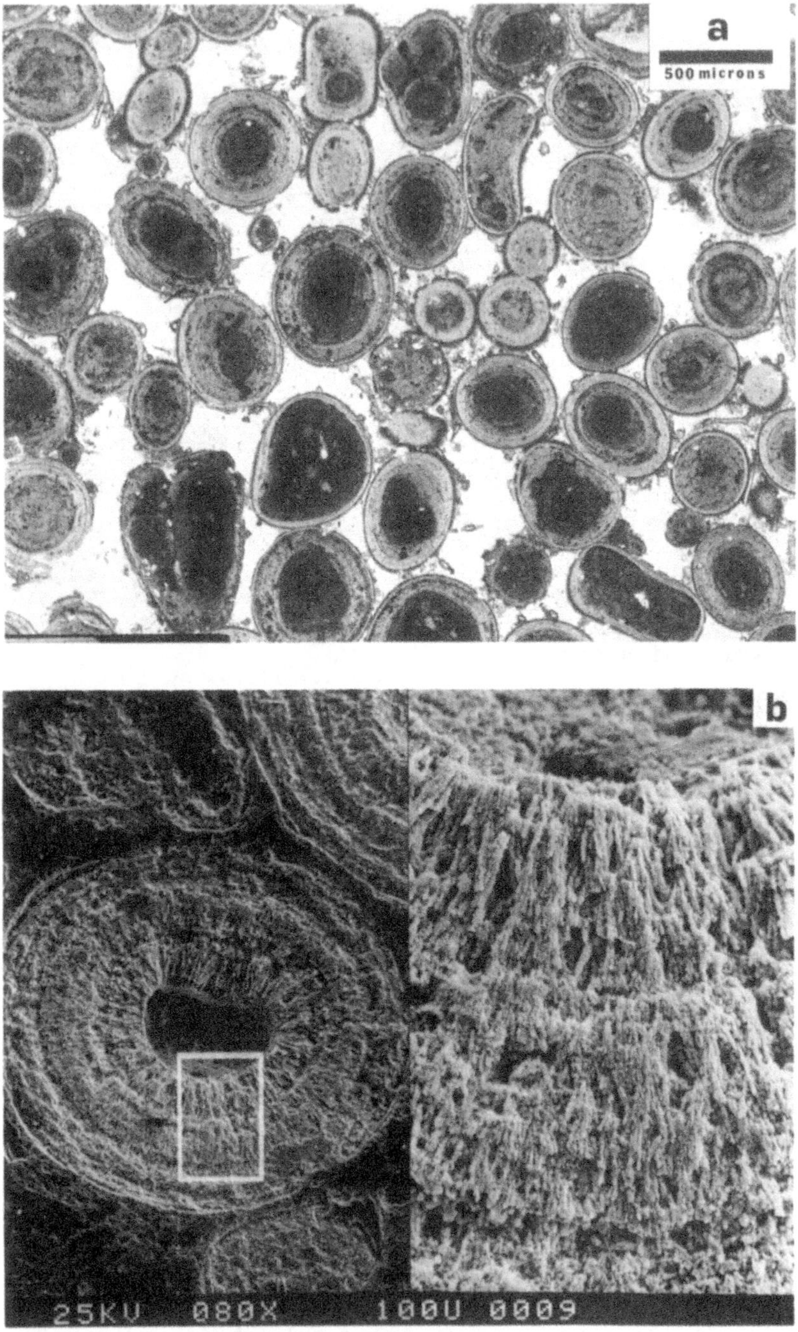

Figure 1. Marine ooids. a. Tangential ooids from the Bahamas. b. Radial ooid from the Lower Inferior Oolite, Jurassic, Selsey Common, Gloucestershire, U.K., with 5X radial orientation. Base of square is 100 microns.

these grains. A brief treatment of these topics is presented in this chapter. The reader is directed to the referenced literature for more complete discussions.

Typical marine ooids are spherical to ellipsoidal carbonate grains, less than 2 mm in diameter composed of a cortex of concentric laminae which surrounds a detrital nucleus (Newell *et al.*, 1960; Teichert, 1970; Bathurst, 1975). The nucleus is commonly a quartz grain, a peloid, a skeletal,or an ooid fragment. The cortex is made up of calcite, Mg-calcite, or aragonite crystals arranged either tangentially, radially, or randomly into concentric layers (Fig. 1). Grains with these same characteristics, but larger than 2 mm, are termed pisolites. Mineralogy and grain fabric have been ascribed to both environmental and diagenetic control (see below). An oolite is a sediment or sedimentary rock containing a significant proportion of ooids (Newell *et al.*, 1960).

Modern ooids from a wide variety of geographical settings have been described in the literature. These include marine ooids from the Bahamas (Illing, 1954; Newell *et al.*, 1960); Cuba (Pavlidis *et al.*, 1972); the Great Barrier Reef (Marshall & Davies, 1975); the Dead Sea (Garber *et al.*, 1981); the Red Sea (Friedman *et al.*, 1973); the Persian Gulf (Loreau & Purser, 1973); Baffin Bay, Texas (Land *et al.*, 1979); and non-marine ooids from the Great Salt Lake, Utah (Eardley, 1938); caves (Donahue, 1969); and low-salinity lakes (Wilkinson *et al.*, 1980). Non-marine ooids, because they in no way are indicative of sea level (SL), will be ignored in this chapter.

The ooid-producing regions of the Bahamas are the best known of those listed above (Illing, 1954; Newell *et al.*, 1960; Purdy, 1961; Ball, 1967; Buchanan, 1970; Hine, 1977; Harris, 1979; Dravis, 1979). These regions are restricted to the physically-active areas of the Bahama Banks, such as the shallow margins, local areas of tidal acceleration (tidal inlets, inter-island gaps, bedrock highs and the ends of closed basins), and shallow nearshore zones. In these areas, large-scale bedforms with superimposed smaller-scale ripples commonly form in response to strong, persistent tidal currents, and build within 1-3 m of the sea surface (Fig. 2).

At Brown's Cay in the Bahamas, Newell *et al.* (1960) noted that the percentage of oolitically-coated grains increased as water depth decreased. A marked change at about 2 m led the authors to conclude that this is the maximum depth of the 'chief zone of ooid accretion'.

Studies of other regions (listed earlier) corroborate the observation from the Bahamian investigations that ooid production is a shallow-water phenomenon. However, it is important to remember that ooids do not form *at* SL.

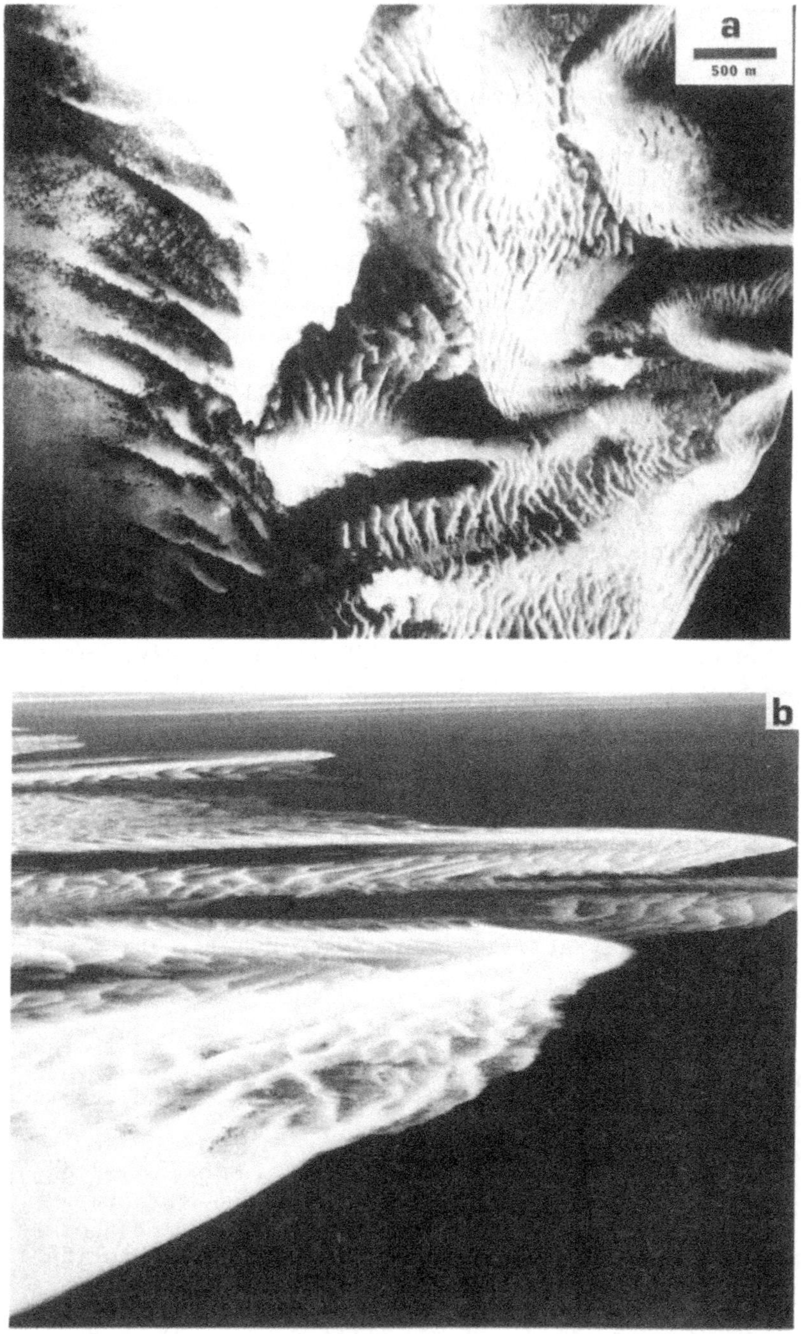

Figure 2. a. Aerial photograph of a physically-active ooid shoal on Little Bahama Bank (DMA photo). b. Oblique photograph of the same ooid shoal, showing spillover-lobes which have migrated onto the shelf-lagoon surface. Airplane altitude was 800 feet.

Investigators of ancient oolites, armed with the classical notion that most ooids formed in Bahamian-like settings, have assigned a very shallow depth of accumulation to these sediments. Quaternary SL curves have been generated in part from the distribution of ooids on continental shelves. Below are a few examples. Later, the arguments of those who stress the likelihood of transport of ooids will be discussed.

Pilkey *et al.*, (1966) collected ooids from the shelf-edge off the S.E. United States at water depths of from 35 to 150 m. A composite sample from 79 m and 104 m gave a radiocarbon age of 26 000 BP, which indicated to the authors that the ooids were not growing at the present time. Because they did not have single-sample dates they were not able to delineate paleo-shorelines. Nevertheless, the authors felt that the presence of ooids in outer shelf sediments should reflect the location of the ancient shoreline, and dates on these grains should yield precise ages for these deposits. Problems with the dating of ooids are discussed later.

This idea was utilized extensively by Milliman & Emery (1968) in constructing a SL curve for the last 35 000 years. Many of their 80 samples, especially the older samples, were oolites. Other indicators used were shallow-water molluscs, beachrock, and peat. Milliman & Barretto (1975) again used oolites to plot a relative SL curve for the Amazon shelf. These sediments were collected at 80 to 150 m water depth, and the ooids contained in these sediments were relict (radiocarbon ages of 14 430 to 38 000 BP). When plotted on the eustatic curve of Milliman & Emery (1968) the older Amazon ooids consistently indicated a lower position for SL. From this interesting result they concluded that the Amazon shelf underwent a period of significant subsidence (140 m in 5000 years) between 21 000 and 16 000 years ago. Younger ooids fell on the curve, indicating that little if any subsidence has occurred in the last 14 000 years.

Finally, in their study of the origin of Bahamian grapestone, Winland & Matthews (1974) concluded that a SL rise was responsible for the developmentof grapestone from on oolitic sediment. The grapestone consisted of recrystallized ooids cemented together by Foraminifera and worm tubes. Carbon^{14} dates for these recrystallized ooids were significantly older than those for presently-forming ooids. Thus, the alternative interpretation, that the grapestone was composed of modern ooids transported to a quieter environment and then cemented together, was rejected, and a SL rise was inferred. The transport of sand across and off the Bahama Banks was more active in the past (3000-5000 BP) when water depths were shallower and storm waves were more effective in suspending and transporting sediment (Hine, in press). Large scale, inactive bedforms on the Bahama Banks are indicative of an earlier, higher-energy period.

OTHER OOID-LIKE GRAINS

When using ooids in SL reconstructions, the researcher must first determine that indeed the ooids are of shallow-marine origin. Spherical grains with concentric structure can form in settings and under conditions quite different from those presented earlier for the formation of true marine ooids. For example, sediment trapping by planktonic algae may have been responsible for the pelagic 'oolite' of the Tethyan Jurassic (Jenkyns, 1972). The sediment in which these 'ooids' accumulated was replete with planktonic Foraminifera, calcitized Radiolaria, and calpionellids. The cortexes of these grains even contained nannofossils. Clearly a shallow-water, agitated environment would be a poor choice for paleo-reconstruction of this depositional setting.

Ooids also form in non-marine environments, and these could be confused with marine ooids, especially the quiet-water variety. Fortunately, other indicators of non-marine conditions, including fresh-water fossils and unique facies relationships, identify the appropriate paleo-environment of these grains.

Grains once interpreted as the oldest (silicified) carbonate ooids (Reimer, 1975), from the Overwacht Group, South Africa (∿3.3 billion years old) were later reinterpreted to be volcanic accretionary lapilli. Close examination by Lowe & Knauth (1978) revealed that the shells of these 'ooids' were in fact constructed of volcanic clastic debris.

Diagenetic processes may alter the appearance of other grains so that they may be easily confused with ooids. This explanation was invoked by Siesser (1973) to account for the presence of ooid-like grains in caliche layers on top of or within Pleistocene eolianites from South Africa. He rejected the more obvious explanation, that the grains were true ooids transported into the subaerial realm during a storm, because of the complete lack of ooids in the adjacent marine facies.

Examination of the grains in thin section allows the distinction of oncolites, rhodoliths and 'pseudoolithes' from true ooids. Oncolites and rhodoliths are unattached, spherical stromatolites which grow by the cohesion of sediment to blue and red algal filaments as these grains roll along the seafloor. These grains lack the uniform crystal arrangement seen in true ooids. Pseudoolithes (Cayeux, 1935) are apparently either fecal in origin or are the micritized remains of skeletal grains (Bathurst, 1975). They are uniformly micritic and thus distinguishable from marine ooids.

An awareness of the possible presence of 'fraudulent' ooids in conjunction with good stratigraphic and facies analysis should prevent the misinterpretation of these grains as true marine ooids.

THE FORMATION OF OOIDS

The environmental controls on ooid growth may not be sufficiently stringent to allow a precise interpretation of the ooid's depth of origin. Several factors may be important to the growth process, including the supersaturation of sea water with respect to calcium carbonate, water depth and degree of agitation (Berner, 1971), algal and bacterial control of precipitation (Rothpletz, 1892; Vaughn, 1914) poisoning by other ions such as Mg^{2+}(Davies *et al.*, 1978), aggregation of colloidal particles (Bucher, 1918), availability of nuclei, interactions with organic molecules (Mitterer, 1971, 1972; Suess & Futterer, 1972; Davies *et al.*, 1978), and the ability of the ooid-forming environment to retain grains for a sufficient length of time to develop oolitic coatings (Bathurst, 1975). The history of oolite theory is presented in Bathurst (1975). Simone (1981), provides an up-to-date review of the literature on ooid formation.

No definitive explanation of ooid growth has appeared in the literature to date (Bathurst, 1975; Simone, 1981). Laboratory experiments involving saturated solutions of $CaCO_3$ (Monaghan & Lytle, 1956; Donahue, 1965; Deelman, 1978) have not yielded Bahamian-type ooids. The discovery of ooids forming in quiet-water environments indicates that agitation may not always be required for ooid development (Bathurst, 1967; Kahle, 1974; Davies & Martin, 1976; Halley, 1977; Davies *et al.*, 1978; Land *et al.*, 1979). These quiet-water ooids are composed of aragonite or Mg-calcite crystals with radial orientation.

Loreau & Purser (1973) and Loreau (1982) have investigated the relationship between grain structure and environment for Persian Gulf ooids. They found that the dominant structure of an ooid's youngest cortical layer was determined by the physical energy of the grain's immediate surroundings. Radial ooids were restricted to very protected areas such as lagoons and depressions between ridges, while tangential ooids were predominant in agitated settings such as ridge crests. Loreau and Purser proposed that the aragonite crystals of an ooid are arranged radially during growth in a quiescent setting. If transported to an agitated environment, the radial crystals are flattened and abraded and a tangential structure results.

In areas of weak agitation, the development of organic substrates may promote the precipitation of oolitic coatings. Davies *et al.* (1978) synthesized ooids under both agitated and non-agitated conditions in laboratory solutions containing humic acids. Membranes formed from the humic acids presumably enveloped potential nuclei and provided a substrate for carbonate precipitation. This process was suggested by Land *et al.* (1979) as an explanation for the formation of Baffin Bay, Texas, ooids.

Very shallow water depth thus does not appear to be a decisive factor in ooid genesis. This is another of the many uncertainties in the assignment of paleo-depths to ooid accumulations.

PROBLEMS OF TRANSPORT

Because many ooids are formed under high-energy conditions, they are often transported to the platform interior, the lee of a barrier, a deep-shelf basin, or up onto a beach/dune complex, where they are buried. Sometime after burial, these grains may be re-exposed to waves and currents and be rejuvenated as relict particles back into the active sedimentary domain.

Investigators working in the Bahamas and the Persian Gulf, two areas of high-energy ooid formation, have noted that these grains can be found in areas quite different from those in which they formed. Newell *et al.*, (1960) reported that ooids formed a quantitatively important fraction of the sediment 50 km bankward of the zone of optimum formation at the edge of the Bahama Banks. Harris (1979) found that while the region of active ooid formation was quite restricted on the margin of Great Bahama Bank bordering Tongue of the Ocean north of Andros Island (~25 km^2), the region of accumulation, Joulters Ooid Shoal, was built almost to SL over a vast area (~400 km^2). Harris concluded that 'the vast sand flat formed by reworking of mobile sands and spit-like tidal bar accretion.'

The position of the carbonate bank margin with respect to the dominant wind direction determines whether sand transport is bankward or offbank. On windward coasts, such as the eastern margins of the Bahama Banks, transport is generally towards the bank interior. Along leeward margins, sediment transport is generally offbank to great depths (Mullins & Neumann, 1979; Hine *et al.*, 1981; Hine & Steinmetz, 1984, where gravity-flow processes and deep-ocean currents may move the sediment 100's of km away from their zone of production (Schlager & Chermak, 1979; Mullins *et al.*, 1980; Fig. 3). At the moment, it is unknown how much of this offbank transported/periplatform sand consists of ooids.

Loreau & Purser (1973) documented the wide distribution of ooids in the Persian Gulf. Holocene ooids were not only found in coastal regions where they are currently forming but also in regions as diverse as the deep basin of the Gulf (80-120 m) and sabhka and desert environments.

Similar relocation of Quaternary ooids was reported by Fabricius *et al.*, (1970) near the South Tunisia coast and by Kumar *et al.*, (1977) who reinterpreted the work of Milliman & Barretto (1975) on the subsidence of the Amazon shelf. Citing the lack of evidence that the ooids had not been transported, and what they felt was an excessive subsidence rate based on an *in situ* interpretation for the ooids, Kumar *et al.*, (1977) argued that the ooids had been redeposited. As the shoreline receded at the end of the last glaciation, ooids were transported out on the shelf to what is now a

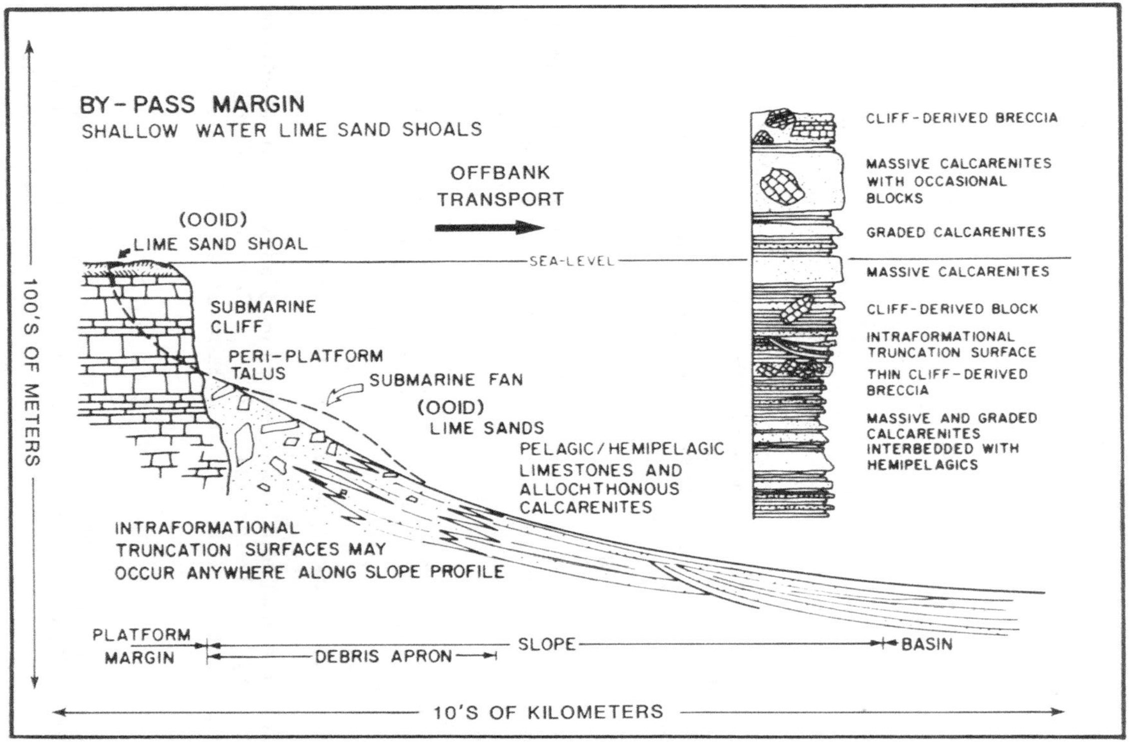

Figure 3. Hypothetical model of a leeward by-pass margin in which shallow-water lime sands, including ooids, accumulate in deep-water slope environments. (From McIlreath & James, 1979; used with permission.)

deep-water environment. In reply to these comments, Milliman (1977) suggested that the stratigraphic sequence which exists in the shelf sediments would not have been maintained during an extensive relocation of a large oolitic zone. Thus he again opted for subsidence as an explanation of the ooids whose age and depth plot below the eustatic curve.

The wide dispersal of ooids in at least some of these regions occurred in a geologic 'instant' of less than 20 000 years. With this in mind the geologist must be wary of precise locations of paleo-shorelines and SLs based solely on the occurrence of oolites.

Likewise, the rejuvenation of older ooids into younger sediments must be considered in the interpretation of ancient oolites. Sediment-starved areas are especially prone to reworking. One such area which has been extensively studied is the east coast of the United States. Holocene SL rise has flooded the coastal region creating estuaries where much of the sediment supplied by the rivers is trapped. Therefore the contribution of Recent sediments to the shelf is minimal, and storms and currents are able to erode older sedimentary deposits (Swift *et al.*, 1972). Shoreline indicators in these older sediments such as shells of the oyster *Crassostrea virginica*, may have been relocated, and this reduces their usefulness in SL reconstructions (MacIntyre *et al.*, 1978).

In an attempt to determine what the net effect of the redistribution of shelf sediments on the east coast of the United States may be, Pilkey & Field (1972) investigated, among other things, the distribution of ooids off the coast of central Florida. They found that ooids in beach sands near Cape Kennedy (Canaveral) likely were derived from submarine Pleistocene outcrops 10-20 km offshore. No coastal source of ooids was present in this region. This, they argued, was evidence for the landward transport of relict shelf sediments, namely ooids, in this area.

These findings and those mentioned above cast a shadow of uncertainty on the use of ooids as SLIs once advocated by Milliman & Emery (1968) and Pilkey *et al.* (1966).

THE DIAGENESIS OF OOIDS

During and after the deposition of an ooid there are at least four additional influences on the eventual appearance of the grain and its host rock: (*a*) micritization of the grain by boring algae; (*b*) a change in mineralogy; (*c*) the breakage of the grain, especially through overburden pressure; and (*d*) the infilling of pore space by cements. The first three processes may complicate the correct identification of the grain as an ooid, precluding its use as a SLI. The fourth may in fact aid in SL determinations.

a) Microborings. The original fabric of an ooid may be partially or totally obscured by the action of endolithic algae. These plants bore into the surface of the grain, producing cavities which later are infilled with microcrystalline aragonite or Mg-calcite (Bathurst, 1975; Gaffey, 1983).

b) Mineralogy. Ancient ooids are commonly composed of radially oriented calcite crystals, while their most widely-used modern analogs from the Bahamas have a tangential aragonite composition. To explain this discrepancy, researchers have proposed complicated diagenetic pathways by which the mineralogy and fabric of these grains were modified. Sandberg (1975) undermined these arguments when he determined that the Great Salt Lake oolite, long considered to contain prime examples of original radial aragonite ooids altered to radial calcite, are in fact relatively unaltered, radial aragonite ooids. He now recognizes an oscillating Phanerozoic trend in non-skeletal carbonate mineralogy which appears to be related to fluctuations in eustatic SL and climate (Sandberg, 1983). At certain times in the geologic past, inorganic calcite precipitation was favored over that of aragonite, and radial calcite ooids were produced.

The primary radial ooids recently recognized in several modern environments - the Gulf of Aqaba, Red Sea (Friedman *et al.*, 1973); protected lagoons of the Persian Gulf (Loreau & Purser, 1973); and Shark Bay, Great Barrier Reef, Australia (Davies & Martin, 1976) - may represent ooids more closely resembling ancient ooids at their time of deposition.

c) Breakage. Broken ooids have been described from various localities and geologic ages (see Table 3 in Halley, 1977) although their occurrence is considered rare. The tendency for post-depositional breakage has been attributed to a nucleus more soluble than its cortex (Wilkinson & Landing, 1978) while syndepositional breakage may occur selectively in ooids with a weaker radial fabric (Halley, 1977). Wilkinson & Landing (1978) noted that Jurassic ooids from the Twin Creek Formation in Wyoming, which had molluscan aragonitic nuclei, were more often fractured than those that had crinoidal, foraminiferal or molluscan calcitic debris as nuclei. The aragonite nuclei had dissolved away during early diagenesis, leaving an 'eggshell' structure that was easily crushed by the weight of overburden. The cortexes of both types of ooids were originally calcitic and thus less susceptible to dissolution.

Halley (1977) determined that radial ooids from the Great Salt Lake, Utah, were much more easily fractured than tangential Bahamian ooids of similar size and shape. He proposed that the radial structure is inherently weaker, so that a high proportion of broken ooids in a sample may be indicative of an environment of formation conducive to radial growth.

d) Cementation. The high porosity of ooid accumulations usually is reduced during diagenesis by intergranular calcite cement. Differences in cement type can be indicative of the diagenetic environment during cementation. Especially pertinent to the discussion of SLIs is the ability to distinguish vadose (above the water table) from phreatic (below the water table) cements[1]. For low-relief islands, the water table is a rough indicator of SL. Halley & Harris (1979) described these cements from the South Joulters Cay, Bahamas oolite. They found that while the degree of contemporary cementation was equivalent above and below the water table, the style of cementation was markedly different. Vadose cements exhibited uneven blocky-calcite meniscus cements primarily at grain contacts. Upper phreatic cements were characterized by an even rim of calcite rhombs perpendicular to and surrounding ooid grains. The lower phreatic zone was not cemented, probably because of the influence of marine waters. Because the transition between vadose and upper phreatic cements was abrupt, the authors were able to locate the water table within one meter.

Pre-Holocene diagenetic environments are more difficult to delineate because of the blurring of the transition zone through SL fluctuations (Harris, pers. comm.). However, cements might be used to indicate how fast SL changed. A sharp drop could leave a water table line 'suspended' in a sedimentary unit (A.C. Neumann, pers. comm.).

Cement stratigraphy is a useful tool in determining the depositional and burial history of an oolite. For example, acicular aragonite or micritic Mg-calcite cements followed by blocky or drusy calcite cements would suggest that submarine stabilization of the sediment occurred before the onset of meteoric diagenesis. See Bricker (1971) for a collection of papers on carbonate cements.

DATING OOIDS

Pleistocene and younger ooids can be dated using modified ^{14}C techniques. Standard methods using bulk samples are unable to distinguish the range of ages represented by old nuclei, younger cortexes, and rejuvenated grains. With this method, Hine (1977, Fig. 24) found an apparent 'stratigraphic inversion' in ooids collected from the Bahamas; 'older' ooids appeared to overlie the younger ones. Martin & Ginsburg (1965) investigated the range of ages in the modern Bahamian ooid sands. Bulk analyses yielded an age of 900 years. Nuclei were much older, dating back 2000 years, while their cortexes were dated as contemporary.

The following is an outline of the method of Martin & Ginsburg (1965) for dating Quaternary ooids presented at the Sixth International Conference on Radiocarbon and Tritium Dating.

The procedure for dating Quaternary non-cemented ooids is as follows:

[1]See Coudray & Montaggioni, this volume

1) Select a narrow size range to insure more even dissolution of particles (Martin & Ginsburg used .75 and 1.5 ϕ).

2) Collect ten times as much $CaCO_3$ as normally used in radiocarbon age determinations.

3) Employ the usual caution in preventing contamination by modern carbon.

4) Separate ooids from skeletal material by rolling the particles down a vibrating, inclined ramp.

5) A bulk age for the entire sample can be obtained by collecting all of the carbon dioxide produced during the complete dissolution of the sample (Martin & Ginsburg used 20 grams of sample).

6) Incremental collection of carbon dioxide during dissolution of the sample allows one to determine ages for successively older layers of the cortex and nucleus.

Incremental analysis is preferable to whole ooid analysis because much older detrital carbonate grains may serve as nuclei for younger ooids. However, there are also several analytical and environmental complications for even this preferred method. One is the non-uniformity of ooid sizes, nucleus sizes, and cortical thicknesses. Smaller grains, and those with large nuclei or thin cortexes, may add old nucleus-derived carbon to the younger cortical carbon released during the dissolution of the majority of the grains. Another problem is the varying degree of microboring of ooids. These cavities are filled with sediment or cement which would have a different age. A final complication arises from the fact that ooid-producing environments, by necessity, have the ability to retain grains for great lengths of time. Ooids may spend over 95% of their time buried within an active bedform (Bathurst, 1975). Thus, a given ooid may have just precipitated its first coating, or it may have done so at a much earlier time.

For these reasons, the accurate position of Quaternary SL, based on the age of oolites, is a difficult task without even considering the effects of transport.

RECOMMENDATIONS

The discussions above have considered the uncertainties of ooid formation, diagenesis, transport, reworking, and dating. Ooids are forming in both quiet-water and turbulent environments. The differences in these two quite dissimilar environments are manifested in their fabrics: radial ooids are found in quiet-water environments while tangential ooids form in turbulent environments. It is not known whether ancient ooids, now mostly radial and calcitic, have retained their original fabric and mineralogy or if they have undergone diagenetic transformation from an original tangential aragonite crystal arrangement. The problem of the transport of ooids from their sites of formation to sites of accumulation is especially acute if we suppose that ancient ooids were primarily of the Bahamian type.

What then is the utility of ooids in SL reconstruction? Do the unknowns and complications described above render ooids useless to this task? Certainly, to tacitly assume that any ooid-bearing rock was deposited within perhaps two meters of SL would be incorrect. But what of the often-encountered ooid grainstone? Does this texturally-mature, predominately oolitic rock indicate an extremely shallow environment of accumulation?

A common conclusion is that this type of rock must have been deposited in very shallow water near to where it formed, because if it were transported any distance it would have been diluted with skeletal grains, non-carbonate grains, deep-water skeletal indicators, or mud. This conclusion may be valid if it is supported by an analysis of adjacent facies of the type carried out by Carozzi (1971). He showed that for the Lower Carboniferous St. Genevieve Limestone of the Illinois Basin there is a marked decrease in the contribution of ooids and a concomitant increase in mud content for the facies adjacent to the ooid grainstone. The lateral association of facies for this unit bears strong similarities to that of the ooid-forming areas of the Bahama Banks, and his assignment of a shallow depth of accumulation to the ooid grainstone is probably correct.

However, not all oolitic grainstones accumulated in shallow water. Bosellini *et al.* (1981) have described an oolitic grainstone (the Dogger Vajont Limestone) from the Venetian Alps which was deposited in deep water. The authors drew an analogy with modern sedimentation in the Tongue of the Ocean, Bahamas (see Schlager & Chermak, 1979), and suggested that the ooids formed on the bank top, cascaded off the side of the platform and accumulated in deep-water fans.

Ooids provide a wealth of information about the environments of origin, deposition, and diagenesis of a particular sedimentary deposit. Unfortunately, the complicated story they present often makes them unreliable indicators of SL when used alone. Careful facies analysis based on correct analogues from the Recent and the coincidence of other SLIs may strongly suggest that a particular oolite may have accumulated at or near SL. For such an interpretation, however, it is the supporting evidence rather than the presence of the oolite itself that is most conclusive. Marine ooids may be indicators of shallow water but are not precise indicators of SL even under conditions where little or no transport has occurred.

ACKNOWLEDGEMENTS

We wish to express our gratitude to R.G. Bathurst, P.M. Harris, W.L. Miller, H.T. Mullins and A.C. Neumann for critically reviewing the manuscript, and P.A. Sandberg and E.A. Shinn for providing helpful comments and photomicrographs.

REFERENCES

Ball, M.M., 1967. Carbonate sand bodies of Florida and the Bahamas. *Journal of Sedimentary Petrology*, 37, 556-591

Bathurst, R.G.C., 1967. Oolitic films on low energy carbonate sand grains, Bimini Lagoon, Bahamas. *Marine Geology*, 5, 89-109

Bathurst, R.G.C., 1975. *Carbonate sediments and their diagenesis*. (Elsevier Scientific Publishing Company, Amsterdam)

Berner, R.A., 1971. *Principles of chemical sedimentology*. (McGraw-Hill, New York)

Bosellini, A., Masetti, D. & Sarti, M. 1981. A Jurassic 'Tongue of the Ocean' infilled with oolitic sands: the Belluno Trough, Venetian Alps, Italy. *Marine Geology*, 44, 59-95

Bricker, O.P. (ed), 1971. *Carbonate cements*. Johns Hopkins Press, Baltimore)

Buchanan, H., 1970. *Environmental stratigraphy of Holocene carbonate sediments near Frazers Hog Cay, British West Indies*. Ph.D. thesis (Columbia University)

Bucher, W.H., 1918. On oolites and spherulites. *Journal of Geology*, 26, 593-609

Carozzi, A.V., 1971. Interpretation of ancient oolitic environments and petroleum exploration. *Journal of the Geological Society of the Philippines*, 25, 9-32

Cayeux, L., 1935. *Les roches sédimentaires de France; roches carbonatées*. (Masson, Paris)

Davies, P.J. & Martin, K., 1976. Radial aragonite ooids, Lizard Island, Great Barrier Reef, Queensland, Australia. *Geology*, 4, 120-122

Davies, P.J., Bubela, B. & Ferguson, J. 1978. The formation of ooids. *Sedimentology*, 25, 703-730

Deelman, J.C., 1978. Experimental ooids and grapestones: carbonate aggregates and their origin. *Journal of Sedimentary Petrology*, 48, 503-512

Donahue, J.D., 1965. Laboratory growth of pisolite grains. *Journal of Sedimentary Petrology*, 35, 251-256

Donahue, J.D., 1969. Genesis of oolite and pisolite grains: an energy index. *Journal of Sedimentary Petrology*, 39, 1399-1411

Dravis, J., 1979. Rapid and widespread generation of Recent oolitic hardgrounds on a high energy Bahamian platform, Eleuthera Bank, Bahamas. *Journal of Sedimentary Petrology*, 49, 195-208

Eardley, A.J., 1938. Sediments of the Great Salt Lake, Utah. *American Association of Petroleum Geologists Bulletin*, 22, 1305-1411

Fabricius, F.H., Berdau, D. & Munnich, K.-O. 1970. Early Holocene ooids in modern littoral sands reworked from a coastal terrace, south Tunisia. *Science*, 169, 757-760

Friedman, G.M., Amiel, A.J., Braun, M. & Miller, D.S., 1973. Generation of carbonate particles and laminites in algal mats - example from a sea-marginal hypersaline pool, Gulf of Aqaba, Red Sea. *American Association of Petroleum Geologists Bulletin*, 57, 541-557

Gaffey, S.J., 1983. Formation and infilling of pits in marine ooid surfaces. *Journal of Sedimentary Petrology*, 53, 193-208

Garber, R.A., Friedman, G.M. & Nissenbaum, A. 1981. Concentric aragonitic ooids from the Dead Sea. *Journal of Sedimentary Petrology*, 51, 455-458

Halley, R.B., 1977. Ooid fabric and fracture in the Great Salt Lake and the geologic record. *Journal of Sedimentary Petrology*, 47, 1099-1120

Halley, R.B. & Harris, P.M., 1979. Fresh-water cementation of a 1,000-year old oolite. *Journal of Sedimentary Petrology*, 3, 969-987

Harris, P.M., 1979. Facies anatomy and diagenesis of a Bahamian ooid shoal. *Sedimenta VII*. (University of Miami)

Hine, A.C., 1977. Lily Bank Bahamas: history of an active oolite sand shoal. *Journal of Sedimentary Petrology*, 47, 1554-1581

Hine, A.C., Wilber, R.J., Bane, J.M., Neumann, A.C. & Lorenson, K.R. 1981. Offbank transport of carbonate sands along open leeward bank margins: Northern Bahamas. *Marine Geology*, 42, 327-348

Hine, A.C., 1983. Relict sand bodies and bedforms of the northern Bahamas: evidence of extensive early Holocene sand transport. In: *Coated Grains*, ed. T. Peryt, (Springer-Verlag, Heidenberg)

Hine, A.C. & Steinmetz, J.C., 1984. Cay Sal Bank Bahamas - a rapidly submerged and partially drowned carbonate platform. *Marine Geology*, 59, 135-164

Illing, L.V., 1954. Bahamian calcareous sands. *Bulletin of the American Association of Petroleum Geologists*, 38, 1-95

Jenkyns, H.C., 1972. Pelagic 'oolites' from the Tethyan Jurassic. *Journal of Geology*, 80, 21-33

Kahle, C.F., 1974. Ooids from the Great Salt Lake, Utah, as an analogue for the genesis and diagenesis of ooids in marine limestones. *Journal of Sedimentary Petrology*, 44, 30-39

Kumar, N., Damuth, J.E. & Gorini, M.A., 1977. Relict magnesian calcite oolite and subsidence of the Amazon shelf: discussion. *Sedimentology,* 24, 143-148

Land, L.S., Behrens, E.W. & Frishman, S.A., 1979. The ooids of Baffin Bay, Texas. *Journal of Sedimentary Petrology,* 49, 1269-1278

Loreau, J.-P., 1982. *Sédiments aragonitiques et leur genèse.* (Memoires du Muséum National d'Histoire Naturelle XLVII, Paris)

Loreau, J.-P. & Purser, G.H., 1973. Distribution and ultrastructure of Holocene ooids in the Persian Gulf. In: *The Persian Gulf,* ed. G.H. Purser, (Springer-Verlag, New York), 279-328

Lowe, D.R. & Knauth, L.P., 1978. The oldest marine carbonate ooids reinterpreted as volcanic accretionary lapilli, Overwacht Group, South Africa. *Journal of Sedimentary Petrology,* 48, 709-722

MacIntyre, I.G., Pilkey, O.H. & Stuckenrath, R., 1978. Relict oysters on the United States Atlantic continental shelf: a reconsideration of their usefulness in understanding late Quaternary sea-level history. *Geological Society of America Bulletin,* 89, 277-282

Marshall, L.F. & Davies, P.J., 1975. High-magnesium calcite ooids from the Great Barrier Reef. *Journal of Sedimentary Petrology,* 45, 285-291

Martin, E.L. & Ginsburg, R.N., 1965. Radiocarbon ages of oolitic sands on Great Bahama Bank. *Proceedings of the Sixth International Conference on Radiocarbon and Tritium Dating.* (U.S. Atomic Energy Commission Report, conference 650652), 705-719

McIlreath, I.A. & James, N.P., 1979. Facies models 12, carbonate slopes. In: *Facies Models,* ed. R.G. Walker, (Geoscience Canada Reprint Series 1), 133-143

Milliman, J.D., 1977. Relict magnesian calcite oolite and subsidence of the Amazon shelf: reply. *Sedimentology,* 24, 149-151

Milliman, J.D. & Emery, K.O., 1968. Sea levels during the last 35,000 years. *Science,* 162, 1121-1123

Milliman, J.D. & Barretto, H.T., 1975. Relict magnesian calcite oolite and subsidence of the Amazon shelf. *Sedimentology,* 22, 137-145

Mitterer, R.M., 1971. Influence of natural organic matter on $CaCO_3$ precipitation. In: *Carbonate Cements,* ed. O.P. Bricker, (Johns Hopkins, Baltimore)

Mitterer, R.M., 1972. Biogeochemistry of aragonite mud and oolites. *Geochimica et Cosmochimica Acta,* 36, 1407-1422

Monaghan, P.H. & Lytle, M.L., 1956. The origin of calcareous ooliths. *Journal of Sedimentary Petrology,* 26, 111-118

Mullins, H.T. & Neumann, A.C., 1979. Deep carbonate bank margin structure and sedimentation in the northern Bahamas. In: *Geology of Continental Slopes,* eds L.J. Doyle & O.H. Pilkey, (Society of Economic Paleontologists and Mineralogists Special Publication No.27, Tulsa, Oklahoma), 165-192

Mullins, H.T., Neumann, A.C., Wilber, R.J., Hine, A.C. & Chinburg, S.J., 1980. Carbonate sediment drifts in the northern straits of Florida. *American Association of Petroleum Geologists Bulletin,* 64, 1701-1717

Newell, N.D., Purdy, E.G. & Imbrie, J., 1960. Bahamian oolitic sand. *Journal of Geology,* 68, 481-497

Pavlidis, Y.A., Ionin, A.S., Igantov, Y.I., Avello-Suarez, O.A. & Luis-Riera, M., 1972. Oolite sedimentation conditions in the shallow water of a tropical sea, island of Cuba, Golfo de Batabaño. *Oceanology,* 12, 705-713

Pilkey, O.H., Schnitker, D. & Pevear, D.R., 1966. Oolites on the Georgia continental shelf edge. *Journal of Sedimentary Petrology,* 36, 462-467

Pilkey, O.H. & Field, M.E., 1972. Onshore transportation of continental shelf sediment: Atlantic Southeastern United States. In: *Shelf Sediment Transport,* eds D.J.P. Swift, D.B. Duane & O.H. Pilkey, (Dowden, Hutchinson & Ross, Pennsylvania), 492-498

Purdy, E.G., 1961. Bahamian oolite shoals. In: *Geometry of sandstone bodies,* eds. J.A. Peterson & J.C. Osmond, (American Association of Petroleum Geologists, Tulsa), 53-62

Reimer, T.O., 1975. Paleogeographic significance of the oldest known oolite pebbles in the Archean Swaziland supergroup (South Africa). *Sedimentary Geology,* 14, 123-133

Rothpletz, A., 1982. On the formation of oolite. *American Geologist,* 10, 279-282

Sandberg, P.A., 1975. New interpretation of Great Salt Lake ooids and of ancient non-skeletal carbonate mineralogy. *Sedimentology,* 22, 497-537

Sandberg, P.A., 1983. An oscillating trend in Phanerozoic carbonate mineralogy. *Nature,* 305, 5929, 19-22

Schlager, W. & Chermak, A., 1979. Sediment facies of platform basin transition, Tongue of the Ocean, Bahamas. In: *Geology of Continental Slopes,* eds. L.J. Doyle & O.H. Pilkey, (Society of Economic Paleontologists and Mineralogists Special Publication No.27, Tulsa, Oklahoma), 193-208

Siesser, W.G., 1973. Diagenetically formed ooids and intraclasts in South African calcretes. *Sedimentology,* 20, 539-551

Simone, L., 1981. Ooids: a review. *Earth-Science Reviews*, 16, 319-355

Suess, E. & Futterer, D., 1972. Aragonitic ooids: Experimental precipitation from seawater in the presence of humic acid. *Sedimentology*, 19, 129-139

Swift, D.J.P., Kofoed, J.W., Saulsbury, F.P. & Sears, P., 1972. Holocene evolution of the shelf surface, central and south Atlantic shelf of North America. In: *Shelf Sediment Transport*, eds. D.J.P. Swift, D.B. Duane & O.H. Pilkey, (Dowden, Hutchinson & Ross, Pennsylvania), 499-574

Teichert, C., 1970. Oolite, oolith, ooid: discussion. *American Association of Petroleum Geologists Bulletin*. 54, 1748 - 1749

Vaughn, T.W., 1914. Preliminary remarks on the geology of the Bahamas, with special reference to the origins of the Bahamian and Floridian oolites. *Papers from the Tortugas Lab.* Carnegie Institute of Washington Publication, 183, 47-54

Wilkinson, B.W. & Landing, E., 1978. 'Eggshell diagenesis' and primary radial fabric in ooids. *Journal of Sedimentary Petrology*, 48, 1129-1138

Wilkinson, B.W., Pope, B.N. & Owen, R.M., 1980. Nearshore ooid formation in a modern temperate region marl lake. *Journal of Geology*, 88, 697-704

Winland, H.D. & Mathews, R.K., 1974. Origin and significance of grapestone, Bahama Islands. *Journal of Sedimentary Petrology*, 44, 921-927

Added in proof:
Coudray, J. & Montaggioni, L., (this volume). The diagenetic products of marine carbonates as sea-level indicators

8

Corals and reefs as indicators of paleo-sea levels with special reference to the Great Barrier Reef

D. HOPLEY

ABSTRACT

Coral reefs are built by and contain a large range of plants and animals, many of which have a specific relationship to sea level. However, the main framework builders, scleractinian corals, have a wide range of depths down to about 40 m in which they can grow. Their upper limit, however, is determined by emersion and is normally close to mean low water springs. Where reef tops are moated behind shingle ramparts or algal ridges this level may be raised, even above mean sea level. The micro-atoll colonial morphology is indicative of a moated situation. Corals also contain internal evidence of their growth environment which may be determined by careful analysis. Similarly, with careful choice and preparation of samples, corals provide excellent material for a variety of radiometric dating techniques.

INTRODUCTION

The importance of Quaternary sea-level (SL) changes to the formation of coral reefs has long been a central theme of coral-reef study. Belt (1874), Upham (1878) and Penck (1894, 1896) were amongst the first to realise that the major SL oscillations would cause migration of the vertical zone in which coral growth could occur and Daly (1910, 1915, 1916, 1917, 1919, 1934, 1948) formulated a model of atoll development based on these oscillations. More recently suites of emerged reef terraces have been used to reconstruct Quaternary SL histories (eg. Bloom *et al.*, 1974). However, although a gross relationship between SL and reef growth is quite clear, the use of coral reefs to establish paleo-SLs has a number of reservations and difficulties. This paper examines these problems and concludes that, with acknowledgement of these reservations, fossil reefs can provide excellent dateable material for the establishment of former SLs.

A coral reef is a complex organo-genetic structure dependent upon a wide range of calcium carbonate secreting plants and animals each of which may have contrasting environmental requirements. As it is largely from skeletal material that the reef is built, samples suitable for several radiometric dating techniques are readily available, although, as indicated below, care must be taken because of the complex chemistry of the reef system. Superimposed upon the net results of biological construction are purely physical and chemical processes which also have vertical ranges related to SL within which they can operate. Corals may thus form the initial reef framework but many other contributors construct, maintain or modify the total structure. Because of the complexity of the reef several other chapters in this volume should be studied for specific details. Mainly because of light requirements, reef building is limited to depths above 40 m below which the rate of calcium carbonate deposition may be at least matched by chemical and biological degradation processes. Reef maintenance therefore requires that a coral reef remains within the upper photic zone. Although the majority of reefs appear to have grown at rates below that at which the Holocene SL rose, the presence of older reefal foundations of presumably Pleistocene age (which on the Great Barrier Reef lie at depths of between 5 and 25 m below modern reef-flat levels) has meant that few reefs have been outstripped to the extent that they lie below the photic zone by the rate of SL rise (see Hopley, 1982, for discussion).

THE RELATIONSHIP OF REEF BUILDING CORALS TO SEA LEVEL

1. Coral ecology and calcification

The Scleractinia (Wells, 1956), together with other anthozoans such as the Stoloniferan *Tubipora*, the Coenothecalian *Heliopora* (Bayer, 1956) and the hydrozoan *Millepora* (Boschma, 1956), are the major contributors to reef framework. They produce calcareous external skeletons largely in colonial form which are constructed of aragonite, except for the majority of octocorals which are composed of magnesium calcite. The polyp occupies the outer part of the coral skeleton and contains, in a symbiotic relationship, numerous unicellular algae, the zooxanthellae. These perform important functions related to nutrient conservation, waste removal and primary production. However, their most important function is a fundamental role in the inorganic growth of the coral skeleton, aiding in the calcification by removing carbon dioxide. Precipitation of the skeleton is from a supersaturated solution of calcium carbonate in a cavity between skeleton and tissue. As the zooxanthellae aid calcification by removing carbon dioxide, light increases the supersaturation of the solutions beneath the polyp by promoting the photosynthetic processes of these small plants. Thus whilst temperature and oxygen levels in the upper 50 m or so of water depth show little variation, decrease in illumination is so important that rates of calcification are significantly reduced (eg. Baker & Weber, 1975). Diurnal growth bands in the coral skeletons result from the link between photosynthesis and calcification (Wells, 1963, 1969) and these are superimposed on growth banding grouped into lunar cycles (Jell, 1974). Most distinctive, however, are the high- and low-density bands laid down seasonally and detected by x-radiographic techniques (Buddemeier, 1978; Buddemeier & Kinzie, 1976). This distinctive banding of one high- and one low-density band per year, the low-density band being by far the wider and laid down during optimal growth conditions, allows the application of dendrochronological techniques to corals. Annual linear radial growth rates reported by these techniques range between 4 and 20 mm/yr for massive species, 10-12 mm being about the average; and about 100 mm/yr extension for branching species such as *Acropora* (Buddemeier & Kinzie, 1976).

2. Zonation and colonial morphology

In spite of the rigid appearance of the calcareous skeletons, the Scleractinia are highly plastic organisms with a considerable range of variations in their skeletal characteristics (Veron & Pichon, 1976). In the Caribbean, where there are probably no more than 70 species (Glynn, 1974) a distinctive species zonation exists which allows the identification of fossil relationship between older reef tracts and paleo-SLs (eg. Mesolella, 1967; Mesolella *et al.*, 1970). In the Indo-Pacific, with not less than 500 species representing about 80 genera (Wells, 1956) and the Great Barrier Reef specifically, with at least 350 species representing about 60 genera (Wells, 1955), no clear species relationship has been demonstrated. This has made reconstruction of paleo-reef geometries based on species alone extremely tentative. More reliable may be the morphologies of coral colonies that have adaptive growth strategies to minimise hydraulic stress (Graus *et al.*, 1977) and respond to light (Graus & Macintyre, 1976) or combinations of environmental stresses (Chappell, 1980). On the Great Barrier Reef, Done (1977), Veron & Done (1979) and Done & Pichon (1981) have shown that prevailing hydrodynamic conditions are prime influences on zonation with crustose and digitate forms dominating in high-energy locations (windward reef crest), branching forms in moderate-energy locations and massive forms where energy is low. Colonial morphology may thus be a guide to the paleo-reef geometry and thus an indicator of the relationship with an older SL (Fig. 1).

3. In situ position

Unfortunately corals do not always remain in this growth position but during storms may be removed by wave action and deposited as rubble in sediment accumulating zones. Very large proportions of coral found in reef cores appears

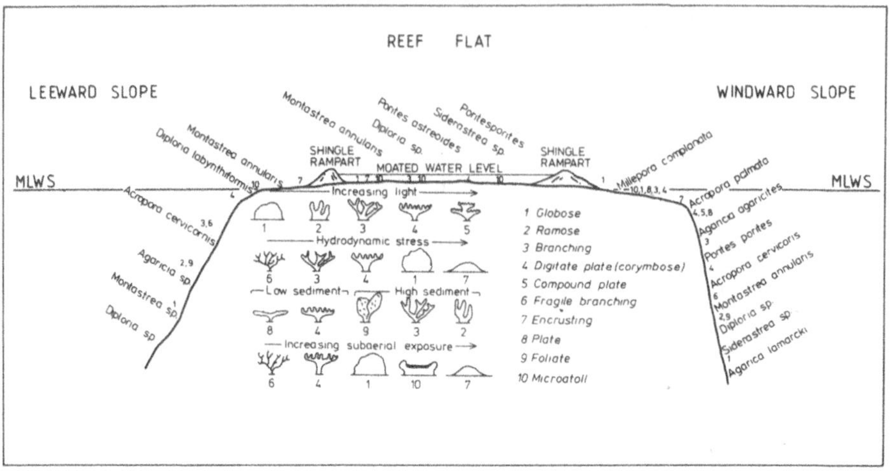

Figure 1. Variation of colonial morphology in response to environmental stress (after Chappell, 1980) and actual species variation with reef location for typical Caribbean reef. Note that some forms and species occur in more than one area of the reef. In particular there is an overlap between the reef flat and the deeper fore-reef.

to be detrital. Experience of the author in drilling through the Holocene veneer of reefs in northern Queensland suggests that less than 5% of the reef structure is formed from *in situ* coral. In exposed reefs the determination of an *in situ* position is relatively simple as the majority of branching corals will be branching upwards whilst head corals, particularly in section, will show radiation of corallite extension from a basal point of origin. In reef cores, however, the problem is more difficult and it is not always possible to determine an *in situ* position although again, the direction of branching and corallite extension may act as a guide.

4. The upper limit of open-water coral growth

The majority of coral species can withstand short periods of emergence on low tides, but after about 3 hours of exposure in high day-time temperatures the mucus, with which the polyps protect themselves from desiccation, will tend to dry and simultaneously the polyps may commence to expel the symbiotic zooxanthellae (Yonge, 1963). As the algae are responsible for much of the pigmentation in corals, as well as aiding in calcification, the net result is an initial paling of the colour of the living coral on exposed upper surfaces and subsequently a cessation of further upward growth. The colony may continue to grow horizontally, forming a micro-atoll morphology with the near horizontal upper surface dying and frequently colonised by crustose coralline algae (see Scoffin & Stoddart, 1978; Stoddart &

Figure 2. The reef front of Gable Reef, central Great Barrier Reef, with low corymbose forms of *Acropora* sp. typical of the high-energy reef front close to the maximum level of coral growth (MLWS).

Scoffin, 1979). In general terms this uppermost level of living coral growth on the Great Barrier Reef approximates mean low water springs (MLWS) whatever the tidal range (2 m to 6 m). A variation of up to 25 cm may occur due to year to year variation in tidal behaviour, varying exposures to wave action, which may slow down the desiccation process, or varying tolerances to exposure of different species. In general on the Great Barrier Reef *Porites* sp. appear most tolerant of exposures but some Favids may also grow to levels slightly higher than MLWS. The end result is a reef surface composed of living corals growing to within (+ or -) 25 cm of MLWS (Fig. 2). However, this is not the absolute height of reef-flat development due to moating effects (see below) and the throwing up of debris banks in storms which may be subsequently colonised by crustose coralline algae (see Adey, this volume). These have the ability to cement coral debris to levels as high as mean sea level (MSL). The resultant reef rock may be difficult to distinguish from a reef rock composed of reef-crest corals.

5. The problem of moating

In some circumstances, water may be held back at low tide and not drain to the same level as in the open ocean. Moating may occur at any intertidal level and some lower level moated pools may come into being only on the lowest spring tides. Smaller moated pools form in depressions within the reef flat or, on continental islands, within crevices in shore platforms. Much larger portions of reef

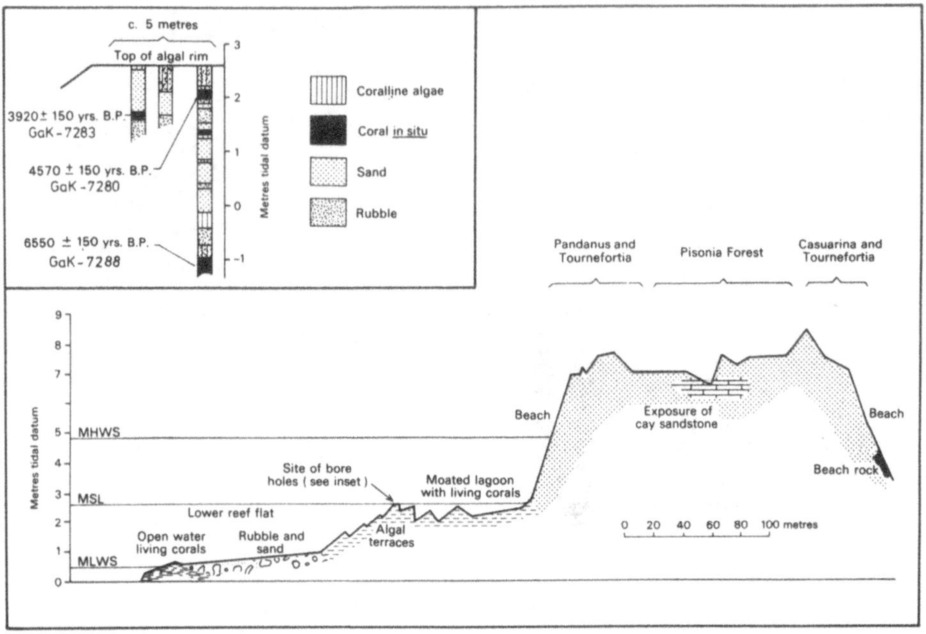

Figure 3. Redbill Reef, central Great Barrier Reef. The section is through the high algal rim near the cay (Bushy Island). Tidal range is about 5 m and the algal rim moats the lagoon to MSL, a level to which the lagoon corals grow and 2 m higher than open-water corals. Drilling into the rim established that it has a base of *in situ* corals of mid-Holocene age, the location of which is unlikely to have been moated at time of growth. Thus an emergence of 1 to 1.5 m is suggested by the corals beneath the algal rim.

flat are moated by a variety of damming agents. Most commonly these are rubble banks or ramparts of coral shingle deposited by storm waves around the reef-flat margins. Although small gaps may occur in the dam, flow of water off the reef flat is retarded and the subsequent lowering of the water level on the reef flat may be very slow. Under such circumstances the water level on the reef flat may be at least 0.5 m higher than the open sea at MLWS. The situation is very common on the low wooded island reefs of the northern Great Barrier Reef. Given sufficient time the efficiency of the dam may be enhanced by basal cementation of the shingle banks and/or filling of interstitial spaces by compaction and finer sediments. Water levels then drop to the level of the dam and show no further fall on the low tide. this too is seen on many low wooded island reefs with levels as high as MHWN (neap) being maintained. In the higher tidal range area of the central Great Barrier Reef, high algal rims commonly form dams around pools varying in size from a few m^2 to several km^2. Typical is Redbill Reef (Fig. 3) which is completely encircled by a massive algal rim

Figure 4. *Porites* sp. micro-atolls on Three Isles Reef, northern Great Barrier Reef, coalescing and effectively raising the level of the reef flat. The moat dam is formed by shingle ramparts.

terraced on its seaward side. MLWS here is 0.5 m and MHWS 4.8 m but the algal rim which rises to over 2.6 m, maintains low water levels within 1.6 cm of 2.6 m for up to 5 hours on each low tide. This is only 8 cm below MSL and on spring tides, up to 2.3 m above the adjacent open-water level. Ludington (1979) has described similar moating within the reef crest of One Tree Reef where lagoon-water levels are also maintained near MSL for 5 to 6 hours of each tide. On the majority of lagoonal reefs the water is impounded to some extent but normally the level of the rim is much lower and less effective than the massive and relatively impervious algal rims of the central province. A more unusual form of moat on the Great Barrier Reef is formed by emerged platform rocks and associated emerged corals on the low wooded islands. Nymph Island and Turtle II Island near Lizard Island are examples. Both are low wooded islands containing large central lagoons with narrow exits through which water drains at low tide, but not below the sill level which in both examples is between MHWN and MHWS.

The most important aspect of moating is that behind the dam the tide consistently falls to the same level on every tide, unlike in the open-water situation where variations between spring and neap tides may produce low water levels more than 2 m apart on the central Great Barrier Reef. The upper level of coral growth is thus firmly fixed and the result is the typical micro-atoll form (Fig. 4). In their simplest form micro-atolls are flattened, disc-shaped coral colonies, consisting of living vertical sides and dead centres often covered by crustose coralline algae. The majority are formed by the massive rounded corals such as *Porites* or *Goniopora*, but Wells (1957) has described large less compact micro-atolls formed by branching corals such as *Acropora palifera*. On the Great Barrier Reef branching corals such as *Porites andrewsi*; *Helipora coerulae* also adopt the micro-atoll morphology.

Micro-atolls develop from evenly growing hemispherical coral colonies whose upward growth is interrupted by a specific level of subaerial exposure (Scoffin & Stoddart, 1978; Stoddart & Scoffin, 1979). Upward growth is limited to no more than 5 cm above moated tide levels, a height at which complete desiccation of the upward part of the colony is prevented by the lapping of small waves, protective mucus coatings and by capillary action. The surface of a classical micro-atoll assumes a slightly raised rim above a flat or dished centre which many believe to be an erosional feature (eg. Fairbridge, 1968). As the micro-atoll develops the leading edge of the rim grows outwards whilst the trailing edge dies. Under-cutting of the inner edge may result from both solution and bioerosion aided by filamentous algae and bacteria. After the erosional phase, an algal succession may take place and protective calcareous algae prevent further rapid attack, allowing a sub-horizontal inner surface to form whilst the original width of the rim is maintained (Fig. 5).

This hypothesis assumes that the moated water level is constant throughout the life of the micro-atoll. Where this is not so, distinctive variations in surface morphology occur. A rise in the moated level will cause the living rim to grow upwards producing, in time, a distinctive ring with deeper inner pool. If the raised level is sufficiently high the inner pool may provide a suitable environment with sufficient circulation and depth to allow the inner side of the rim to grow inwards (Fig. 5). A lowering of moated water level will cause the reverse effect and given that the new level provides a pool of depth greater than a minimum threshold value, a new lower surface will develop in the micro-atoll producing a 'top hat' morphology (Fig. 5). A sudden drop in level may be due to breaching of a former shingle barrier during a tropical cyclone. A gradual lowering of level may be the result of a slower widening or deepening of trickle zones in the ramparts. A sudden rise may be a response to a new moating rampart system being thrown up during a cyclonic storm, whilst a slower rise may be associated with upgrading algal rims, or slow cementation of the shingle ramparts. Significantly, micro-atolls displaying rapid changes of level are associated with moated conditions provided by *shingle ramparts*. Such changes are not seen in pools dammed by the more massive algal rims. Similar modification to micro-atoll forms takes place when it is the micro-atoll itself which is moved, rather than the moat level. On many reef flats micro-atolls disturbed by cyclonic seas may be seen with new horizontal rims growing at an angle to the older central tilted disc.

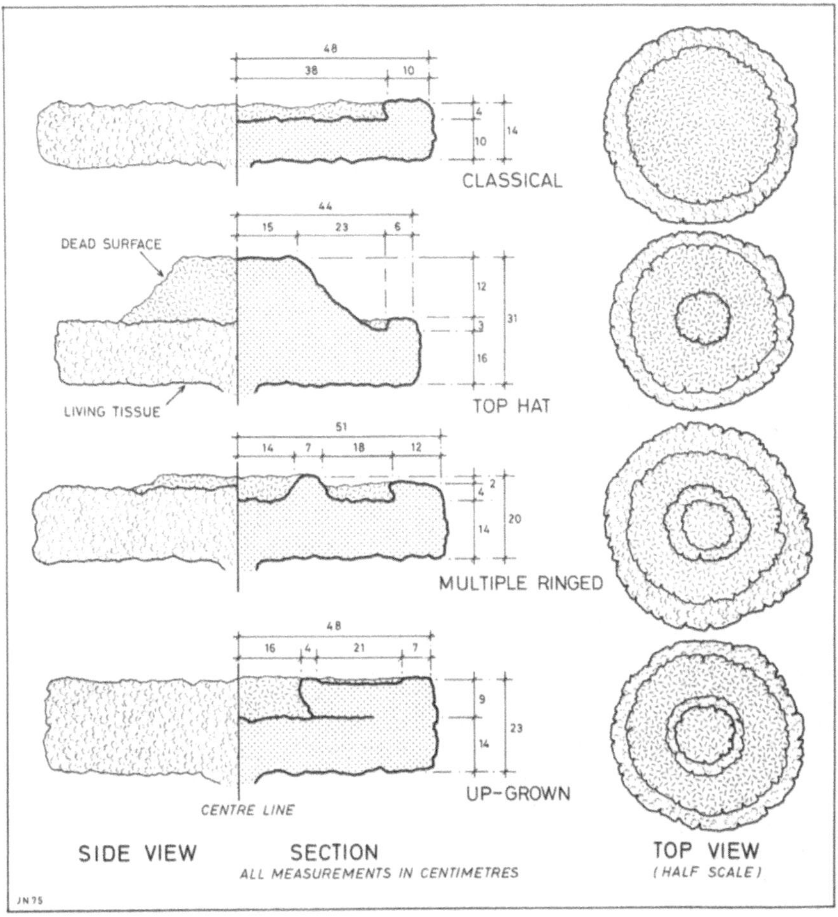

Figure 5. Changing morphology of micro-atolls in response to changing moat levels. The classical micro-atolls grow at an unchanging moat level; the top hat forms in response to a falling moat level; the multiple ring form is the result of a stable moat level but with periods of retarded growth possibly as the result of heavy rainfall and fresh-water lens development; and the up-grown form in response to a raised moat level.

Where the date of disturbance of either micro-atoll or moat level is known and can be identified with a specific cyclonic event, rates of growth of reef-flat corals subsequent to that event can be calculated. An example is seen on the fringing reef of Holbourne Island near Bowen (Hopley & Isdale, 1977) where a cyclone in 1918 breached a former rampart dam, lowering moated water levels over most of the reef flat by nearly 40 cm. This has left a series of the pre-1918 micro-atolls completely emerged and these were mistaken by Marshall *et al.* (1925) as part of a 'raised' reef.

However, radiocarbon assay (GaK-5217) of the centre of a *Goniastrea* colony from this 'raised' reef produced a date of 60 ± 70 BP. Taylor & Jouannic (1981) have similarly used micro-atoll levels to evaluate vertical movements in the tectonically active New Hebrides.

Not all moats contain micro-atolls. Some, because of their small size, may be rapidly heated on the low tide to temperatures which at times may be lethal to corals. Others may be so high that their waters are not sufficiently renewed or provided with enough circulation for coral growth. The pools behind the raised platforms on low wooded islands do not contain living corals for this reason. However, small living corals have been noted close to MHWN behind cemented rampart dams on the Turtle Islands, and at the same level within pools impounded in a mid-Holocene beach conglomerate on the windward side of Stone Island near Bowen. It is more usual to find micro-atolls extending only up to MSL and quite frequently below this level. Nevertheless, there is a distinct variation between the upper limit of growth and that of their open-water counterparts. Given sufficient stability of moat levels, the reef flat may be considerably elevated above the level suggested by open-water coral growth by coalescence of micro-atoll forms into a reef-flat pavement (Fig. 4).

Micro-atolls obviously produce problems for determination of past SLs, particularly in areas of large tidal range. However, whilst individual colonies may be found at levels up to MHWNs such examples are very unusual and restricted. Large fields of moated micro-atolls do not normally grow to levels in excess of MLWN and, in emerged reef exposures, where the former geometry of the reef top can be established with micro-atolls extending over 100 m^2 or more, their upper level can be conservatively taken as MLWN (for discussion see McLean *et al.*, 1978).

OTHER INDICATORS OF PALEO-SEA LEVELS ON CORAL REEFS

1. Other reef organisms

Ideally an organic indicator of a paleo-SL should have a distinctive upper and lower limit of growth, be easily identifiable and either its skeletal remains or its impact on its habitat in the form of casts or burrows have some permanency. Unfortunately, a large number of plants and animals living on coral reefs have a similar type of distribution pattern to the corals, ie. they have distinctive upper limits in which they will grow but highly variable lower limits. As with corals, the presence of *in situ* material merely indicates that the upper limit of growth (MLSW, MSL, MHWS etc.) lay above that position at time of growth.

Distinctive organisms with known intertidal distributions (see Endean *et al.*, 1956 and Stephenson *et al.*, 1958 for description of Great Barrier Reef organisms) and which may be retained in the geological record include the boring clams (especially *Tridacna crocea*), the encrusting rock oyster(*Crassostrea amasa*) and chitons (*Acanthozostera gemmata* and *Onithochiton quercinus*) but many of the distinctive intertidal Littorinids, barnacles, limpets and gastropods which have strict vertical limits are unlikely to be retained in growth position except in unusual circumstances. Bioeroding organisms, however, may be indicated by the distinctive shape of their burrows which may also retain the hard parts of the organism after death (for a general review of burrow shapes see Risk & MacGeachy, 1978; for Great Barrier Reef borers see Otter, 1937). However, the observation made by Otter in 1937 (p. 343) is still pertinent:

> '... an animal in a stone burrow is not nearly so suspect to changes in its external environment as animals in other habitats; for instance, water can be held in the burrow at low tide, or the burrow may be partially closed by the shell ... or a portion of the body ... may act as an operculum. Most of the burrowers were found to begin at a certain vertical limit and appeared to extend downwards without a break. No information was obtained as to the lower limit of any of the animals concerned.'

Suggestions made, for example by Scoffin (1977), that *Lithophaga* sp. are limited to the intertidal environment are demonstrably incorrect. Certainly many of the bioeroding organisms reach their maximum population numbers in the intertidal zone and the concentration of their attack is responsible for the notch and overlying visor found in many tropical emerged reefs, cemented platform rocks and reef blocks (Trudgill, 1976). However, individuals may extend below the absolute low tide level. This unfortunately limits their use as SLIs (indicators) in reef cores, but uppermost levels ascertained over wide areas in emerged reefs may be useful and accurate indicators of paleo-SLs.

A further complicating factor in the use of current reef-flat organisms is the probable plasticity in choice of habitat of those plants and animals during the post-glacial transgression. The rate of SL rise was greater than the rate of reef growth (with the possible exception of some rapidly growing Acroporid or Pocilloporid reefs). As Stoddart (1976) has noted, at the full glacial, the low SL shoreline would have been placed against steeply sloping shoulders of continental shelves. Until the reefs caught up with the stabilised or only slowly rising late-Holocene SL, reef-flat environments would have been very limited or even non-existent. Nevertheless, reef-flat inhabitors survived and until the exact mode of survival is known, then the possibility of these organisms occupying ecological niches different to those they occupy today, has to be recognised in attempts to use their fossils in SL reconstructions.

Figure 6. Typical caliche stringers formed in the soil horizons of Pleistocene reefs, Barbados. The caliche is a clear indicator of exposure and is readily identified in reef cores.

2. Geochemical indicators

Diagenetic changes, and in particular a reversion of the aragonitic corals and high-Mg calcite of coralline algae to low-Mg calcite, occur when a reef mass is subaerially exposed on lowering of SL. Accompanying changes include: marked reductions in primary porosity and increase in secondary solution pores, iron staining and reduction in strontium content (see for example Thom *et al.*, 1978; Hopley *et al.*, 1978). Such changes are indicators of major stratigraphic unconformities in reef cores but the length of time available for recognisable diagenetic changes to take place during minor falls within the Holocene SL rise has been insufficient. Particularly diagnostic of major periods of exposure is the development of caliche (Fig. 6), secondary $CaCO_3$ deposited as horizontal stringers near the upper surface of the weathering profile (Harrison, 1977). Steinen *et al.* (1973) attempted to use such an exposure in Barbados to establish a minimal low SL stand during the last glacial period, although later Fairbanks & Matthews (1978) suggested a revision to the age assigned to the exposure (see also Coudray & Montaggioni, this volume).

Cementation levels may also be used as SLIs in the reef environment. Two types of cemented materials are limited to an intertidal location, namely beachrock *sensu stricto* and rampart rocks formed from the cementation of coarse shingle deposits usually on the windward sides of reefs (see Scoffin, 1977; McLean *et al.*, 1978; Hopley, 1980, 1982, for full discussion). Unfortunately, other forms of cementation can take place both above and below the intertidal level and the resulting deposits may be difficult to distinguish from the intertidal rocks. Submarine cements, largely aragonitic in shallow water are important in Holocene reefs (see for example Shinn, 1969; Land & Goreau, 1970; Goreau & Land, 1974; Macintyre, 1977). Similarly, subaerially cemented or exposed phosphatic cay sandstones (guano rock), dune calcarenites, *in situ* emerged coral reefs and Pleistocene humic sandstones (developed as soil B horizons) have been misinterpreted as former intertidal beachrocks in the literature. Diagnostic features which suggest an originally intertidal origin for cemented reef detrital deposits include horizontality of the upper surface of cementation, the nature of the detrital fabrics, the nature of the cementing matrix and the possible retention of distinctive erosional features (see Hopley & Mackay, 1978). Only with great care, therefore, can these cemented deposits be used as indicators of past SLs (see Hopley, this volume).

3. Detrital deposits

Reefs, like most other coastal environments, have a variety of mechanically derived beach deposits associated with them. Unfortunately, the distribution of coral reefs coincides very closely with the distribution of tropical cyclones, (typhoons or hurricanes). It is during these high-energy events that most of the coarse reef debris, consisting largely of fragments of branching corals from the reef front, is formed. Not only are wave heights up to 4 times those normally experienced but also water levels may be raised by up to 6 m by the wind and pressure induced storm surge (see Hopley & Harvey, 1979, for review in Australian region). It is possible during these events that coral shingle can be thrown up to 8 m above the level of MLWS. In the Central equatorial Pacific where cyclones are rare or unknown, similar effects may result from tsunami waves. Thus the decline northwards along the Great Barrier Reef of various types of beaches (Fig. 7) is the result of declining cyclone incidence, prevailing wind speeds and tidal range as well as a variation in the regional pattern of mid-Holocene SLs (see Hopley, 1982, for further discussion).

Comparison of the heights of modern and dated mid-Holocene beach materials has been used as a basis for the estimation of mid-Holocene SLs (eg. McLean *et al.*, 1978). On the Great Barrier Reef a higher inner sand terrace on many cays contrasts with a lower and younger marginal terrace whilst shingle ridges are often in a declining sequence seawards. Both these situations may result from higher relative SLs (RSLs) in the past but in the absence of corroborating evidence may also result from the declining supply of materials for beach construction as a reef flat becomes stable at its maximum elevation, or from a decline in effective wave energy reaching the inner reef flat as

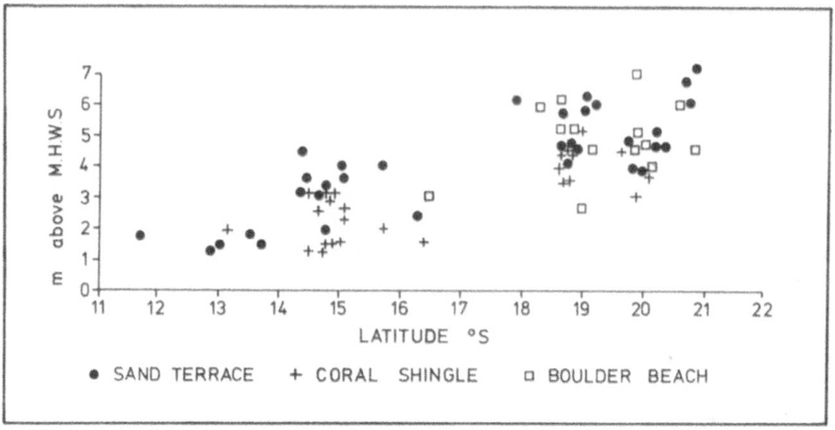

Figure 7. The general decline in elevation of surveyed heights of various beach deposits in north Queensland (plotted by latitude) is the result of a combination of exposure, wind strength variation, and cyclonic surge incidence as well as variation in mid-Holocene sea levels (see Hopley, 1982 for further discussion).

Figure 8. Large reef block with subaerial fluting on upper surface, Grub Reef, central Great Barrier Reef.

the result of increasing amount of protective reef-flat development with time, the closure of the 'Holocene energy window' of Neumann (1972).

One feature of reefs which may be helpful in the reconstruction of minor SL oscillations in the late-Holocene are reef blocks. These are massive slabs of coral rock or single colonies up to 5 m high and 40 m circumference generally thought to be thrown onto the outer reef flat from the reef front during cyclones (Fig. 8). Blocks up to 4310 ± 100 ^{14}C yrs in age have been reported from the Great Barrier Reef (Polach *et al.*, 1978) and have been used to calculate rates of reef extension (Scoffin *et al.*, 1978). This latter calculation assumes that the blocks furthest from the reef edge are oldest, a dubious assumption as during cyclone Althea in 1971 near Townsville new blocks were thrown well back on a number of reefs. However, if the assumption is made that the age of the corals in the reef block dates the event which emplaced it (which may be checked by also dating the reef flat beneath the block) then the contact between block and reef flat, which is normally close to LWN, is fossilised for the time that the reef block remains. Once established on the reef flat, 'mega blocks' become cemented to the flat and are attacked by bio-eroding organisms to produce the typical notch and visor profile (Fig. 9)[1]. However, appearances can be misleading due to the rapidity of the notching process and observations on the Great Barrier Reef and an ongoing research programme suggest that all three of the possible situations shown in Figure 9 related to stable, rising or falling SLs may be found.

4. Internal indicators

Corals and other reef organisms may retain records of the water-depth environment in which they grow in their internal growth structures as indicated by X-radiography. Two examples illustrate the potential of such methods. First, a relationship between water depth and rate of growth has been clearly demonstrated in at least the massive corals (eg. Baker & Weber, 1975; Graus & Macintyre, 1976). For example, over a 25 m depth range in Belize annual skeletal growth increment decreased from 11 to 3 mm/yr. However, as environmental fluctuations which affect coral growth (daily, seasonally, annually) are greatest in near surface waters (*c.* upper 5 m), surface corals not only display greatest mean growth rates but also greatest variance year to year, this latter factor probably being the more diagnostic (Fig. 10). However, site to site variations do exist in growth rates and, as Isdale (in prep.) shows, both absolute growth rate and variance decline with distance from the mainland in the Great Barrier Reef which may require that the paleo-geography of a reefal area needs to be considered when interpreting the growth patterns.

Similar caution is needed in the use of stable isotopes, although, like growth patterns, they can give an indication of water depth at time of coral growth. Weber

[1]See Pirazzoli, this volume.

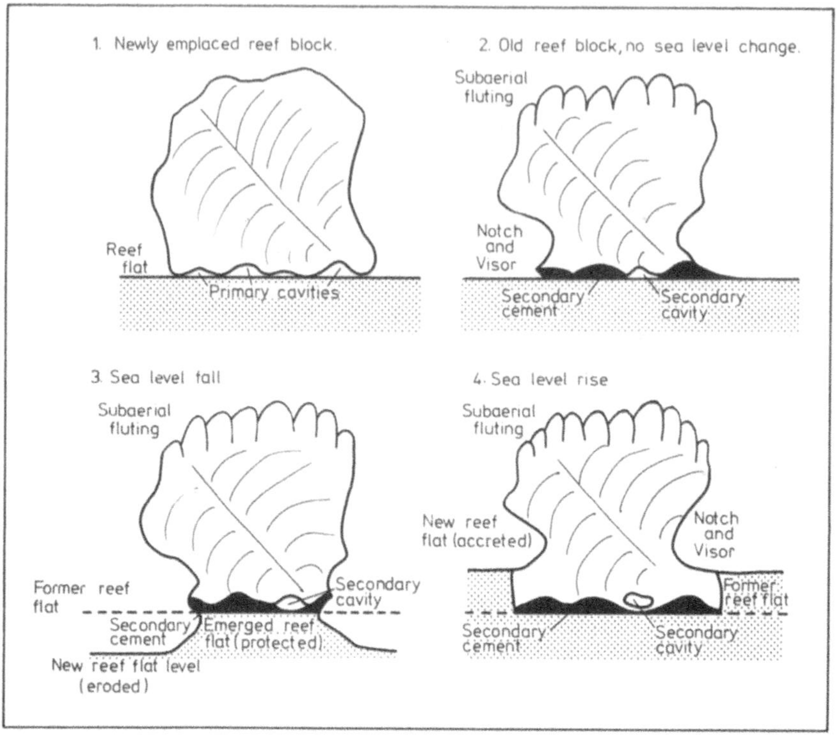

Figure 9. Three possible post-emplacement changes to reef-block/reef-flat relationships dependent on the nature of relative sea-level change.

& Woodhead (1970) showed that $^{13}C/^{12}C$ decreases with increasing depth; the average decrease in $\delta^{13}C$ between 18 m and the surface being about 1.2%. They noted that the depth effect is about twice as great for corals constructing massive forms than for branching or encrusting types. The $^{18}O/^{16}O$ ratio was also correlated with depth, $\delta^{18}O$ increasing with depth. However, the average difference in $\delta^{18}O$ between surface corals and those living at 18 m was small, about 0.10%.

Further work is needed on these internal indicators of SL. At present they require careful corroboration of modern measurements in the local area on the coral species being used for SL reconstruction. Further, there is no guarantee that temperature/salinity gradients and other environmental parameters were similar to present patterns in the paleo-environment. However, they do provide a guide for the interpretation of water depths beyond the basic assumption that MLWS lay above the *in situ* position of the coral at its time of growth.

Figure 10. X-radiographs of two *Porites* sp. coral heads. The upper example is *Porites lutea* from the reef flat of Three Isles, northern Great Barrier Reef: 6 years growth, mean = 1.40 cm/yr, σ = 0.08 cm. The lower example, *P. mayeri,* is from -8 m on Kangaroo Reef: 13 years growth, mean = 0.63 cm/yr, σ = 0.04 cm. (X-radiographs, P.J. Isdale)

ESTABLISHING THE AGE OF CORAL REEFS

1. Radiometric dating of coral reefs

As large carbonate structures of biogenic origin, coral reefs apparently provide a wealth of material for radiometric dating. However, the complex chemistry of the reef, especially when it is subaerially exposed, requires that great care and caution be applied to the choice, preparation and interpretation of materials. Reviews of radiometric dating of reefs have been given by Chappell *et al.* (1974, 1978) and Veeh & Green (1977). Comments by Thom (1973) are also pertinent to the problems.

Many of the problems result from calcium carbonate not behaving as a simple salt (which the equilibrium concentration of both anion and cation is simply the square root of the solubility product). Further, the various crystalline forms of calcium carbonate each behaves in an independent manner. The most important phases in the reef are aragonite, high-Mg calcite and low-Mg calcite and each of these phases has its own solubility. Matthews (1974) has shown that in the reef system, containing two or more solid phases, the water can never be at equilibrium with the rock. If the water is near saturation with respect to aragonite it will be supersaturated with calcite and should precipitate calcite. Conversely, if the water is near saturation with respect to calcite it should still be capable of dissolving more aragonite. Equilibrium is unlikely to be achieved without completely dissipating one of the solid phases, although a steady state may exist in which mineral matter is being removed from aragonite crystals and precipitated onto calcite crystals.

Geochemical changes during early diagenesis thus produce problems for dating of reef materials by, for example, introducing new ^{14}C into the system. Chappell *et al.* (1974) noted that different geochemical changes are associated with different modes of recrystallisation of the skeleton. In summary they state:

> 'One mode is recrystallisation to sparry calcite, and involves free exchange of carbon isotopes with the surroundings. Another recrystallisation mode is subtle coarsening of skeletal aragonite fibres to fine laths, mainly of aragonite, but sometimes of calcite. Carbon isotope exchange with the surroundings in this second mode is shown to be very low, and in *Tridacna* sometimes to be zero, over tens of thousands of years. Only completely unaltered specimens can be regarded as reliable for both C^{14} and Th^{230}/U^{234} dating. Of those samples affected only by 'subtle coarsening', the giant clam *Tridacna* appears reliable, with a lesser degree of confidence, for C^{14} dating, while coral is still satisfactory for Th^{230}/U^{234} dating but not so good for C^{14}. Any recrystallisation to sparry calcite indicates an unreliable sample.'

The care which needs to be taken in the choice and preparation of reef samples for radiometric dating is illustrated by Polach *et al.* (1978). At the very least, physical removal of parts of a sample thought to be contaminated (such as the outer parts of molluscs or secondary carbonate fillings in corals) should be undertaken in the laboratory. Use of dilute hydrochloric acid and ultrasonic cleaners together with a small dentist drill allow such preparation. Mineralogical analysis to determine the aragonite/calcite content and Mg calcite phase is necessary to confirm the removal of contaminants. With such careful preparation, and where possible, by dating stratigraphic sequences of materials, reliable radiocarbon and uranium series dates can be obtained in corals and related materials. In addition under favourable circumstances, indirect dating in coral reefs is possible by potassium-argon dating of interbedded volcanic material (Veeh & Green, 1977).

2. Other dating methods

Only one absolute dating method other than radiometric is currently available for corals. This is sclerochronology based on the annual growth rings described above (see Buddemeier, 1978). The techniques used are similar to those of dendrochronology, with each pair of low- and high-density bands representing a year's growth. As individual coral colonies up to 8 m in diameter are found in many back-reef areas (mainly *Porites* sp.), sclerochronology provides the opportunity of obtaining up to 1000 years of record. Unfortunately, as Isdale (1977) has shown, corals growing even in close proximity to one another may not show similar year to year growth patterns thus eliminating the possibility of extending the record back to older dead colonies by correlation of the overlapping time period in a currently growing coral. Problems may also arise in the record retained in the older parts of massive colonies because of intense bioerosion. Nonetheless, the technique has been, applied to short-term changes in RSL or moat level produced for example, by cyclonic influences (Hopley & Isdale, 1977) or tectonics (Taylor & Jouannic, 1981).

Other comparative dating methods are far less accurate and can be applied only on a coarse time scale. This is particularly true of the fossil record (see for example Frost, 1977) although occasional local variations between the Holocene and last interglacial may allow a differentiation between emerged reefs of these two periods. For example, *Fungia* is present in the Pleistocene but absent in the modern reefs of Formosa, whilst *Acropora* is present in the Pleistocene and absent in the Holocene reefs of Hawaii (Stoddart, 1976). Similarly, progressive diagenesis in reefs is time dependent and the progressive removal of aragonite and strontium in particular from the reef system has been suggested as a potential age indicator by Friedman & Bremmer (1977) and Gvirtsman & Friedman (1977) in the Red Sea. However, as research in Barbados has shown (eg. James, 1974) the rate of diagenesis is equally dependent on climatic considerations and if there is a possibility of climatic change since a reef assemblage has been deposited, its diagenetic stage can be only a gross indicator of age.

APPLICATIONS AND EXAMPLES

Data on SL changes provided by coral reefs come from three sources, each of which provides its own problems. The three sources are from emerged reefs, currently above the level of modern reef growth, submerged reefs which are either in areas where coral growth is no longer possible or where the reefs appear to have been left behind by the rise of SL, and reef cores provided by drilling into the reef structure.

1. Emerged reefs

Emerged reefs provide by far the best evidence for SL changes. Having the whole reef geometry available, together with biological and morphological zonation allows for a recognition of facies types and an accurate determination of a paleo-SL. Similarly, emerged reefs usually provide a wide range of dateable materials from which a choice can be made, and in which cross checking is possible. Particularly in reefs emerged only a few meters above modern reef-growth levels, problems may arise from the differentiation of moated and open-water corals but comparison with the modern height ranges of both open-water corals and micro-atolls should allow resolution of these problems.

Suites of emerged reef terraces are associated with tectonically rising areas and form essentially as fringing reefs. Beach calcarenites or deltaic lobes from coastal streams provide evidence which parallels that of the reefs. The level of the sea at the time of reef formation is thus determined by a variety of features as indicated by work in Barbados (Mesolella, 1968; Broeker *et al.*, 1968; Mesolella *et al.*, 1969; James *et al.*, 1971; Steinen *et al.*, 1973; Fairbanks & Matthews, 1978), in the Ryukyu Islands (Konishi *et al.*, 1970, 1974), Huon Peninsula, New Guinea (Chappell, 1974; Bloom *et al.*, 1974; Chappell & Polach, 1976) and East Timor (Chappell & Veeh, 1978). What is clear from examination of all these works is that there is no one standard structure, zonation, coral formation or associated feature which can be used as a SLI. Local differences occur even within the same bio-geographical region as indicated by Chappell & Veeh (1978) in comparing the emerged reefs of Atauro and Huon. Variations in the rate of reef growth in response to RSL rise and differences in exposure conditions are critical and necessitate that in any new study area any interpretation of emerged reefs requires detailed examination of the modern reef morphology, ecology and zonation.

Where the emergence is more recent (mid-Holocene) and of an amount which is possibly less than the local tidal range, then the problems are compounded. It is then important to examine the modern equivalent close to the situation of the emerged reef. Exposures at Rattlesnake Island near Townsville illustrate the relationships between moated and open-water corals (Fig. 11). An emerged reef consists purely of open-water corals (mainly Acroporids) which rise to 1.6 m (Tidal Datum), almost exactly to MSL, thus suggesting an emergence of 1.2 m in comparison with modern MLWS. Dated core recovered from beneath this emerged open-water reef shows that it had grown to or above the level of modern open-water growth by 6480 ± 170 BP (GaK-9295) and was up to 1 m higher by 5380 ± 120 BP (GaK-9294). Higher

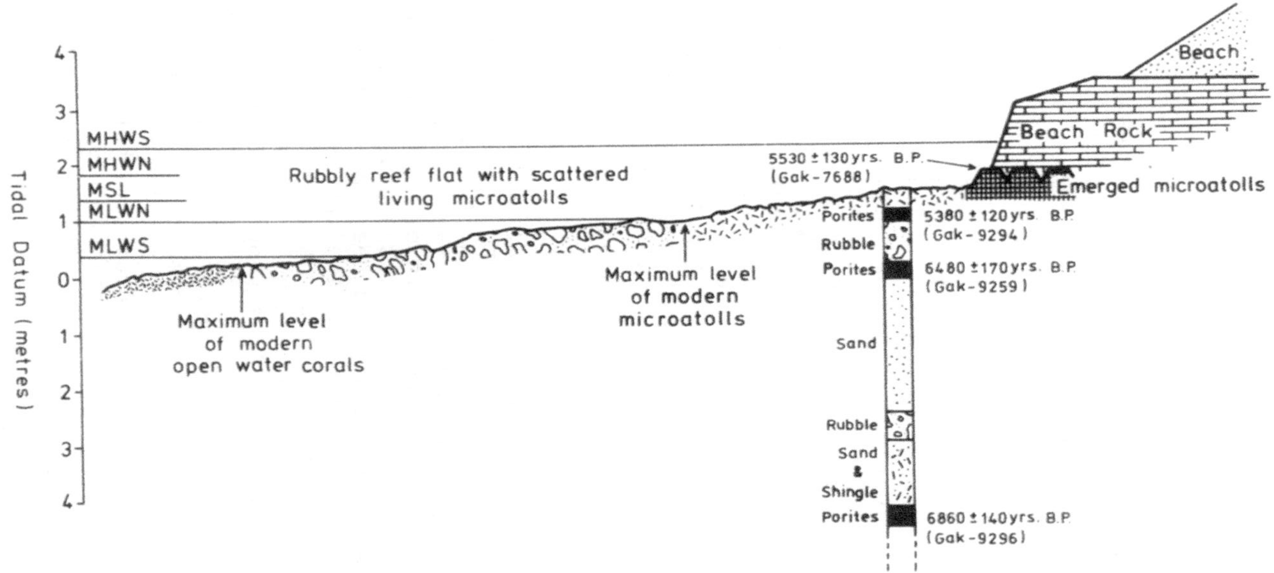

Figure 11. Cross section of Rattlesnake Island fringing reef showing the modern levels of open-water and moated corals and their relationship to the emerged mid-Holocene features. Only the top section of the borehole through the emerged open-water reef is shown.

on the reef, about 20 m landwards, a field of micro-atolls (mainly *Favids* and *Goniastrea*) rise to 1.86 m, suggesting a minimum emergence of 0.86 m compared to modern MLWN. This is less than the emergence suggested by the open-water assemblage but illustrates the problems of using micro-atoll heights to establish paleo-SLs. The amount of emergence can be made compatible if the micro-atoll field were moated to a level in relation to the paleo-tidal regime 0.36 m below the maximum level of their modern counterparts. A radiometric date from a micro-atoll of 5530 ± 130 BP (GaK-7688) confirms the age of this reef and the paleo-SL. It is overlain by two suites of beachrock, a higher level rising to 3.78 m above MHWS and a lower level to 0.96 m above MHWS unconformably lain against the higher level. An age of 3240 ± 100 BP for whole beachrock from the upper level is probably a minimal age for a paleo-SL of about +3.0 m (Hopley, 1971).

The modern reef-flat biological zonation is superimposed over the zonation formed at mid-Holocene times, producing a pattern which can only be understood in terms of a SL change (Hopley, 1982). Where even smaller SL changes are involved, especially on open-shelf reefs or atolls where an adjacent land mass is not present to provide confirmatory evidence, such lack of correlation between the biological communities presently occupying a reef flat and those which form or underlie the reef flat may provide an indicator of a SL change. Such an anomaly led Buddemeier *et al.* (1975) to conclude that the coral-algal pavements and Halimeda pavements of Enewetak reef flat had formed at a SL more than 1 m higher than present.

2. Submerged reefs

Numerous submerged reefs, usually in a shelf-edge location, have been described from the Caribbean (eg. Macintyre, 1972; Lighty, 1977 a, b; Lighty *et al.*, 1978, 1979 a, b; Adey *et al.*, 1977). These relict shelf-edge reefs typically form a low (10 m relief) ridge, the crests of which occur at depths of 15 to 30 m. Present-day fauna consists of alcyonarians, sponges and scattered deep-water coral heads but the internal structure is dominated by *Acropora palmata*, the typical reef-crest coral of the Caribbean. Similar submerged shelf-edge reefs have been recognised only recently in the Great Barrier Reef (Hopley, 1982).

The explanation given by Lighty (1977a, b) and Lighty, *et al.*, 1978, 1979 a, b) lays prime emphasis on the rapid increase in turbidity by reworking of soils and lagoon deposits during the transgression over the shelf behind reefs established on the shelf edge in early-Holocene times. A subsidiary factor is temperature, as the presence of tropical bryozoans in the early-Holocene reefs suggest warmer waters at this time. After transgression over the shelf, shallow waters would have been more susceptible to winter cooling. However, further south at St. Croix, the similar cessation of growth of an early-Holocene reef at about the same time is unlikely to be associated with water temperatures, reinforcing the explanation of increased turbidity as waters first flooded the shallow shelf (Adey *et al.*, 1977). Adey (1978) has stressed that inundation of surfaces of carbonate platforms during deglaciation results in deterioration of water quality, impeding growth of marginal reefs.

If it can be established that the submerged reef crest developed at or close to SL by recognition of a reef-crest facies or by other means, a SLI is established. These reefs have not been used in this way in the Caribbean. Rather they have been compared to previously established SL curves to confirm their position near SL at time of formation, but it seems possible that, with reservations, the reverse argument can be made in areas where no former SL curve has been established.

3. Reef drilling

Although W.M. Davis referred to reefs as 'inscrutable' (1928, p. 14) and largely ignored internal structure, reef drilling has subsequently provided a wealth of information on the way reefs evolve and has been particularly valuable in confirming the broad outline of antecedent platform hypotheses of MacNeil (1954) and Purdy (1974).

There is little doubt that a reef borehole can indicate very clearly the nature of reef growth at a particular point on a reef (Fig. 11). Sequential dating of carefully selected and prepared samples is necessary. Unfortunately, with the possible exception of some highly porous low energy Acroporid and Pocilloporid reefs from the Caribbean and eastern Pacific (Adey, 1975, 1978; Glynn & Macintyre, 1977) the upward growth rate of reefs appears to be slower than the rise of SL, even during the latter part of the Holocene transgression when the foundations of the modern reefs were being inundated. *The reef growth curve therefore is not a SL curve*. However, Lighty *et al*. (1982) working in the Caribbean, have been able to use the distinctive reef-crest *Acropora palmata* facies as a reliable indicator of SL. Modern calcification studies (eg. Kinsey & Davies, 1979) show that different zones on reefs may calcify at contrasting rates. The rapid transport of sediment in the reef system may also complicate the pattern of reef growth. Different parts of the reef may have accreted at highly contrasting rates (Fig. 12). This has been confirmed by the drilling of reefs in the Bunker-Capricorn Group (Davies & Marshall, 1979, 1980). Davies & Marshall (1979) have prepared a hypothetical growth curve for a reef, which shows an early period of slow accretion when difficulties in colonising the Pleistocene surface were encountered, a period of accelerated growth, and a period of decreasing growth as the reef approaches a stabilised SL in which erosion in the surf zone would decrease actual accretion rates. Thus even the basal dates of the Holocene reef may not represent SL at that time as a delay may have occurred between the inundation of the antecedent surface when high turbidity may have existed and coral colonisation. However, comparison of

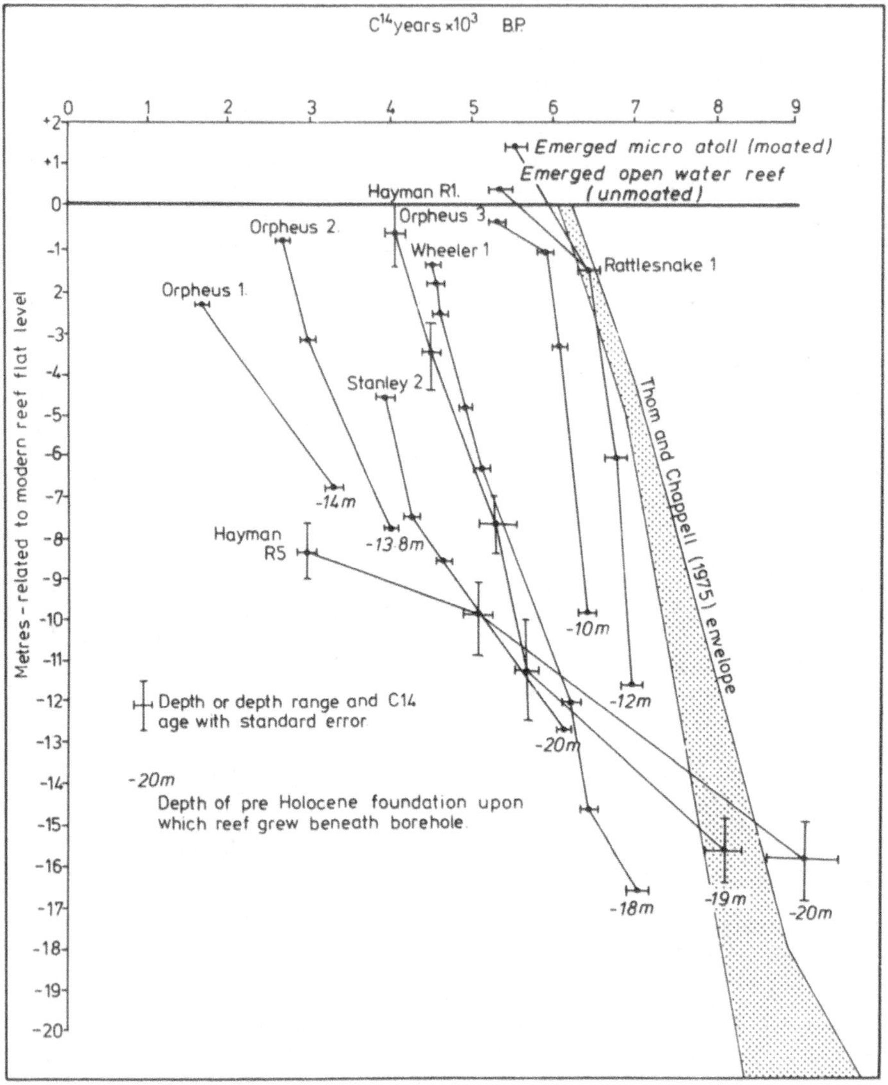

Figure 12. Accretion curves for specific reef cores, Great Barrier Reef, with the depth of the pre-Holocene platform from which the reef was growing beneath the core indicated. The Thom & Chappell (1975) sea-level curve is shown and is probably close to the actual SL rise in this area. Note the relationship of emerged open-water and emerged moated corals on Rattlesnake Island.

dated sequences from both Australia and the Caribbean with SL curves previously established by other means suggest that this delay may be minimal (eg. Hopley *et al.*, 1978). Therefore it may be possible to use basal Holocene reef dates to establish a SL curve in a particular area in a way similar to the use of basal peat dating described by Bloom (1964). A relative relief of up to 14.5 m has been established on the pre-Holocene surface of individual reefs of the Great Barrier Reefs (Harvey & Hopley, 1982) whilst the total spread of depths ranges from surface exposures in the Torres Strait to in excess of -30 m in the central Reef. Theoretically at least, this should mean that dated SL may be established for this total depth range. Fringing reefs, growing from bedrock foundations may similarly provide a range of basal dates from different depths up the bedrock slope to establish a rate of SL rise.

Refinement in the interpretation of techniques involving X-radiographic densitometry and stable isotopes (^{18}O and ^{13}C) may overcome the problems associated with an unknown water depth overlying a coral recovered from a core. This is an area of research currently being undertaken by the author but even if successful minor problems may still remain, not the least of which is the clear establishment that a coral-core sample is indubitably from an *in situ* coral head.

CONCLUSION

Corals and coral reefs are not the clear indicators of past SLs that they are sometimes suggested to be. Problems arise associated with the depth range over which corals and many related or symbiotic organisms grow and from the delayed response of a reef system to the Holocene SL rise. Nevertheless, careful observation and measurement can overcome these problems for emerged reefs at least, and a wide range of techniques are available for the interpretation of paleo-SLs associated with submerged reefs and from reef-borehole samples. The coral reef is a major habitat for numerous plants and animals which themselves may have more specific relationships to SLs than do the corals. For that reason, several other chapters in this volume are pertinent to the use of coral reefs in SL research. Runcorn (1966) has suggested that corals may act as paleontological clocks. With increasing application of stable isotope and geochemical research to coral reefs (eg. Weber & Woodhead, 1970; Weber, 1977; Woodhead & Weber, 1973; Dodge & Thomson, 1974; Baker & Weber, 1975; Land *et al.*, 1975; Goreau, 1977; Fairbanks & Dodge, 1979; Smith *et al.*, 1979) it may be possible to prove that corals retain records not only of SL changes but also of other interesting environmental fluctuations such as water temperature and salinity.

ACKNOWLEDGEMENT

Funds for research by the author on the Great Barrier Reef have been provided by the Australian Research Grants Committee and the Australian Marine Sciences and Technologies Advisory Panel - Funding Advisory Panel. Acknowledgement is also given for assistance in the field, particularly by Frazer Muir and Clive Grant, but also by many other individuals.

REFERENCES

Adey, W.H., 1975. The algal ridges and coral reefs of St Croix, their structure and Holocene development. *Atoll Research Bulletin*, 187, 1-67

Adey, W.H., 1978. Coral reef morphogenesis: a multi-dimensional model. *Science*, 202, 4370, 831-837

Adey, W.H., Macintyre, I.G., Stuckenrath, R. & Dill, R.F., 1977. Relict barrier reef system off St Croix: its implications with respect to late Cenozoic coral reef development in the western Atlantic. *Proceedings of the Third International Coral Reef Symposium*, (Miami, 1977), 2, 15-21

Adey, W.H., (this volume). Coralline algae as indicators of sea level

Baker, P.A. & Weber, J.N., 1975. Coral growth rate: variation with depth. *Earth and Planetary Science Letters*, 27, 57-61

Bayer, F.M., 1956. Octocorallia. In: *Treatise on Invertebrate Paleontology*, ed. R.C.Moore, F, Coelenterata, F166-F233, (Geological Society of America, New York)

Belt, T., 1874. An examination of theories that have been proposed to account for the climate of the glacial period. *Quarterly Journal of Science*, 11, 421-464

Bloom, A.L., 1964. Peat accumulation and compaction in a Connecticut coastal marsh. *Journal of Sedimentary Petrology*, 34, 599-663

Bloom, A.L., Broeker, W.S., Chappell, J., Matthews, R.K. & Mesolella, K.J., 1974. Quaternary sea-level fluctuation: new Th^{230}/U^{234} dates from New Guinea. *Quaternary Research*, 4, 185-205.

Boschma, H., 1956. Milleporina and Stylasterina. In: *Treatise on nvertebrate aleontology*, ed. R.C.Moore, F, Coelenterata, F90-F106, (Geological Society of America, New York)

Broeker, W.S., Thurber, D.L., Goddard, J., Ku, T., Matthews, R.K. & Mesolella, K.J., 1968. Milankovitch hypothesis supported by precise dating of coral reefs and deep sea sediments. *Science*, 160, 297-300

Buddemeier, R.W., 1978. Coral growth: retrospective analysis. In: *Coral eefs: esearch ethods*, eds. D.R. Stoddart & R.E. Johannes, 551-571, (UNESCO, Paris)

Buddemeier, R.W. & Kinzie, R.A., 1976. Coral growth. *Oceanography and Marine Biology Annual Review*, 14, 183-225

Buddemeier, R.W., Smith, S.V. & Kinzie, R.A., 1975. Holocene windward reef flat history, Enewetak Atoll. *Geological Society of America Bulletin*, 86, 1581-1584

Chappell, J., 1974. Geology of coral terraces, Huon Peninsula, New Guinea: a study of Quaternary tectonic movements and sea-level changes. *Geological Society of America Bulletin*, 85, 553-570

Chappell, J., 1980. Coral morphology, diversity and reef growth. *Nature*, 286, 249-252

Chappell, J. & Polach, H.A., 1976. Holocene sea-level change and coral reef growth at Huon Peninsula, Papua-New Guinea. *Geological Society of America Bulletin*, 87, 235-240

Chappell, J. & Veeh, H.H., 1978. Late Quaternary tectonic movements and sea-level changes at Timor and Atauro Island. *Geological Society of America Bulletin*, 89, 356-368

Chappell, J., Broeker, W.S., Polach, H.A. & Thom, B.G., 1974. Problems of dating upper Pleistocene sea levels from coral reef areas. *Proceedings of the Second International Coral Reef Symposium*, (Brisbane, 1973), 2, 563-571

Chappell, J., Thom, B.G. & Polach, H.A., 1978. Radiometric dating of coral reefs. In: *Coral Reefs: Research Methods*, ed. D.R. Stoddart & R.E. Johannes, 81-91, (UNESCO, Paris)

Daly, R.A., 1910. Pleistocene glaciation and the coral reef problem. *American Journal of Science*, Series 4, 30, 297-308

Daly, R.A., 1915. The glacial control theory of coral reefs. *Proceedings of the American Academy of Arts and Science*, 1, 155-251

Daly, R.A., 1916. A new test of the subsidence theory of coral reefs. *Proceedings of the National Academy of Science*, 2, 664-670

Daly, R.A., 1917. Origin of the living coral reefs. *Scientia*, 22, 188-199

Daly, R.A., 1919. The coral reef zone during and after the glacial period. *American Journal of Science*, Series 4, 48, 136-159

Daly, R.A., 1934. *The Changing World of the Ice Age*, (Yale University Press, New Haven)

Daly, R.A., 1948. Coral reefs - a review. *American Journal of Science*, 246, 193-207

Davies, P.J. & Marshall, J.F., 1979. Aspects of Holocene reef growth - substrate age and accretion rate. *Search*, 10, 276-279

Davies, P.J. & Marshall, J.F., 1980. A model of epicontinental reef growth. *Nature*, 287, 37-38

Davis, W.M., 1928. *The Coral Reef Problem*. American Geographical Society Special Publication, 9, 596 pp

Dodge, R.E. & Thomson, J., 1974. The natural radiochemical and growth records in contemporary hermatypic corals from the Atlantic and Caribbean. *Earth and Planetary Science Letters*, 23, 313-322

Done, T.J., 1977. A comparison of units of cover in ecological classifications of coral communities. *Proceedings of the Third International Coral Reef Symposium*, (Miami, 1977), 1, 9-13

Done, T.J., 1983. Coral zonation, its nature and significance. In: *Perspectives on Coral Reefs*, ed. D.J. Barnes, Australian Institute of Marine Science, Townsville, 107-147

Endean, R., Stephenson, W. & Kenny, T., 1956. The ecology and distribution of intertidal organisms on certain islands off the Queensland coast. *Australian Journal of Marine and Freshwater Research*, 7, 317-342

Fairbanks, R.G. & Dodge, R.E., 1979. Annual periodicity of the $^{18}O/^{16}O$ and $^{13}C/^{12}C$ ratios in the coral *Montastrea annularis*. *Geochemica et Cosmochemica Acta*, 43, 1009-1020

Fairbanks, R.G. & Matthews, R.K., 1978. The marine oxygen isotope record in Pleistocene coral, Barbados, West Indies. *Quaternary Research*, 10, 181-196

Fairbridge, R.W., 1968. Microatoll. In: *Encyclopaedia of Geomorphology*, ed. R.W. Fairbridge, 701-705, (Reinhold, New York)

Friedman, G.M. & Brenner, I.B., 1977. Progressive diagenetic elimination of strontium in Quaternary to late Tertiary coral reefs of the Red Sea: sequence and time scale. In: *Reefs and Related Carbonates - Ecology and Sedimentology*, eds. S.H. Frost, M.P. Weiss & J.B. Saunders, *American Association of Petroleum Geologists, Studies in Geology*, 4, 353-355

Frost, S.H., 1977. Cenozoic reef systems of the Caribbean - prospects for paleo-ecologic synthesis. In: *Reefs and elated arbonates - Ecology and Sedimentology*, eds. S.H. Frost, M.P. Weiss & J.B. Saunders, *American Association of Petroleum Geologists, Studies in Geology*, 4, 93-110

Glynn, P.W., 1974. Aspects of the ecology of coral reefs in the western Atlantic region. In: *Biology and Geology of Coral Reefs II*, eds. O.A. Jones & R. Endean, Biology I, 271-324, (Academic Press, New York)

Glynn, P.W. & Macintyre, I.G., 1977. Growth rate and age of coral reefs in the Pacific coast of Panama. *Proceedings of the Third International Coral Reef Symposium*, (Miami, 1977), 2, 251-259

Goreau, T.J., 1977. Coral skeletal chemistry: physiological and environmental regulation of stable isotopes and trace metals in *Montastrea annularis*. *Proceedings of the Royal Society, London*, B, 196, 291-315

Goreau, T.J. & Land, L.S., 1974. Fore-reef morphology and depositional processes, north Jamaica. In: *Reefs in Time and Space*, ed. L.F. Laporte, Society of Economic Paleontologists and Mineralogists Special Publication 18, 77-89

Grauss, R.R. & Macintyre, I.G., 1976. Light control of growth form in colonial reef corals: computer simulation. *Science*, 193, 895-897

Grauss, R.R., Chamberlain, J.A. & Baker, A.M. 1977. Structural modification of corals in relation to waves and currents. In: *Reefs and Related Carbonates - cology and Sedimentology*, eds. S.H. Short, M.P. Weiss & J.B. Saunders, *American Association of Petroleum Geologists, Studies in Geology*, 4, 135-153

Gvirtsman, G. & Friedman, G.M., 1977. Sequence of progressive diagenesis in coral reefs. In: *Reefs and Related Carbonates - Ecology and Sedimentology*, ed. J.H. Frost, M.P. Weiss & J.B. Saunders, *American Association of Petroleum Geologists, Studies in Geology*, 4, 356-380

Harrison, R.S., 1977. Caliche profiles: indicators of near-surface subaerial diagenesis, Barbados, West Indies. *Bulletin of Canadian Petroleum Geology*, 25, 123-173

Harvey, N. & Hopley, D. 1982. The relationship between modern reef morphology and a pre-Holocene substrate in the Great Barrier Reef province. *Proceedings of the Fourth International Coral Reef Symposium*, (Manila, 1981), 1, 549-554

Hopley, D., 1971. The origin and significance of north Queensland island spits. *Zeitschrift für Geomorphologie*, N.F., 15, 371-389

Hopley, D., 1980. Mid-Holocene high sea levels along the coastal plain of the Great Barrier Reef province: a discussion. *Marine Geology*, 35, M1-M9

Hopley, D., 1982. *The geomorphology of the Great Barrier Reef: Quaternary evolution of coral reefs*, (John Wiley-Interscience, New York)

Hopley, D., (this volume). Beachrock as a sea-level indicator

Hopley, D. & Harvey, N., 1979. Regional variation in storm surge characteristics around the Australian coast: a preliminary investigation. In: *Natural Hazards in Australia*, eds. R.L. Heathcote & B.G. Thom, 164-185, (Australian Academy of Science, Canberra)

Hopley, D. & Isdale, P., 1977. Coral micro-atolls, tropical cyclones and reef flat morphology: a north Queensland example. *Search*, 8, 79-81

Hopley, D. & Mackay, M.G., 1978. An investigation of morphological zonation of beachrock erosional features. *Earth Surface Processes*, 3, 363-377

Hopley, D., McLean, R.G., Marshall, J. & Smith, A.S., 1978. Holocene-Pleistocene boundary in a fringing reef: Hayman Island, north Queensland. *Search*, 9, 323-325

Isdale, P.J., 1977. Variations in growth rates of hermatypic corals in a uniform environment. *Proceedings of the Third International Coral Reef Symposium*, (Miami 1977), 2, 403-408

Isdale, P.J. 1981. Geographical variation in the growth rate of the Hermatypic coral *Porites* in the Great Barrier Reef Province, Australia. Unpublished PhD theses, James Cook University of North Queensland

James, N.P., 1972. Holocene and Pleistocene calcareous crust (caliche) profiles: criteria for subaerial exposure. *Journal of Sedimentary Petrology*, 42, 817-836

James, N.P., 1974. Diagenesis of scleractinian corals in the subaerial vadose environment. *Journal of Paleontology*, 48, 785-799

James, N.P., Mountjoy, E.W. & Omura, A., 1971. An early Wisconsin reef terrace at Barbados, W.I. and its climatic implications. *Geological Society of America Bulletin*, 82, 2011-2018

Jell, J.S., 1974. The microstructure of some scleractinian corals. *Proceedings of the Second International Coral Reef Symposium*, (Brisbane, 1973), 2, 301-320

Kinsey, D.W. & Davies, P.J., 1979. Inorganic carbon turnover, calcification and growth in coral reefs. In: *Biogeochemistry of Mineral Forming Elements*, eds. P. Trudingar & D. Swaine, 131-162, (Elsevier, New York)

Konishi, K., Schlanger, S.O. & Omura, A., 1970. Neotectonic rates in the central Ryukyu Islands derived from Th^{230} coral ages. *Marine Geology*, 9, 225-240

Konishi, K., Omura, A. & Nakamichi, O., 1974. Radiometric coral ages and sea-level records from the late Quaternary reef complexes of the Ryukyu Islands. *Proceedings of the Second International Coral Reef Symposium*, (Brisbane, 1974), 2, 595-613

Land, L.S. & Goreau, T.F., 1970. Submarine lithification of Jamaican reefs. *Journal of Sedimentary Petrology*, 40, 457-462

Land, L.S., Lang, J.C. & Barnes, D.J., 1975. Extension rate: a primary control on the isotopic composition of West Indian (Jamaican) scleractinian reef coral skeletons. *Marine Biology*, 33, 221-233

Lighty, R.G., 1977a. Relict shelf-edge Holocene coral reef: south-east coast of Florida. *Proceedings of the Third International Coral Reef Symposium*, (Miami, 1977), 2, 215-221

Lighty, R.G., 1977b. Submarine diagenesis in relict Holocene coral reef, south-east Florida - diagenetic control on porosity and stratigraphy. *American Association of Petroleum Geologists Bulletin*, 61, 808

Lighty, R.G., Macintyre, I.G. & Stuckenrath, R., 1978. Submerged early Holocene barrier reef, south-east Florida shelf. *Nature*, 275, 59-60

Lighty, R.G., Macintyre, I.G. & Stuckenrath, R., 1979a. Holocene reef growth on the edge of the Florida shelf. *Nature*, 278, 281-282

Lighty, R.G., Macintyre, I.G. & Stuckenrath, R., 1979b. Shelf temperatures and reef growth on the south-east Florida coast: reply. *Nature*, 278, 670

Lighty, R.G., Macintyre, I.G. & Stuckenrath, R., 1982. *Acropora palmata* reef framework: a reliable indicator of sea level in the Western Atlantic for the past 10 000 years. *Coral Reefs*, 1, 125-130

Ludington, C.A., 1979. Tidal modifications and associated circulation in a platform reef lagoon. *Australian Journal of Marine and Freshwater Research*, 30, 425-430

Macintyre, I.G., 1972. Submerged reefs of the eastern Caribbean. *American Association of Petroleum Geologists Bulletin*, 56, 720-738

Macintyre, I.G., 1977. Distribution of submarine cements in a modern Caribbean fringing reef, Galeta Point, Panama. *Journal of Sedimentary Petrology*, 47, 505-516

McLean, R.F., Stoddart, D.R., Hopley, D. & Polach, H., 1978. Sea-level change in the Holocene on the northern Great Barrier Reef. *Philosophical Transactions of the Royal Society, London*, A, 291, 167-186

MacNeil, F.S., 1954. The shape of atolls: an inheritance from subaerial erosion forms. *American Journal of Science*, 252, 404-427

Marshall, P., Richards, H.C. & Walkom, A.B., 1925. Recent emergence at Holbourne Island, Great Barrier Reef. *Reports of the Great Barrier Reef Committee*, 1, 29-34

Matthews, R.K., 1974. A process approach to diagenesis of reefs and reef associated limestones. In: *Reefs in Time and Space*, ed. L.F. Laporte, Society of Economic Paleontologists and Mineralogists Special Publication 18, 234-256

Mesolella, K.J., 1967. Zonation of uplifted Pleistocene coral reefs on Barbados, West Indies. *Science*, 156, 638-640

Mesolella, K.J., 1968. The uplifted reefs of Barbados: physical stratigraphy, facies relationships and absolute chronology. Unpublished PhD thesis, Brown University, Providence RI

Mesolella, K.J., Matthews, R.K., Broeker, W.S. & Thurber, D.L. 1969. The astronomical theory of climatic change: Barbados data. *Journal of Geology*, 77, 250-274

Mesolella, K.J., Sealy, H.A. & Matthews, R.K., 1970. Facies geometries within Pleistocene reefs of Barbados, West Indies. *American Association of Petroleum Geologists Bulletin*, 54, 1890-1917

Neumann, A.C., 1972. Quaternary sea-level history of Bermuda and the Bahamas. *American Quaternary Association Second National Conference Abstracts*, 41-44

Otter, G.W., 1937. Rock-destroying organisms in relation to coral reefs. *Scientific Reports of the Great Barrier Reef Expedition 1928-9*, (British Museum, Natural History), 1, 323-352

Penck, A., 1894. *Morphologie der Erdoberfläche*. (Engelhorn Stuttgart), 2 volumes

Penck, A., 1896. Das grosse australische Wallriffe. *Vorträge Verein zur Verbr. Naturar. Kenntnisse, Wien*, 36, 13, 1-23

Polach, H.A., McLean, R.F., Caldwell, J.R. & Thom, B.G., 1978. Radiocarbon ages from the northern Great Barrier Reef. *Philosophical Transactions of the Royal Society, London*, A, 231, 139-158

Purdy, E.G., 1974. Reef configurations, cause and effect. In: *Reefs in Time and Space*, ed. L.F. Laporte, Society of Economic Paleontologists and Mineralogists Special Publication, 18, 9-76

Risk, M.S. & MacGeachy, J.K., 1978. Aspects of bioerosion of modern Caribbean reefs. *Revista de Biologia Tropical*, 26 (Supplement), 85-105

Runcorn, S.K., 1966. Corals as paleontological clocks. *Scientific American*, 215, 26-33

Scoffin, T.P., 1977. Sea-level features on reefs in the northern province of the Great Barrier Reef. *Proceedings of the Third International Coral Reef Symposium*, (Miami, 1977), 2, 319-334

Scoffin, T.P. & Stoddart, D.R., 1978. The nature and significance of microatolls. *Philosophical Transactions of the Royal Society, London*, B, 384, 99-122

Scoffin, T.P., Stoddart, D.R., McLean, R.F. & Flood, P.G., 1978. Recent development of reefs in the Northern Province of the Great Barrier Reef. *Philosophical Transactions of the Royal Society, London*, B, 284, 129-139

Shinn, E.A., 1969. Submarine lithification of Holocene carbonate sediments in the Persian Gulf. *Sedimentology*, 12, 109-144

Smith, S.V., Buddemeier, R.W., Redalje, R.C. & Houck, J.E., 1979. Strontium-calcium thermometry in coral skeletons. *Science*, 204, 404-407

Steinen, R.P., Harrison, R.S. & Matthews, R.K., 1973. Eustatic low stand of sea level between 105 000 and 125 000 BP: evidence from the subsurface of Barbados, W.I. *Geological Society of America Bulletin*, 84, 63-80

Stephenson, W., Endean, R. & Bennett, I., 1958. An ecological survey of the marine fauna of Low Isles, Queensland. *Australian Journal of Marine and Freshwater Research*, 9, 261-318

Stoddart, D.R., 1976. Continuity and crisis in the reef community. *Micronesica*, 12, 1-9

Stoddart, D.R. & Scoffin, T.P., 1979. Micro-atolls: review of form, origin and terminology. *Atoll Research Bulletin*, 224, 1-17

Taylor, F.W. & Jouannic, C., 1981. Coral heads: natural recorders of changes in relative level of the land and sea. *Fourth International Coral Reef Symposium*, (Manila, 1981), Abstracts

Thom, B.G., 1973. The dilemma of high interstadial sea levels during the last glaciation. *Progress in Geography*, 5, 170-246

Thom, B.G. & Chappell, J., 1975. Holocene sea levels relative to Australia. *Search*, 6, 90-93

Thom, B.G., Orme, G.R. & Polach, H.A., 1978. Drilling investigations of Bewick and Stapleton Islands. *Philosophical Transactions of the Royal Society, London*, A, 291. 37-54

Trudgill, S.T., 1976. The marine erosion of limestones on Aldabra Atoll, Indian Ocean. *Zeitschrift für Geomorphologie*, Supplement Band, 26, 164-200

Upham, W., 1878. Geology of New Hampshire. *Concord*, 3, 18, 329

Veeh, H.H. & Green, D.C., 1977. Radiometric geochronology of coral reefs. In: *Biology and Geology of Coral Reefs*, eds. O.A. Jones & R. Endean, IV, Geology, 11, 183-200, (Academic Press, New York)

Veron, J.E.N. & Done, T.J., 1979. Corals and coral communities of Lord Howe Island. *Australian Journal of Marine and Freshwater Research*, 30, 203-236

Veron, J.E.N. & Pichon, M., 1976. Scleractinia of Eastern Australia 1, Thamnasteriidae, Astrocoeniidae, Pocilloporidae. *Australian Institute of Marine Science Monograph Series*, 1, 86 pp

Weber, J.N. & Woodhead, P.M.J., 1970. Carbon and oxygen isotope fractionation in the skeletal carbonate of reef-building corals. *Chemical Geology*, 6, 93-117

Weber, J.N. & Woodhead, P.M.J., 1972. Temperature dependence of Oxygen-18 concentration in reef coral carbonates. *Journal of Geophysical Research*, 77, 463-473

Weber, J.N., 1977. Use of corals in determining glacial-interglacial changes in temperature and isotopic composition of seawater. In: *Reefs and Related Carbonates - Ecology and Sedimentology*, eds. S.H. Frost, M.P. Weiss & J.B. Saunders, *American Association of Petroleum Geologists, Studies in Geology*, 4, 289-295

Wells, J.W., 1955. A survey of the distribution of coral genera of the Great Barrier Reef region. *Reports of the Great Barrier Reef Committee*, 6(2), 21-29

Wells, J.W., 1956. Scleractinia. In: *Treatise on Invertebrate Paleontology*, ed. R.C. Moore, F, Coelenterata, F328-F440, (Geological Society of America, New York)

Wells, J.W., 1957. Coral reefs. In: *Treatise on Marine Ecology*, ed. J. Hedgpath, *Geological Society of America Memoir*, 67, 609-631

Wells, J.W., 1963. Coral growth and geochronology. *Nature*, 197, 948-950

Wells, J.W., 1969. The formation of dissepiments in Zoantharian corals. In: *Stratigraphy and Paleontology: Essays in Honour of Dorothy Hill*, ed. K.J.W. Campbell, 17-26 (Australian National University Press, Canberra)

Woodhead, P.M.J. & Weber, J.N., 1973. The evolution of reef building corals and the significance of their association with zooxanthellae. In: *Proceedings of the Symposium on Hydrogeochemistry and Biogeochemistry*, ed. E. Ingerson, 2, 280-304

Yonge, C.M., 1963. The biology of coral reefs. In: *Advances in Marine Biology*, ed. F.S. Russell, 1, 209-260

Added at proof:
Coudray, J. & Montaggioni, L., (this volume). The diagenetic products of marine carbonates as sea-level indicators.

Pirazzoli, P.A., (this volume). Marine notches.

9

Coralline algae as indicators of sea-level

W. H. ADEY

ABSTRACT

Calcified crustose coralline algae occur from the arctic to the tropics in the ocean's photic zone. However, in warmer seas, particularly supersaturated waters, growth is more rapid and secondary lithification more extensive leading to a greater frequency of bioherm and biostrome development. The generic and species composition of tropical and colder seas differ considerably and a depth zonation of genera and species is characteristic of all waters. Depth characterization of fossil material can be carried out on the basis of community structure. However, precision of determination is likely to be within the ±5-10 m range.

In tropical and temperate waters, and particularly in supersaturated seas, coralline bioherms (algal ridges and trottoir) develop in close association with sea level and can be precision (±10 cm) indicators of that level. Wave action control of grazing, and temperature and salinity/pH enhancement of growth and secondary mineralization are factors determining the extent of bioherm development. ^{14}C dating along with consideration of the above factors and coralline community structure can lead to great confidence in the use of crustose corallines for sea-level determination.

Basic techniques and information needed for preparation and identification of crustose corallines is provided.

INTRODUCTION

The crustose coralline algae (Melobesiodeae) occur along most normal salinity (30-40 parts per thousand, ‰) coasts of the world ocean not dominated by large stretches of shifting sand. Although generally reduced in relative abundance, they have been found at salinities as low as 20‰ and as high as 50‰. In the tropics, either alone or with stony corals and/or calcified worms or molluscs, crustose corallines can form considerable reef frameworks; in subtropical to temperate regions, the reef frameworks developed by these algae are considerably smaller but still potentially preservable in the fossil record. Although corallines are abundant in boreal and arctic waters, lithification of framework rarely occurs and destructive elements exceed constructive elements. Nevertheless, in those colder waters, the coralline algae often provide a sub-continuous crust over rocky substrate, and under some circumstances, this crust can be preserved in place. On the other hand, abundant attached coralline crusts in colder waters can lead to or participate in the development of significant sediments, rich in coralline skeletal fragments, that are called rhodoliths or maerls. These coralline sediments dominated by unattached, or free, calcareous plants also occur throughout the tropics as well as in colder areas.

The coralline skeleton is high-magnesium calcite (typically 8-20% Mg). While moderately stable, this skeleton sometimes suffers enough alteration to render ^{14}C dates unreliable. Nevertheless, structural integrity sufficient to allow generic identification is often preserved and the possibilities for using corallines as fossil environmental indicators based on taxonomic characters are considerable. The fine, closed and open porous structure of the magnesium-rich calcitic coralline skeleton leads to considerable development of submarine cementation in the pores. It is this secondary calcification that eventually improves the structural integrity of the skeleton and makes it more likely that it will withstand scraping and boring to become a geological structure. The statement of Alexandersson (1977): '...it must be stressed that even biochemic cementation is dependent on the ambient state of carbonate saturation. In supersaturated seas, like the Mediterranean and the Caribbean, the cement is accumulated in frameworks that last indefinitely in their environment of formation. On the other hand, in undersaturated regions, like the Skagerrak and the North Sea, the algae must work against an unfavorable chemical gradient. They manage to produce small amounts of cement in restricted frames only, and often on the death of the organisms both skeleton and cement are attacked by dissolution' is generally true. However, coralline skeletons, sometimes rather massive in dimension, are produced in colder, undersaturated seas, and under some conditions (maerl and rhodoliths are striking examples) these skeletons can be preserved in the fossil record. On the other hand, in tropical, saturated seas, massive coralline frameworks usually develop only where boring and scraping is limited by some means - usually intensive wave action. Removing the latter will generally insure the beginning of a destructive process that may eventually be complete.

Although strongly calcified, coralline algae are fully autotrophic, and physiologically plants in every respect.

Light limitation is a primary ecological factor in coralline distribution and rarely do living plants occur in abundance below 100 m. In the tropics, 80 m is a more typical lower limit for abundant living material, while in colder climates 20-40 m is more characteristic. Also, most species, genera and subfamily groupings demonstrate a depth zonation which is much narrower than the overall depth range of the group. Coralline fragments are often transported some distance, dead and alive, by wave and current action. Whenever reef framework or substrate heterogeneity is present, cryptic and low-light plants can occur close to high-light plants. Single plants in a fossil environment would provide little secure information regarding sea level (SL) position at the time of growth. However, an analysis of community structure, where corallines are abundant in a Quaternary reef framework or sediment, could provide quite reliable information as to paleo-water depth.

The following discussion is based on rather extensive ecological studies carried out by several workers during the past twenty years. The principal field research to which this analysis applies was carried out in the Caribbean, the Hawaiian chain and the North Atlantic by numerous workers (see eg. Adey & Adey, 1973; Adey, Townsend & Boykins, 1982; Adey, 1978a, b; Adey & Vassar, 1975; Bosselini & Ginsburg, 1971; Ginsburg & Schroeder, 1973; Focke, 1977 and Focke & Gebelein, 1978); and in Atlantic Europe and the Mediterranean by Laborel, 1961; Peres & Picard, 1964; Bosence, 1976; and Alexandersson, 1977. Less extensive ecological work has also been carried out in the North Pacific (Adey, Masaki & Akioka, 1976; Lebednik, 1974; Doty, 1974), and the Great Barrier Reef of Australia (Steneck, in ms). The uniformity of results from one area to another suggests that this analysis applies to the entire world ocean except perhaps for the coast of Antarctica. The latter coast remains unstudied ecologically with regard to corallines and the systematic data are weak. Since the Antarctic climate has been polar for at least 15 million years, evolution has probably developed systematic/ecological patterns among corallines that we are not presently equipped to recognize.

In this paper, I make no attempt to present a thorough treatment of crustose coralline ecology. To do so, many systematic/ecological groups and patterns would have to be covered which are very unlikely to appear in the fossil records. Since the intention of this volume is to provide as much information as possible, in compact form, to assist workers in identifying the relative position of late Quaternary or Holocene SLs, I have chosen to concentrate on the more preservable ecological/paleoecological situations.

The following treatment of corallines as SL indicators is structured for convenience of discussion around the following biogeographic/ecologic units:

Table 1. A comparison of the dominant coralline elements of the Indo-Pacific and Caribbean. About two thirds of the species of each region have similar morphologic and ecologic elements in the other region. (Adey, 1979)

Genus	Caribbean: No obvious relationship	Caribbean: Similar	Caribbean: Very similar	Indo-Pacific: Very similar	Indo-Pacific: Similar	Indo-Pacific: No obvious relationship
Porolithon		*P. antillarum*	*P. pachydermum*	*P. onkodes*	*P. craespedium*	*P. gardineri*
Paragoniolithon	*Pg. solubile*	*Pg. "typica"*				*Pg. "conicum"*
Neogoniolithon	*N. caribaeum* *N. "ramusculum"* *N. mamillare* *N. "simplex"* *N. "moricatum"*	*N. "megacarpum"* *N. "clavynodum"* *N. strictum* *N. "lenormandi"* *N. accretum*			*N. fosliei* *N. "clavycymosum"* *N. frutescens* *N. rugulosum*	*N. rufus*
Hydrolithon		*H. børgesenii*			*H. reinboldii*	*H. breviclavium* *H. "laeve"* *H. "megacystum"*

		Lithoporella			
		L. melobesioides	*L. melobesioides*		
		Lithophyllum			
		L. congestum	*L. kotschyanum*		*L. pallescens* *L. "ganeopsis"* *L. "insipidus"* *L. mollucense*
L. intermedium				*L. punctatum*	
		Tenarea			
T. bermudense	*T. prototypum*			*T. tessellata*	
		Archeolithothamnium			
	A. dimotum *A. episporum*			*A. erythraeum* *A. "episoredion"*	
		Lithothamnium			
L. ruptile	*L. occidentale*			*L. australe*	*L. pulchrum*
		Mesophyllum			
	M. erubescens *M. "friaselis"* *M. "levolaminum"*			*M. madagascariensis* *M. prolifer* *M. mesomorphum*	*M. purpurascens* *M. "syrphetodes"* *M. siamense* *M. "fluatum"*
		Total number of species			
8 3	13	3	3	13 3	13
Listed Caribbean species-27				Listed Indo-Pacific species-32	

Tropical Environments

1. Systematic patterns; 2. Ridges and trottoir; 3. Coral reefs; 4. Rhodolith and 'maerl' bottoms.

Arctic, Subarctic and Boreal Environments

1. Systematic patterns; 2. Rhodolith and 'maerl' bottoms.

Temperate and Subtropical Environments

1. Systematic patterns; 2. Trottoir and subtropical-temperate bioherms; 3. 'Maerl' bottoms.

To assist the inexperienced worker, a brief discussion of field and laboratory techniques, methods of systematic analysis and a critical literature survey is also included.

TROPICAL ENVIRONMENTS

1. Systematic patterns

Little detailed ecological study of crustose corallines has been carried out in the Tropical East Pacific and Tropical East Atlantic. Area-wise, these biogeographic regions form only a small part of the tropics, and carbonate structures in general are only weakly developed. Nevertheless, this lack of information should be taken into account if SL analyses using corallines are to be carried out in those areas.

Crustose coralline algae in the Indo-Pacific and Caribbean are characterized at the generic level by similar populations (Adey, 1979; Adey *et al.*, 1982). Most genera occupy equivalent ecological situations in both areas, and over half of the species are 'pairs', most of which occupy similar niches (Table 1). This minimal evolutionary divergence has occurred over 3 to 20 m.y. of isolation and is considerably less than that to be found in the other major group of Tertiary bioherm builders, the scleractinian corals

Many coralline species, particularly in the tropics, are very specific in their requirements for light and wave action. Using coralline assemblages, while taking into account the mosaic of exposed and cryptic habitats within the reef framework, and the morphology of the species involved (extent and size of branches, thickness of crust, and so on), it is generally possible to determine the depth of a fossil coralline framework to within roughly ±2 m near mean low water, to within ±4 m from 5 to 20 m, and to within ±15 m at depth greater than 30 m. In addition, island shore, reef flat, reef-crest, upper fore-reef, and lower fore-reef zones, and shelf less than 40 m and shelf drop-offs, each have characteristic assemblages that it would be possible to recognize in the fossil record (eg. see Adey, 1979).

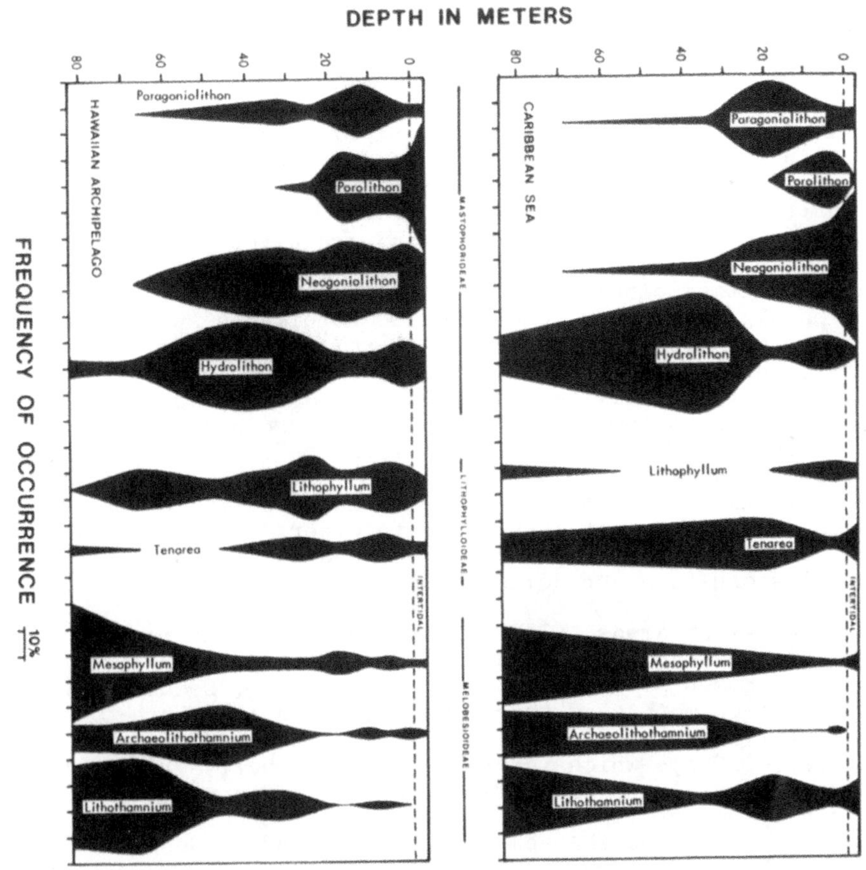

Figure 1. Patterns of crustose coralline depth distribution, by genus, in the Caribbean Sea and the Indo-Pacific (Hawaiian Archipelago). Only dominant framework or maerl-building genera are shown. (After Adey, Townsend & Boykins, 1982.)

Figure 1 compares the patterns of depth distribution, by genus, in the Caribbean area and the Hawaiian Archipelago. Considerable correlation can be seen in the major characteristics of generic dominance: *Porolithon* and *Neogoniolithon* dominate in the intertidal and uppermost subtidal to about 10 m; from that point to about 20-25 m *Paragoniolithon* tends to replace the shallow-water genera; from 20-30 m *Hydrolithon* begins to replace *Porolithon, Neogoniolithon* and *Paragoniolithon; Lithothamnium, Mesophyllum,* and *Archaeolithothamnium* dominate at depths greater than 50 m.

Thus, the assemblage of coralline genera in a carbonate framework can give a good indication of depth at the time of formation. The major genera involved in this level of interpretation are determinable in ground sections and scanning electron microscope (SEM) preparations, using only easily recognizable characters such as placement of heterocysts, type of hypothallium and the presence of multipored conceptacles and/or sporangial sori.

More precise determinations of depth are possible by examining the occurrence and extent of development of certain species that have well-defined ecological niches (Fig. 2). For example, *Porolithon pachydermum* and *craespedium* in the Pacific (Littler & Doty, 1975), as well as their Atlantic pairs *Porolithon pachydermum* and *Porolithon antillarum*, form massive structures virtually independent of other corallines only on intertidal and wave-beaten algal ridges. Some of these *Porolithon* species can form considerable framework within a few meters below low water, under very rough conditions, but species of other genera, particularly *Neogoniolithon* and *Paragoniolithon* species, then become intermixed with the *Porolithon*. *Porolithon gardineri* can also be found on the more protected parts of algal ridges, and it also occurs to some depth in lagoonal and back-reef situations. However, while the anatomy of this plant is similar to other *Porolithon* species, the growth habit (densely-branched) easily distinguishes it from its crustose and columnar relatives.

Reef-flat areas, although as shallow as the range dominated by the crustose and columnar *Porolithon* species on the exposed side of a reef or algal ridge, seldom have appreciable amounts of those plants. Such flats are often characterized by a red boring alga (Melobesioid 'C' of Littler, 1973a), which is probably a destructive rather than constructive agent. In this situation, often heavily grazed by parrot fish and urchins, crustose corallines are present usually on the shaded sides and undersides of rubble. In this basically cryptic environment, the most common corallines are *Hydrolithon reinboldii* and the easily distinguishable *Archaeolithothamnium erythraeum* in the Pacific and the pair species *Hydrolithon borgesenii* and *Archaeolithothamnium dimotum* in the Caribbean. These species are very useful as ecological indicators; except for *H. borgesenii* they are sharply restricted to shallow water and have distinguishing morphological characteristics making identification relatively simple. The large cells of *H. reinboldii* and *H. borgesenii* set them off from the other species of the genus; the unique sori of the *Archaeolithothamnium* is easily distinguishable in fossilized material, and the difference in their relative size is often great enough to distinguish them from the plants of the same genus associated with deeper water. However, the presence of abundant *Hydrolithon* and *Archaeolithothamnium* in a geological sample, even if not determinable to species, would indicate reef-flat environments only if *Lithothamnium* and *Mesophyllum* species were not present. On more protected reef flats and in shallow lagoons, the elongate and slender type of branching seen in *Porolithon gardineri* and *Neogoniolithon* species may be encountered. It is not likely that either of these would be found outside the reef at any great depth. It is possible to find representives of species characteristic of much deeper water in shallow areas, but these occurrences are limited to cryptic situations where light conditions are similar to those existing at depth. If the structural orientation and low abundance of these species does not give away their cryptic nature, the near proximity of shallow-water species in the framework would likely do so.

Abundant coarse-branch *Hyrolithon*, if mixed with a branching *Lithophyllum*, would indicate deeper reefs or island slopes (15-20 m) in the Pacific (Littler, 1973b; Adey *et al.*, 1982). This morphological form and ecological situation appears to be lacking in the Caribbean where shallow island slopes tend to be rather barren of coralline crusts in general. Where present, they are either plants of the wide-depth-ranging *Hydrolithon borgesenii* or species from shallow or deeper water. Large amounts of *Lithothamnium* and *Mesophyllum*, in either ocean, especially if in rhodolith (nodule) form, can only indicate bank depths of greater than 50-60 m. Although members of both of these genera do occur in shallow water, they are not nodule formers, but are represented by thin crusts in cryptic reef and ridge environments. Shade-loving *Neogoniolithon* species usually dominate even in those areas. One rather rare exception to this rule of thumb is the branching *Mesophyllum madagascariensis* and its Caribbean pair *M. erubescens*. This plant forms branching heads and sometimes nodules in shallow reef environments. The remainder of the coralline association should expose this case, however.

2. Algal ridges and trottoir

Algal ridges, massive carbonate frameworks built largely by crustose coralline algae, develop only within a few meters plus or minus of MSL in tropical-trade wind belts. It has been known for some time that crustose coralline algae can be major framework builders in Indo-Pacific reefs. Howe (1912), Weins (1962), and Doty (1974) briefly describe Indo-Pacific algal ridges and, in the latter cases, refer to the literature of the 1950s and 60s. Emery, Tracey & Ladd (1954) relate the height and distribution of the algal ridge at Bikini Atoll to wave action, and Doty (1974) discusses the major elements of the coralline flora that are involved in Indo-Pacific algal ridges. With the possible exception of the coralline-rich reef described by Easton & Olson (1976) for Oahu in the Hawaiian Islands, no Indo-Pacific algal ridge has been cored and its internal structure examined. Thus, this description relies heavily on Caribbean algal ridges (Adey, 1978a; Bosence, 1984). Figure 3 shows generalized diagrams of algal ridges of the Caribbean area based on mapping and surface observations and on numerous core holes drilled in the Lesser Antilles. Photographs of typical ridges are shown in Figures 4-7.

Caribbean algal ridges are often young structural caps (less than 5000 years old) on deeper coral-reef features (Adey, 1975), although other rocks can also serve as foundations (Pleistocene reef rock - Guadeloupe, Adey &

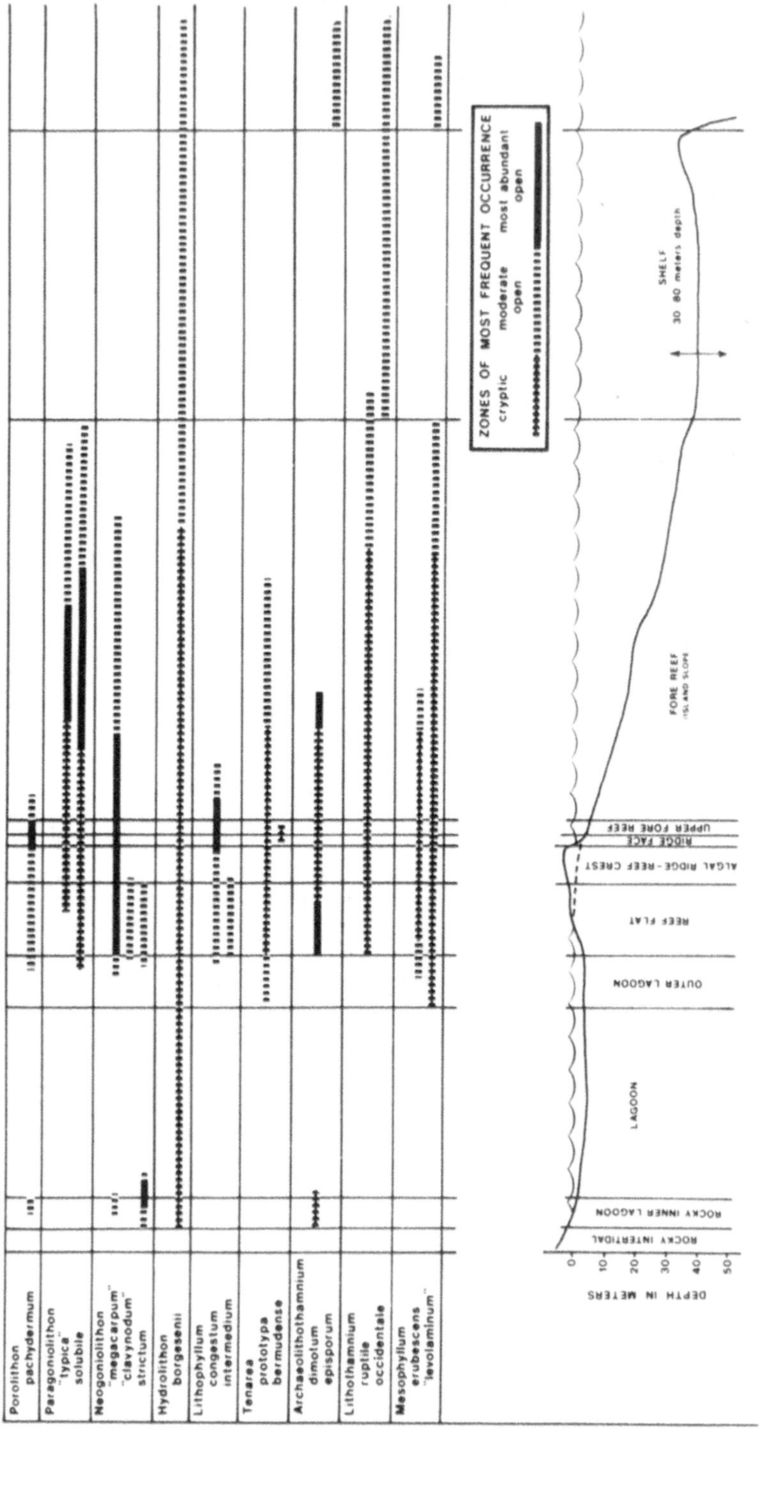
Porolithon pachydermum
Paragoniolithon "typica" solubile
Neogoniolithon "megacarpum" "clavynodum" strictum
Hydrolithon borgesenii
Lithophyllum congestum intermedium
Tenarea prototypa bermudense
Archaeolithothamnium dimotum episporum
Lithothamnium ruptile occidentale
Mesophyllum erubescens "levolaminum"
ZONES OF MOST FREQUENT OCCURRENCE
cryptic
moderate open
most abundant open
ROCKY INTERTIDAL
ROCKY INNER LAGOON
LAGOON
OUTER LAGOON
REEF FLAT
ALGAL RIDGE-REEF CREST
RIDGE FACE
UPPER FORE REEF
FORE REEF
ISLAND SLOPE
SHELF
30 80 meters depth
DEPTH IN METERS
0
10
20
30
40
50

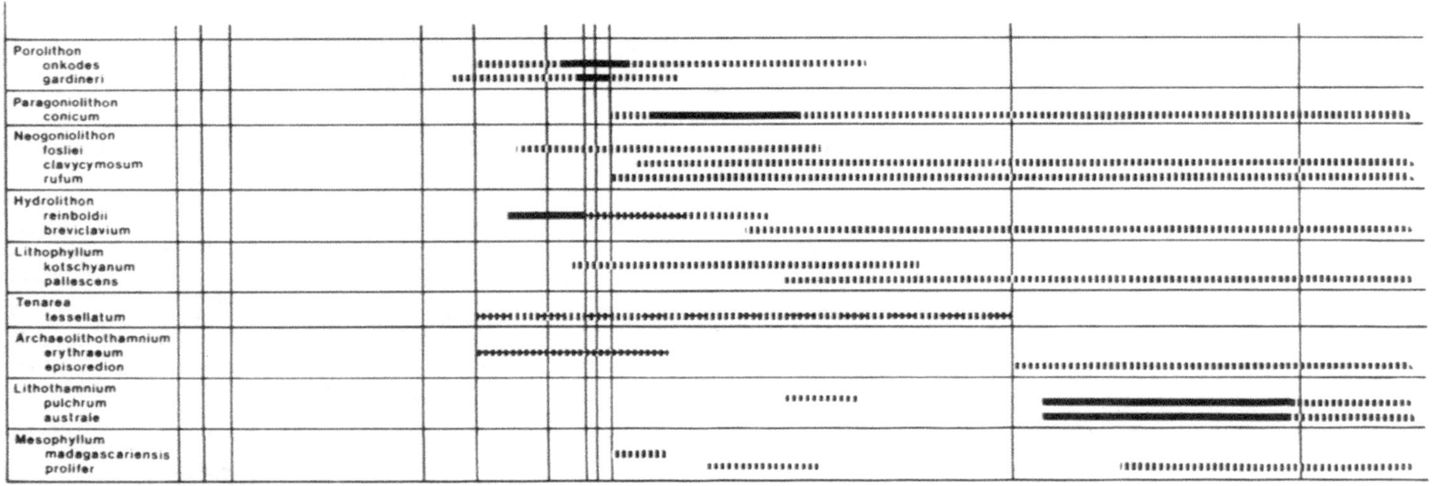

Figure 2. Major reef-building or maerl-forming species in the Caribbean and Indo-Pacific. Patterns of distribution from shore-line to drop-off on typical reef shore. Caribbean-above; Pacific-below. (After Adey, 1979.)

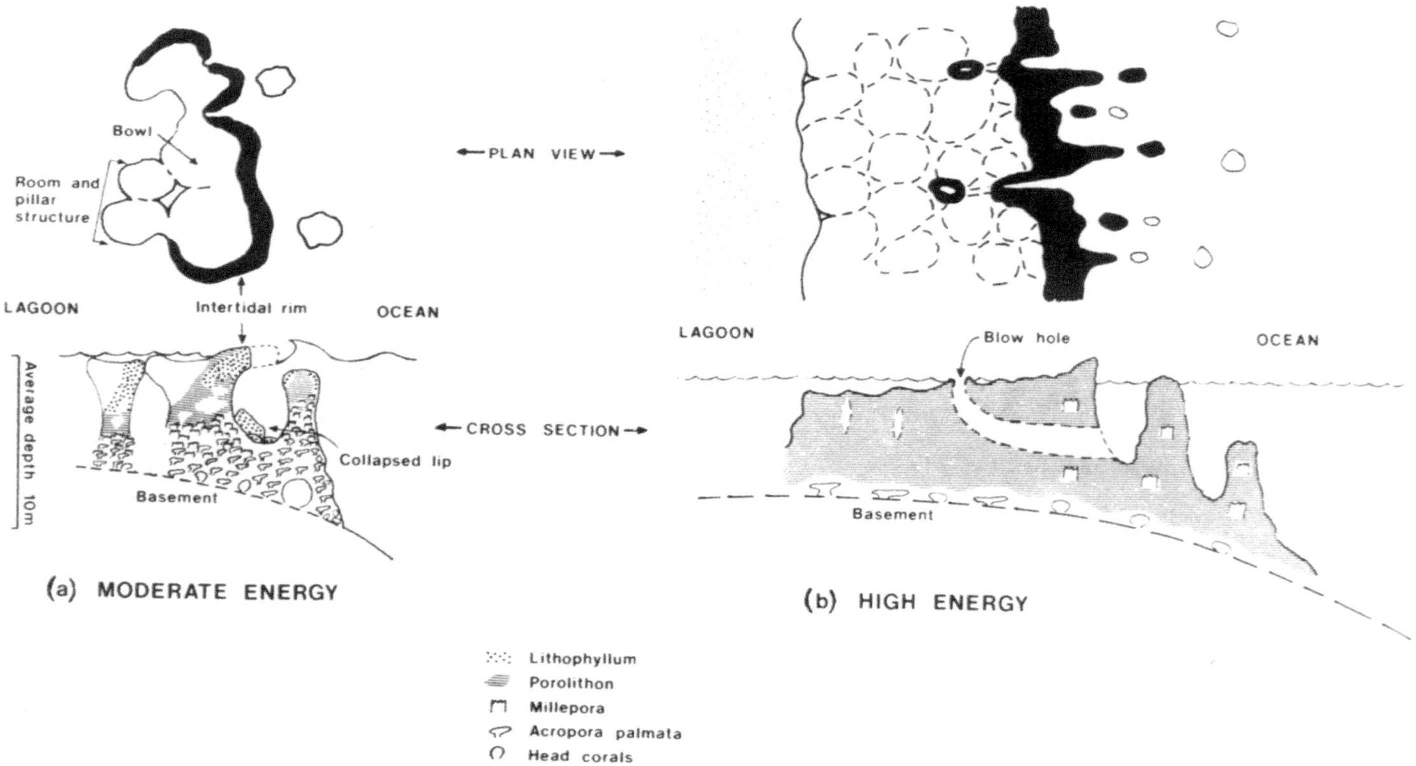

Figure 3. Morphology of typical moderate-energy (a) and high-energy (b) Caribbean algal ridges. A complete gradation of development occurs ranging from a single or few fused boilers to extensive ridge-like structures with minimal evidence of the existence of the original boiler units. (After Adey, 1978. See Doty,1974 for brief description of Pacific algal ridges.)

Figure 4. Front face of Corre-Corse algal ridge on St. Eutatius, Caribbean Sea on an exceptionally quiet day. The shear, lip-less seaward face of this feature is characteristic of high energy Caribbean algal ridges. (See Adey & Burke, 1976)

Figure 5. Lobate margin of Pinsonnelle algal ridge on Martinique. (See Adey *et. al.*, 1977)

Figure 6. **Surge channel on fore side of Pinsonnelle** algal ridge on Martinique

Figure 7. **Protected section of Pinsonnelle algal** ridge rim on exceptionally quiet day. Note mounds of *Porolithon pachydermum* projecting through stubby cover of *Laurencia* and *Sargassum* spp.

Burke, 1976, 1977; Holocene beachrock - Brazil, Kempf & Laborel, 1968, aeolianites - Bermuda, Laborel, pers.comm.). The thickness of the 'cap' is a function of basement depth (relative to SL rise) and wave action. Coring in the Lesser Antilles has shown the thickness of the coralline caps to range from roughly 1-8 m. In areas of medium energy, such as St. Croix, *Lithophyllum congestum* can be the major framework element of the algal ridges, typically between overlying and underlying *Porolithon pachydermum* strata. In the Caribbean *Lithophyllum congestum* appears to be the ecological equivalent of *Porolithon gardineri* in the Indo-Pacific. However, in ridges found in areas of very high energy, *Porolithon pachydermum* is always the dominant frame builder. In earlier (incipient) stages of development, *Millepora* is also an important contributor. Vermetid gastropods are generally consistent but minor elements in Caribbean ridges, although in and above corniches they can be considerably more abundant (see below). A number of cryptic calcifiers, including several *Neogoniolithon* species and the encrusting foraminifer *Homotrema*, along with secondary sediment and submarine cementation can considerably modify the original structure (Ginsburg & Schroeder, 1973; Bosence, 1984). Bosence (1984) describes a situation in St. Croix algal ridges where only a small percentage of the original coralline framework remains. Nevertheless, he felt that the signature of the algal ridge remained intact and could still be identified as a fossil structure. On the other hand, Bosence also regarded St. Croix ridges as in the process of complete degeneration due to the slowing of SL rise. This is unlikely the cause since the massive features of Barbuda, Guadeloupe, Marie Galante and Martinique are still actively growing without signs of extensive degeneration. Perhaps most interesting in this respect, however, is the relative abundance of secondary cryptic coralline framework (*Tenarea bermudense, Lithothamnium ruptile Mesophyllum* sp.) that could result in a misinterpretation of the relative depth of the structure.

Lithophyllum congestum is a major element of moderate-energy algal ridges. It achieves its highest growth rates at and slightly above the low water line. Since these growth rates greatly exceed those of both the same species and *Porolithon pachydermum* above and below that level, a tendency to develop 'mushroom-shaped' structures or boilers is inherent. Individual boilers often fuse to develop ridge-like structures, as shown in Figure 3, where large cavities and tunnels remain and growth of secondary framework can be considerable. In high-energy situations, the abundance of *Lithophyllum congestum* is reduced. Also, where cantilevered projections might otherwise occur, they tend to be broken off by the intense wave action. High-energy ridges tend to be built by mound fusion rather than 'mushroom' fusion and are considerably more solid. Also, since chambers tend to remain more open to light until fully closed, cryptic secondary coralline frameworks are probably much less important in these ridges. The extensive and well-developed algal ridge on the northeast side of Grande-Terre, Guadeloupe is a good example of this situation and type of construction. It would appear that the most difficult fossil situation to interpret, is one in which the high-energy environment has been changed, usually due to additional reef or ridge growth.

Although several coral species, especially *Porites astreoides* and *Diploria strigosa,* and a number of corallines are able to grow in the lower and landward 'bowl' sections of Caribbean algal ridges, these lagoonward sections are usually heavily bored, mainly by the sea urchin *Echinometra lucunter*. The intertidal rim tends to be more uniformly encrusted by *Porolithon pachydermum* and *Neogoniolithon* spp. and often bears a dense coating of fleshy algae. Several species of *Sargassum, Laurencia, Jania, Hypnea, Gracilaria* and *Turbinaria* are the major components. On medium-energy ridges fleshy algae can reach standing crops of 1-5 kg/m^2 (Connor & Adey, 1977; Adey *et al.*, 1977); however, in zones of very high energy, the upper surfaces can be quite bare of fleshy algae, presumably due to a combination of direct removal by wave action and drying on occasional calm days.

There seems to be little question that the development of algal ridge caps on Caribbean reefs is directly related to wave action. Within the trade wind belt of easterlies, both a moderately high wind strength and a directional constancy are necessary to build large algal ridges. The coralline community that is building and maintaining these structures in the face of considerable biological erosive elements requires freedom from intensive grazing on the one hand and minimum desiccation on the other (see Steneck & Adey, 1976). The maximum depth of development of massive coralline framework in the subtidal (ie. incipient algal ridge) is between 2-6 m (Adey & Burke, 1976, 1977). The maximum intertidal height is about 2 m. Thus, an extensive coralline structure dominated by *Porolithon,* with some *Lithophyllum, Homotrema* and *Millepore,* probably developed at least within +2 m to -6 m of the then existing SL. More typically, the range of elevation relative to SL is less than those values. Although scattered small plants of *Lithophyllum congestum* can extend to depths of several meters, extensive framework even on a small scale (ie. greater than a cubic decimeter) can only mean a situation very close to or slightly above mean low water.

Algal ridges occur in exposed waters that are usually non-turbid and in which the lower limit of *Acropora palmata* is typically 5-15 m (see however, Lighty, Macintyre and Stuckenrath, 1982). Under those conditions, they are considerably sharper indicators of SL than the coral framework they often cap.

Over most of the Caribbean/West Indian area, where algal ridges occur (see Adey, 1978a & b), the mean tidal range is less than 25 cm, and therefore microtidal in character (Kjerfve, 1981). While spring tides in these areas are from roughly 30-50 cm, the average trade wind sea in the calmest season in the same areas is greater than 1 m. These relationships are diagrammed in Figure 8. Rarely are

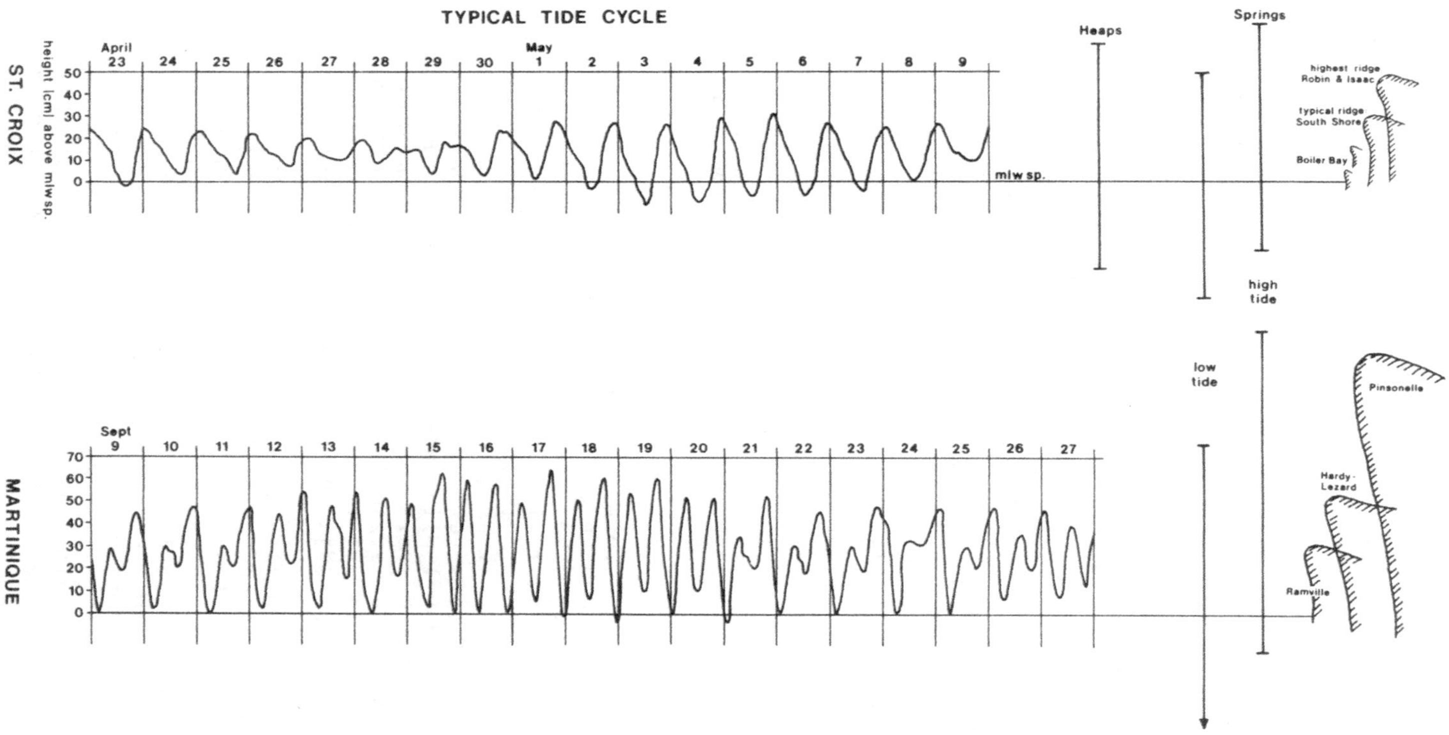

Figure 8. Relationship between tidal range and character, wave peaks and troughs (in moderate weather) and algal ridge height for St. Croix and Martinique, Caribbean. (Partly after Adey, 1975 and original data for Martinique.)

algal ridges not heavily wave washed; only on low-tide springs in very calm weather, perhaps ten to twenty days a year will wave crests not be breaking constantly over the crests of well-developed ridges. Even at those times, an unusually high wave every few minutes will usually pour over the ridge crest. It would appear that Windward Island algal ridges are higher relative to calm seas at low water springs than those at St. Croix. This may result from the more semidiurnal character of the Windward Island area and a more limited low-tide period for ridge drying in exceptionally calm weather.

Below, we will briefly discuss the 'Trottoir a *Tenarea*' or corniches; intertidal ledge-like structures similar to algal ridges in framework that are abundant in the Mediterranean. These features also occur scattered throughout the tropics (see eg. Focke, 1977 - Caribbean; Kempf & Laborel, 1968 - Brazil). In the Caribbean, several cases are known where these 'trottoir' grade into true algal ridges (Figs. 9-11), and they are obviously the equivalent of algal ridges on steeply-cliffed shores not blocked by reefs or algal ridges. While trottoir sharply delineate MSL, they are probably well-developed only in microtidal seas

A friable, sub-trottoir structure, constructed primarily by *Neogoniolithon strictum,* is known from the northern Caribbean/West Indian area. It is discussed in more detail under subtropical-temperate bioherms.

3. Coral reefs

Coralline algae are certainly abundant elements of tropical coral reefs. They have been widely regarded as cementing agents, welding together the carbonate framework of stony corals, although there is little real evidence to support this contention. More typically, these encrusting plants simply coat the basal coral structure to varying degrees. Thick encrustations over coral (ie. greater than several cm) generally develop in water depths of less than 6 m. Depths less than 6 m are also characterized by massive crusts with little marginal growth. Deeper crusts are often thin and 'leafy' in habit, often projecting off the substrate and onto adjacent coral or other objects.

A fossil carbonate substrate of growth, with consistently thin encrustations of coralline probably developed its crustose coating at depths over 10 m. However, as discussed above, generic, and in some cases specific, identification can greatly narrow the potential range, relative to then-existing SL. For example, community structure dominated by *Porolithon* probably developed at less than 10 m. *Hydrolithon* and *Neogoniolithon* species are often present in this shallow reef community, but they tend to be thin cryptic elements. Communities dominated by *Paragoniolithon* and *Neogoniolithon* with perhaps some *Hydrolithon* as cryptic elements, grew in the 10-30 m range relative to the then existing SL. These plants are often

Figure 9. Atlantic, *Lithophyllum congestum* 'trottoir' and upper notch at low tide. Southeastern Antigua.

Figure 10. Detail of Figure 9 'trottoir' showing well-developed *Lithophyllum congestum* heads.

Figure 11. Gradation of 'trottoir' into algal ridge at Indian Town Point, Antigua. (See Adey & Burke, 1976.)

leafy crusts with a coaxial hypothallium. Rarely does significant fixed (non nodular) coral-reef framework develop at depths greater than 30 m in tropical seas. Where framework does develop at those depths, *Hydrolithon* species are likely to be the dominant encrusting corallines.

4. Rhodolith and maerl bottoms

Under conditions of episodic wave or current movement, some species of tropical crustose corallines are capable of producing considerable growth and carbonate production in the free or unattached state (Parker & Curray, 1956; Bosellini & Ginsburg, 1971; Focke & Gebelein, 1978; Adey *et al.*, 1982). In some cases these growths develop from an encrustation of a core fragment of shell or coral. In other cases they represent continued growth of a detached coralline fragment. Unattached coralline nodules, showing continued growth in an environment of occasional movement, occur on a wide variety of substrate types from a silty mud to cobble bottoms made up largely of the nodules themselves. Although the terminology has not been stabilized in the literature, I prefer the terms 'rhodolith' for the larger, more massive types (2-5 cm, and more or less spherical) and 'maerl' for the smaller, more irregular types (see eg. Adey & McKibbin, 1970).

In the tropics both maerl and rhodoliths occur at depths ranging from low water to the limit of coralline algae, depending upon exposure (Weber van Bosse, 1901; Adey & Macintyre, 1973). However, they are perhaps more typical of deeper banks. Coralline nodules are capable of building sediment and subsequently rock formations. They are known from the fossil record, and could be critical in the interpretation of Quaternary SLs (eg. Schlanger & Johnson, 1969).

Depth interpretation of rhodoliths and maerl generally requires more care than that of the encrusting habit. While the majority of *Hydrolithon* and *Mesophyllum* species are nodule-formers in deep water, a few species occur in shallow seas. For example, the *Mesophyllum* species, *M. madagascariensis* in the Pacific and *M. erubescens* in the Atlantic, are both full-branching types, a relatively rare habit type for the genus. In some cases (Zaneveld, 1958; Adey *et al.*, 1982), they can be nodule formers in shallow water. Also, a single species of *Hydrolithon* occurs abundantly (primarily as a back-reef rhodolith former) in shallow water in both the tropical Pacific (*H. rheinboldii*) and Atlantic (*H. borgesenii*). However, in these cases the plants are usually accompanied by species of the genera *Porolithon* and *Neogoniolithon*, which only occur in shallow water. *Lithothamnium* in the rhodolith and maerl habit is only known from deep water.

ARCTIC, SUBARCTIC AND BOREAL ENVIRONMENTS

1. Systematic patterns

The often conspicuous development of crustose corallines in tropical reef environments has led to the often-stated misapprehension that crustose corallines are uniquely characteristic of such environmental conditions. However, crustose coralline algae thrive in polar as well as tropical conditions, and they commonly form the dominant component of benthic communities from low water down to the lower limit of the photic zone (Figs. 12-14; Adey & Macintyre, 1973; Adey, 1966, 1968, 1971).

The dominant form of crustose corallines in colder waters of the northern hemisphere is the encrusting type on rock substrate. These range up to 10-30 cm in thickness, but are usually not mineralized and often break off in sheets. The maximum depths of occurrence are more in the 20-40 m range, and as in the tropics a marked depth zonation of genera and species occurs. On the other hand, while *Clathromorphum* and *Phymatolithon* are the primary shallow genera with *Lithothamnium* and *Leptophytum* occurring in mid to deeper water, the depth ranges are generally wider than in the tropics and interpretation requires more care. Also, because of the general lack of heterocysts, the genera can be more difficult to identify with certainty in fossil form.

Figure 12. Ledge-boulder bottom at 10 m in southern Newfoundland. The branching heads are *Lithothamnium* spp. and range to about 15 cm in thickness.

Figure 13. Close up of portion of surface shown in Figure 12. Crustose species primarily *Clathromorphum* spp., branched forms *Lithothamnium* spp.

Figure **14. Coralline-coated cobble-pebble bottom** at 20 m in the northern Gulf of St. Lawrence.

Considerable information on the ecology of cold water corallines has been developed in the last 20 years and most of it applies to the rock encrusters. While this ecotype would be rarely preserved, even in the latest Quaternary fossil record, corallines on morainal boulders laid down in sea water have been found. Rather than using space in this treatment to deal in detail with the depth distribution of the rock encrusters in cold water, I simply refer to the major literature (Adey, 1966, 1968, 1971, 1973; Adey *et al.*, 1976).

2. Rhodolith and maerl bottoms

Rhodoliths and maerls can be extraordinarily abundant in the colder waters of the northern hemisphere and numbers of square miles of bottom occupied by them have been reported in Spitzbergen, Novoya Zemlya, Kara Sea, East Greenland, Norway, Newfoundland and Labrador (see eg. Figs. 15-17; Adey & Macintyre, 1973). The author has observed quarries established in elevated beds of rhodoliths in North Norway. It seems likely that such deposits are abundant on glacial rebound coasts, particularly in areas of fjord topography.

In arctic, subarctic and boreal waters, rhodoliths are primarily, species of the genus *Lithothamnium*. However, in the European boreal, *Phymatolithon polymorphum* (the dominant rock-encruster) occasionally develops rhodolith beds on shell/silt/sand bottoms at the base of the rock zone.

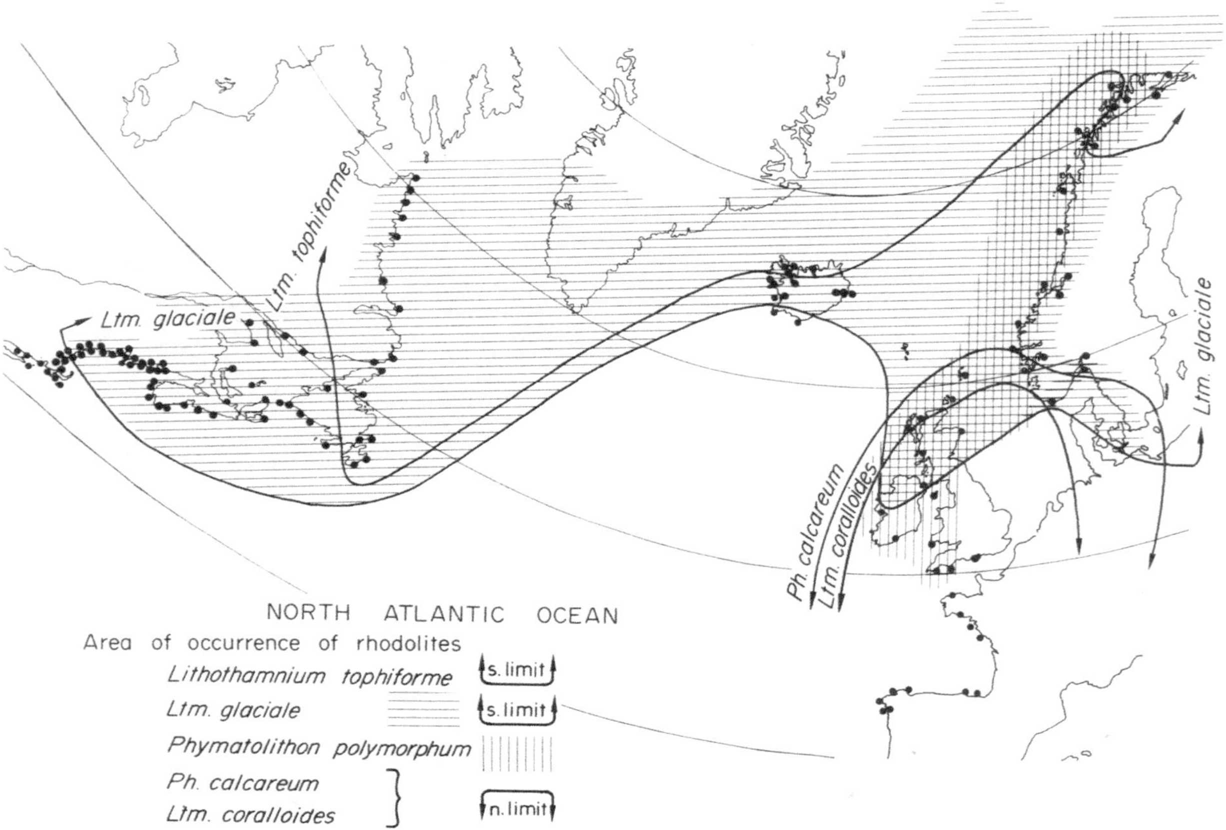

Figure 15. Areas of occurrence of maerl and rhodoliths in boreal-subarctic waters of the North Atlantic. *Lithothamnium tophiforme* range widely throughout the Arctic Ocean. *Phymatolithon calcareum* and *Lithothamnium corallioides* extend into the Mediterranean. (After Adey & Adey, 1973).

Figure 16. Maerl bottom of *Lithothamnium glaciale* at 6 m in White Bear Arm, Labrador. Scallop shells are about 12 cm diameter.

Figure 17. Maerl bottom of *Lithothamnium tophiforme* at 20 m in protected, strong current sound near Trans in North Norway.

In the British Isles the temperate species *Phymatolithon calcareum* and *Lithothamnium corallioides* are important maerl-formers.

In colder waters rhodoliths generally occur at the base of steeply-sloping coralline encrusted bedrock or boulders. However, their geographic pattern of occurrence is rather complex, depending upon light, current and wave exposure. Where some knowledge of the paleotopography is available, considerable depth interpretation is possible. Generally, extensive rhodolith deposits in arctic-boreal waters occur in bays and sounds, less than 10-30 m in protected areas and from 20-50 m in more exposed areas. In open-ocean areas in northern seas, extensive rhodoliths are to be expected only in the clearest waters and then in the 40-60 m range.

TEMPERATE AND SUBTROPICAL ENVIRONMENTS

1. Systematic patterns

At this time, relatively few ecological studies including corallines are available from temperate and subtropical climates. The most extensive work has been carried out in western Europe, the Mediterranean and in Japan (Peres & Picard, 1964; Adey *et ál.*, 1976; Adey, unpublished), and this discussion relies heavily on those studies.

Some of the more typically cold water and tropical genera of crustose corallines extend into temperate regions. From colder regions *Lithothamnium* and *Phymatolithon* often extend into temperate regions but primarily in mid to deeper water. From the tropical side *Neogoniolithon* spp. (in shallow water) and *Mesophyllum* spp. in mid depths are important. However, of the heavily calcified plants, the single characteristic genus of temperate waters is that of *Lithophyllum*. In both crustose and branching forms and with a closely-related genus (*Tenarea*), which adds the epiphytic habit, this genus of many species dominates mid-latitudes of both northern and southern hemispheres.

Unfortunately, in temperate waters, the depth distribution patterns of crustose corallines are not as sharply defined, on the generic level, as they are in the tropics and in boreal-arctic waters. *Lithophyllum* tends to be the most abundant genus in shallow water, but several species also occur at depth in moderate quantities. *Lithothamnium* species are almost always present but are generally restricted to the lower end of the photic zone. With abundant collections of living material, there would be little difficulty in estimating depth ±10 m based on the abundance of *Lithothamnium* alone.. Unfortunately, in fossil material the highly distinctive cover cells of this genus may not be preserved. Differentiation from the generally abundant *Phymatolithon* which ranges widely across mid-depths, would be difficult. At the species level, where taxonomic discrimination is often based on gross morphology, depth differentiation can be more narrowly defined. However, the

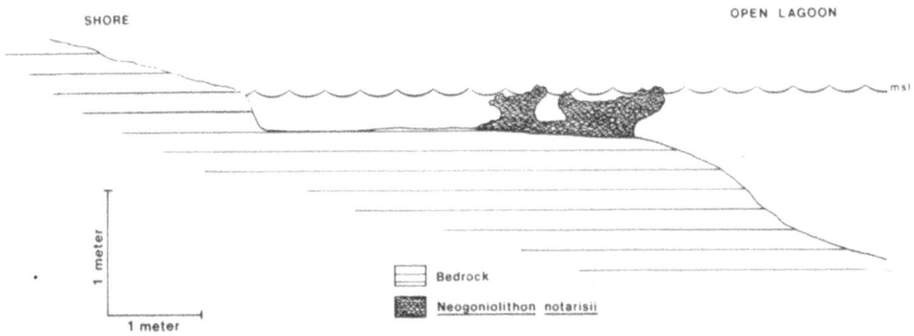

Figure 18. Idealized section across algal ridge at Bahiret el Bibane, Tunisia. (After Thornton *et al.*, 1978.)

appropriate taxonomic/ecologic studies for temperate regions in general are not adequate at this time to allow extensive use of species for depth discrimination. In temperate regions, depth discrimination is best carried out with reference to biohermal structures and the occurrence of rhodoliths and maerls, as discussed below.

2. Trottoir and subtropical-temperate bioherms

Reef-building stony corals do not extend into temperate seas with sufficient abundance or vigor of growth to construct bioherms. Coralline algae, on the other hand, extend from the tropics to arctic seas and are capable of developing considerable carbonate structure at any latitude. (Coralline framework, as most biogenic carbonate structure, is subject to the scraping and boring of animals, to physical damage by waves and to dissolution). In supersaturated seas, extensive secondary calcification can significantly improve the integrity of coralline skeleton and make it more likely to withstand erosive effects. Nevertheless, a positive balance between erosive and constructive elements is required to build coralline frameworks. The Mediterranean is a supersaturated sea and coralline framework is more or less abundant. However, while extensive algal ridges are known (eg. Thornton, Pilkey & Lynts, 1978), they are considerably more limited than tropical algal ridges in vertical height and in the overall mass of carbonate developed. The algal ridge at Bahiret el Bibane in Tunisia, although roughly 30 km long, has from 1/10 to 1/30 the vertical development of Caribbean ridges (Fig. 18). This could be a function of SL rise and antecedent platform position, although the relative rarity of algal ridges in the Mediterranean and their small size suggests that there is more to the story. Unlike tropical algal ridges, the Bahiret el Bibane algal ridge occurs in a protected, though rather large lagoon. The Mediterranean lies in the horse latitudes, rather than a trade wind belt, and one suspects that the same wind consistency and strength factors that control ridge

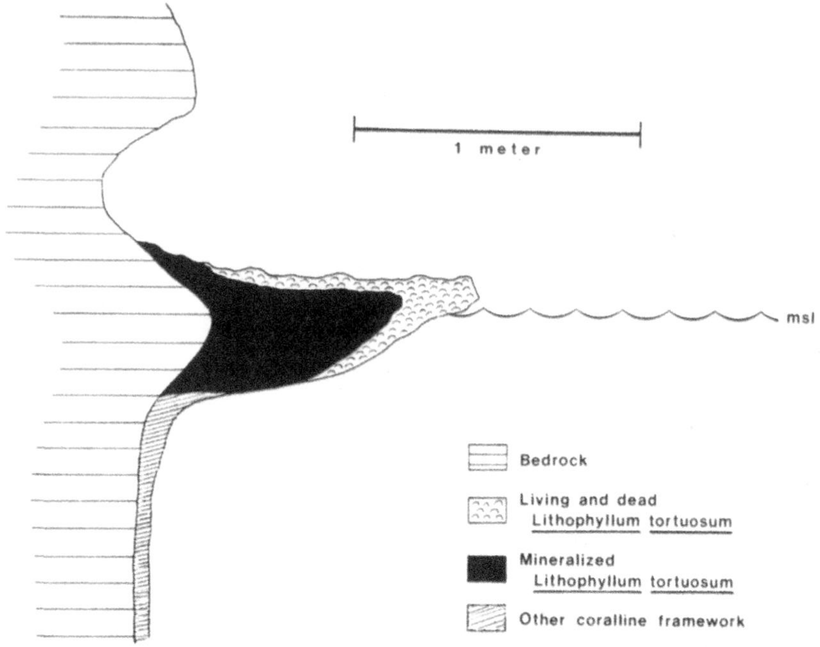

Figure 19. Generalized and well-developed Mediterranean 'Trottoir' of *Lithophyllum tortuosum* with upper notch. (After numerous authors.)

development in tropical seas relative to the grazing and boring of animals applies here as well. It is to be expected that the moderately high salinity in the lagoon with the Tunisian ridge is a primary factor limiting the activity of grazers but not that of the corallines (Thornton *et al.*, 1978). *Neogoniolithon notarisii* is the primary constructing element of the Tunisian ridge. Although a *Tenarea* species (sic.) is also present in small quantities, *Lithophyllum tortuosum*, vermetid gastropods, as well as urchins and other borers, are either absent or very limited.

Marked, crustose coralline, tide-level rims or bioherms, variously called 'trottoir', 'trottoir a *Tenarea*' or 'corniches' (Figs. 19-21), are abundant throughout the Mediterranean and have been described by many authors (eg. Peres & Picard, 1964). These are excellent SL indicators, and have been recently used to follow Holocene eustatic SL rise (Laborel, Delibrias & Boudouresque, 1983) and to determine patterns of tectonic activity (Thommeret *et al.*, 1981).

Neogoniolithon notarisii, along with vermetids is the primary framework element of trottoir in the southern and eastern Mediterranean, while *Lithophyllum tortuosum* and vermetids dominate trottoir in the northern and western parts of this sea. Using care to identify species and spatial relationships in fossil 'trottoir', at least in the

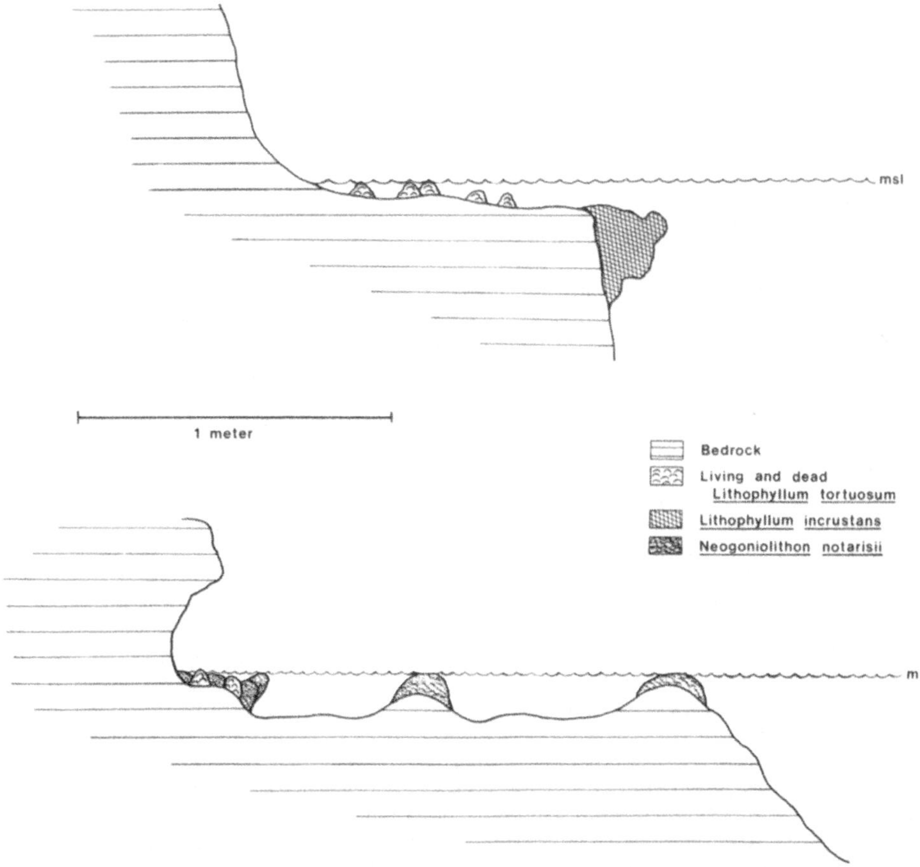

Figure 20. Examples of 'trottoir' of *Neogoniolithon notarisii* (below) subtidal *Lithophyllum incrustans* framework (above). (After Peres & Picard, 1964).

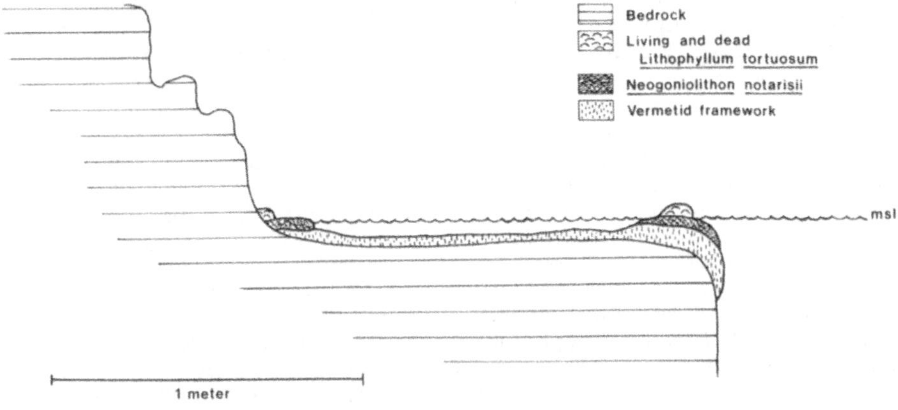

Figure 21. Complex framework of 'trottoir' including most major framework builders. (After Peres & Picard, 1964).

Mediterranean, an accuracy of determination of ±20 cm of the SL existing at time of growth can probably be determined.

In the western Atlantic, particularly in the Florida-North Bahamian area, a 'trottoir' of *Neogoniolithon strictum* is quite common. Usually, this micro-bioherm occurs in lagoons large enough to have a moderate local sea developed but not having open-ocean waves. These are relatively small structures, in the order of several meters wide one-half meter thick and in a rather broken form up to 1 km long. One of the best examples known to the author occurs in the southwest reef lagoon on the island of Mayaguana in the southeastern Bahamas. This 'fringing reef' is about 3 km long, along the northwest shore of the lagoon, which in summer is characterized by temperatures over 30°C and salinities over 38‰. Although much more friable in nature than the Bahiret el Bibane ridge, it is exposed only on extreme low tides and is generally similar in character.

The 'trottoir' of *Neogoniolithon strictum* grades into a maerl of the same species which is quite common in the subtropical West Atlantic on quiet shores. In some cases, mixed with the sea-grass *Thallasia testudinum,* it can form extensive, coarse sediments along lagoon shores.

As with algal ridges, a rule of thumb with regard to any fossil coralline biohermal framework is: any massive coralline carbonate structure, not constructed by rhodoliths or maerls was built in close approximity with SL in a narrow tidal regime. A further generalization that has considerable application in this context is: any extensively branched *Neogoniolithon* forming a framework (or for that matter, an extensive and tight agglomeration of maerl) has formed very close (within a meter or two) to mean low water.

In tropical seas, at depths greater than 2 m to perhaps 6-8 m, crustose corallines produce major accumulations only as rhodoliths or maerls. While these accumulations can be geologically quite significant, they are at best biostromal in character and probably would not be mistaken for 'in place' framework. This situation is apparently not the case in the Mediterranean, where a 'coralligene' bottom extends from the upper subtidal to at least 20 m and probably in some cases to well over 100 m. Based on diving at several stations, Alexandersson (1969) stated that 'The lithified algal framework, which was found down to 20 m, occurs only on hard bottoms, normally as irregular ridges or patches with a thickness between 5 and 20 cm.' Laborel (1961), applying the term 'coralligene' to coralline encrustations at depths over about 30 m, and based on information from many stations, has classified numerous types of coralligene bottom. Based on observations of numerous workers he states that one metre is a very common thickness for that formation and that thicknesses of more than 8 metres are known (Laborel, pers. comm., 1983). Some of this framework, particularly at greater depth, may be early Holocene and late Pleistocene. However, the primary question, in this context, is whether or not 'coralligene' bottom could be mistaken for a SL feature.

Figure 22. Typical maerl of Rio de Vigo, Spain at 15 m depth.

It seems likely that most coralligene framework, would not be taken for an intertidal or mediolittoral feature because of its 'sheet' or 'mound-like' character. In the event of confusion, examination of crustose corallines involved in the framework (and many animals as well), should establish the depth of growth. While *Lithophyllum tortuosum* and *Neogoniolithon notarisii* are the primary elements in the SL features, almost immediately below these 'trottoir' they are replaced by *Lithophyllum incrustons* and *Phymatolithan lenormandii*. Beginning at several meters depth, depending on the locality, *Mesophyllum lichenoides, Phymatolithan polymorphum,* the leafy and very distinctive *Lithophyllum (Pseudolithophyllum) expansum* and calcified *Peyssonelia* spp. become the dominant coralline, framework builders.

3. Maerl bottoms

Maerl or fine-grained rhodolith bottoms consisting of *Lithothamnium corallioides* and *Phymatolithon calcareum* occur abundantly from the British Isles and western France southwards into the Mediterranean (Fig. 22). In the Mediterranean, several other genera and species (eg. *Lithothamnium futiculosum* and *Lithothamnium valens*) also become elements of this extensive and well-known but little studied bottom type. In deeper areas a calcified member of the red algal family Squamariaceae (*Peyssonnelia polymorpha)* also contributes to the maerl (see eg. Jacquotte, 1962;

Peres & Picard, 1964; Bosence, 1976; Adey & McKibbin, 1970). Milliman (pers. comm.) reports abundant maerl in the western Mediterranean with modern dates up to 80 m depth. He found maerls up to 250 m deep, but below 80 m, dates of 7000-11,000 years were tallied.

In the turbid, high-energy waters of western Europe, maerl occurs primarily at depths of 1-10 m (but ranging to 20 m), generally within more or less protected bays. In the quieter and clearer Mediterranean, on the other hand, depths up to 40 m are more characteristic. However, live maerl is occasionally reported to 100 m and Peres & Picard (1964) report a record 180 m in very clear water near Crete. Considerable depth and SL interpretation is possible in Quaternary deposits of maerl. But just as with cold water rhodoliths, a precision interpretation is dependent upon some additional knowledge of environment, particularly exposure and water clarity. Several species of *Lithophyllum*, eg. *L. incrustom* and *L. expansum* are commonly found in the Atlantic/Mediterranean maerls. These are likely to be absent at depths below 30-50 m, depending on water clarity.

In the biohermal section above, I briefly discussed the Floridean/Bahamian 'trottoir' of *Neogoniolithon strictum*. Associated with this trottoir, generally extending for a short distance into the lagoon from the base of the structure, is a maerl of the same species. It is apparent that under the right wave conditions *Neogoniolithon strictum* is capable of continued growth in the free state, and rather fragile maerls of 10-15 cm in diameter are common in the region. Whether in the maerl or the biohermal state, *Neogoniolithon strictum* and its related Caribbean branching species of the genus only occur in lagoonal situations very close (1 m) to SL conditions existing at the time of growth. Where a *N. strictum* trottoir is well-developed extensive sand beaches dominated with skeletal fragments of the same species can be developed.

TECHNIQUES FOR USING CORALLINES AS INDICATORS OF SEA LEVEL

Under some circumstances coralline algae are excellent and very precise indicators of past SLs. Thommeret *et al.*, (1981) and Laborel (1983) have shown this in precision determinations of Holocene tectonic movements in Crete (Fig. 23) and late Holocene SL rise in the Mediterranean (Fig. 24). Both ^{14}C dating and species determination using gross morphology were applied in these cases.

In general, however, and particularly outside of the Mediterranean, one should be careful of significant skeletal diagenesis when taking C^{14} dates. While the age pattern shown in Figure 23 for Crete was quite consistent with a constantly-rising SL, several younger anomalies appeared in the Port Cros data (Fig. 24), suggesting either contamination with younger crusts or later re-mineralization and inclusion of young carbon. Also, as shown in the carbon-14 data for Caribbean algal ridges (Fig. 25), a

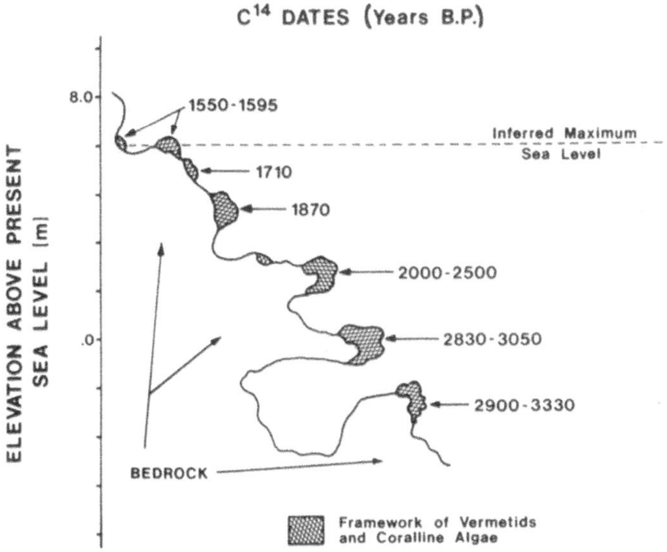

Figure 23. Fossil 'trottoir' on raised shore at Moni Khrisoskalitisas, Crete. The carbon-14 ages are consistent with a eustatic rise of sea level from approximately 3000 to 1500 BP, after which a rapid elevation of the coast exposed the entire series. (After Thommeret *et al.*, 1981.)

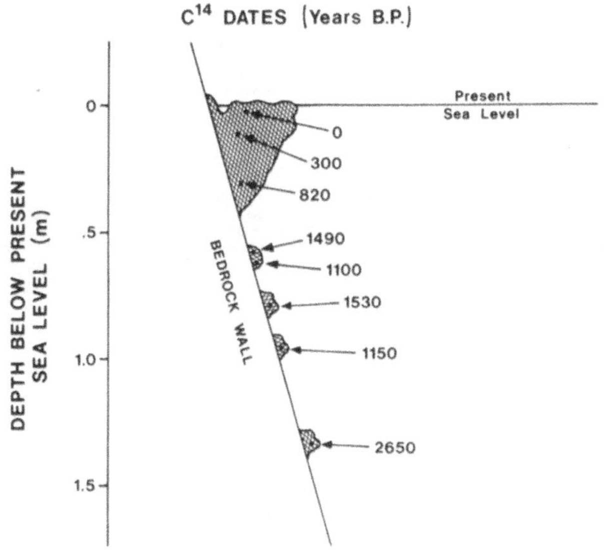

Figure 24. Carbon-14 dates on a 'trottoir' of *Lithophyllum congestum* and preexisting remnants at Port Cros, Mediterranean France showing pattern of sea-level rise over the last 2600 years.

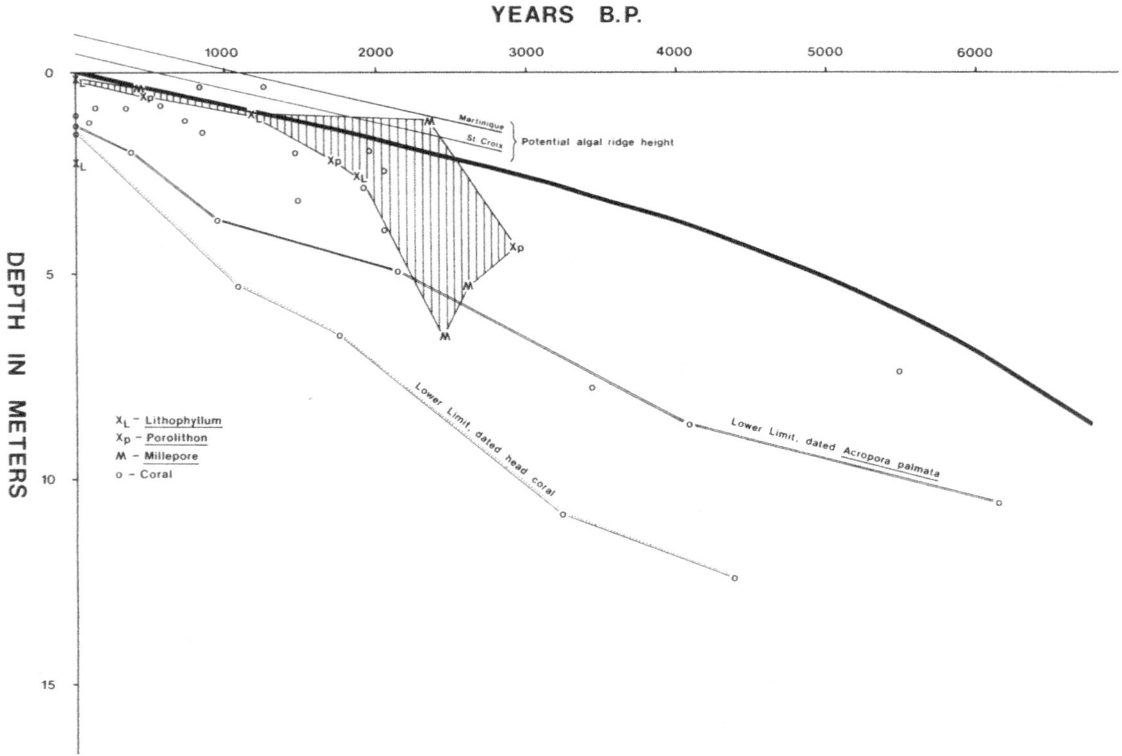

Figure 25.

Carbon-14 dates from Caribbean algal ridge framework on Martinique and St. Croix. The vertical hachuring encloses dates taken on coralline algae and *Millepore* framework that one would expect to lie a maximum of 1-2 m from low water. An apparent tendency to relatively younger dates with time is seen (Note one young date - Modern - at 2.2 m is assumed to be due to hole collapse during drilling.) Sea-level curve after Lighty *et al.*, 1982.

possibility for younger contamination exists there as well. Routine ground sectioning and examination, where there is a question, should allow judgement as to whether or not there is a re-mineralization problem. Also, one should take great care to separate primary framework from cryptic, secondary framework, as many thousands of years difference may result. Sectioning should also solve this problem as long as one avoids using typical cryptic species and genera.

While bioherm morphology will almost always indicate whether or not a framework is a SL feature or not, under some conditions corallines can develop significant carbonate accumulations at depth. As a general rule microscopic examination of sections and routine species identifications is particularly desirable to avoid potential errors.

Classically, slide-mounted, ground-sections were used to examine coralline anatomy and morphology. Unfortunately, it is difficult to use such sections unless they are less than 20 μm in thickness, and the preparation technique is time-comsuming at best. SEM examination of relative vertical fracture sections provides an alternate technique which can provide as much information as ground sections for most fossil material.

Orientation in a framework, relationship to cavities and examination of the interrelationships of individual specimens can be critical to interpretation. In some cases genera and species normally occurring near the local light limits of crustose corallines can occur in cryptic situations less than a meter from high-light, typical shallow-water taxa. Usually only a few deep water species will exhibit this shallow-water 'shift' and examination of community structure and habitat relationships will provide an appropriate answer to the depth question.

CRUSTOSE CORALLINE SYSTEMATICS

This is not the place to provide an in-depth treatment of the identification of crustose corallines. The recent literature treating the subject is extensive. However, because of the mystique that has developed surrounding the group, particularly with regard to the difficulty of identification, a brief discussion is in order. Most important, the crustose corallines are no more difficult to identify than species of most large groups of marine algae and invertebrates. Particularly now with the general availability of the SEM to replace the more time-consuming methods of paraffin and ground sectioning, routine identification of much material by the non-specialist or with an occasional check of critical specimens by the specialist is well within the realm of possibility. A brief working key and pictoral glossary is presented below. For a more in-depth treatment, and reference in detail to the literature, see: Adey *et al.* (1982), Adey & Adey (1973), and Adey & Macintyre (1973).

The basic phylogenetic differentiation in crustose corallines is the development of thick-walled, asexual sporangial caps in the Melobesioideae (that is, multipored asexual conceptacles) (Table 2). This differentiation has perhaps led to some of the hesitancy of nonspecialists to work with corallines, as the statement that asexual conceptables or microtome sections are required before any generic identification can be attempted, is common in the literature. This may be true in some cases, and conceptacles and sectioning is always desirable, but frequently not necessary. The working key (Table 3) is based on the possible absence of asexual conceptacles, and is aimed toward generic breakdown wherever possible with minimum preparation. A 30-50X dissecting microscope or SEM preparations is critical for this work.

Table 2

Suprageneric classification
of Recent
crustose coralline algae
family Corallinaceae

SPORANGIAL CAPS PRESENT (Conceptacles or sori multi-pored)
Subfamily Melobesioideae (J. Areschoug) Yendo 1902

Meristem growth, Lithothamnium cover cells, conceptacle primordia from meristem
Tribe Melobesieae Areschoug 1852
Lithothamnium, Archeolithothamnium, Mesophyllum, Clathromorphum

Progressive growth, rounded cover cells, adventitious primordia
Tribe Phymatolitheae Adey and Johansen 1972
Phymatolithon, Leptophytum

SPORANGIAL CAPS ABSENT (Conceptacles single-pored)

Secondary pits present
Subfamily Lithophylloideae Setchell 1943

Tribe Lithophylleae (Setchell) Mason 1953
Lithophyllum, Tenarea

Secondary pits absent
Subfamily Mastophoroideae (Svedelius) Setchell 1943

Without heterocysts
Tribe Pseudolithophylleae Adey and Johansen 1972
Pseudolithophyllum, Lithoporella

With heterocysts
Tribe Porolitheae
Porolithon, Paragoniolithon, Neogoniolithon, Hydrolithon

Note: does not include genera which are only parasitic, epiphytic or otherwise unlikely to be preserved in the fossil record.

Table 3. Working key to the genera of crustose coralline algae. Does not include genera which are only parasitic, epiphytic or otherwise unlikely to be preserved in any significant quantity in fossil record.

D-dissecting microscope or SEM fractures required
M-sections or SEM fractures required

D1.	Multipored asexual conceptacles present	
D2.	Hypothallium coaxial on radial fracture	
D2.	Hypothallium noncoaxial	Mesophyllum
D3.	Large sunken meristem obvious on vertical fracture	Clathromorphum
D3.	Sunken meristem obvious	
M1.	Lithothamnium cover cells, large meristem	Lithothamnium
M1.	Non-Lithothamnium cover cells-small celled meristem	
M2.	Pore cells present, conceptacles strongly raised	Leptophytum
M2.	Pore cells absent, conceptacles sunken or weakly raised	Phymatolithon
D1.	Sporangia in sori, sporangial caps visible	Archeolithothamnium
D1.	No multipored asexual conceptacles sunken or weakly raised	
D4.	Individual crusts extremely thin (if thick, by overlapping growth, primarily hypothallium)	
D5.	Large cells (10 to 15μ diam)	Lithoporella
D5.	Smaller cells	Tenarea
D4.	Thicker crusts, developing perithallium	
D6.	Heterocysts present	
D7.	In fields	Porolithon
D8.	Hypothallium plumose	
D8.	Hypothallium coaxial	Paragoniolithon
D9.	Hypothallium obviously coaxial or multilayered on fractured section	Neogoniolithon
D9.	Not coaxial or if multilayered thin	
D10.	Perithallium granular in appearance	Hydrolithon
D10.	Nongranular	Neogoniolithon
D6.	Heterocysts absent (may have to verify with sections or SEM)	
D11.	Surface distinctly glossy and generally deep brownish-red or buried sporangial sori obvious on vertical fracture	Archeolithothamnium
D11.	Surface dull, buried sori absent	
D12.	Hypothallium coaxial on radial fracture	Mesophyllum
D12.	Hypothallium noncoaxial	
D13.	Large sunken meristem obvious on vertical fracture	
M3.	Perithallium non-layered, secondary pits absent	Clathromorphum
M3.	Perithallium layered, secondary pits	Lithophyllum
M4.	Secondary pits	
D14.	Palisade hypothallium	Tenarea
D14.	Nonpalisade hypothallium	Lithophyllum
M4.	No secondary pits (fusions)	
M5.	Lithothamnium cover cells	Lithothamnium
M5.	Non-Lithothamnium cover cells	Leptophytum

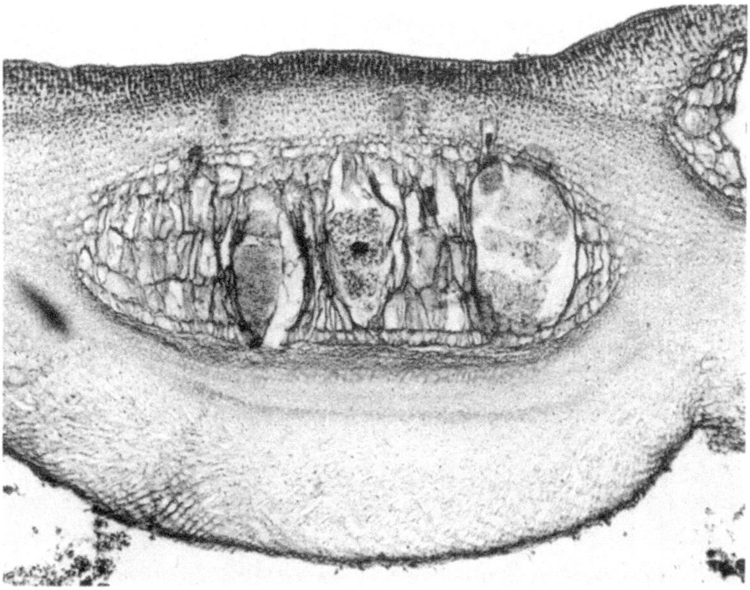

Figure 26. Tetrasporangial conceptacle with sporangial caps, ie. multipored. *Mesophyllum purpurasiens*. Conceptacle 500 µm diameter.

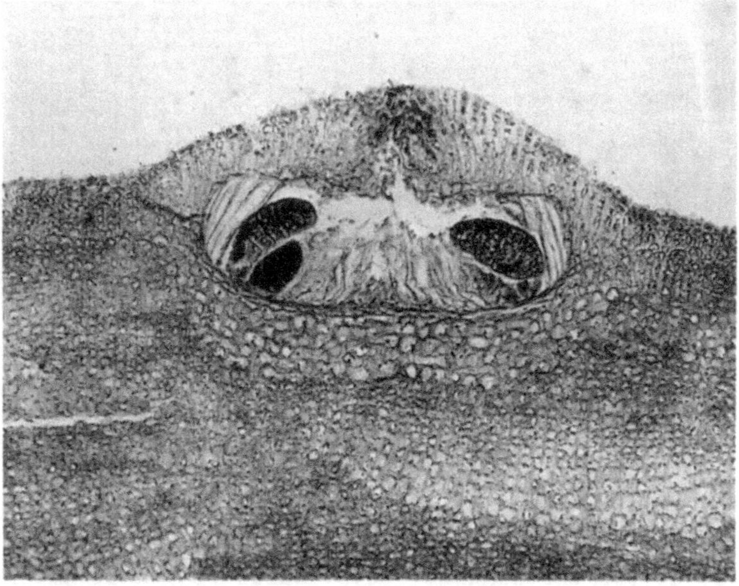

Figure 27. Single-pored tetrasporangial conceptacle. *Paragoniolithon conicum*. Conceptacle 250 µm diameter.

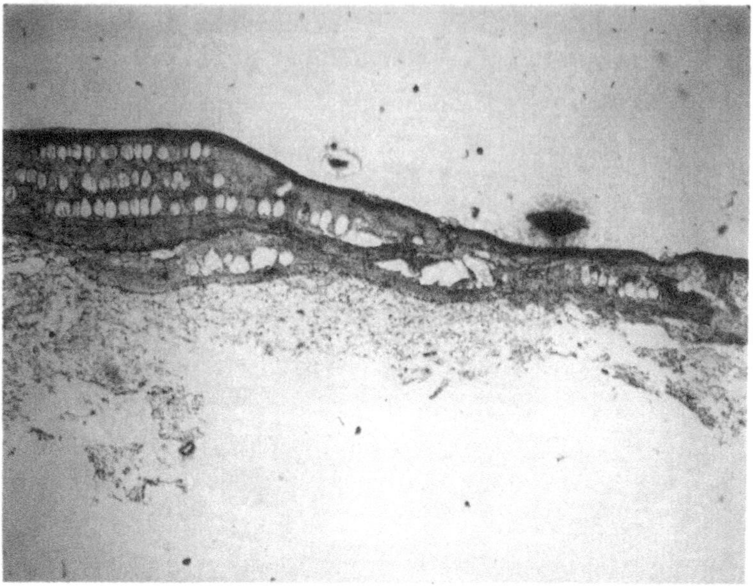

Figure 28. Sporangial sori in crustose *Archeolithothamnium episoredioni*. Sporangia approximately 150 μm long.

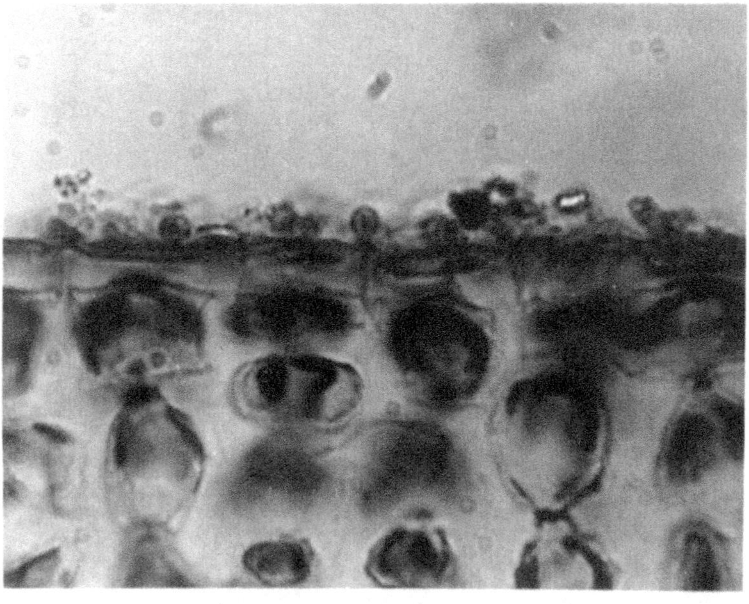

Figure 29. *'Lithothamnium'* cover cells (epithallium) in *Lithothamnium australe*.

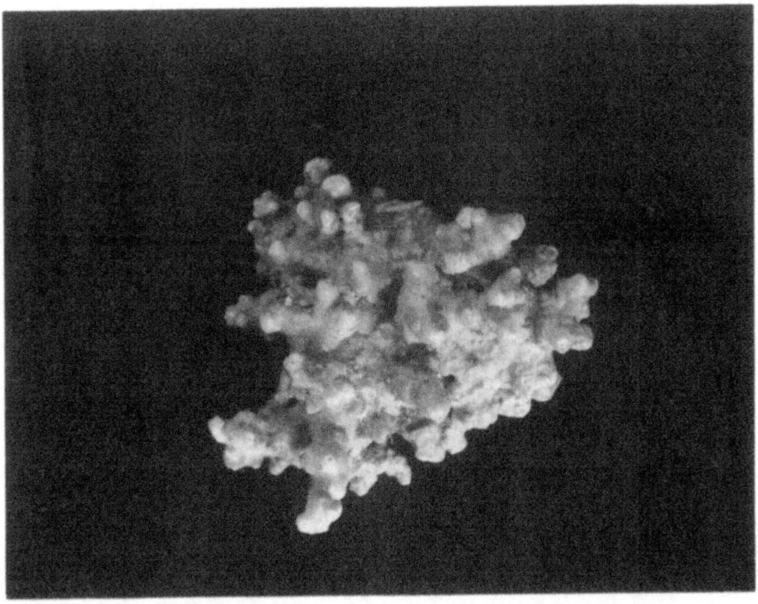

Figure 30. Branching maerl fragment of *Lithothamnium australe*. Natural size.

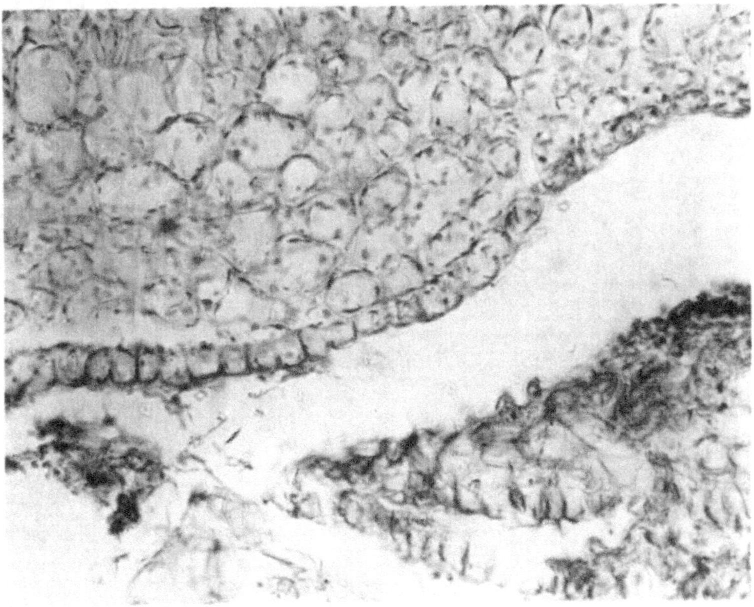

Figure 31. Single-layer hypothallium in ***Hycholithon reinboldii*****.** Note 'spongy', large-celled perithallium. Hypothallial cells 20 µm long (high).

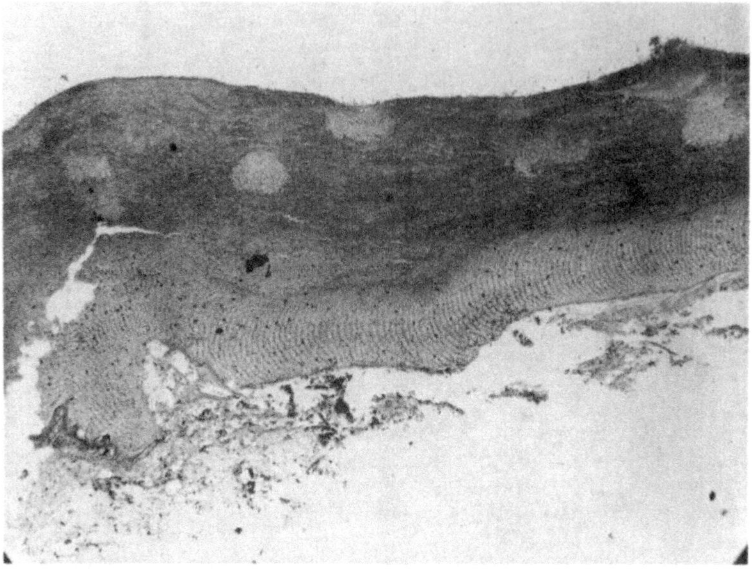

Figure 32. Coaxial hypothallium in *Paragoniolithon conicum*. Hypothallial thickness approximately 300 µm.

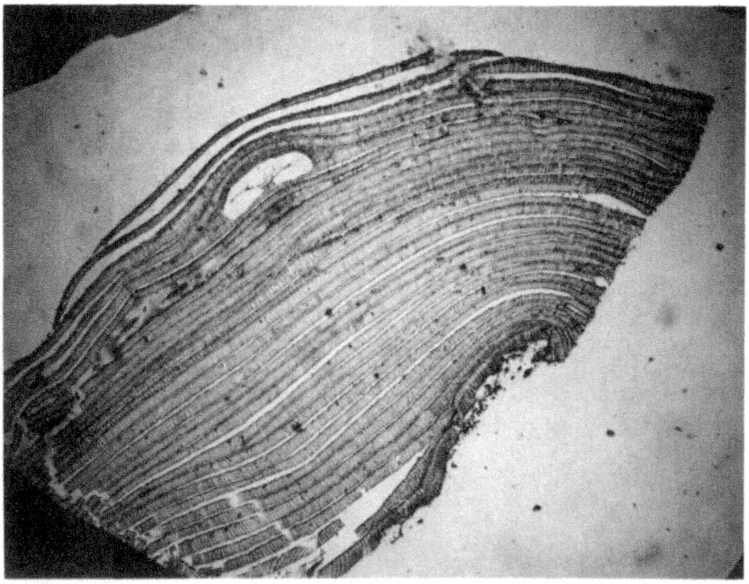

Figure 33. Palisade hypothallium of overlapping plants in *Tenarea tessellatum*. Cell length (ie. height) 40 µm.

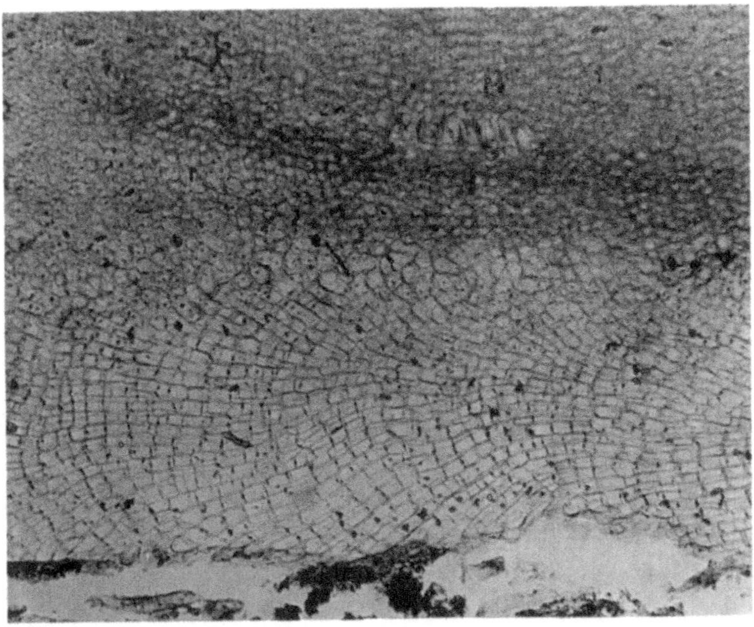

Figure 34. Plumose hypothallium in *Porolithon ankodes*. Note heterocyst field, upper middle. Hypothallial thickness about 200 µm.

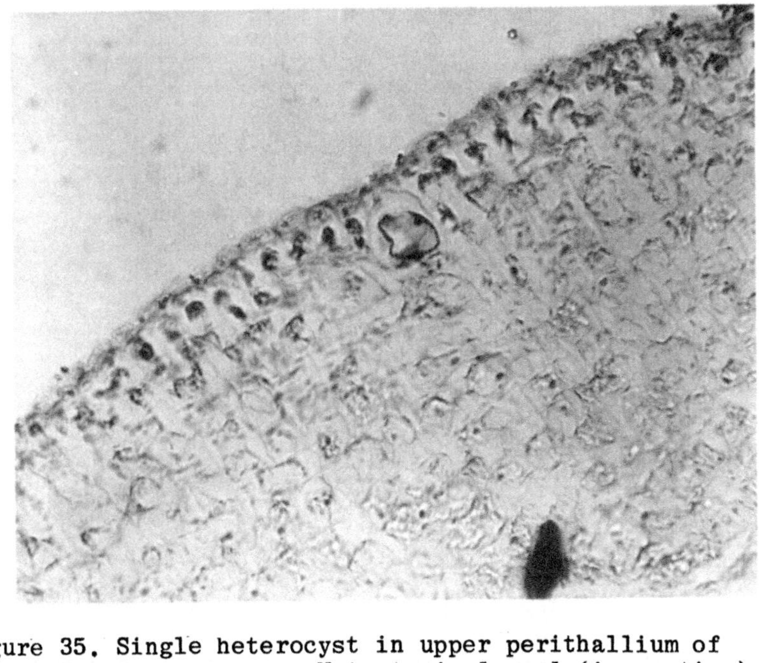

Figure 35. Single heterocyst in upper perithallium of *Hydrolithon laeve*. Note typical oval (in section) epithallial (cover) cells. Heterocyst about 20 µm in diameter.

Figure 36. Loose heterocyst fields characteristic of *Paragoniolithon*. Heterocysts 10-20 μm in diameter.

Figure 37. Secondary pits in *Lithophyllum lotschyanum*. Cells approximately 8-10 μm in diameter.

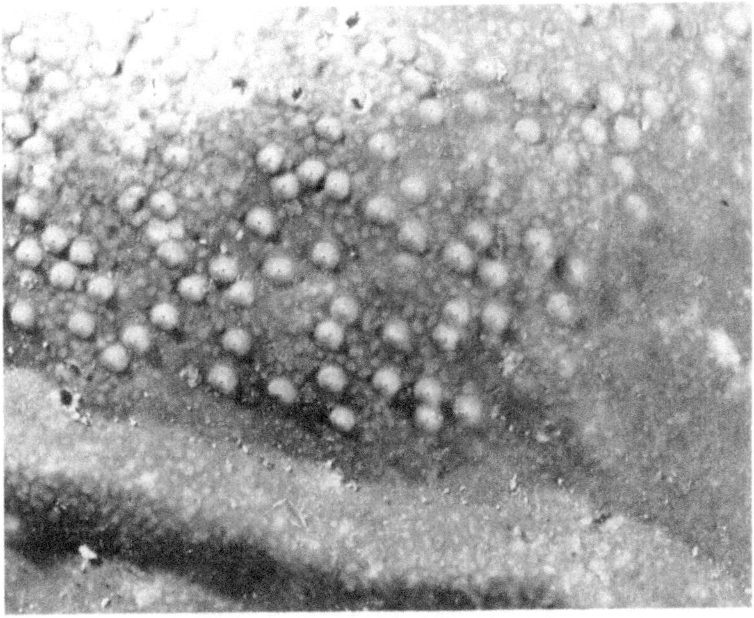

Figure 38. Single-pored tetrasporangial conceptacles and 'indistinct' heterocyst fields of *Paragoniolithon conicum*. Conceptacles approximately 400 μm diameter.

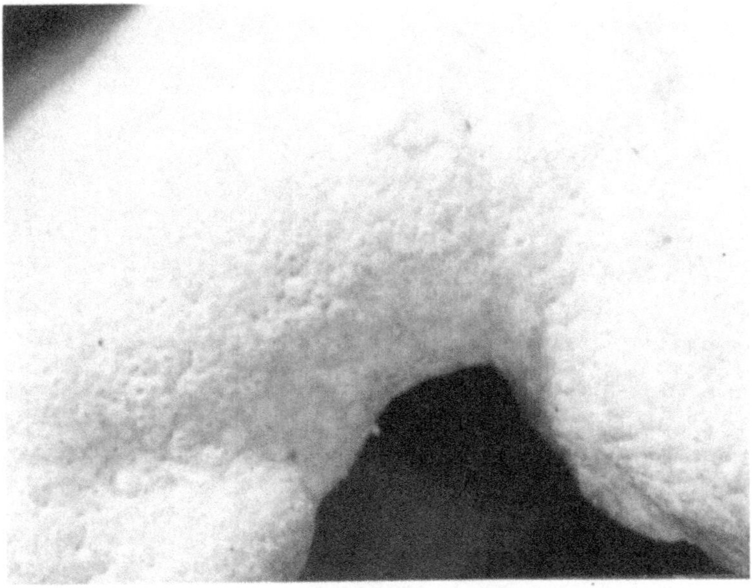

Figure 39. 'Pustulous' heterocyst fields of *Porolithon ankodes*. Fields approximately 60 μm in diameter.

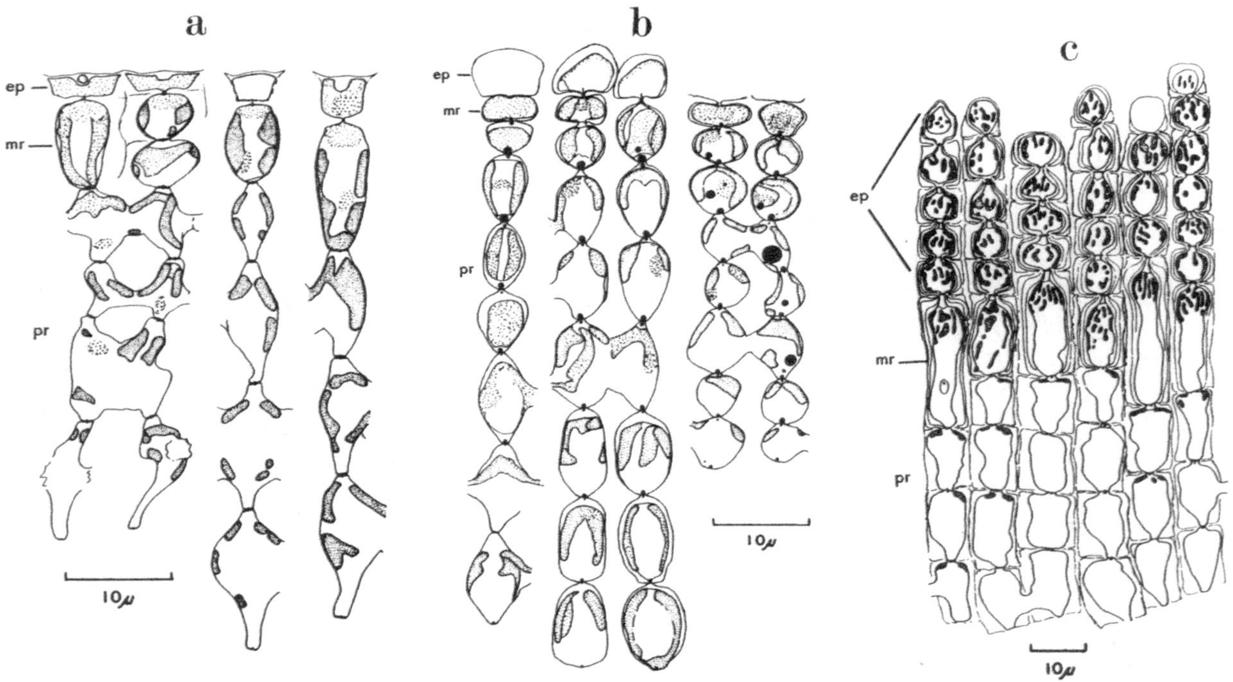

Figure 40. ep-epithallium; mr-meristem; pr-perithallium. a. Upper perithallium and epithallium in *Lithothamnium coralloides* (Adey & McKibbin, 1970). b. Upper perithallium and epithallium in *Phymatolithon calcareum* (Adey & McKibbin, 1970). c. Upper perithallium and epithallium in *Neopolyporolithon reclinatum* (Adey & Johansen, 1971)

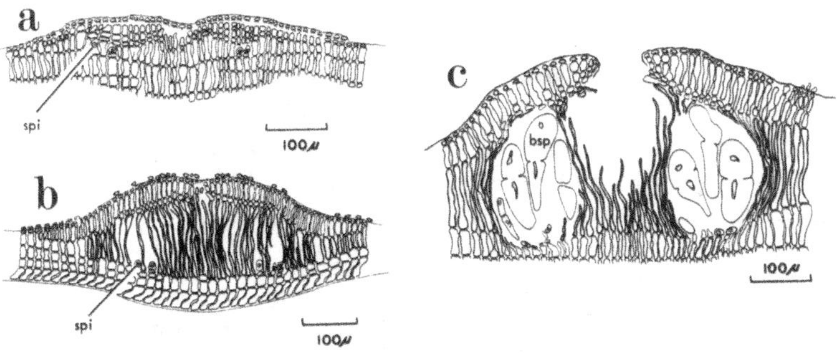

Figure 41. Development of asexual conceptacles stage, a. mature conceptacle. spi-sporangia initial; in *Tenarea* sp. (Adey, 1965a); b. primordial; c. mid-bsp-bispofds.

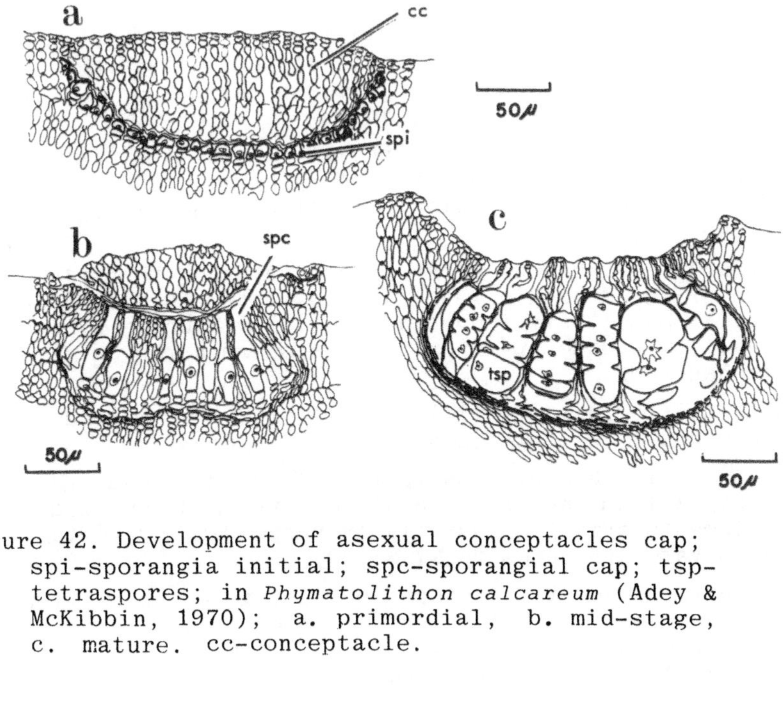

Figure 42. Development of asexual conceptacles cap; spi-sporangia initial; spc-sporangial cap; tsp-tetraspores; in *Phymatolithon calcareum* (Adey & McKibbin, 1970); a. primordial, b. mid-stage, c. mature. cc-conceptacle.

WORKING GLOSSARY

(see Figures 26 to 42 for pictorial explanation)

branching: development of crust into elongate and sometimes multiple-forked protuberances. (Figure 30)

conceptacle: a cavity in which the reproductive structures are developed. (Figures 26,27,28,41,42)

conceptacle cap: layer of varying thickness of vegetative cells above developing conceptacle, typically breaking away as it matures. (Figure 42)

conceptacle primordium: layer of cells from which conceptacle develops. In most genera in meristem, in Phymatolitheae, adventitious in perithallium. (Figure 42)

epithallium: generally the tissue lying outside the lateral meristem and being produced from it. Equivalent to cover cells. (Figure 40)

fusion (of cells): a relatively wide passage through the cell wall connecting cells of adjacent filaments, apparently allowing the free communication of cytoplasm. (Figures 29,31,40)

heterocyst (trichocyte): perithallial cell that is generally larger than surrounding cells. Commonly does not show the original hairlike projection. (Figures 35,36,39)

hypothallium: the basalmost tissue in which the filament(s) are oriented parallel to the substratum (basal hypothallium). Coaxial hypothallium: any kind of hypothallium tissue in which the cells are arranged in arched rows.

meristem: region or cell layer in which most of the vegetative cell division occurs. (Figure 40)

perithallium: thin or massive tissue lying between the hypothallium and the lateral meristem and composed of filaments oriented essentially perpendicular to the substrate. (Figures 26,27,28,29)

primary pit: generally closed or restricted connections between cells of a filament, ie.: cells that have separated by cell division in the formation of a filament, in most red algae including corallines.

secondary pit: pit connecting adjacent cells not belonging to the substrate. Only Lithophylleae, perhaps occasionally *Archeolithothamnium*. (Figure 37)

sorus: group of single sporangial compartments. Only *Archeolithothamnium*. (Figure 28)

sporangium: a cell which eventually produces one or more spores. (Figures 26,27,28,41,42)

sporangial cap: mucilaginous plug formed as part of the sporangial wall and eventually filling the pore in the conceptacle roof above the sporangium. (Figures 26,43)

ACKNOWLEDGEMENTS

Many people have helped over the years in developing the background that made this paper possible. I owe special thanks to Jacques Laborel for assistance with the Mediterranean 'trottoir' with which I have little personal experience, to Orson van de Plassche for very helpful editorial comments, to Charlotte Johnson for the drafting effort and to Diane Cloyd for typing.

REFERENCES

Adey, W., 1966. The distribution of saxicolous crustose corallines in the northwestern North Atlantic. *Journal of Phycology*, 2, 49-54

Adey, W., 1968. The distribution of crustose corallines on the Icelandic Coast. *Scientia Islandica*, 1, 16-25

Adey, W., 1971. The sublittoral distribution of crustose corallines on the Norwegian Coast. *Sarsia*, 46, 41-58

Adey, W., 1975. The algal ridges and coral reefs of St. Croix. *Atoll Research Bulletin*, 187, 1-67

Adey, W., 1978a. Algal ridges of the Caribbean Sea and West Indies. *Phycologia*, 17, 361-367

Adey, W., 1978b. Coral reef morphogenesis: a multidimensional model. *Science*, 202, 831-837

Adey, W., 1979. Crustose coralline algae as micro-environmental indicators for the Tertiary. In: *Historical Biogeography*, eds. J. Gray & A. Boucot, Oregon State University, 459-464

Adey, W. & Adey, P., 1973. Studies on the biosystematics and ecology of the epilithic crustose corallinacean of the British Isles. *British Phycological Journal*, 8, 343-407

Adey, W. & Burke, R., 1976. Holocene bioherms of the eastern Caribbean. *Bulletin of the Geological Society of America*, 87, 95-109

Adey, W. & Burke, R., 1977. Holocene bioherms of the Lesser Antilles - geologic control of development. *Bulletin of the American Association of Petroleum Geologists*, 4, 67-81

Adey, W. & Macintyre, I., 1973. Crustose coralline algae: a re-evaluation in the geological sciences. *Bulletin of the Geological Society of America*, 84, 883-904

Adey, W. & McKibbin, D., 1970. Studies on the maerl species *Phymatolithon calcareum* and *Lithothamnium coralloides* in the Rio de Vigo. *Botanica Marina*, 13, 100-106

Adey, W. & Vassar, M., 1975. Colonization, succession and growth rates of tropical crustose coralline algae (Rhodophyta, Cryptonemiales). *Phycologia*, 14, 55-69

Adey, W., Adey, P., Burke, R. & Kaufman, L., 1977. The Holocene reef systems of eastern Martinique. *Atoll Research Bulletin*, 218, 1-40

Adey, W., Masaki, T. & Akioka, H., 1976. The distribution of crustose corallines in eastern Hokkaido and the biogeographic relationships of the flora. *Bulletin Faculty of Fisheries,* (Hokkaido University), 26, 303-313

Adey, W., Townsend, R. & Boykins, W., 1982. The crustose coralline algae of the Hawaiian Islands. *Smithsonian Contributions to Marine Science,* 15, 1-74

Alexandersson, E.T., 1969. Recent littoral and sublittoral high-Mg calcite lithification in the Mediterranean. *Sedimentology,* 12, 47-61

Alexandersson, E.T., 1977. Carbonate cementation in recent coralline algal constructions. In: *Fossil Algae,* ed. E. Flugel, (Springer Verlag), 261-269

Bosence, D., 1976. Ecological studies on two unattached coralline algae from western Ireland. *Palaeontology,* 19, 365-395

Bosence, D., 1984. Paleoecological studies of two coralline algal reefs, St. Croix, Caribbean. *Palaeontology,* 27(3), 549-574

Bosellini, A. & Ginsburg, R., 1971. Form and intertidal structure of recent algal nodules (Rhodolites) from Bermuda. *Journal of Geology,* 79, 669-682

Connor, J. & Adey, W., 1977. The benthic algal composition, standing crop and productivity of a Caribbean algal ridge. *Atoll Research Bulletin,* 211, 1-40

Doty, M., 1974. Coral reef roles played by free-living algae. *Proceedings of the Second International Coral Reef Symposium,* (Brisbane, 1973), 1, 27-33

Easton, W. & Olson, E., 1976. Radiocarbon profile of Hanauma reef, Oahu, Hawaii. *Bulletin of the Geological Society of America,* 87, 711-719

Emery, K., Tracey, J. & Ladd, H., 1954. Geology of Bikini and nearby atolls. *United States Geological Survey Professional Paper,* 260-A, 1-265

Focke, J., 1977. The effect of a potentially reef-building vermetid/coralline algal community on an eroding limestone coast. *Proceedings of the Third International Coral Reef Symposium,* (Miami, 1977), 1, 239-245

Focke, J. & Gebelein, C., 1978. Marine lithification of reef rock and rhodolites at a fore-reef slope locality off Bermuda. *Geologie en Mijnbouw,* 57, 163-171

Ginsburg, R. & Schroeder, J., 1973. Growth and submarine fossilization of algal cup reefs, Bermuda. *Sedimentology,* 20, 575-614

Howe, M., 1912. The building of 'coral' reefs. *Science,* 35, 837-842

Jacquotte, R., 1962. Etude des fonds de maerl de Méditerranée. *Recueil des Traveaux de la Station Marine d'Endoume,* 26, 141-235

Kempf, M. & Laborel, J., 1968. Formation de vermets et d'algues calcaires sur les côtes du Brésil. *Recueil des Traveaux de la Station Marine d'Endoume*, 43 59 325-435

Kjerfve, B., 1981. Tides of the Caribbean Sea. *Journal of Geophysical Research*, 86, 4243-4247

Laborel, J., 1961. Le concrétionnement algal 'coralligene' et son importance géomorphologique en Méditerranée. *Recueil des Traveaux de la Station Marine d'Endoume*, 23, 37-60

Laborel, J., Delibrias, G. & Boudouresque, C., 1983. Variations récentes du niveau marin à Port-Cros mises en évidence par l'étude de la corniche littorale à *Lithophyllum tortuosum*. *Comptes Rendus Acádémie des Sciences Paris*, 297, 157-160

Lebednik, P., 1974. The genera *Clathromorphum* and *Mesophyllum* from the northeast Pacific Ocean. Ph.D. thesis, University of Washington, Seattle

Lighty, R., Mactintyre, I. & Stuckenrath, R., 1982. *Acropora palmata* reef framework: a reliable indicator of sea level in the western Atlantic for the past 10,000 years. *Coral Reefs*, 1, 125-130

Littler, M., 1973a. The population and community structure of Hawaiian fringing reef crustose corallinaceae. *Journal of Experimental Marine Biology and Ecology*, 11, 103-120

Littler, M., 1973b. The distribution, abundance and communities of deep water Hawaiian crustose Corallinaceae. *Pacific Science*, 27, 281-289

Littler, M. & Doty, M., 1975. Ecological components structuring the seaward edges of Pacific Reefs. *Journal of Ecology*, 63, 117-129

Parker, R. & Curray, J., 1956. Fauna and bathymetry of banks on continental shelf, northwest Gulf of Mexico. *Bulletin of the American Association of Petroleum Geologists*, 40, 2428-2439

Peres, J. & Picard, J., 1964. Nouveau manuel de bionomie benthique de la Méditerranée. *Recueil des Traveaux de la Station Marine d'Endoume*, 31, 5-137

Schlanger, S. & Johnson, C., 1969. Algal banks near LaPaz, Baja, California - Modern analogues of source areas of transported shallow water fossils in pre-alpine flysch deposits. *Palaeogeography, Palaeoclimatology and Palaeoecology*, 6, 141-157

Steneck, R. & Adey, W., 1976. The role of environment in control of morphology in *Lithophyllum congestum*, a Caribbean algal ridge builder. *Botanica Marina*, 19, 197-215

Thommeret, J., Thommeret, Y., Laborel, J., Montaggioni, L. & Pirazzoli, P., 1981. Late Holocene shoreline changes and seismo-tectonic displacements in western Crete. *Zeitschrift für Geomorphologie*, Supplement Band, 40, 127-149

Thornton, S., Pilkey, O. & Lynts, G., 1978. A lagoonal crustose coralline algal micro-ridge: Bahiret el Bibane, Tunisia. *Journal of Sedimentary Petrology*, 48, 743-750

Weber van Bosse, A. & Foslie, M., 1904. The Corallinaceae of the Siboga Expedition. In: *Siboga Expeditie*, 61, 110 pp. (Leiden)

Weins, H., 1962. *Atoll environment and ecology*, (Yale University Press, New Haven, London)

Zaneveld, J., 1958. A *Lithothamnion* bank at Bonaire (Netherlands Antilles). *Blumea Supplement*, 4, 206-219

10

Vermetid gastropods as sea-level indicators

J. LABOREL

ABSTRACT

In tropical and subtropical waters, several genera of marine Gastropods belonging to the family *Vermetidae* may, under certain circumstances, build conspicuous benches or corniches on the littoral zone of rocky coasts that are subject to moderate or heavy surf action. These formations always grow at a constant level in the biological zonation, which is the upper limit of the infralittoral zone (Peres & Picard, 1964) or the lower limit of the midlittoral zone (Stephenson & Stephenson, 1949). As this level, according to the distribution of intertidal organisms (and not by any tidal reference), is the height at which periodic emergence at low tide (or in the trough between two waves) is of minor significance as an ecological factor, it may be accepted as a *biological zero level* (even though it does not correspond exactly to the 'mean sea level' as defined by hydrographers).

The shape and size of vermetid formations is subject to variations, according to the nature of the substratum and the relative and absolute importance of tide and surf.

Vermetid formations fossilize easily when raised above their normal growth level, provided they are protected from rain and the humic acids of the soil. When submerged, their preservation is generally not so good due to biological erosion.

The measurement of the height of an elevated vermetid-zone remain must be made, whenever possible, in relation to the summit of the living Vermetids of the present shore in the same place. In tideless seas, and especially when Vermetid lines depend upon erosional features such as *notches* or marine platforms, they constitute almost ideal sea-level indicators with a range of error as small as

±0.1 m. In less favourable conditions (strong surf and tide, absence of erosion notch), their interpretation may be more difficult, requiring a thorough survey of present zonation and environmental conditions, but they are still quite useful even if the range of error reaches ±0.5 m or even ±1 m in some cases.

Vermetid shells are wholly aragonitic, and are therefore little prone to contamination and are easily dated by ^{14}C techniques, which makes them all the more desirable as sea-level indicators.

INTRODUCTION

Since the early 1960s the little known group of *Vermetidae*, a family of uncoiled and fixed Gastropods, has acquired a certain importance as a sea-level indicator (SLI). The aim of this chapter is to present that group to non-biologists, explain how it can build biostromes closely linked to mean sea level (MSL) and give some practical recommendations regarding interpretation in the field of vermetid remains, possibilities of error, reliability of their datation by the ^{14}C method and so on. First, however, it is necessary to describe the group, its principal genera, the way they live and the places where they grow how they can be linked to MSL and the various types of bioconstructions they can build.

PRESENTATION OF THE PRESENTLY LIVING VERMETIDS

1. Origin of the group

The Gastropod family *Vermetidae* is related to the family *Turritellidae* from which it differs mainly by a complete uncoiling of the whorls of the adult shell and a sessile way of life, the individuals being generally cemented to a substrate without any possibility of moving themselves. They look somewhat like tubeworms with which they have been long mistaken.

The first fossil forms to be related to this family are some dubious species from the Upper Cretaceous, not to mention some much older Paleozoic fossils, the exact nature of which is not known. The first true Vermetids occur in lower Eocene beds (Cuisian) and look like modern *Serpulorbis* (big, solitary forms). Then, in Miocene times the important colonial, biostrome-building genus *Petaloconchus* appears. It represents a major component of Pleistocene and Holocene reefs. This genus has undergone a very strong and brutal decline for unknown reasons, notably in tropical Atlantic waters where it had nearly disappeared at the turn of the century (Laborel, 1977), and seems to have been superseded by another important reef-building vermetid genus: *Dendropoma*. This genus is presently the more common Vermetid on tropical shores.

2. Systematics

Recent works by Keen (1961), the leading specialist of this difficult group, divide the family into five genera and ten subgenera. Since the ecology of these forms differs widely we must limit our discussion to the genera and species which may be used successfully as SLIs: that is, those which, have a colonial way of life in order to build solid biostromes that fossilize easily and, most important, live at a constant biological level (see discussion on biological zonation below).

For these reasons solitary living forms such as *Serpulorbis* spp., many *Vermetus* and so on, which may be found within a wide depth range, will not be discussed. Suffice to say that the study of their remains, together with other sessile forms such as corals or calcareous algae, may be of importance in some cases.

Fortunately, the great majority of reported vermetid biostromes are built by a limited number of species belonging to two genera (and subgenera) which are: *Dendropoma* Mörch subgenus *Novastoa* Finlay, and *Petaloconchus* Lea subgenus *Macrophragma* Carpenter. Some species of *Vermetus* Daudin may also build biostromes in some parts of the world.

A brief description of these two most important vermetid genera, quoted from Keen (1961) (Fig. 1) follows.

Genus *Dendropoma* Mörch 1861, subgenus *Novastoa* Finlay 1927

Genus comprising solitary and colonial forms, generally able to entrench themselves into the calcareous substrate.

Coiling of the shell planorboid for the first whorls, then becoming lax with a tendency for bending upwards at right angles and for rising above the surface of the substratum. The colour of the adult is mostly white, variously stained with dark brown inside. Sculpture of lamellar growth striae intersected or not by longitudinal lines, sometimes rising into a crest near the outer edge of the whorl. Tube often chambered within by convex calcareous septa isolating the inner portions. Operculum well developed, as large as the aperture of the tube, bright reddish brown, flat or convex, showing a button like calcareous protuberance on its lower side. Nuclear whorls two, dark brown, inflated.

Genus *Petaloconchus* Lea 1843, subgenus *Macrophragma* Carpenter 1857

Coils forming a hollow, flattened cylinder, at least in the young parts. Solitary or colonial. Median whorls possessing a more or less complex internal structure of spiral laminae projecting from the columellar whorl. Sculpture when present evenly cancellate but with the longitudinal ribs tending to be predominant and

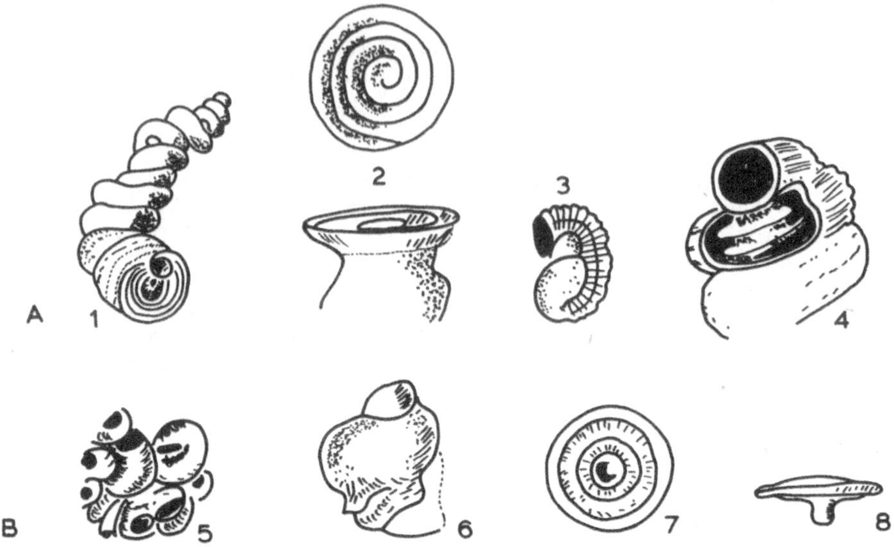

Figure 1. Morphology and anatomy of the two main reef-building vermetid genera.
A: *Petaloconchus (Macrophragma)*. 1) coiling of adult tube; 2) operculum; 3) embryonary shell; 4) inner laminae protruding inside the shell cavity.
B: *Dendropoma (Novastoa)*. 5) colony showing typical adult coiling; 6) embryonary shell; 7) and 8) operculum with its typical inner central mamilla.
(From Keen, 1960, Figures 22-25, 30-33)

subcarinate. Nuclear whorls two to four, ivory white to waxy yellow, conic to cylindrical. Adult shell medium to dark brown. Operculum concave, narrower than the aperture with an upstanding spiral lamina on the upper side, of chitinous origin. The shell may give subvertical feeding tubes which are progressively abandoned and replaced by the animal.

3. Shell structure

The shell of Vermetid Gastropods may be distinguished rather easily from the calcareous tubes of many Serpulid Polychaete worms by the following criteria:

a) it is generally made of three different layers of aragonite crystals:

- *an outer layer:* often coloured in buff or brown and bearing radial or longitudinal striations (or sometimes both),
- *a middle layer:* generally thicker, and
- *an inner layer:* usually thin and glossy, white or brown in colour and which in *Petaloconchus* bears sharp inner columellar laminae projecting inside the tube;

b) typical structures like embryonary shells (nuclear whorls) and the operculum are sometimes found preserved. It is nevertheless not rare that erosion and chemical dissolution may render any identification at the genus level impossible, but microscopic study nearly always allows an identification at the family level.

These shell characteristics are sufficient to avoid confusion with other tube-inhabiting organisms, most especially tube-worms (Polychaetes), the tube of which is always one-layered, calcitic and chalky, contrasting with the gloss of the inside of vermetid tubes; operculae, when they exist, are also of a completely different form. Reef-building Polychaetes are rare and either limited to brackish environments (estuarine or lagoonal), which are unsuitable for vermetid growth, or have a very different structure (reef-building *Phragmatopoma*, the tube of which is made of sand cemented with mucus). Some cases exist, however, where ill-preserved Vermetids could be confused at first glance with Serpulid Polychaetes and it is strongly recommended to have material checked by a specialist.

4. Biology

The fixed way of life of the Vermetids has led them to a certain number of biological adaptations which cannot be studied in detail here. Since the species which are of interest for our purpose all live in high-energy environments on rocky substrate I shall briefly outline the adaptations linked with that biotope. First and most important for us is gregariousness, which in certain cases can be a true coloniality (Hughes, 1979). This has various advantages regarding mating and fertilization of the eggs. Free larval life is generally very short and most species have only crawling larvae which have to fix rapidly to a substrate (often a tube of the same species). Fixation is generally helped by the possiblity of entrenching the substrate by rasping it with the radula; the animals may thus cut through their own tubes or those of neighbouring individuals. Many species, notably in the genus *Dendropoma*, grow in mixed stands with calcareous algae (*Porolithon*, *Lithophyllum*, *Neogoniolithon*) which tend to cement the tubes together and contribute to the construction, giving it a smooth surface on which the vermetid tubes appear as small black holes. A strong competition between Vermetids and algae may be observed, leading to different types of equilibrium following the local conditions. When surf is too strong for vermetid growth calcareous algae tend to overgrow and replace them. In contrast, in calmer sites the proportion of Vermetids is greater and they tend to grow in pure stands (Focke, 1977). It is interesting to notice that a different type of equilibrium is observed in the Eastern Mediterranean between *Dendropoma petraeum* and *Neogoniolithon notarisi*, the latter being dominant in very calm waters. Food is collected either by filtration or by means of a floating lamina of mucus which acts as a trap, or by these two methods together.

5. Geographical distribution

Colonial Vermetids are warm water animals which occur all

over the world in tropical or subtropical waters. In temperate regions they are generally absent: for example, they are not found in the Northwestern Mediterranean but grow abundantly on the coasts of Corsica, Sardinia, Southern Italy and Sicily, Greece, and on the coasts of the Middle East and North Africa. In tropical waters vermetid biostromes have been described in many localities along the Atlantic, including the coasts of West Africa and Brazil and their presence in manyundescribed localities is probable.

They have also been found in the Red Sea, Madagascar and tropical East Africa down to the Cape of Good Hope, in many Australian localities as well as on many tropical islands of the Pacific, including Hawaii and New Zealand. They are also present on the coasts of Southern California and Baja California. Since no complete monograph exists for these biostromes there are still many places from where they have not been described and one can expect to find them approximately from 35°N to 35°S and sometimes even 40° depending on local hydrographic conditions, with a minimum winter temperature of about 15°C. One should nevertheless bear in mind that in some places Pleistocene or Holocene biostromes may be observed while living Vermetids are absent from the present marine zonation due to climatic modifications. This is, for example, the case for the coasts of Brazil south of Cabo Frio. Some vermetid species recently also appear to have undergone a strong recession, both in range and species diversity, the cause of which is still unknown (Laborel, 1977). Some non-building species of *Vermetus* and *Serpulorbis* have a wider range of distribution.

6. Nature of the substrate

Vermetid biostromes may develop on any kind of rocky substrate: crystalline or metamorphic (Corsica, Molinier, 1960; Brazil, Kempf & Laborel, 1968), volcanic (Islands of Cape Verde, Crossland, 1905; Fernando de Noronha, Kempf & Laborel, 1968; Hawaii islands, Hadfield *et al.*, 1972), compact limestone of various age and origin (Tunisia, Molinier & Picard, 1954; Lebanon, Fevret & Sanlaville, 1965; Crete, Pirazzoli & Thommeret, 1977), Pleistocene or Holocene coral rock (Florida, Curaçao, Focke, 1978b, c; Brazil Madagascar, Battistini *et al.*, 1976), eolianites (Bermuda, Aggassiz, 1895; Verrill, 1906; Israel, Safriel, 1966; Sicily, Molinier & Picard, 1953a, b) or even, though rarely directly upon sandy bottom (Florida, Shier, 1969). While there seems to be no preference for one particular substrate, it appears that the morphology of the biostrome is strongly influenced by the nature of the bedrock. In effect, the final pattern of the shore is a resultant of complex interactions between erosive forces (particularly strong on eolianites and limestones, but very weak or non-existent on volcanic or crystalline rocks) and constructive forces and there are no two similar conditions. So the variety of forms is extremely great and has been abundantly described by various authors, varying from simple linear corniches on hard rock to complex platforms and rimmed pools sometimes associated with erosion notches. For a detailed approach of these problems see Focke (1978a,b) Kempf & Laborel (1968) Safriel (1975) and Figure 3.

7. Influence of other ecological factors

Vermetids are stenohaline, open-ocean animals and as such are not found in estuarine and lagoonal environments. Although precise evidence is yet lacking we can assume that their sensitivity to accidental flooding by fresh water is limited by their capacity to take refuge in their operculated tubes. Temporary burial on the contrary has severe effects and may kill them.

POSITION OF VERMETIDS IN THE BIOLOGICAL ZONATION ON ROCKY COASTS

1. 'Sea level': myths and realities

For many years SL was taken for granted, notably among geomorphologists and geographers. Papers dealing with its variations used a great variety of datums such as 'high tide', 'low tide', 'mean sea level', or more refined mareographic notions such as 'extreme high water spring'. It was not until recently that general criticism of these unqualified datums arose among geographers, notably thanks to cooperation with marine biologists. I shall not enter into a detailed account of the controversy; interested readers are referred to the papers by Kidson and Heyworth (1979) and Pinot (1979). However, to summarize briefly the main points of it:

- The level of the sea at a given place and time is not a measurable entity but the statistical result of highly complex movements of the actual water surface under the influence of a host of factors such as: tides, winds, atmospheric pressure, currents, water temperature and salinity, shoreline orientation and morphology, bottom topography, geoid deformations and so on. These movements may be periodic (with periods ranging from some seconds to several millenia) or aperiodic.
- When afield, no one has the material possibility of taking all these factors into account, all the more if one is surveying little known areas with severe time limitations that are the rule in fieldwork.
- Mere use of mareographic data as supplied by mareographic stations and charts is insufficient since it does not take important factors such as surf into account and because of the great varieties of tidal irregularities introduced by local factors and atmospheric pressure, even in well-known regions of the world.

2. Marine zonation

It is a fact of general observation that on rocky shores the vertical distribution of marine fixed organisms is not random but they instead grow in horizontal belts or *zones*,

Table 1. A comparison of worldwide marine zonation schemes in the littoral region based on the vertical repartition of marine organisms and showing the place of the Vermetids in the littoral zonation

Stephenson & Stephenson (1961)	Lewis (1961)	Peres & Picard (1964)	Prevailing conditions and marine organisms
SUPRALITTORAL ZONE	MARITIME ZONE	Not covered	Terrestrial, not marine but strongly influenced by sea brines. Terrestrial salt-loving vegetation.
Supralittoral fringe	SUPRALITTORAL FRINGE (or LITTORAL FRINGE)	ETAGE SUPRALITTORAL	Spray zone: wetting by surf or highest tides only. Very poor populations of Littorinid Gastropods and rock boring blue Algae.
MIDLITTORAL ZONE – Upper	EULITTORAL ZONE	ETAGE MEDIOLITTORAL	Surf and tide zone: mainly wetted by the daily movement of tide or by surf in tideless seas. Rich populations of Algae (notably fucoids), limpets and barnacles with clear belt patterns. Exposure to air progressively diminishes downwards.
MIDLITTORAL ZONE – Lower		ETAGE INFRALITTORAL	Vermetid and calcareous Algae bioherms.
Infralittoral fringe		ETAGE INFRALITTORAL	No more drying out except in rare opportunities and in the upper part of the zone only. Laminarians, sea grasses, reef corals. This zone extends down to the limit of light loving species which is variable with latitude and turbidity.
SUBLITTORAL ZONE	SUBLITTORAL ZONE		

whatever the local mareographic pattern may be. This type of vertical distribution, generally known under the term of *zonation*, was described by marine biologists from all over the world. This phenomenon is the resultant of complex interactions between the organisms and the local ecological factors, the latter varying along gradients perpendicular to the shoreline and the former growing only at the precise height where the value of the gradient is favourable for their biological requirements.

Observations of these belts over the years reveals a great stability if local mareographic and hydrodynamic parameters remain constant. Seasonal variations in belt height and composition have been described in some places by various authors but they are generally linked to algal forms with a short life cycle which are not part of this discussion

The systematic composition of the littoral zones or belts shows an interesting worldwide pattern: similar zones occur in regions of comparable climate, some invertebrate genera having a very wide biogeographical range and being always found at the same level. So a worldwide classification of the littoral zones was rapidly worked out by marine biologists, the most used by English-speaking authors being that of Stephenson & Stephenson (1949). A general scheme of littoral zonation, in order to be valid for any region of the world, cannot be based on mareographic parameters only. The phenomenon of tides is too variable on a world scale and other factors such as surf intensity may locally have a far greater biological influence than tides. Marine biologists are therefore now basing their schemes of littoral zonation on the distribution of marine organisms in the field.

Two schemes of zonation are now widely used the world over: English-speaking biologists use the scheme of Stephenson & Stephenson or the modification by Lewis (1961). French-speaking authors and some specialists around the world use the parallel scheme of Peres & Picard (1964).

A table of correspondence between these zonations is presented in a simplified form in Table 1 compiled from various sources notably the Stephensons and Lewis (1961).

From this table of correspondence one can see that there is a marked discrepancy between these three authors, and by way of consequence between French and English speaking specialists, regarding the separation of the tide and surf zone from the sublittoral/infralittoral zone: for Peres & Picard this limit is higher than for Stephenson & Stephenson, a situation which stems mainly from slight differences in the theoretical concepts of these authors. Hence the field usage varies between two main linguistic groups. However, this is not a crucial point; just a problem of definition which could be settled easily since there is no argument about the biological zonation itself.

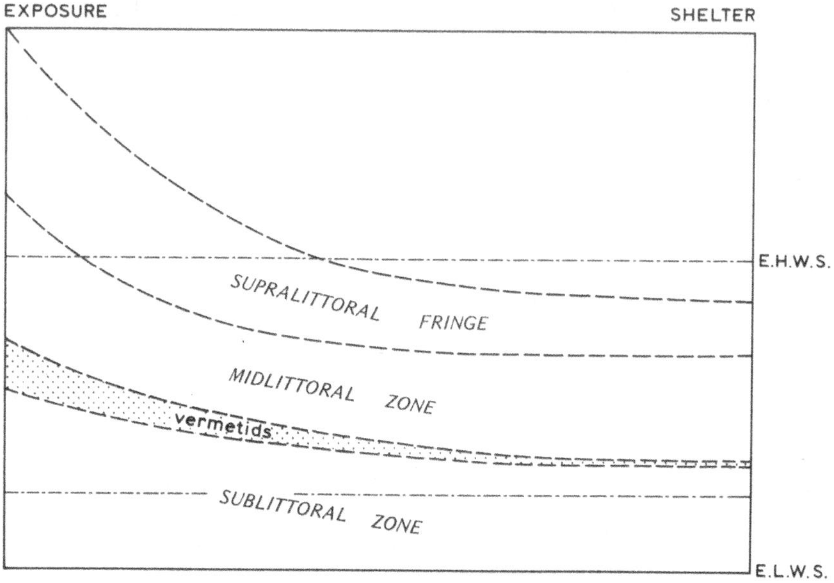

Figure 2. Diagram showing the upward displacement and widening of the various littoral sub-zones as a function of surf conditions (tidal range being constant). Modified from Lewis (1961, Figure 2) to include the vermetid zone in tropical and subtropical waters.

Contrary to widespread belief the biological zones so defined are not always horizontal; the influence of local factors such as surf tends to warp the limits upwards in higher energy segments, even if tide factors are constant. A good discussion of this phenomenon is found in the paper by Lewis (1961) from which Figure 2 is taken and modified to include the vermetid zone.

3. Place of Vermetids in the biological zonation

Many genera and species of Vermetids are solitary and live at any depth in the infralittoral zone, so they generally cannot be used as SLIs. This is the case for all species of *Serpulorbis*, the largest Vermetids, which are found in tropical and subtropical waters around the world. On the contrary, many species of the two most important genera, *Petaloconchus (Macrophragma)* and *Dendropoma (Novastoa)*, are strictly limited to the upper boundary of the 'étage infralittoral', or lower midlittoral zone of English authors, as indicated by various observations in many parts of the world (Keen, 1960; Kempf & Laborel, 1968; Molinier 1955; Peres & Picard, 1952; Safriel, 1966).

Vermetids are often found growing in close association with calcareous algae: these are either growing in competition with the former or have slightly different depth ranges. Safriel (1975) points out that while in the

Mediterranean Sea conspicuous corniches of *Lithophyllum tortuosum* may develop *above* the Vermetids and do not mix with them, this is not so in tropical waters, where the calcareous algae of the 'algal ridge' are found more or less mixed with the Vermetids, according to surf intensity. This is the case in the Caribbean region (Glynn, 1973; Focke, 1978c) and in the Indo-Pacific (Emery, 1962). Quite often also in tropical waters a biostrome of mixed Vermetids and calcareous algae may be found growing immediately over the upper part of a living coral reef, capping and smothering live corals. It is stressed that all types of littoral 'corniches' or 'lips', whether of mixed algal-Vermetid or purely algal composition, are good SLIs and that all discussions and arguments presented here apply equally to these as well as to pure vermetid bioconstructions.

4. Influence of surf and tide on the height of the vermetid zone

Vermetids, like many littoral invertebrates, need a certain amount of water movement, but how much exactly is difficult to tell, since there is no objective scale for measuring it. While tides are relatively easy to define and to quantify this is not true for surf. In spite of these difficulties it is important to study the relative influence of these two phenomena on the development of vermetid biostromes and their altitudinal relation to mareographic MSL.

a) Influence of the surf

Surf is an important factor. As a rule, living species of Vermetids develop best in moderate to strong surf conditions, for example on the windward side of islands in the tradewinds belt. But if the surf is too heavy, Vermetids may give way to organisms more tolerant of high energy such as calcareous algae. Conversely, Vermetids generally are found growing in very quiet waters, although species of *Petaloconchus* seem to thrive better than species of *Dendropoma* in calm conditions. Hence there is an optimum zone marked by the great development of vermetid formations. However, this optimum may occur at different places and heights depending on local hydrodynamic conditions. Strong surf has another important effect: the plant and animal populations living on the shore tend *to grow at higher levels* and *the stronger the surf the greater is the upward shift*. So when one compares marine zonation at the tip of a cape to that inside a protected bay, altitudinal differences ranging to more than one metre may be apparent (Fig. 3). This upward displacement is stronger in the supralittoral and midlittoral zones than in the sublittoral. Thus vermetid biostromes grow thicker and higher with increasing exposure conditions; they may even be composite and show the appearance of a staircase of rimmed pools which may be one or two metres high (Fernando de Noronha, Curaçao). It is of course an unfavourable effect for our purpose and as a rule it is advisable to *select the more sheltered sectors of the coast bearing Vermetids* for the study of SL variations. One should also always bear in

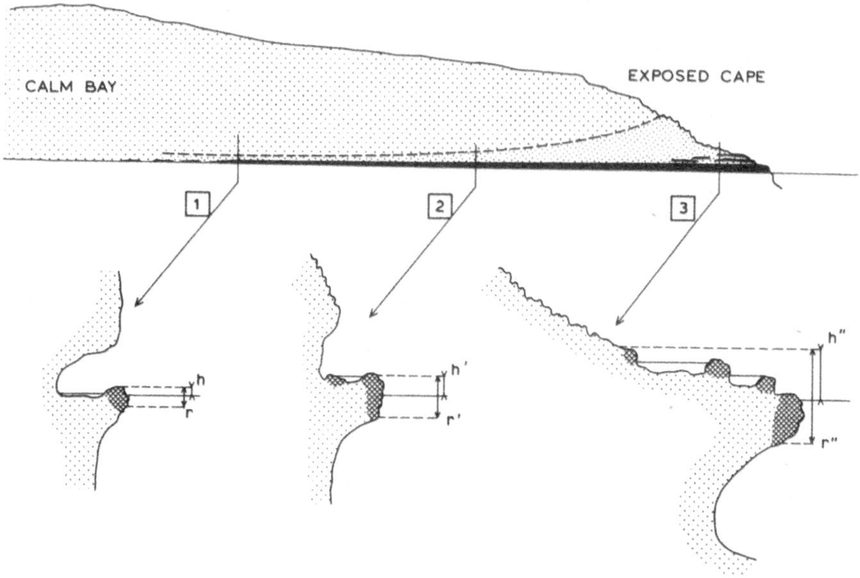

Figure 3. Changes in height and morphology of vermetid constructions and erosion notches going from a calm, sheltered bay to the tip of an exposed cape. Based on Focke (1978a and 1978b) and Kempf & Laborel (1968).

mind that the *altitude of a vermetid remain must always be measured in relation to that of living Vermetids at the same place of the coast,* in similar conditions of exposure to surf.

b) Influence of tide

Vermetid biostromes are found in regions where the tidal range is small or negligible (such as the Mediterranean or Caribbean seas or the Central Pacific Ocean). Some exceptions exist such as the coasts of Northeastern Brazil (with tidal range between 2 and 6 m). A comparison between these different parts of the world shows that the development and the morphology of the biostromes do not seem to be affected in any recognizable manner by strong tides, so it is reasonable to say that tides are ecologically unimportant compared to surf intensity.

Nevertheless, the effect of tides is an important one if we are concerned by the relative height of the Vermetids in relation to MSL. In effect, let us compare, as did Safriel (1975), vermetid formations in the near-tideless Mediterranean and in regions of stronger tides: we can see that in the Mediterranean the snails will develop more or less at MSL, whereas they will grow at a lower relative level in the regions of stronger tides. The explanation is that in tideless seas no marked difference exists between mareographic MSL and 'biological MSL', the level where

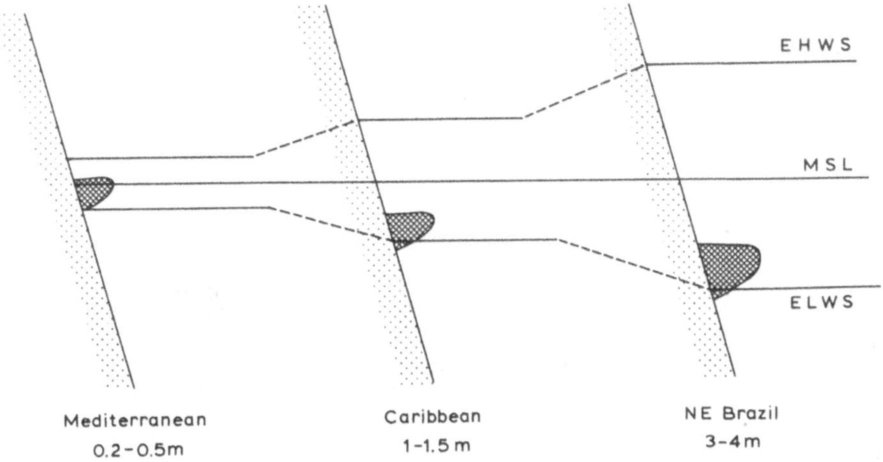

Figure 4. Variation in height and thickness of vermetid formations in function of changing tidal range. Surf is supposed constant. 1: case of a tideless sea; 2: sea with small to medium tidal range such as the Caribbean region; 3: case of relatively large tidal range (Northeastern Brazil). From Safriel (1975), modified to include personal observations.

flooding is a dominant ecological factor. On the other hand, when tidal range increases the mareographic MSL corresponds to the mid-tide level which is flooded and dried out twice daily (as a rule). In this case the biological MSL is much lower, and the wider the tidal range the greater is the difference between these two concepts of MSL. Thus, the wider the tidal range the stronger the downward effect on the level of the Vermetid growth (Fig. 4).

So if we consider the zone where Vermetids develop as the equivalent of a 'biological MSL' we must not forget that this notion does not correspond either with the mareographic 'MSL' or with the zero of the marine charts (which differs from country to country).

If we compare now the Brazilian coasts near Recife (tidal range a little over 3 m) with the Bermuda islands (1 m) and Eastern and Western Mediterranean (0.5 m and 0.2 m respectively) we can see that the respective levels of Vermetid growth for these regions are: 0.4-0.9 m below mid-tide for Recife, 0.3-0.2 m below mid-tide for Bermuda and approximately 0.1 m below mid-tide for the Mediterranean. *One can state that Vermetids grow in the lower quarter of the tidal range,* ie. between *low water neap tides and low water spring tides,* if surf action does not interfere as stated above. A change in the tidal range (at constant SL) will thus involve a vertical movement of the Vermetids: a smaller tidal range leading to an upward movement and a greater tidal range to a downward movement. This may be represented by the following rule of thumb:

$$dH \# \frac{A1 - A2}{4}$$

where dH is the variation of altitude of the Vermetids and A1, A2 the initial and final values of the tidal range; dH is positive if A1 is greater than A2 and conversely.

METHODS OF FIELD STUDY

Vermetids were first used as paleo-SLI on the coasts of Northeastern Brazil by the present author (Van Andel & Laborel, 1964; Delibrias & Laborel, 1971) and by Fevret & Sanlaville (1965, 1966) on the coasts of Lebanon, then on the coasts of West Africa (Laborel & Delibrias, 1976) and Madagascar (Battistini *et al.*, 1976). More or less complete discussions of the problems of field study have appeared in some earlier papers (Fevret & Sanlaville, 1966; Delibrial & Laborel, 1971; Laborel, 1979a, b; Laborel & Delibrias, 1976; Sanlaville, 1972), so I shall present here a synthesis with practical recommendations for the non-biologist field researcher.

1. Favourable types of shorelines

As already stated, reef-building Vermetids may develop on any kind of rocky shoreline. Since they are intolerant to low salinities and to turbid waters they shun estuaries and brackish environments. They are not often found either in the close proximity of beaches or sand spits.

These preferences are all the more interesting when fossil biostromes are found in places which now seem to be inhospitable for the snails since this is generally a very valuable indication of a change in the local environment. For example in the region of Sao Paulo, Suguio *et al.* (1977) have found vermetid rock half covered by a present sand beach which, of course, could not have been present when the animals were alive. I shall discuss in a further paragraph the difficulties of interpretation and measurement inherent to that special type of location.

2. Conditions of preservation; erosion

Vermetid biostromes are generally rather compact and resistant to erosion, especially when the proportion of encrusting calcareous algae is high. But this is not always so; some slabs of *Petaloconchus* are composed of loosely intermingled tubes without any biological cementation and are thus prone to rapid erosion after the death of their components. Even when they are quite resistant, vermetid formations develop in a zone where erosive forces at work are considerable: in the mid-littoral zone (just above the Vermetids), the biological action of boring Cyanophytes and scraping limpets, combined with the mechanical action of waves leads to extensive erosion of littoral limestones with the most frequent development of a notch. Conversely, just below the Vermetids, biological erosion is at work with two predominant agents: the Clionid sponges and the sea urchins. So when a biostrome dies, with or without any modification of the SL, it is immediately the prey of biological erosion. I shall try to briefly present the most usual possibilities.

a) Case of a falling SL

If the movement is large and rapid enough, the biostrome may be raised above the level of marine erosional agents. Its preservation is then a function of the intensity of terrestrial erosional agents such as rain and humic acids of the soils. If the climate is quite dry the preservation may be extraordinary, as it is in Crete (Laborel *et al.*, 1979). More generally the influence of the rains may lead to rapid erosion, so that only scarce remnants will be found. In such circumstances (tropical humid climate such as that of the West African or Brazilian coasts) the best remains will generally be found in caves or under rocky ledges or overhangs. Sometimes also fine remnants may be preserved under sand dunes or buried under slumped material. In many cases however, the fall in SL has not been large enough and the Vermetids remain within the range of tides or surf. This case is quite frequent on the northwestern coast of Brazil. The marine erosional agents may then destroy the biostromes, which may also be bored by cyanophytes or covered by barnacles or oysters. This case must always be studied with care and a thorough biological survey of the sector is absolutely necessary. The specimens collected for radiocarbon dating must be carefully cleaned of such secondary boring and encrusting organisms.

b) Case of a rising SL

If the upward movement is slow and gradual (as it was for example in many places during the last millenia) then the vermetid biostrome will simply grow upward, keeping pace with the rising SL, its lower parts being attacked by Clionids or capped by calcareous algae: such an evolution was described by Ginsburg & Schroeder (1973) from some Bermudian vermetid reefs. Growth rates ranging from 1 cm/ 8.5 years to 1 cm/38 years have been measured by these authors and greater rates up to 1 cm/5 years have been measured by Jindrich (unpublished) on the vermetid reefs of Fernando de Noronha. If the rise of SL is rapid Vermetids may die and the biostrome will be attacked by boring organisms while its outer surface is generally densely covered with *Sargassum* or other forms of algal growth. So the characteristic shape of the bioherm will be quickly altered and its remains will be difficult to locate for the underwater observer: only systematic breaking of the rock will eventually permit their discovery.

c) Case of complex movements of SL

They are generally difficult to resolve. It is therefore

interesting to note that vermetid biostromes may in certain circumstances yield useful information. Recent field work in Crete (Thommeret *et al.*, 1981) has shown that elevated vermetid lines were marked by traces of an intensive erosion by Clionid sponges and had been covered by deeper living organisms such as corals (*Balanophyllia*) or solitary Vermetids (*Serpulorbis*) and calcareous algae. Studies of thin slices have also given further details on the complex phenomena of cementation and diagenesis leading to the recognition of a series of downward movements of the shore, followed by a strong and rapid uplift.

3. How to make sure that vermetid remains are *in situ*

It is of course a most important problem, when dealing with any type of (biological) SLI, to ascertain whether they are in place or have been reworked and transported by erosional agents. Fortunately, vermetid biostromes are generally strongly cemented to the bedrock and are difficult to break. Moreover they are often found associated, on calcareous shores, with such morphological features as marine notches at the base of which they develop and which of course cannot be displaced. But this optimal case is by no means common. Pieces of vermetid rock may have been broken from the biostrome and then cemented into a beachrock or a storm beach. I have quoted elsewhere (Laborel & Delibrias, 1976) the case of some big slabs of *Petaloconchus*: this genus is now nearly extinct, at least in the Atlantic (Laborel, 1977), so any comparison of living and fossil formations is not possible. The scanty living individuals now found (Brattstrom, 1980; Lewis, 1960) never build reef rock and seem to have a larger vertical distribution and a greater tolerance to calm environments than *Dendropoma*. However, recent field work on Brazilian and Caribbean vermetid reefs has shown, by coring, that layers of dead *Petaloconchus* are overlain by the presently active *Dendropoma*. This is an indication that the two forms had similar biotopes and were building very similar types of reef rock. In calm places *Petaloconchus* could nevertheless build large slabs of rock loosely fixed to the substratum, a kind of construction which surf-living *Dendropoma* cannot achieve. Such slabs may have been easy to break and displace in spite of the fact that they apparently developed in rather sheltered environments. When we are faced with such particular formations (Ghana), it is necessary to check the disposition of the tubes: the lower face of the slab must follow exactly the shape of the substrate and show a dominance of *coiled* tubes tightly pressed against one another; the upper part is generally made of twisted and intermingled subvertical feeding tubes. One must look also for points of cementation to the bedrock: it must be done by the tubes themselves or by a layer of calcareous algae, not by any kind of calcareous cement. When Vermetids are present on stone blocks of small to medium dimension great care must be taken to check that these blocks are linked together by an unbroken biostrome: a proof that they have not been reworked. A very favourable case is that of some very thin incrustations, generally made by *Dendropoma* on crystalline rocks: there is no biostrome proper but a thin veneer of tubes, all directly cemented to the rock. Such a formation must be *in situ* (Fort Dauphin, Madagascar, Battistini *et al.*, 1976; Bereby, Ivory Coast, Laborel & Delibrias, 1976).

The presence of biological material which is not normally associated with Vermetids (such as broken valves of Lamellibranchs, sea-urchin spines and so on) is often an indication of reworking.

4. Measuring the height of vermetid remains; precision of measurement

As stated earlier there is only one satisfactory way to measure the height of a vermetid line:

- *The height of a fossil vermetid biostrome above (or under) present SL must be taken between the upper limit of the fossil biostrome and the upper limit of the living biostrome on the same cross section of the littoral:* ie. making sure that conditions of present exposure to surf are the same (or at least not very different) as they were for the fossil biostrome.

- *When no living vermetid formations are growing presently at the place of the study* it is not advisable to make height measurements according to any mareographic criteria such as 'high tide' or other. *One must measure the difference between the summit of the fossil Vermetids and that of the populations which presently replace them on the profile studied,* ie. generally the limit between midlittoral populations of barnacles and limpets and the zone of the big brown algae (*Sargassum, Cystoseira*) or the first Echinids or even the summit of the first living reef corals.

- *Whenever possible one should avoid using vermetid remains corresponding to zones of very strong exposure to surf because of the possibilities of upward displacement,* especially when they grew in places like clefts of the rock where the energy of the surf may have been concentrated.

- *The best suited biostromes are those which grew in zones of relative shelter,* just at the limits before Vermetids disappear: they are thus very thin and, provided they have been preserved, afford the best precision possible, particularly when associated with a notch. A precision of ± 0.1 m is possible in many cases. Where notches are absent and where vermetid remains are more or less fragmented (which is the commonest case on the Brazilian and West African coasts) the precision of height measurement cannot be so good. It is then determined by the value of the total thickness of the living biostromes (or of the fossil one if it can be measured). Thus it is better not to be too optimistic about precision when dealing with fragmentary formations found in sectors of strong exposure to surf and noticeable tidal range, especially if living Vermetids are not present in the modern zonation. It can be evaluated for the examples I have seen personally in Brazil to ± 0.5 m and in very bad cases to ± 1 m. Where the vermetid remains are very eroded it may be interesting to look at the associated fauna in order to learn if one is dealing with the upper or the lower part of the biostrome: the presence of barnacle shells inside the biostrome and mixed with the Vermetids will generally indicate the upper part of the formation. On the contrary, the abundance of calcareous algae (outside the Mediterranean of course) and sessile Foraminifera will point to the lower part.

5. Sample preparation and dating

The percentage of calcite in the shell is very low, not exceeding 2 to 3 percent. Age determination by radiocarbon is generally easy. Thommeret *et al.* (1981) pointed to the necessity of a pretreatment by a concentrated solution of sodium hypochlorite to destroy the remaining organic matter and soften the sample, then breaking into small fragments which are then rinsed with water and immersed for 24 to 48 hours in a solution of H_2O_2 at 130 vol concentration and at 70°C to eliminate algal limestone, then sorting out the vermetid fragments before dissolving them in phosphoric acid. However, such an elaborate treatment does not seem to be mandatory in all cases since dating by Thommeret of treated and untreated fractions of the same sample gave comparable results.

As widely accepted now it is recommended to give all uncalibrated dating in BP notation only.

Contamination by old rock carbonates must be suspected especially of limestone coasts but a preliminary treatment as quoted above is generally sufficient to avoid any problem.

6. Some possible sources of error

In summary, the main causes of error in the estimation of the height of fossil shorelines by the means of vermetid biostromes are:

1 - *The use of any level of reference other than the upper limit of living Vermetids on the same profile of the coast* (or at least the upper limit of the sublittoral plant and animal populations).

2 - *Comparing the height of fossil and living formations in respectively quiet and exposed conditions.*

3 - *Using biostromes in very high-energy environments, notably when Vermetids may have grown in splash pools at some height above their normal living level.*

4 - *Using loose, reworked or transported material.*

To these other factors may be added.

5 - When collecting material for dating, *great care should be given to checking its homogeneity:* taking out the external layers and, most important, any material which could have been deposited at a different level of the sea (for example barnacles overgrowing Vermetids). The same precautions must be taken for the inner structure of the material; one must look for erosion marks or deposition of

secondary layers (either biological or mineralogical). Such study may result in greater precision about the evolution of the biostrome and will avoid dating composite material. Finally one should not forget the possibility of describing as elevated SLs some artifacts due to quite different causes.

6 - *When rimmed pools are built by Vermetids or calcareous algae at some height over their normal growing range,* it is possible that biological or mechanical erosion may break the outer rim of the pool. The pool empties itself and the associated Vermetids and algae will die, giving the appearance of a locally elevated SL. The same may occur on a reef (stone reef or coral reef alike) when the outer rim of a moat is destroyed by a storm. Many such spurious cases have been observed all over the world.

7 - *A final case concerns modifications of local hydrodynamic characteristics at constant SL, changes in the tide range (see page 292) or changes in exposure.* For example, the damming of a bay or an estuary by a natural sand spit or a man-made causeway may convert it into a sheltered environment with a consecutive apparent downward displacement of the biological zonation. In the same manner one cannot overlook the fact that construction or destruction of a spit may have important consequences on the local tidal range inside a bay, thus causing vertical displacement of the biological zones. A special case must also be made for the possible modifications of the wind patterns and intensities, especially of trade winds, during the Holocene. Such variations could give birth to 'elevated' vermetid lines and I feel now that such a phenomenon may have been completely overlooked by myself and other authors, notably on the Brazilian coast. Such changes could well explain some low and recent 'levels'.

CONCLUSIONS

As a whole the fossil remains of vermetid or mixed algal-vermetid biostromes may be considered as a remarkable SLI if properly studied. The precision of the measurements may range from good to quite good when thin biostromes are associated to such morphological features as notches. In only a few well-defined unfavourable circumstances does the precision drop to relatively high figures of possible error. Even in these latter cases it is generally still possible to use them.

The use of fossil vermetid formations necessitates a biological survey of the portion of coast considered. The complexity of the problems involved requires that field and laboratory studies must be done by a team of specialists comprising a geologist (and or a geomorphologist), a marine biologist, various specialists whenever necessary (cementation and diagenesis, algological or zoological problems) and a dating specialist.

Figure 5. Typical coiling of *Petaloconchus*; fossil bioherm at Tema, Ghana

Figure 6. Typical section of *Petaloconchus*; subfossil bioherm at Pointe des Chateaux (La Guadeloupe Island). Note the characteristic inner laminae inside the tubes (arrows).

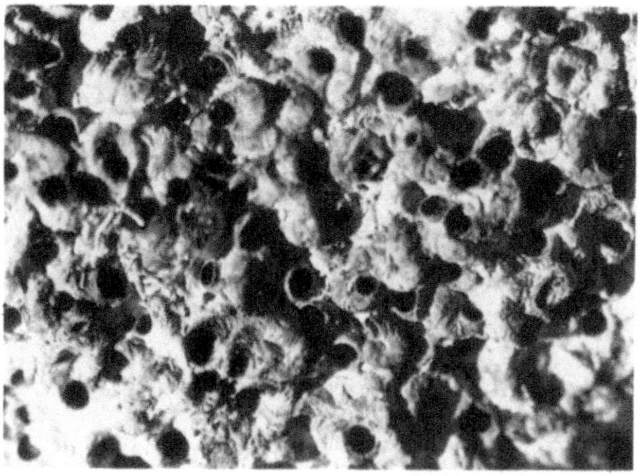

Figure 7. Typical coiling of a *Dendropoma*; living bioherm at Pointe des Chateaux; La Guadeloupe Island.

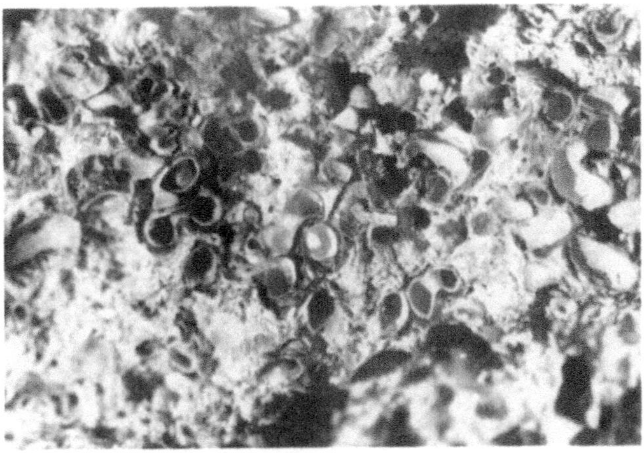

Figure 8. Typical section of *Dendropoma*; living bioherm at Centuri, Corsica.

Figure 9. Algal and vermetid reef now developing at Pointe des Chateaux, La Guadeloupe Island. Thickness of about 0.5 m.

Figure 10. Algal and vermetid reef at Rio Doce (Pernambuco), Brazil, showing an upper fossil layer corresponding to a sea-level stand of 0.5 m above present and a living outer rim at present sea level.

Figure 11. Vermetid remains subject to biological erosion in the tidal zone. They are now covered by *Tetraclita*, a barnacle living just above the vermetid zone. The corresponding (former) water level is 0.5 m above present living Vermetids. Cape Santó Agostinho (Pernambuco), Brazil.

Figure 12. Another view of the vermetid remains shown on the preceding photograph. Notice that the vertical amplitude of the remains is small in spite of a tidal range of about 4 m.

Figure 13. Moni Krysoskalitisas, western Crete. Superposition of fossil vermetid lines corresponding to a succession of downward tectonical movements, followed by strong uplift. The superposed lines are well-preserved on a 45° slope. The highest line is also the youngest. The lower lines exhibit traces of biological erosion which occurred when they were under water.

Figure 14. Moni Krysoskalitisas, western Crete. On a vertical cliff only the highest and youngest line of Vermetids have been preserved. The lower lines have been completely erased by biological erosion, which is stronger on vertical cliffs.

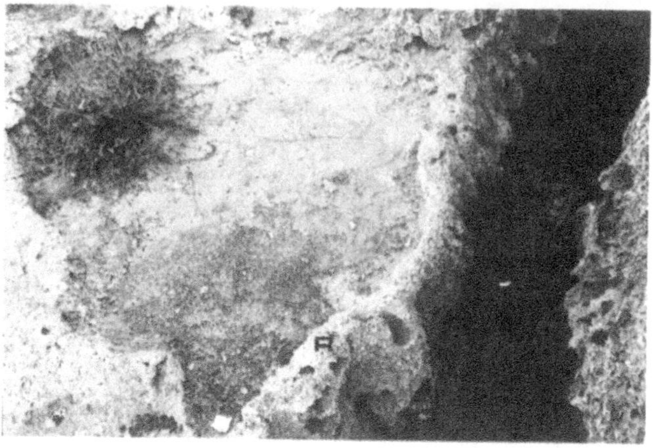

Figure 15. Elevated remains of a vermetid line in western Crete (Moni Krysoskalitisas). Subaerial erosion in the dry climate has been remarkably slight. Even the rim of the pool on the left (R) has been preserved.

Figure 16. Elevated remains of a vermetid line in western Crete (Falasarna). The remains were submerged for more than 2000 years, and then emerged due to uplift. Strong biological erosion (boring sponges and sea urchins) have given the remains their ragged appearance.

Figure 17. A bioherm built by *Petaloconchus* (P), preserved under the soil of a coconut grove, has been exhumed by marine erosion. Itaparica Island, Brazil.

Figure 18. Remains of *Petaloconchus* (P) preserved from dissolution under a rocky ledge which has sheltered them from rainwater. Cabo Frio, Brazil.

ACKNOWLEDGEMENTS

This chapter is a strongly modified rewriting of a paper read at the IGCP-nivmer seminar on fossil SLs organized by P. Pirazzoli in Paris in 1978. It has been completed and augmented as regards both text and illustrations. My best thanks go to Profeesors H. Faure and A.L. Bloom for the stimulating influence which led the author to write this paper. I am grateful also to Drs. D.R. Grant and O. van de Plassche for help and comments.

REFERENCES

Andel, T. van & Laborel, J., 1964. Recent high relative sea level stand near Recife, Brazil. *Science*, 145(3632), 580-581.

Agassiz, A., 1895. A visit to the Bermudas in March 1894. *Bulletin of the Museum of Comparative Zoology, Cambridge*, 26(2), 209-281

Battistini, R., Delibrias, G. & Laborel, J., 1976. Datations des niveaux holocènes à Madagascar. *Comptes Rendus Sommaires de la Société Géologique de France*, 6, 284-285

Blanc, J.J. & Molinier, R., 1955. Les formations organogènes construites superficielles en Méditerranée occidentale. *Bulletin de l'Institut Océanographique, Monaco*, 1067, 26 pp

Brattstrom, H., 1980. Rocky shore zonation in the Santa Marta area, Colombia. *Sarsia*, 65, 163-226

Crossland, C., 1905. The ecology and deposits of the Cape Verde marine fauna. *Proceedings of the Zoological Society, London*, 1, 170-186

DaLongeville, R., 1977. Formes littorales de corrosion dans les roches carbonatées du Liban, étude morphologique. *Méditerranée*, 3, 21-33

Darwin, C., 1841. On a remarkable bar of sandstone off Pernambuco on the coast of Brazil. *London, Edinburgh and Dublin Philosophical Magazine and Journal of Science*, 3, 19, 124, 257-260

Delibrias, G., 1973. Variations du niveau de la mer sur la côte ouest africaine depuis 20 000 ans. *Colloque du CNRS n° 219*, Gif-sur-Yvette, 5-8 juin 1973, 127-133

Delibrias, G. & Laborel, J., 1971. Recent variations of the sea level along the Brazilian coast. *Quaternaria*, 14, 45-49

Emery, K.O., 1962. Marine geology of Guam. *United States Geological Survey Professional Papers*, 403, B, 1-76

Fevret, M. & Sanlaville, P., 1965. Contribution à l'étude du littoral libanais. *Méditerranée*, 2, 117-134

Fevret, M. & Sanlaville, P., 1966. L'utilisation des Vermets dans la détermination des anciens niveaux marins. *Méditerranée*, 4, 357-364

Focke, J.W., 1977. The effect of a potentially reef-building coralline algal community on an eroding limestone coast, Curaçao (Netherlands Antilles). *Proceedings of the Third International Coral Reef Symposium*, (Miami), 1, 239-245

Focke, J.W., 1978a. Subsea (0-40 m) terraces and benches windward of Curaçao (Netherlands Antilles). *Leidse Geologische Mededelingen*, 51(1), 95-102

Focke, J.W., 1978b. Limestone cliff morphology and organism distribution on Curaçao, Netherlands Antilles. *Leidse Geologische Mededelingen*, 51(1), 131-150

Focke, J.W., 1978c. Limestone cliff morphology on Curacao (Netherlands Antilles) with special attention to the origin of notches and Vermetid/coralline algal surf benches. *Zeitschrift für Geomorphologie*, N.F., 22, 329-349

Gessner, F., 1970. Lithothamnium Terrassen in Karibischen Meer. *Internationale Revue der Gesamten Hydrobiologie*, 55, 757-762

Gilet, R., Molinier, R. & Picard, J., 1954. Etudes bionomiques littorales sur les côtes de Corse. *Recueil des Travaux de la Station Marine d'Endoume*, 13, 25-53

Ginsburg, R.N. & Schroeder, J.H., 1973. Growth and submarine fossilisation of algal cup reefs, Bermuda. *Sedimentology*, 20, 574-614

Glynn, P.W., 1973. Aspects of the ecology of coral reefs in the Western Atlantic region. In: *Biology and Geology of Coral Reefs*, 2, eds. O.A.Jones & R. Endean, *Biology*, 1, 9, 574-614

Hadfield, M.G., 1970. Observations on the anatomy and biology of two California Vermetid Gastropods. *The Veliger*, 12(3), 301-309

Hadfield, M.G., Kay, G.A., Gilette, M.V. & Lloyd, M.C., 1972. The Vermetid (Mollusca Gastropoda) of the Hawaiian Islands. *Marine Biology*, 12, 81-98

Hartt, C.F., 1870. *Geology and physical geography of Brazil.* (Boston).

Hoffman, A., Pisera, A. & Studencki, W., 1978. Reconstruction of a Miocene kelp associated macrobenthic ecosystem. *Acta Geologica Polonica*, 28, 377-387

Hughes, R.N., 1979. Coloniality in Vermetidae (Gastropoda). *Biology and Systematics of colonial organisms.* (Academic Press, London), 13, 243-253

Huve, P., 1957. Contribution préliminaire à l'étude des peuplements superficiels des côtes rocheuses de Méditerranée orientale. *Recueil des Travaux de la Station Marine d'Endoume*, 21(12), 50-62

Keen, M.A., 1960. Vermetid Gastropods and marine intertidal zonation. *The Veliger*, 3(1), 1-2

Keen, M.A., 1961. A proposed reclassification of the Gastropod family Vermetidae. *Bulletin of the British Museum (Natural History)*, 7(3), 183-213

Keen, M.A. & Morton, J.E., 1960. Some new African species of *Dendropoma* (Vermetidae Mesogastropoda). *Proceedings of the Malacological Society, London*, 34(1), 36-51

Kempf, M. & Laborel, J., 1968. Formations de Vermets et d'Algues calcaires sur les côtes du Brésil. *Recueil des Travaux de la Station Marine d'Endoume*, 59, 43, 325-435

Kidson, C. & Heyworth, A., 1979. Sea 'level'. In: *Proceedings of the 1978 International Symposium on Coastal Evolution in the Quaternary*, (Sao Paolo), eds. K. Suguio, T.R. Fairchild, L. Martin & J.-M. Flexor, 1-28

Laborel, J., 1969. Les peuplements de Madréporaires des côtes tropicales du Brésil. *Annales de l'Université d'Abidjan*, E, 2, 260 pp

Laborel, J., 1977. Are reef building Vermetids disappearing in the South Atlantic? *Proceedings of the Third International Coral Reef Symposium*, (Miami), 1, 233-237

Laborel, J., 1979a. Fixed marine organisms as biological indicators for the study of recent sea level and climatic variations along the Brazilian tropical coast. *Proceedings of the 1978 International Symposium on coastal evolution in the Quaternary*, (Sao Paolo), eds. K. Suguio, T.R. Fairchild, L. Martin & J.-M. Flexor, 193-211

Laborel, J., 1979b. Les Gasteropodes Vermetidés: leur utilisation comme marqueurs biologiques de rivages fossiles. *Oceanis*, fasc. hors série, 5, 221-238

Laborel, J. & Delibrias, G., 1976. Niveaux marins récents à Vermetidae du littoral ouest africain. *Bulletin Liaison ASEQUA*, 47, 97-110

Laborel, J., Pirazzoli, P., Thommeret, J. & Thommeret, Y., 1979. Holocene raised shorelines in Western Crete. *Proceedings of the 1978 International Symposium on coastal evolution in the Quaternary*, (Sao Paolo) eds. K. Suguiro, T.R. Fairchild, L. Martin & J.-M. Flexor, 475-501

Lewis, J.R., 1960. The fauna of rocky shores of Barbados, West Indies. *Canadian Journal of Zoology*, 38, 391-435

Lewis, J.R., 1961. The littoral zone on rocky shores, a biological or physical entity. *Oikos*, 12(2), 280-301

Loenhoud, P.J. van & Sande, J.C.P.M. van de, 1977. Rocky shore zonation on Aruba and Curaçao (Netherlands Antilles) with the introduction of a new general scheme of zonation. *Proceedings of the Koninklÿke Nederlandse Akademie van Wetenschappen*, C, 80, 437-474

Martin, L. & Suguio, K., 1976. Etude preliminaire du Quaternaire marin. Comparaison du littoral de Sao Paulo et de Salvador de Bahia (Brésil). *Cahiers ORSTOM, serie Geologie*, 8, 33-47

Mikhalski, A.K., 1895. Voprocu o geologischeskoi prirode podolskikh 'toltr' (Sur la nature géologique de la chaîne de collines de Podolie, nommées 'toltry'). *Russia, Izvestiya Geologischeskago Komiteta*, 14, 115-193

Molinier, R., 1954. Première contribution à l'étude des peuplements marins des Iles Pithyuses. *Vie et Milieu*, 5(2), 226-242

Molinier, R., 1955. Les plateformes et corniches récifales de Vermets (*Vermetus cristatus* Biondi) en Méditerranée occidentale. *Comptes rendus des séances de l'Académie des Sciences, Paris*, 240, 361-363

Molinier, R., 1960. Etude des biocénoses marines du Cap Corse. *Vegetatio*, 9, 121-312

Molinier, R. & Picard, J., 1953a. Recherches analytiques sur les peuplements littoraux se développant sur substrat solide. *Recueil des Travaux de la Station Marine d'Endoume*, 9(4), 1-19

Molinier, R. & Picard, J., 1953b. Notes biologiques à propos d'un voyage d'études sur les côtes de Sicile. *Annales de l'Institut Océanographique, Paris*, 28(4), 164-187

Molinier, R. & Picard, J., 1954. Eléments de biologie marine sur les côtes de Tunisie. *Bulletin de la Station Océanographique de Salambo*, 48, 47 pp

Molinier, R. & Picard, J., 1956. Aperçu bionomique sur les peuplements marins littoraux des côtes rocheuses méditerraneennes de l'Espagne. *Bulletin des Travaux de la Station d'Aquiculture et de Pêche de Castiglione*, nouvelle series, 8, 17 pp

Morton, J.E., 1951. The structure and adaptations of the New Zealand Vermetidae. *Proceedings of the Royal Society of New Zealand*, 79, 1-51

Morton, J.E., 1965. Form and function in the evolution of Vermetidae. *Bulletin of the British Museum (Natural History)*, 11, 585-630

Oertel, G. F., 1970. Preliminary investigations of bioconstructional features along the South shore of Bermuda. In: *Seminar on Organism-sediment Inter-relationships*, eds. R.N. Ginsburg & S. Stanley, *Bermuda Biological Station Research Special Publication*, 6, 99-108

Peres, J. M. & Picard, J., 1952. Les corniches calcaires d'origine biologique en Mediterranée. *Recueil des Travaux de la Station Marine d'Endoume*, 4, 1

Peres, J. M. & Picard., 1964. Nouveau manuel de bionomie benthique en Mer Mediterranée. *Recueil des Travaux de la Station Marine d'Endoume*, 31, 47, 137 pp

Picard, J., 1954. Les formations organogènes benthiques méditerranéennes et leur importance géomorphologique. *Recueil des Travaux de la Station Marine d'Endoume,* 13, 55-76

Pinot, J.P., 1979. Méthodes et techniques permettant de déterminer l'altitude relative des niveaux marins anciens en fonction de l'altitude absolue actuelle des divers indicateurs. *Oceanis,* fasc. hors serie, 335-353

Pirazzoli, P., 1976. Les variations du niveau marin depuis 2000 ans. *Mémoires du Laboratoire de Géomorphologie de l'Ecole Pratique des Hautes Etudes,* 30, 421 pp

Pirazzoli, P. & Thommeret, J., 1977. Datation radiométrique d'une ligne de rivage à + 2.5 m près de Aghia Roumeli, Crète, Grèce. *Comptes Rendus des Séances de l'Academie des Sciences, Paris,* D, 284, 1255-1258

Ponder, W., 1967. A new species of *Dendropoma* from New Zealand. *Transactions of the Royal Society of New Zealand,* 10(2), 17-20

Safriel, U., 1966. Recent Vermetid formations on the Mediterranean shore of Israel. *Proceedings of the Malacological Society, London,* 37(1), 27-38

Safriel, U., 1974. Vermetid Gastropods and intertidal reefs in Israel and Bermuda. *Science,* 186, 1113-1115

Safriel, U., 1975. The role of Vermetid Gastropods in the formation of Mediterranean and Atlantic reefs. *Oecologia,* 20, 85-101

Sanlaville, P., 1972. Vermetus dating of changes in sea level. In: *Underwater Archaeology, a Nascent Discipline,* (UNESCO, Paris), 185-191

Shier, D.E., 1969. Vermetid reefs and coastal development in the Ten Thousand Islands, Southwest Florida. *Geological Society of America Bulletin,* 80, 485-508

Stephenson, T.A. & Stephenson, A., 1949. The universal features of zonation between tide marks on rocky coasts. *Journal of Ecology,* 37(2), 289-305

Stephenson, T.A. & Stephenson, A., 1954. The Bermuda Islands. *Endeavour,* 13, 50, 72-80

Stephenson, T.A. & Stephenson, A., 1972. *Life between tide-marks on rocky shores.* (Freeman, San Francisco)

Suguio, K., Marin, L. & Flexor, J.-M., 1977. Sea-level fluctuations during the past 6000 years along the coast of the State of Sao Paulo (Brazil). *Proceedings of the 10th INQUA Congress,* (Birmingham)

Thommeret, J., Thommeret, Y., Laborel, J., Montaggioni, L.F. & Pirazzoli, P.A., 1981. Late Holocene shoreline changes and seismo-tectonic displacements in western Crete. *Zeitschrift für Geomorphologie,* Supplement Band, 40, 127-149

Verrill, A.E., 1906. The Bermuda Islands. *Transactions of the Connecticut Academy of Arts and Sciences,* 12, 348pp

11

The diagenetic products of marine carbonates as sea-level indicators

J. COUDRAY and L. MONTAGGIONI

ABSTRACT

Diagenesis, in this chapter, includes all organic and non-organic changes which affect marine carbonates after deposition but before weathering, within hydrologically active, subsurface nearshore environments (subtidal-marine phreatic, intertidal-marine vadose, supratidal-marine-fresh-water mixing, fresh-water phreatic and vadose). The related products, diagnostic in terms of position of sea level, are of mechanical or physicochemical origin: *burrows*, with significantly distinctive forms; *borings*, and particularly, cyanophyte microborings forming *micritic envelopes*; *desiccation cracks*, typical of supratidal zones; *fenestrae*, well-preserved in upper intertidal and supratidal zones; *intraclasts*, common in coastal sequences; *internal sediments*, each type being characteristic of a particular nearshore zone; *solution pores*, reliable criteria for distinguishing the marine-fresh-water limits; *cements*, morphologically and mineralogically diagnostic of their environment of genesis; *dissolution-reprecipitation*, mainly occurring in the fresh-water zone; *neomorphism*, resulting from supratidal-subaerial conditions; *dolomites* and *evaporites*, common in intertidal-supratidal arid areas; carbon and oxygen isotopic composition, useful for the location of the meteoric-marine water limit.

The reliability of textural, structural, mineralogical and geochemical diagenetic criteria in identification of sea-level changes is illustrated from five regional sequences ranging in age from Pleistocene to Holocene.

Studies of diagenesis require the use of various sophisticated analytical techniques. Absolute dating of diagenetic products may be performed, using the classical radiocarbon, U-Th and amino-acid racemization methods.

Finally, the diagenetic products of ancient marine

carbonates are believed to be useful for roughly defining the location of paleo-sea level, just below or above a given nearshore environment.

INTRODUCTION

The term *diagenesis*, introduced by Von Guembel (1868), was defined as all those physical and chemical changes which a sediment undergoes after deposition at low temperatures and pressures characteristic of surface or near-surface environments. Pettijohn (1957, p 648) emphasized that the primary process of diagenesis is *lithification*, ie. the conversion of a newly deposited sediment into an indurated rock. But limiting the application of diagenesis to lithification is, according to Newell *et al.* (1953), arbitrary, artificial and impractical. As recommended by Chilingar *et al.* (1967) and Fairbridge (1967), it is better to use diagenesis in a wide sense: this phenomenon begins at the moment a sedimentary particle comes to rest on the sea bottom and it continues to a point in history before weathering initiates. Thus diagenetic processes are expressed in a number of different ways, including organic and inorganic changes. Krumbein (1942) mentioned that they may result in textural and structural alterations, and in chemical and mineralogical changes ranging from slight to extensive.

The sophisticated knowledge of the various diagenetic processes of the modern nearshore carbonate environments has opened an important period of understanding in the last two decades and might help towards delineating the distinctive counterparts not only in the Holocene or Pleistocene limestones but even for sequences extending back into the Paleozoic. The nearshore environments usually include three basic zones of deposition based on vertical fluctuations of the tides (Fig. 1): 1) the *subtidal zone*, which encompasses all the deposits lying below mean low-tide level; 2) the *intertidal zone*, the limits of which are defined by mean low- and high-tide levels; 3) the *supratidal zone*, which is defined as that area of shoreline out of reach of the daily tide but not higher than the highest storm tide, and consequently, lying exposed subaerially for long periods of time. Since diagenetic products respond sensitively to changes in degree of water saturation of sediment pore space, rate of pore fluid movement, fluid energy and chemistry and a variety of other factors, each fluctuation in water level can profoundly influence the pattern of diagenesis. Thus the three tidal zones mentioned above can be respectively related to three diagenetically active hydrological zones occurring, according to Longman (1980), in a consistent sequence governed by gravitational laws. From bottom to top, the sequence consists of 1) the *marine phreatic zone* beneath the water-sediment interface, where water saturation is persistent; 2) the *marine vadose zone*, a water and/or air-filled environment, regularly exposed to submergence and emergence; 3) the *marine fresh-water mixing vadose zone*,

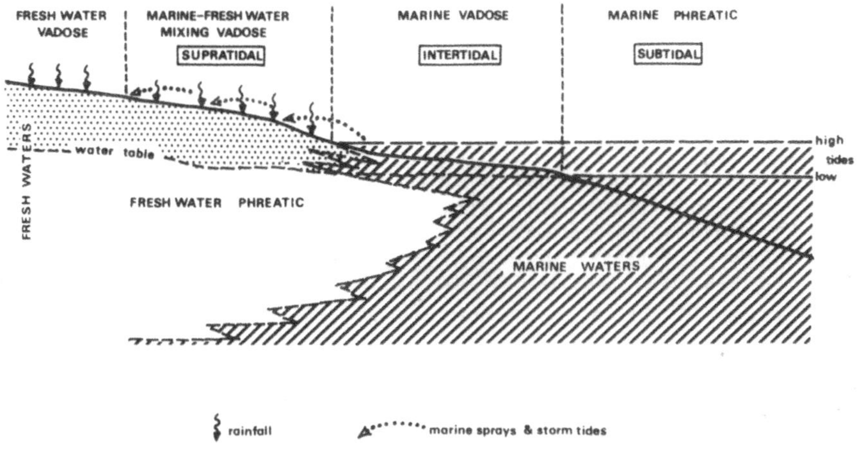

Figure 1. Distribution and relation of major nearshore diagenetic environments, defined on the basis of tidal cycle and of chemistry and distribution of interstitial waters.

an air-filled environment, periodically affected by penetration of percolating waters from sprays, storm tides and rainfall.

In addition to these marine diagenetic zones, two fresh-water related zones occur along or behind the shoreline, within near-surface environments. Termed *fresh-water (meteoric) phreatic and vadose zones,* they lie respectively between the marine phreatic and the mixed vadose zone, and below the land surface. The first exhibits all pore space filled with water, whereas both air and water may be present in the pores of the second. The boundaries between major diagenetic zones need not exactly represent those of the corresponding tidal environments, since they might be held at some position above or below a particular tide level due to local hydrological circumstances. For instance, the upper limits of the marine phreatic zone may be locally raised slightly above the mean low-tide level (effects of capillarity or intergranular retention by superficial tension). Conversely, under strong hydrostatic heads acting on rock frameworks with higher permeability contrasts, fresh water could be forced under the more dense sea water. Under equilibrium conditions, however, it can be assumed that, within the shallow subsurface depositional zone, the diagenetic boundaries crudely define the position of a particular tidal environment.

According to Longman (1980), the typical sequence of diagenetic environments may produce diagnostic sequences in terms of sea-level (SL) fluctuations. For instance, the theoretical sequence of diagenesis for a marine sediment during a relative fall of SL would be to move from the subtidal-marine phreatic zone through the intertidal-marine vadose zone into the supratidal-mixed vadose zone and finally into the meteoric vadose zone. During SL rise, the sequence would be reversed. Thus identification of diagenetic products can play a key role in the interpretation of relative changes in SL.

The scope of this chapter is, firstly, to describe the most diagnostic diagenetic products of the nearshore carbonate environments and, secondly, in order to demonstrate the usefulness of these products as bathymetric and altimetric indicators, to present some applications in identifying shoreline sequences.

DIAGENETIC PROCESSES AND PRODUCTS AND THEIR SIGNIFICANCE IN TERMS OF PALEOBATHYMETRY AND PALEOALTIMETRY

The actual diagenetic processes liable to produce diagnostic features in terms of SL position are divisible as follows: (1) *mechanical processes* both organic or non-organic: burrowing, boring, desiccation, shrinkage, deformation, gas-bubbling, breaking down, reworking, internal sedimentation; (2) *physico-chemical processes*: cementation, recrystallisation, inversion, replacement, micritization, solution, mineralogical and isotopic modifications and alterations, authigenic mineral genesis. The related end-products are relatively easily recognizable in Cenozoic sedimentary sequences.

1. Mechanically produced features

a) Burrows

Most geologists define burrowers as those organisms which excavate unconsolidated sediment particles (Alexandersson, 1972; Milliman, 1974). Impetus to the utilization of burrows as paleobathymetric indicators has come from detailed studies of modern sediments which in the past two decades clearly illustrate the important relations existing between the animals and sediments in a particular environment (Dörjes & Hertweck, 1975; McCall & Tevesz, 1982). Using uniformitarian principles, the recognition of *burrow-fossil associations (ichnofacies)* in marine sequences has opened the way to detailed interpretations in terms of relative water depth (Seilacher, 1964, 1967; Weimer & Hoyt, 1964; Goldring, 1964; Howard, 1972; Frey, 1975; Basan, 1978; Curran, 1984). Unlike most body fossils, burrow fossils are found *in situ* where life activity occurred (Fig. 2). Besides, burrows are one of the most common diagenetic structures of nearshore environments. Incidence of burrowing in such restricted areas ranges from rare or lacking to thorough bioturbation. Two of the most pertinent studies which have proved the value of invertebrate ichnofacies associations as nearshore indicators are Shinn's (1968a) and Farrow's (1971)

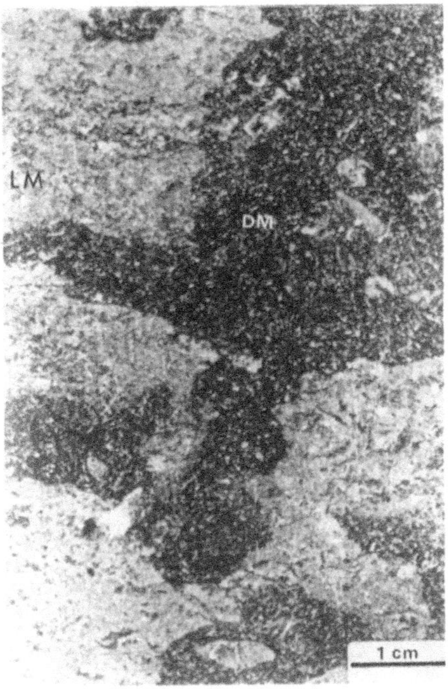

Figure 2.
Burrows of an invertebrate animal - light-coloured micrite matrix (LM) excavated by burrows filled with dark micrite matrix (DM) (Photograph courtesy ELF-Aquitaine).

treatment of burrowing in modern carbonate sediments from Florida (Atlantic Ocean) and Aldabra Atoll (Indian Ocean) respectively. Shinn described diagnostic burrow architecture constructed by five invertebrates. Shrimps, notably *Callianassa* and *Alpheus*, and the anemone *Phyllactis* produce distinctive burrows that are the commonest subtidal types. Intertidal traces are formed mostly by the land crab *Cardisoma* and the fiddler crab *Uca*. Likewise, Farrow showed that each nearshore environment contains diagnostic crustacean burrow assemblages. Subtidal zones reveal a very prolific infauna dominated by the shrimp *Callianassa* and the anomouran *Neaxius*, whereas intertidal environments are characterized by the association of the crabs *Albunea*, *Alpheus* and *Uca*. On supratidal beaches the crab *Ocypode* constructs spiral or 'U'-shaped burrows of a remarkably constant form.

Various aspects of the nearshore paleoenvironment may be determined from a thorough analysis of burrow traces. Takahashi (1932) and Hayasaka (1935) pointed out that whether erect or inclined the base of *Ocypode* burrows always reach groundwater level; thus the higher above high spring tide level the burrow is constructed the deeper it is likely to be. Measurements of penetration depth of burrows in elevated paleo-beaches could provide information on tidal range and position with respect to former high-water levels. Present-day species of *Ocypode* are strictly governed in their distribution by tidal height and the recognition of diagnostic burrow forms of certain species could pinpoint quite precisely the height of any raised

Table 1. List of the main rock-boring organisms occurring within the nearshore environment (from Neumann, 1968, modified).

ORGANISM	FEATURES PRODUCED	MARINE HABITAT
Bacteria	Etched surfaces and enhanced cleavage faces of crystals	ubiquitous
Fungi	Honeycombing of shells, irregular surfaces	ubiquitous
Algae Cyanophytes Chlorophytes	Irregular surfaces, small pits	intertidal and subtidal
Sponges Clionids Spirastrellids	Intensive network of galleries, may penetrate several centimeters	subtidal
Worms Polychaetes Sipunculids Echuroids	Bore holes are long, thin and winding; somethimes closely spaced	intertidal and subtidal
Crustaceans Barnacles	Circular to oval vertical bore holes, 3-5 cm deep, often closely spaced	intertidal
Molluscs Pelecypods Gastropods	Features vary from small pits, made by some species, to long, cylindrical bore holes, often closely spaced	intertidal and subtidal
Echinoids	Deep, cup-shaped holes, circular to oval in plan and variable in cross-section	lower intertidal, subtidal

beach (Farrow, 1971). Stephenson (1965, p 881) observed that numerous present-day *Ocypode Kuhli* along the eastern African coast construct burrows 1 m deep at neap tides, and has found Pleistocene analogues at least 0.70 m deep.

Burrows may easily be confused with a wide range of primary and secondary structures of inorganic origin. They can be related to gas pits (feeding trace of polychaete worms), air escape holes (burrows of amphipod sand-hoppers), and conical fracture patterns (feeding burrows). Nevertheless distinctive burrows and particularly *Callianassa, Alpheus* and *Ocypode* forms which are of very widespread significance at least since the early Pleistocene, are very good guides to the identification of coastline sequences.

b) Borings

Borers, defined as those organisms able to penetrate chemically or mechanically hard substrates (Bromley, 1970; Milliman, 1974, p 253) are widespread at the present time and have an extensive fossil record. According to Bromley (1978), borings of endolithic organisms possess one attribute that renders them potentially useful in paleo-environmental analysis: diagenetically, they are relatively indestructible and therefore are commonly preserved in the geological record.

A wide variety of types of organisms attacks carbonate rocks, including more than three plant and eight animal phyla, in addition to fungi and bacteria. The amount of research done on these various organisms is illustrated by general or regional reviews given by Yonge (1963), Neumann (1968), Milliman (1974, pp 253-258), Warme (1975), Bromley (1978), Risk & MacGeachy (1978) and Golubic *et al.* (1984). Table 1 presents a list of most of the general groups of rock-boring organims occurring within the nearshore belt, the features they produce and the habitat they occupy. Figures 3, 4 and 5 provide some photographic illustration.

Nevertheless, little is at present known of the significance of macroborers as SL markers. Bromley's (1978) careful documentation of boring macroorganisms from Bermuda reefs, living within the depth range 0-20 m, suggests that distribution of related borings is irregular according to water depth; bathymetric control is overprinted by other factors such as hydraulic energy level, particle size and morphology of substrate. Only a few boring Gastropods (vermetid *Dendropoma*) and Crustaceans (barnacle *Lithothrya*) seem to display a depth-dependent distribution.

Conversely, however, several studies of microborers (cyanophytes, green and red algae, fungi) have postulated the possible value of microborings as relative depth tools in recent and possibly ancient carbonates (Ginsburg, 1964; Golubic, 1969; Swinchatt, 1969; Gatrall & Golubic, 1970; Perkins & Halsey, 1971; Rooney & Perkins, 1972; Golubic, Perkins & Lukas, 1975; Le Campion-Alzumard, 1979a; Golubic *et al.*, 1984). For instance, Perkins & Halsey (1971) attempted to use a qualitative and quantitative comparison of fungal and algal borings in Carolina shelf sediments to detect ancient coastline positions. Two offshore zones, paralleling the present coastline, were found to have a high incidence of microborings similar to those in active, nearshore sediments and were interpreted as relic nearshore deposits associated with lower stands of Pleistocene SL. In fact, paleobathymetric application of microbial endoliths necessitates a good knowledge of the morphology and ecology of recent boring microorganims (Golubic *et al.*, 1975). At present, only a few microorganisms and their boring behaviour have been sufficiently studied to permit an identification of paleo-SLs based on boring patterns. According to Le Campion-Alzumard (1970, 1979b) and Golubic *et al.*, (1975), they include: 1) in the lower intertidal

Figure 3. Vermetid shell (VS) partly filled with fibrous aragonite cements (AF) and micrite matrices (Ma), and secondarily perforated by a boring filled with dark micrite (Mb). P = residual pore (Plane polarized light). Emergent organic rim of Holocene age, Western Crete, Greece (Photo L. Montaggioni).

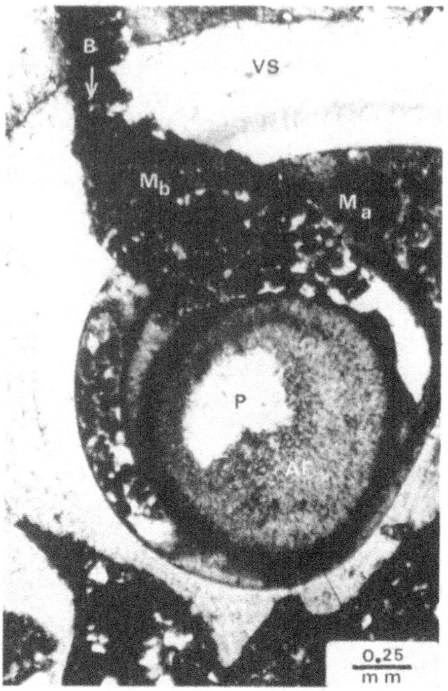

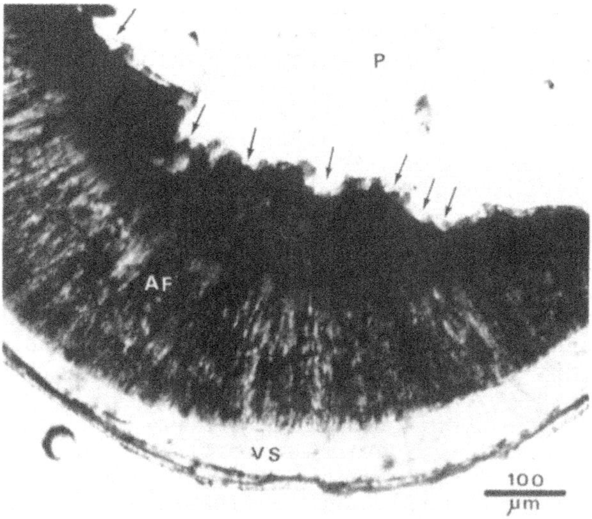

Figure 4. Boring of intragranular fibrous aragonite cement (AF) by sponges (as indicated by the arrows). VS = vermetid shell; P = residual pore (Plane polarized light). Emerged organic rim of Holocene age, Antikythira Island, Greece (Photo L. Montaggioni).

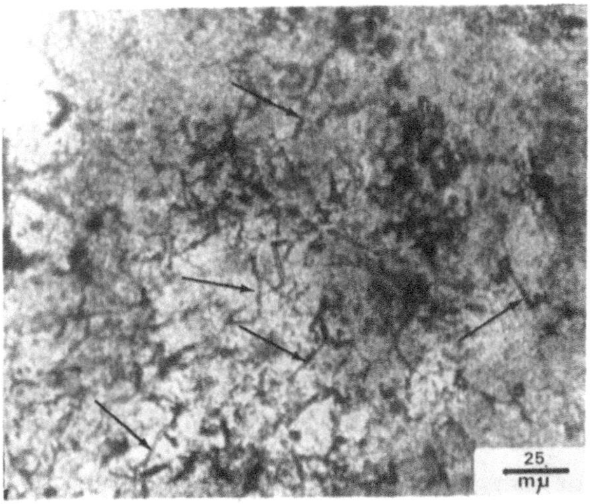

Figure 5. Cyanophyte microborings (arrows) penetrating into a coralline algal nodule. Holocene outer reef slope. Reunion Island (Plane polarized light) (Photo L. Montaggioni).

and subtidal zone (immediately above and below the low-water mark), commonly septate green algae (*Eugomontia, Codiolum*); 2) in the lower intertidal of high-energy coasts, the cyanophytes *Mastigocoleus, Hyella* and *Kyrtuthrix*; 3) in the upper intertidal zone, the blue-green algae *Scopulonema, Hormathonema* and *Solentia*; and 4) in the supratidal zone, coccoid cyanophytes.

Likewise, in order to refine an application of microborings to paleobathymetry, the effects of post-mortem changes prior to fossilization and later diagenetic changes must be taken into account. These processes successively include evacuation of the boring (Alexandersson, 1972); secondary carbonate precipitation (*micritization*, Bathurst, 1971; see this chapter, page 334) or secondary leaching within the boring (Alexandersson, *op.cit.*); diagenetic alteration of host substrates (Chilingar *et al.*, 1967) and filling of microborings by phosphatized chalk (Bromley, 1970) or limonite (Gatrall & Golubic, 1970) to form natural casts.

c) Desiccation mudcracks

As emphasized by Shinn (1983a), probably no single diagenetic feature is better known or more indicative of a depositional environment than polygonally-arranged mudcracks caused by shrinkage of carbonate mud (Fig. 6). Although such well-developed desiccation systems can occur in continental environments (lacustrine, fluvial flood plain), they are more often associated with the marine supratidal environment. By analogy, crack patterns preserved in ancient deposits are frequently interpreted as being diagnostic of a depositional environment periodically subjected to subaerial exposure (Plummer & Gostin, 1981). Shrinkage cracks are also commonly observed in modern

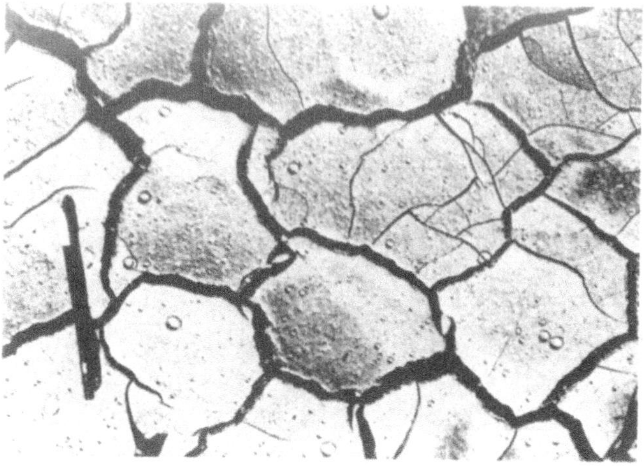

Figure 6. Desiccation cracks in muddy deposits (Photo courtesy ELF-Aquitaine).

intertidal settings, but there is little evidence that they are preserved in such an environment as a result of the severe burrowing activity of organisms. The considerable information available suggests that mudcracks are best preserved in the supratidal and possibly the upper part of the intertidal zone.

Desiccation mudcracks are generally continuous, polygonal and often of several generations with U- or V-shaped cross-sections, from centimetres to metres in diameter. Whilst there is a tendency towards a hexagonal pattern, many polygons are quadrangles or triangles. In rocks they occur in the bedding surfaces of mudstones and are filled by coarse-grained material, usually grainstone. On both present-day surfaces and on bedding surfaces the areas between the cracks are commonly gently concave-upwards. In carbonates, the V-shape often changes; during exposure, the edges of the polygons become rounded from weathering. The result can be a feature that is sausage-shaped in section.

Desiccation cracks may be confused with *synaeresis cracks* resulting from sedimentational compaction and/or foundering induced by the deposition of overlying sediments in subtidal environments (Twenhofel, 1923 in Plummer & Gostin, 1981). Although the synaeresis phenomenon has not been documented in carbonate sediments, it is recommended when using mudcracks as a structure diagnostic of supratidal environment to do so in association with other features typical of subaerial exposure, so as to increase the level of confidence of the interpretation.

d) Fenestrae

Fenestrae (or birdeye, keystone vugs, shrinkage pores, loferites; see Shinn, 1983b) as defined by Tebbutt, Colney & Boyd (1965) are primary penecontemporary holes in rock

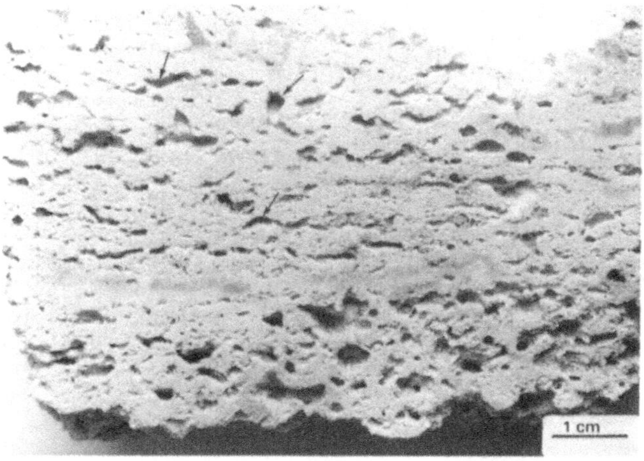

Figure 7. Early Miocene mudstone from Makatea Island (French Polynesia) showing abundant planar-type fenestral voids (arrows) (Photo L. Montaggioni).

framework, larger than grain-supported interstices. Their distinguishing characteristic is that the voids have no apparent support in the framework of primary particles forming the sediment. Fenestrae occur as isolated flattened to spherical 1-5 mm size voids in limestones ranging from mudstone to grainstone (Fig. 7). According to Shinn (1968b) they can be divided into two distinct types, planar-shaped, generally unconnected voids which tend to form parallel to individual laminae, and randomly distributed bubble-shaped voids. In ancient carbonate rocks, these structures are generally completely or partially filled by secondarily introduced sediment or cement (calcite, anhydrite).

Fenestrae represent intergranular voids resulting from shrinkage and expansion, or gas-bubble formation and air space above the low-water mark during the flood tidal cycle (Illing, Wells & Taylor, 1965; Shinn, 1968b; Dunham, 1971). Consequently they are considered reliable indicators of upper intertidal and supratidal deposition. The discovery of fenestrae-like features in the subtidal environment (Shinn, Robbin & Steiner, 1980) may confuse environmental interpretation, but evidence that the latter are preserved in this environment is lacking. On the contrary fenestrae are well-preserved in the upper parts of intertidal deposits and increase in abundance to supratidal sediments. Fenestral features are nevertheless believed to be safely diagnostic of the supratidal only when occurring with other typical diagenetic structures (desiccation mudcracks, evaporites...). As with most geological interpretations, one must use every indicator available in this case.

Subvertical tubular fenestrae occur in some nearshore deposits. Although probably not of the same origin as the

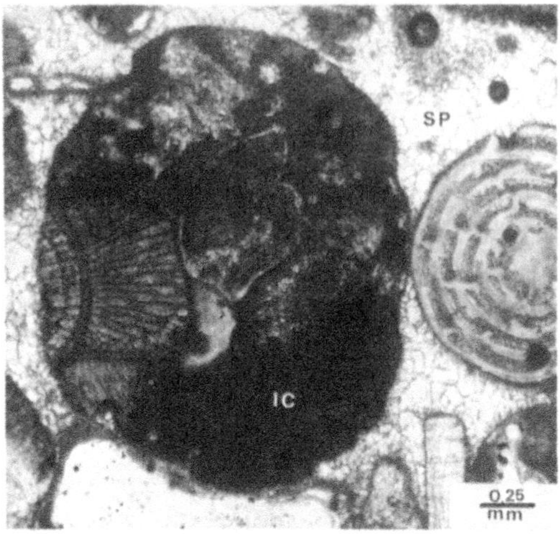

Figure 8. Carbonate intraclast (IC) composed of skeletal fragments (coralline algae, corals) and dark micritic matrix, trapped into low-Mg calcite sparry mosaics (SP). Pleistocene eolianites, Mauritius Island (Plane polarized light). Photo L.Montaggioni.

true planar and bubble-shaped fenestrae, these features are nevertheless common in both modern and ancient supratidal settings. Such pseudofenestral structures are identifiable as burrows or root tubes; since the latter can form in other environments, they alone may not be regarded as reliable criteria for shoreline deposition.

e) Intraclasts

Intraclasts are penecontemporary sediment detritus in rock framework, resulting from a practically *in situ* reworking and sometimes displaying some petrographical characteristics dissimilar to the host rock (Folk, 1959) (Fig. 8).

During storms, nearshore deposits may be subjected to collapse, breakage or reworking into cobble- to pebble-sized clasts before redeposition. Because early hardening or cementation of initially loose sediments is commonly restricted to intertidal and/or supratidal conditions, these zones, therefore, provide most intraclasts. The clasts are recognized in ancient coastline sequences by subtle textural, structural and color contrasts with the trapping deposits (possible occurrence of fenestrae, muddy pebbles composed of redeposited mud polygons, disoriented lamination,...). On a smaller scale, the margins of the clasts exhibit truncated grains and cements and may be penetrated by the borings of various organisms (Inden & Moore, 1983).

A special kind of irregularly shaped intraclast forms in supratidal areas elevated high enough above the marine water table to support flourishing plant life. Here a combination of factors, such as alternate wetting and drying, shrinkage and contraction, disturbance by roots and associated terrestrial burrowers, causes internal shifting of sediment constituents (Shinn, 1983a).

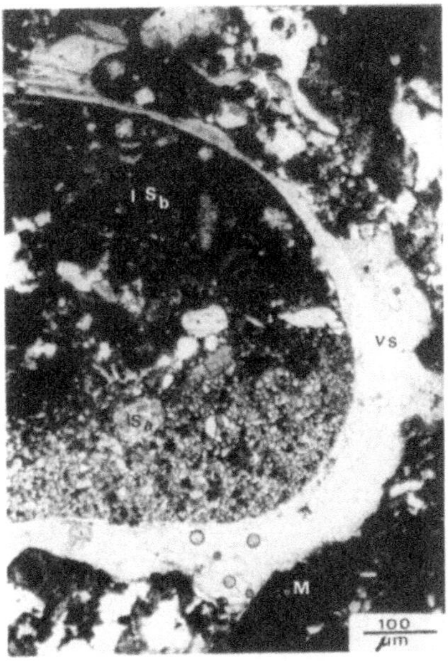

Figure 9.
Vermetid shell completely occluded by normal graded internal sediments. The lower and upper deposits are microbioclastic grainstones (ISa) and dark micritic matrices (ISb). Emerged organic rim of Holocene age, Western Crete, Greece (Plane polarised light). (Photo L. Montaggioni).

f) Internal sediments

In carbonate rocks, cavities of variable origin (primary intergranular voids, fenestrae, burrows, borings, secondary solution voids) frequently are either partially or completely filled with cements and fine-grained, well-sorted detritus (internal sediment). The occurrence of an internal sediment necessitates a hydrodynamic flow sufficient to transport it into the host rock, thus implying near-surface conditions. It exhibits typical textural and structural variations which are clearly related to the environment within which the host rock has developed (Fig. 9). Indeed, within a given cavity, centripetal successions of various cements and internal sediments seem to reflect the lateral evolution of coastline environments; interstitial waters have evolved progressively as the coastline migrated. Thus, the character of the internal sediments which depends on the diagenetic environment (Dunham, 1969) may be a valid clue for interpreting paleo-SL changes.

Aissaoui & Purser (1983) defined four types of internal sediments, each being the product of a particular nearshore environment (Fig. 10).

(1) Geopetal bioclastic micrite.
Unsorted microbioclastic detritus is always disposed geopetally where the cavity is partially filled. It overlies a thin, isopachous layer of early marine cement whose fibrous and regular distribution within the rocks suggests a marine phreatic origin, ie. subtidal or lower intertidal conditions.

(2) Grainstone with inverse graded bedding.
It generally occurs in limestones displaying common

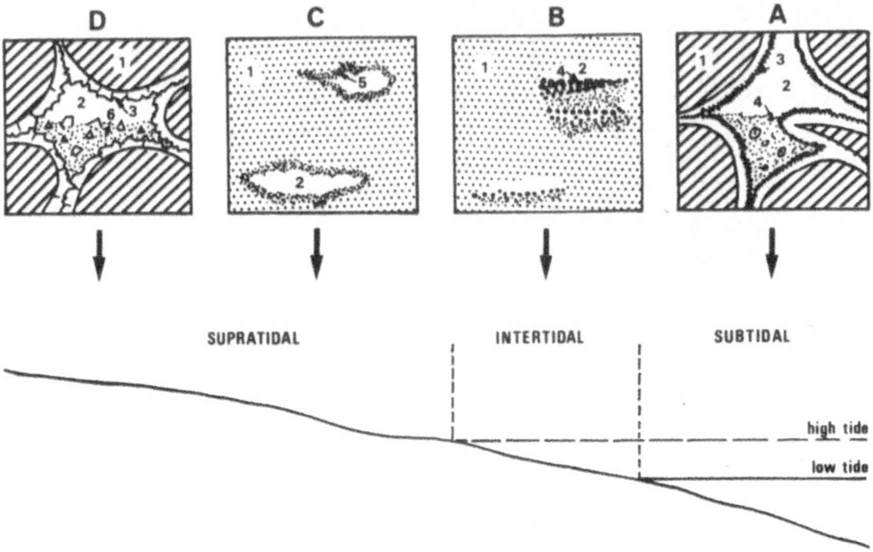

Figure 10. Distribution of the four major types of internal sediment with respect to the tidal environments (from Aissaoci & Purser, 1983). Sediment types: A = geopetal bioclastic micrite; B = grainstone with inverse graded bedding; C = centrifugal micrite; D = crystal silt. Textural characters: 1 = sedimentary grains and/or matrices; 2 = pore spaces; 3 = early cements; 4 = internal sediments; 5 = centrifugal micrite; 6 = crystal silt.

marine vadose criteria (fenestrae, fibrous micro-stalactitic cements). The upward coarsening probably expresses reworking of the heterogeneous internal sediment within microenvironments subjected to certain vadose conditions (Dunham, 1969). The obvious traces of hydrodynamic reworking are logical in an intertidal zone where waters regularly percolated through the sediment to reach the marine water table immediately below.

(3) Centrifugal micrite.
The distribution of this sediment is seemingly independent of gravity; it is plastered both onto the walls and roof, in addition to the more usual geopetal disposition. Such a centrifugal disposition is clearly related to vadose conditions, more particularly to the supratidal zone (Purser & Loreau, 1973; Purser, 1980). In modern supratidal beachrocks, micrite is deposited from muddy waters percolating slowly through the vadose zone; in air-filled cavities, the micrite is presumably retained on the walls and roof by superficial tension of the water film. Moreover micrite may be concentrated at point contacts between adjacent grains as small bridges; this gives

an *alveolar* structure to the host rock, obviously related to supratidal vadose environments (Esteban, 1974).

(4) Crystal silt.
Silt and sand-sized crystals of calcite may occur within certain microcavities. These *vadose silts* (according to Dunham, 1969) display a limpid nature and lack of organic microstructure which suggest a diagenetic origin by erosion of pre-existing sparitic cement. The erosion of this cement and the presence of solution features in the same rock imply the influence of non-marine waters. Thus this kind of internal sediment would seem to be a valuable clue for supratidal and, especially, continental environments.

2. Physico-chemically produced features.

a) Secondary pores

Chemical solution which may affect carbonate material (organic and inorganic grains, cements) leads to the genesis of 'passive cavities' (secondary porosity, Bathurst, 1971), varying in size from a few micrometres to several metres. It almost exclusively occurs within the fresh-water vadose and phreatic zones.

Solution is indeed the result of two factors: the initial undersaturation of rain water with respect to calcium carbonate, due in part to the existence of atmospheric CO_2, and that of CO_2 produced in the soil which increases the partial pressure of CO_2 in the water, thus increasing the amount of calcium carbonate the water can hold in solution.

All the common carbonate mineral species present in modern sediments are found to dissolve. For grains of the same mineralogical composition, solution is controlled by microstructure and origin. Thus biogenic aragonite is less stable than physico-chemically induced aragonite cement (Purser, 1980, p 295).

It is important, but difficult, to differentiate the effects of solution in both the vadose and phreatic zones. Theoretically, the intensity of solution decreases to the lower parts of the vadose zone, as saturation with respect to calcium carbonate increases. However, the most significant solution features (caves) occur at the water-table level, ie. the top of the fresh-water phreatic zone, where saturated vadose waters are cooled or mixed with waters at a lower pressure of CO_2 (Thrailkill, 1968; Ford & Ewers, 1978).

Climate and time play a very important role in fresh-water vadose diagenesis (Harrison, 1975). Solution indeed occurs more rapidly and more extensively in humid climates, partly because of the abundance of fresh water, and partly because a humid climate favors the development of a soil, if the time of emergence is sufficiently long.

So meteoric water often may pass through the vadose zone without becoming saturated with respect to $CaCO_3$, particularly when permeability is high (presence of open fractures or caves). Once under the water table, in the fresh-water phreatic zone, the meteoric water may continue dissolving until it reaches saturation. Either *vuggy* or *moldic porosity* may be produced, that is to say, voids similar to those produced in the vadose zone, depending on the crystal size and mineralogy of the components in the rock.

Vuggy porosity is formed by diagenetic solution of indistinct areas, principally in mudstones or cystalline carbonates (Fig. 11). Moldic porosity is formed by diagenetic solution of a former sedimentary consituent, such as shell, ooid or non-carbonate crystal(gypsum, halite, anhydrite) (Fig. 12). In limestones, it commonly affects aragonite particles while micritic envelopes (see page 334) if existing, remain intact. In dolomites, solution of calcitic shells may likewise form moldic porosity (Friedman, 1964, 1975; Friedman & Sanders, 1978).

In some cases such as calcareous crusts (see page 332) the solution of coarse elements which takes place at the lower layers may be a valid indicator of former meteoric vadose conditions (Esteban, 1974). Such a geopetal solution is due to the retention of pendant acid water drops which leach the substrate instead of inducing precipition of gravitational cement (see page 326).

In conclusion, products of chemical solution are believed to be useful as reliable criteria for distinguishing the limits between the fresh-water, and marine environments; at the present state of the art, it seems impossible to define the fresh-water vadose-phreatic limit from such an investigation.

Identification of secondary pores furnishes less ambiguous information in cases where water-table level intermingles with SL. However that may be, it is preferable to use the distributional pattern of secondary cavities in association with that of cements, thus providing the fundamental basis of understanding.

b) Cements

Cementation refers to the process of precipitation of a binding material around grains or materials and inside primary and secondary pore spaces in sediments and rocks. By aggradation, it tends to progressively occlude open-space structures.

The morphological and compositional characteristics of individual cements depend upon environmental conditions. Numerous studies of cementation were compiled in comprehensive books and articles (Chilingar *et al.,* 1967; Bathurst, 1971, 1980, 1983; Bricker, 1971; Longman, 1980; Purser, 1980).

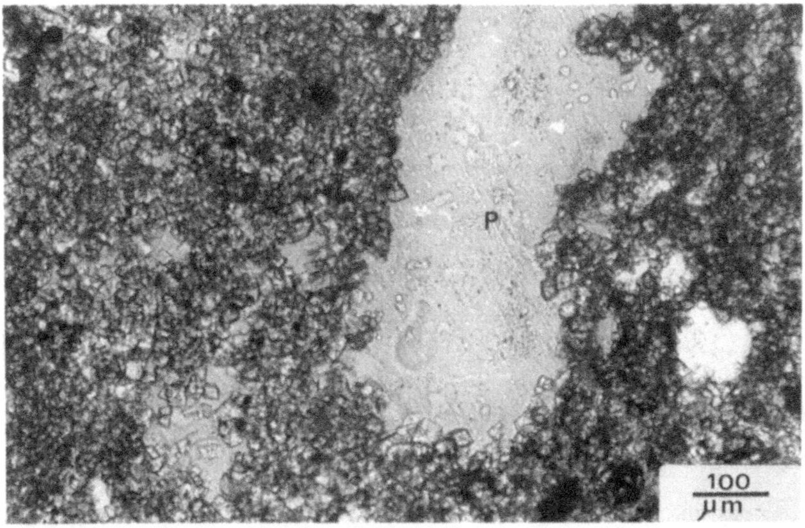

Figure 11. Pleistocene dolomitic micrite showing a vuggy porosity (P = solution pore) (Plane polarized light). Borehole 132.50 m deep, Tenia Island, New Caledonia (Photo J. Coudray).

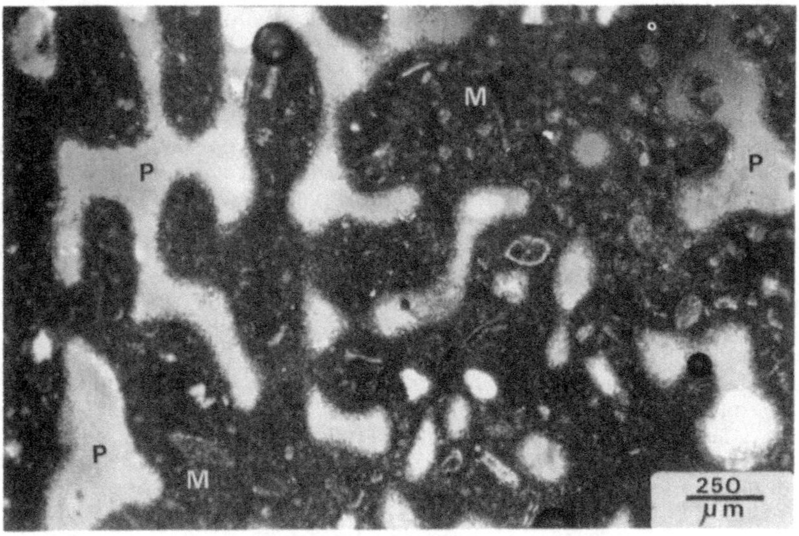

Figure 12. Pleistocene coral skeleton showing a moldic porosity (P = solution pores). Initially, aragonite skeleton is completely dissolved, while internal micrite matrices (M) are preserved (Plane polarized light). Borehole 109.8 m deep, Tenia Island, New Caledonia (Photo J. Coudray).

In the marine phreatic zone, principally near the sediment-water interface, where water does actively move into the sediment, cementation occurs in various forms. The dominant cements precipitated in this environment are both high-Mg calcite, which contains more than 5 mole % $MgCO_3$ (Chave, 1954) and aragonite.

High-Mg calcite cements commonly exhibit the habit of micritic, finely crystalline and semi-opaque rhombic crystals (Alexandersson, 1972; James *et al.*, 1976; Macintyre, 1977) distributed as isopachous crusts around grains (Figs. 13 and 14). In some cases, continued growth of the upper faces of the rhombs produces thick rims of fibrous isopachous cements. In other cases the micritic high-Mg calcite clusters to form pseudopellets which may totally fill pore spaces.

Such textures are common in recent marine-cemented reef rocks (Land & Goreau, 1970; Coudray, 1976; Thomassin & Coudray, 1981; Macintyre, 1977; Montaggioni, 1978) and are present in marine phreatic beachrocks and conglomerates (Moore, 1977; Montaggioni & Pirazzoli, 1984).

Aragonite cement typically is present as clear and acicular crystals in a variety of textures: isopachous fibrous fringes which can radiate outward from grains and/or inward into vugs are the most common (Fig. 15) but in some places aragonite forms also a loose meshwork of needles (Longman, 1980) or coarse botryoids (Ginsburg & James, 1976). Like high-Mg calcite and generally associated with it, marine aragonite may form in all kinds of carbonate sediments, including reefs, beachrocks, ooïd shoals or certain sediments, generally deeper, that are slowly deposited (hardgrounds).

In the stagnant marine phreatic environment, where tides, waves and currents are absent or not strong enough to move through the sediments, diagenesis occurs slightly as micritization (see below) and minor intergranular cementation.

Finally, the fibrous texture, which may be preserved in ancient rocks in spite of neomorphism into low-Mg calcite (see page 335) is the best petrographic criterion to distinguish former marine cements. Crystal morphology also permits recognition of former aragonite, after neomorphism into calcite: it commonly has square-ended fibres (Loucks & Folk, 1976). For fibrous Mg-calcite cements, it is more difficult, one of the strongest lines of evidence being the presence of microdolomite inclusions in the neomorphosed calcite (Lohmann & Meyers, 1977). Internal sediments and encrusting organisms accumulated within, or on the top of, cements (see page 323) and the stable isotopes of carbon and oxygen (see page 337) can be used successfully also to recognize the phreatic marine diagenesis.

In the fresh-water and marine vadose environments, where both air and water (meteoric or marine) may be present

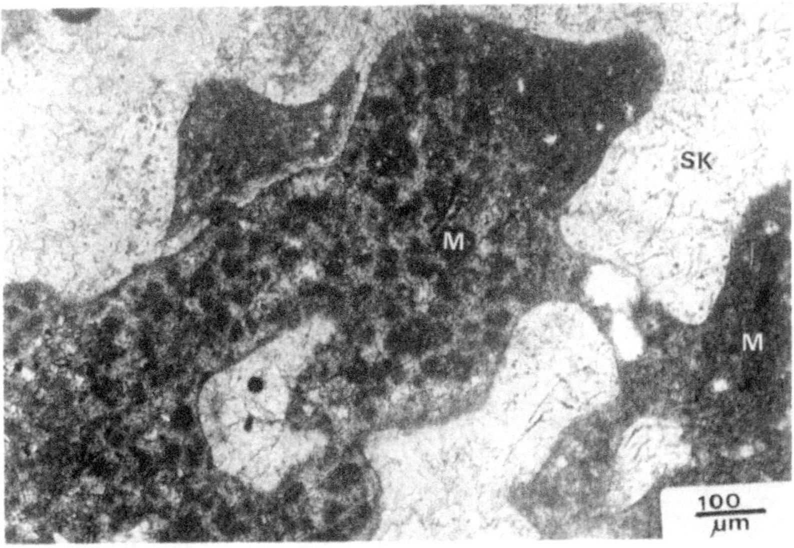

Figure 13. Microcrystalline (micrite) pelleted cement of high-Mg calcite (M) filling internal cavities of a coral skeleton (SK) (Plane polarized light). Pleistocene reef flat, Mauritius Island (Photo L. Montaggioni).

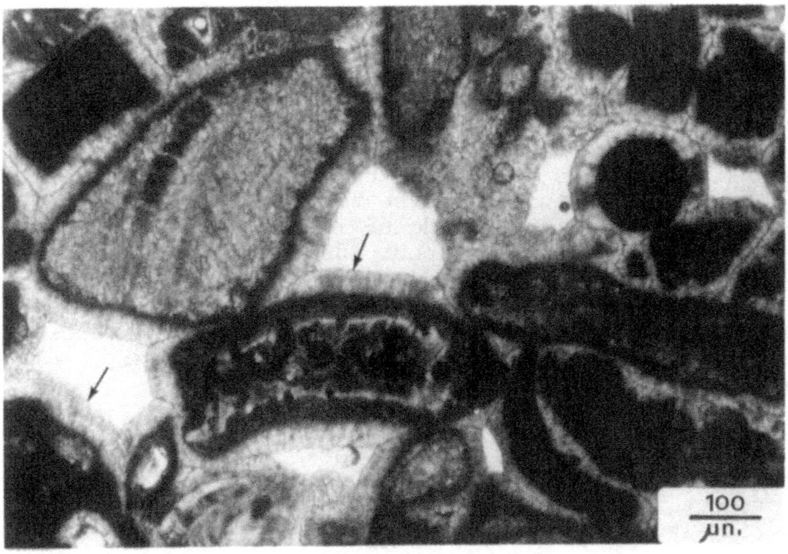

Figure 14. Isopachous fringes of high-Mg calcite (arrows) around skeletal grains (plane polarized light). Emerged beachrock of Holocene age, Mauritius Island (Photo L. Montaggioni).

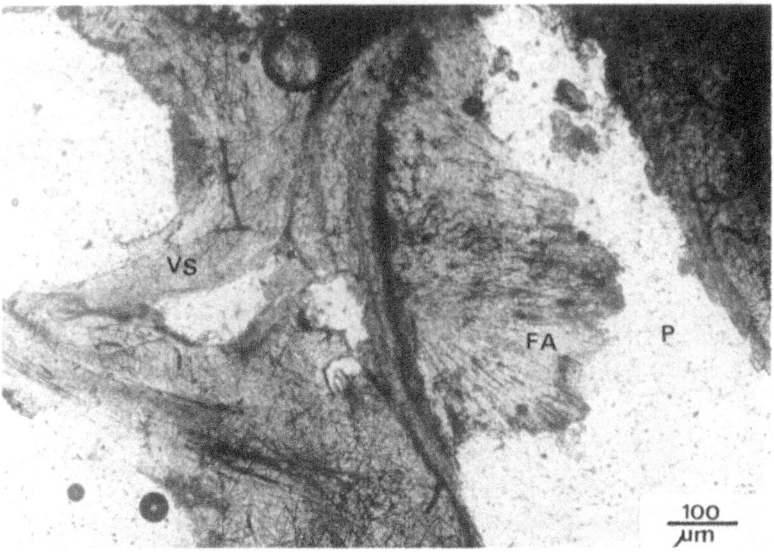

Figure 15. Acicular aragonite (FA) partly filling a vermetid shell (VS) (P = residual cavity) (Plane polarized light). Modern reef flat, Reunion Island (Photo L. Montaggioni).

in pores, cementation is a relatively minor process, but may develop rapidly below the zone of solution (see page 325) once vadose waters reach saturation.

The clearest examples of the intertidal cements are found in beachrocks which represent, like the beach itself, a transitional situation, sometimes bathed in marine water, and in other localities at times affected by fresh water (Bricker, 1971). The global result of the penecontemporaneous beach cementation is the development of large seaward dipping slabs, described by many authors, in modern beach sediments around the world, between latitudes 35°N and 35°S (Friedman & Sanders, 1978). The origin of such cements, discussed for about 100 years, is still debated.

Cement morphology in the vadose zone tends to reflect the distribution of pore water. Thus at least two types of cement, recognized both in modern and ancient rocks, characterize the vadose zone:

(i) a *dissymmetric* or *microstalactitic* (Taylor & Illing, 1969; Purser, 1969), *gravitational* (Müller, 1971) or *pendant* Longman, 1980) type forming when water is present as droplets on the underside of grains (Fig. 16).

(ii) a *meniscus* type (Dunham, 1971) forming when only capillary water is present at, and very close to, grain contacts (Fig. 17).

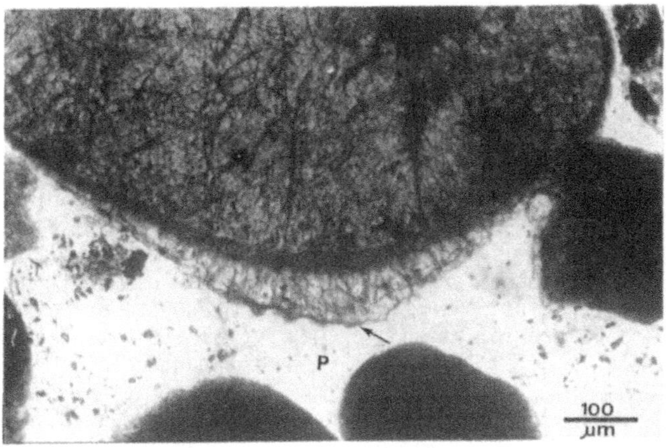

Figure 16. Microstalactitic cement of low-Mg calcite (arrow) hanging from the lower side of a skeletal grain (P = residual pore) (Plane polarized light). Pleistocene eolianites, Rodriguez Island (Photo L. Montaggioni).

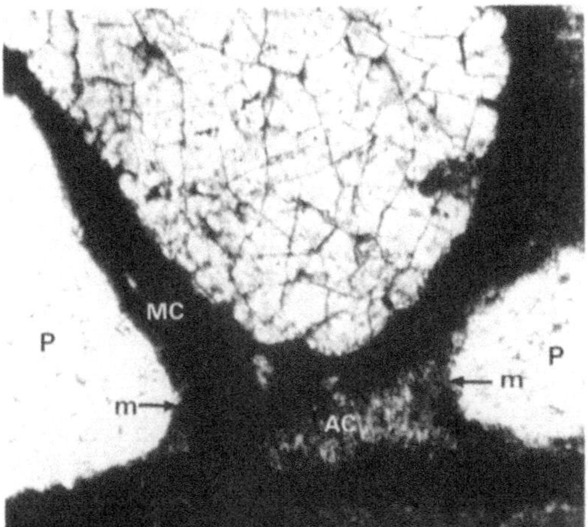

Figure 17. Thin rims in high-Mg calcite micrite (MC) and aragonite fibers (AC) locally of meniscus form (m, arrows) at grain contact (Plane polarized light). Emergent coral conglomerates within intertidal environment, Mataiva Atoll, French Polynesia (Photo L. Montaggioni).

These cements tend to be patchy and concentrated in some parts of sediments. They are not isopachous and increase the roundness of pores. If precipitation continues, they may fill pores completely, with, according to Schroeder (1973), a coarsening of the crystals toward pore centers.

Mineralogically, cements tend to be calcite, commonly as very fine equant crystals without terminations, when the main precipitating fluid is meteoric water, that is to say principally in the fresh-water vadose zone. This is not true if magnesium ions inhibit the precipitation of calcite

Table 2. Idealized zonation in fresh-water phreatic environment (from Longman, 1980)

		Zone	Characteristics
Increasing saturation with $CaCO_3$	Increasing depth	1. Zone of solution of calcite and aragonite	Formation of vuggy and moldic porosity
		2. Zone of solution of aragonite	Formation of moldic porosity
		3. Zone of aragonite solution and calcite precipitation	Rapid neomorphism of aragonite grains to equant calcite cementation by equant calcite
		4. Zone of no solution precipitation of calcite	Rapid cementation by equant calcite
		5. Stagnant zone, saturated with $CaCO_3$	Little or no cementation; slow neomorphism of aragonite; grains with preservation of some structures

and facilitate the precipitation of aragonite or even dolomite, commonly occurring in caves (Thrailkill, 1971; Cabrol & Coudray, 1982).

Within marine beaches and intertidal flats extensive cementation by aragonite and/or high-Mg calcite is common. Cements tend to be fibrous or micritic (Purser, 1969, 1980; Schroeder, 1973; Bathurst, 1974; Moore, 1977; Hanor, 1978; Montaggioni, 1978).

Another unusual type of cement which may form in the vadose zone is 'whisker calcite' formed by randomly distributed straight fibres (1 to 4μm wide and hundreds of μm long), as described by Supko (1971) and Steinen (1974). However, it seems rare and its origin is not definitively sure (Harrison, 1974; Badiozamani, Mackenzie & Thorstensen, 1977).

At the top of the fresh-water vadose zone, where solution is the most important diagenetic process, the only cementation that occurs is that of evaporitic and/or biologic crusts (caliche, nari, calcrete) at or near the air-rock interface in arid and semi-arid regions. Such crusts, characterized by an ooidal micritic texture have been described in detail (James, 1972; Coudray, 1976; Montaggioni, 1978). Generally, they do not provide an interesting tool for identification of ancient strandlines, because they originate in supratidal zones as well as high continental areas.

Finally, one of the best recorders of vadose

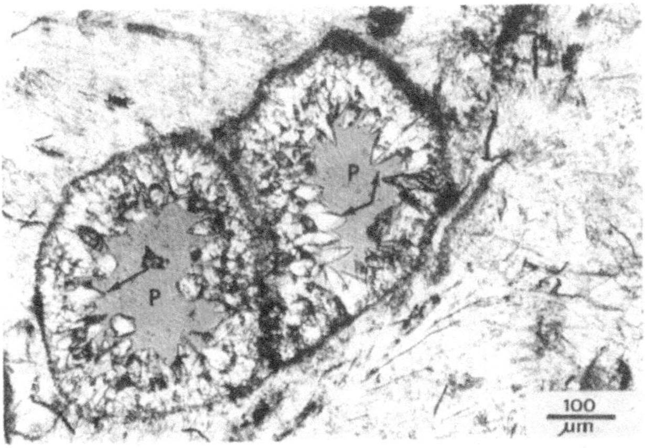

Figure 18. Dog-tooth-shaped sparry cement of low-Mg calcite (arrows) lining two internal cavities of a coral skeleton (P=residual pores) (Plane polarized light). Borehole 38.4 m deep, Tenia Island, New Caledonia (Photo J. Coudray).

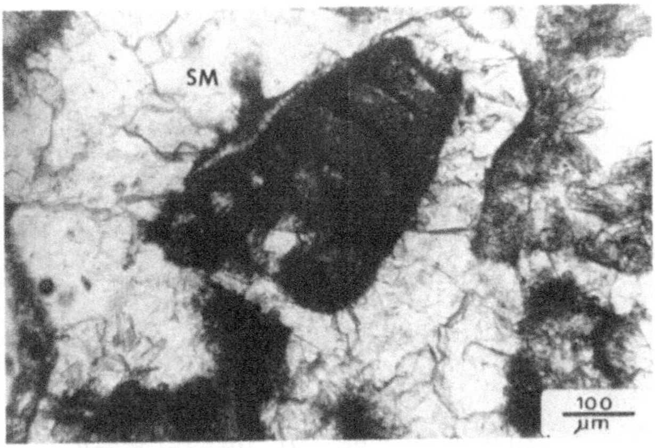

Figure 19. Mosaic of low-Mg calcite blocky cement (SM) (Plane polarized light). Pleistocene eolianites, Mauritius Island (Photo L. Montaggioni).

environment is microstalactitic cement which is easily recognizable in ancient rocks, because of its distinct morphology and texture (Purser, 1969). Fossil meniscus cement is easier to recognize when not obscured by later cementation, but is frequently observed in Quaternary carbonate rocks (Schroeder, 1973). However, rocks in which vadose cements form are very susceptible to erosion because of their position above SL and the preservation of these cements is likely only if a rise of SL follows soon after their formation.

In the fresh-water phreatic environment, five diagnostic products of cementation may be recognized on the basis of saturation of waters with respect of $CaCO_3$ (Longman, 1980; Table 2). The more common and widespread cement is equant calcite, distributed equally around all grains and filling all pores with possible coarsening of crystal towards pore centres (drusy, sparry or blocky mosaic, Bathurst, 1971; Friendman, 1975; interlocking mosaic, Longman, 1980) (Figs. 18 and 19). Other common fabrics are syntaxial overgrowths (or rim cements): the original echinoderm fragment and the

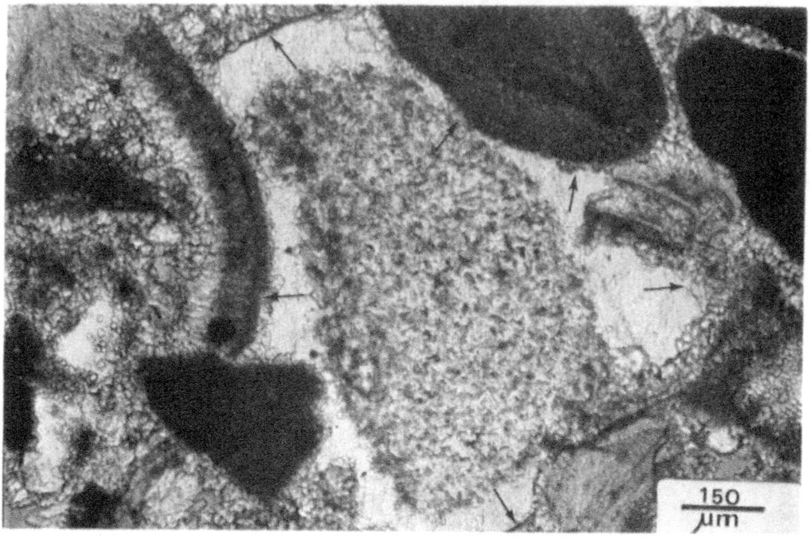

Figure 20. Overgrowth of an echinoderm fragment (delineated by arrows) (Plane polarized light). Borehole 48.20 m deep, Tenia Island, New Caledonia (Photo J. Coudray).

overgrowth form a single crystal (Evamy & Shearman, 1965) (Fig. 20). However, such a product may also form from saturated vadose water and sea water (Bathurst, pers. comm.) but usually does not completely surround grains.

c) Micritic envelopes

Bathurst (1966, 1971) described a process of intragranular cementation occurring inside grain borings. By repeated boring, followed by vacation of the bore and filling of it with cryptocrystalline material (micrite), carbonate grains are gradually and centripetally replaced by micrite (Fig. 21). In this process (called micritization), complete replacement of the grain by micrite can take place, or a rind of micrite develops within the grain periphery. Micritization has been recognized as one of the most significant geologic effects of boring algae (filamentous cyanophytes, chlorophytes and rhodophytes), and is present to some extent in practically every modern and ancient shallow-water marine carbonate environment (Winland, 1968; Swinchatt, 1969; Loreau, 1970; Alexandersson, 1972; Golubic *et al.*, 1975). Endolithic boring algae being dominantly photosynthetic and showing a vertical species zonation from the supratidal to the light-compensation level, the related process of micritization may have some paleobathymetric application (see page 319).

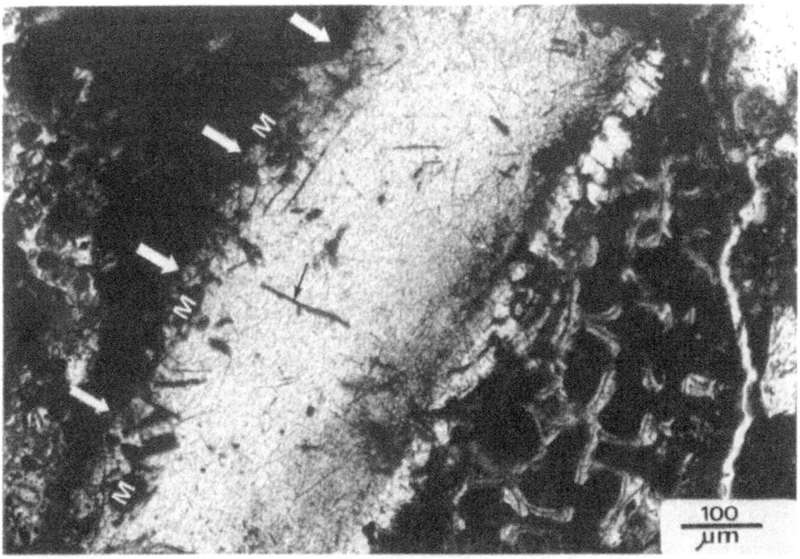

Figure 21. Micritization of a molluscan test by microbial infestation: white arrows delineate the front of microboring; a black arrow indicates an algal microboring (M = internal micrite) (Plane polarized light). Modern reef flat, Reunion Island (Photo L. Montaggioni).

d) Mineralogical and chemical changes

Among the numerous mineralogical changes which affect the carbonate sediments during early diagenesis, named by Folk (1965) 'neomorphic processes', and largely described and discussed by Bathurst (1971), the more significant in terms of recognition of SL are the dissolution-reprecipitation phenomena which tend to transform aragonite into low-Mg calcite (Folk's 'inversion') and high-Mg calcite into low-Mg calcite, and also aggrading neomorphism in the form of crystal enlargement (recrystallization), with or without 'replacement' of a mineral by another.

Considering the dissolution-reprecipitation phenomena, both particles (organic and inorganic) and early diagenetic cements formed in shallow tropical seas consist mainly of two carbonate polymorphs: aragonite, characterized by a high Sr content (more than 7 parts per thousand according to Kinsman, 1969; Mercadier *et al.*, 1975; Friedman & Sanders, 1978) and high-Mg calcite, containing more than 5 mole percent $MgCO_3$ (Chave, 1954). Relatively stable in their original marine environment, they are quite unstable under the influence of meteoric waters, being colder and poorer in Mg ions. So either they are completely dissolved, giving passive cavities (mainly aragonite; see page 325) or they progressively have been transformed by *in situ* wet processes, into low-Mg calcite (from high-Mg calcite especially).

All the data (Stehli & Hower, 1961; Friedman, 1964; Gross, 1964; Land, 1967; Land, Mackenzie & Gould, 1967; Purdy, 1968; Gavish & Friedman, 1969) show that the first mineralogical change under fresh-water regime is the loss of Mg^{2+} from high-Mg calcite, without destroying the apparent structure. The newly formed low-Mg calcite, mainly from high-Mg calcite grains, by incongruent dissolution of Mg as well as replacement of Mg by Ca is only, but clearly, distinguishable with the help of a magnesium spot test and by X-ray diffraction. According to Friedman & Sanders (1978), high-Mg calcite can change into low-Mg calcite within a time span of 7000 to 10 000 years.

The polymorphic transformation of aragonite to calcite occurs later, mainly in the fresh-water phreatic environment. It generally affects sediments which have a low permeability, where the composition of the local pore fluid probably is between the solubility product of aragonite and calcite (Füchtbauer, 1975). Between the aragonite and the replacing calcite a solution film is present thin enough to enable *in situ* preservation of a ghost of the original structure; for example original lamellar structure of the shell wall (Land, 1967; Pingitore, 1970; James, 1974) or concentric structure of ooids (Shearman, Twyman & Karimi, 1970). In this case, the transformation is hetero-axial (no optical and textural relationship between aragonite and calcite crystals). The products may be microcrystalline calcite, mainly in the supratidal zone (James, 1972), but more commonly, mosaics of sparry calcite. In Pleistocene limestones of Bermuda, Schroeder (1973) showed each crystal, with plane boundaries, is in optical continuity with sparry calcite crystals which occlude the nearby voids. Purdy (1968), in the Bahamas, showed a partial recrystallization of aragonite tests into sparitic crystals of variable size, the smaller located near the periphery of particles. This looks like Barthurst's drusy mosaic (see page 333) which results from void's filling by a sparitic cement. These two examples well illustrate a convergent phenomenon between dissolution-precipitation and cementation products which together mainly occur in the fresh-water phreatic environment. In other cases, the aragonite - calcite transformation may be homoaxial (transformation of fibrous aragonite into fibrous calcite; Füchtbauer, 1975) and then, very difficult to recognize. In all cases, finally, this process is accompanied by Sr^{2+} loss, falling down from more than 7000 ppm in aragonite particles to less than 600 ppm in the replacing calcite (Kinsman, 1969; Mercadier, Coudray & Joseph, 1975).

Aggrading neomorphism (recrystallization and replacement; Folk, 1965) is the process whereby finer crystal mosaics are replaced by coarser crystal mosaics of the same mineral (recrystallization) or its polymorph (replacement). It includes the classical diagenetic alterations of aragonite or calcite micritic mud or micrometre-sized skeletal fabrics (crystal diameters: 0 to 4μm) into microsparry (4 to 10μm or even 50μm) and sparry calcite (more than

10-50µm). These alterations, which are a complex process combining the *in situ* processes described above, are most frequently associated with non-marine water conditions, for example in the fresh-water phreatic zone for skeletal grains (Longman, 1980) and in the supratidal zone for carbonate oozes (Gebelein, 1969; Purser, 1980).

Early dolomitization may occur in all kinds of tropical carbonate environments (Coudray, 1976) and generally is not a valid criterion for identification of SL if occurring alone. Nevertheless, syngenetic dolomicritic muds and crusts frequently occur in upper intertidal to supratidal areas as in Bahamian tidal flats and the Florida cays (Shinn, 1968a, 1983a; Shinn, Ginsburg & Lloyd, 1965; Bourrouilh-Le Jan, 1973) or in the Persian Gulf sabkhas (Illing *et al.*, 1965; Kinsman, 1964; Purser, 1973). Similar to fresh-water vadose caliche crusts, they frequently contain pisolites (Purser & Loreau, 1973; Shinn, 1973) and are associated with mechanically and biologically produced features as desiccation mud-cracks, birdeyes, stromatolitic laminations, algal mats and with authigenic minerals as halite, anhydrite, aragonite, celestite, magnesite, huntite.

In all present-day examples, the site of formation of these dolomite hypersaline caliches (Shinn, 1983a) is in the seaward portion of the tidal-flats exposed to sea spray and spring or storm-tide flooding. If their origin in ancient counterparts is similar, then one can assume that their presence may be a useful indicator for predicting the position of a former shoreline.

Many authors have studied isotope ratios $^{16}O/^{18}O$ and $^{12}O/^{13}O$ in carbonate minerals as indicative of sedimentary and diagenetic environments (Craig, 1953; Friedman, 1964; Land, 1967, 1970; Choquette, 1968; Land & Epstein, 1970; Hudson, 1975; Magaritz, 1983).

^{18}O and ^{13}C are indeed enriched in marine waters compared with non-marine waters owing to fractionation during evaporation and precipitation in equilibrium with the atmosphere. The isotopic compositions of carbonate minerals formed or modified in marine or fresh-water environments reflect these variations as shown in Figure 22, where δ represents the deviation from a universal standard sample, given in parts per thousand. So, particles and cements produced in a marine environment have mean $\delta^{13}C$ and $\delta^{14}O$ near zero, whereas those formed in a fresh-water environment become systematically more negative ($\delta^{13}C = -5$ to -15; $\delta^{18}O = -5$ to -10). These last values also indicate a post-depositional exchange of isotopes with meteoric waters.

Thus the carbon and oxygen isotopic composition of primarily marine carbonates, which have been collected at close spacing within vertical sequences, may prove to be an important data base which can be used to develop predictive models for the location of the marine - meteoric water limits in ancient carbonates.

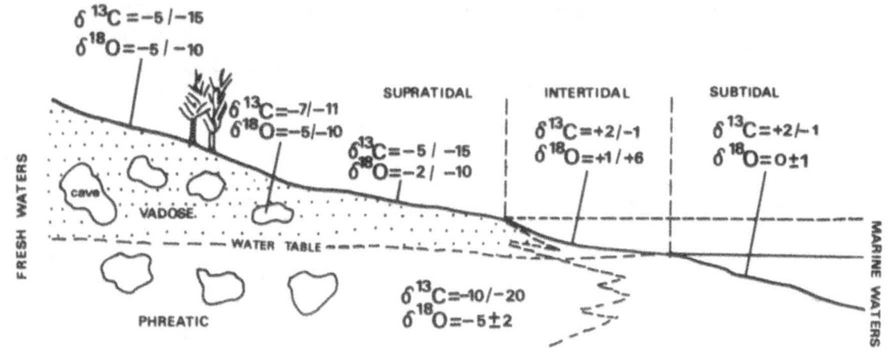

Figure 22. Range of carbon and oxygen isotopic composition of carbonates formed in various modern nearshore and subaerial environments (from Magaritz, 1983, modified).

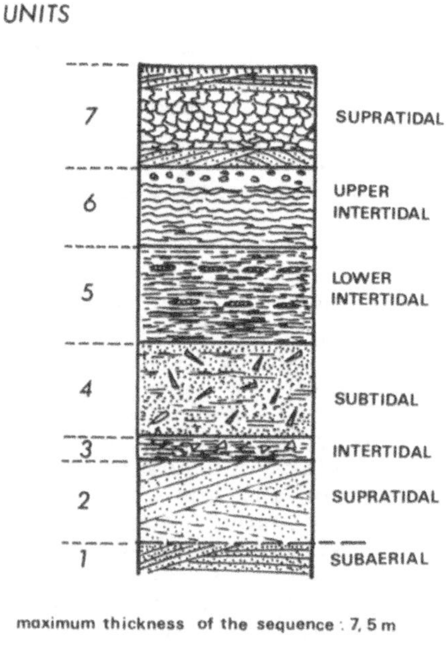

Figure 23. Idealized vertical profile of sabkha sediments, Persian Gulf; see text for detailed explanation (from McKenzie *et al.*, 1980, modified).

EXAMPLES OF DIAGNOSTIC DIAGENETIC SEQUENCES

Mention is made of some regional diagenetic sequences as a reliability test of the vertical mosaic of diagenetic features mentioned previously in the recognition of ancient shorelines and interpretation of SL changes.

1. Use of structural and mineralogical criteria.

An idealized offlap profile representing a complete vertical sequence of Holocene tidal-flat sediments from the Persian Gulf, Abu Dhabi, is given in Figure 23. The sequence is based on a compilation of data from various works (Evans *et al.*, 1969; McKenzie, Hsu & Schneider, 1980; Shinn, 1983). The upper complex includes three units (numbered 5,6,7). Throughout units 5 and 6, diagenetic alteration is manifested by the occurrence of gypsum mush, algal laminations with fenestral voids and desiccation cracks, grading downward into burrowed muds. These two units have been referred to an intertidal zone, whereas unit 7, characterized by the presence of nodular anhydrite and cross-bedded sands with root marks, is typical of a supratidal facies. The upper complex is commonly separated from unit 4 by a lithified, bored grainstone crust. The unit 4 is composed of bioturbated, gray-green carbonate sand with varying mud composition, gastropods, bivalves and gypsum discs, subtidally displaying algal laminations, shrinkage and birdeye pores; it was probably deposited mainly in an intertidal environment. The lowest units (2 and 1) reflect supratidal to subaerial deposition: occurrence of cross-bedded (eolian) sands with intraclasts and root tubes.

Examination of this profile shows that a shallowing-upward sequence (units 4 to 7) overlaps a formerly exposed formation (units 2 and 1), indicating an early relative rise or SL, then probably a rapid sediment accumulation under relatively stable SL.

2. Use of textural and mineralogical criteria (under exclusive marine control).

Montaggioni & Pirazzoli (1984) demonstrated that exposed coral conglomerates from French Polynesia (Pacific Ocean), ranging in age from the present to about 6000 years BP can be utilized as valuable indicators of former SL. Reaching a maximum elevation of more than 1 metre above present mean high-tide level, the rubble outcrops occur mainly as sub-horizontal rampart platforms of skeletal detritus deposited by storm surges. With the aim of clarifying their reliability as markers of recent SL changes, a determination of the environments of lithification has been attempted. For each outcrop, two sequences of marine cementation have been defined: 1) a *lower sequence* displaying several generations of intergranular cements. The earlier generations are characterized by high amounts of high-Mg calcite pelleted micrite normally completely occluding interparticle pore spaces (Fig. 24). The later generation is characterized by the presence of irregular rims of truncated aragonite fibres and pendant microstalactites (Fig. 25). Residual interparticle porosity is relatively low; 2) and *upper sequence* mainly exhibiting blunt aragonite needles and microstalactites; residual porosity is relatively high (Fig. 26). The diagenetic limit between these two sequences defines the position of a former marine water table that is regarded as having been

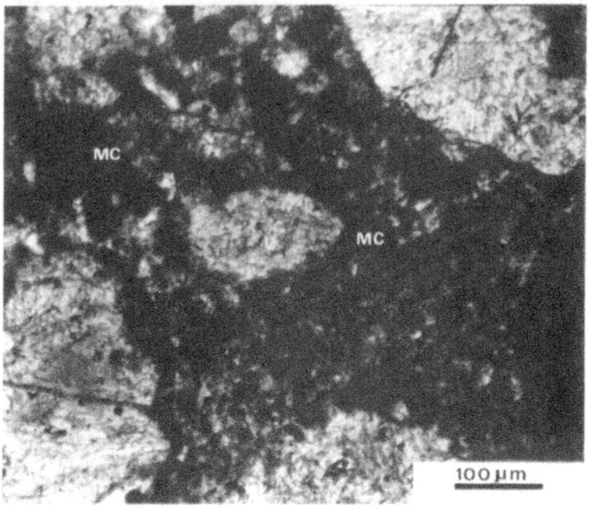

Figure 24. Marine phreatic cementation: occurrence of high-Mg calcite-pelleted micrite (MC) trapping bioclastic elements and completely filling pore spaces (Plane polarized light). Emergent coral conglomerates, lower sequence, Rangiroa Atoll, French Polynesia (Photo L. Montaggioni).

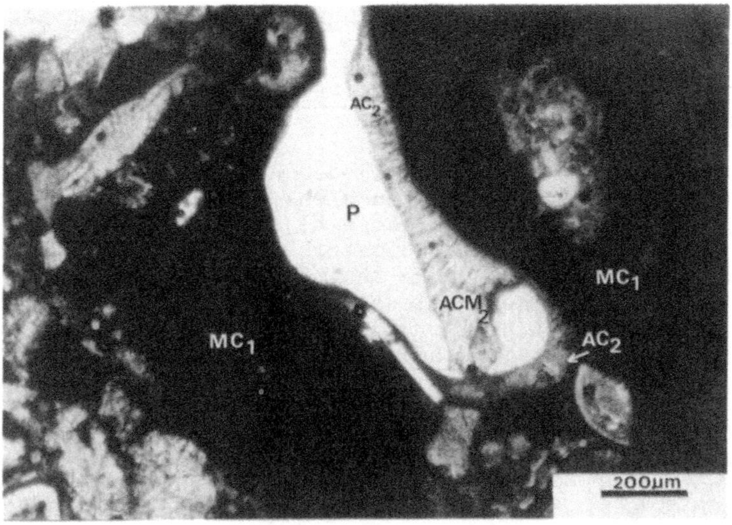

Figure 25. Marine phreatic and vadose cementation: occurrence of two generations of cement. The earlier generation reveals micrite matrix (MC_1). The later one includes irregular fringes of blunted aragonite needles (AC_2) locally forming microstalactites (ACM_2) within a residual cavity (P) (Plane polarized light). Emergent coral conglomerates, lower sequence, Moorea Island, French Polynesia (Photo L. Montaggioni).

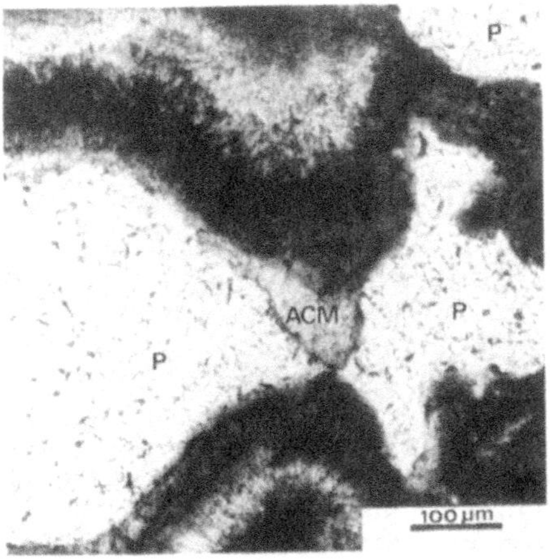

Figure 26.
Marine vadose cementation: occurrence of an aragonite microstalactite (ACM). Notice the high residual intergranular porosity (P) (Plane polarized light). Emergent coral conglomerates, upper sequence, Apataki Atoll, French Polynesia (Photo L. Montaggioni).

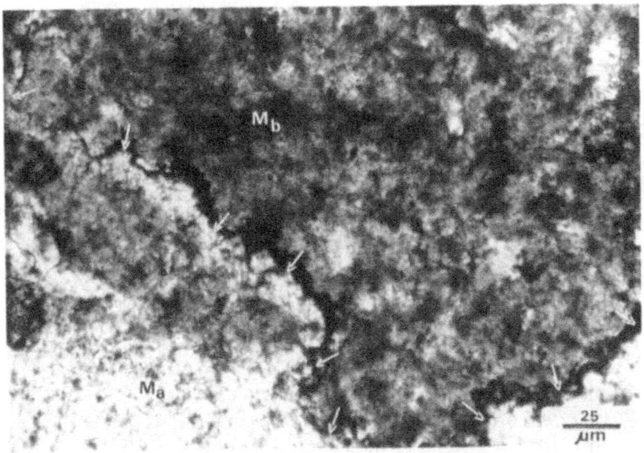

Figure 27. Coralline algal thallus bored by sponges and subsequently filled with marine cement. The diagenetic sequence is as follows: 1) after settlement of a coralline algal buildup, transformation of initial calcite micrite (Ma) into microspar during early emergence; 2) boring of the carbonate substrate by sponges (white arrows delineate the rim of organic matter marking the bioeroded front), then precipitation of micrite cements (Mb) during the following stage of submergence; 3) ultimate emergence (Plane polarized light). Emerged organic rims of Holocene age, Western Crete, Greece (Photo L. Montaggioni)

closely related to an ancient mean low-tide level. All this seems to indicate that the lower sequence has spent time in two distinctly different diagenetic environments (marine phreatic overlain by marine vadose), whereas the upper sequence has always been within one diagenetic zone (marine vadose). Thus it is clear that these conglomerates have recorded a relative drop of SL after the deposition and early lithification of the lower deposits.

3. Use of textural and mineralogical criteria (under both marine and subaerial control).

a) Holocene age

In Western Crete (Greece), the type and sequence of diagenetic changes which elevated organic-built rims of Holocene age have undergone since their formation, have been correlated with environmental changes (submergence - emergence phases) from examination of thin sections and X-ray mineralogical analysis (Thommeret *et al.*, 1981).

Thus a cycle including a period of submergence - emergence, then a period of new submergence - new emergence was manifested over the last millennia. The petrogenetic and diagenetic sequence is as follows (Fig. 27).

- during and following the settlement of the building organisms, early marine cements were precipitated in the same single generation as micrite or fibrous high-Mg calcite and acicular aragonite.

- first emergence: the most significant diagenetic alterations are those of the formation of microspar, which is a recrystallization of the primary microcrystalline cements into a granular crystal mosaic (microsparitization) and the transformation of fibrous high-Mg calcite into equant granular crystals (pseudospar). This neomorphism is the result of leaching by meteoric waters of Mg ions trapped in calcite network and which impedes the growth of microcrystals in the marine environment. Consequently, the $MgCO_3$ content of marine calcites varies markedly from rim to rim (8-16 mole percent $MgCO_3$; initial content: about 20 mole percent). Subaerial diagenesis is also evident from the development of low-Mg calcitic rinds or sparry mosaics.

- new submergence: the biogenic rim is bored by various organisms (*Cliona* especially), then vacated borings are filled by micrite-size, high-Mg calcite material.

- final emergence: strong microsparitization of the first

Figure 28 (see opposite). Diagenetic patterns and reconstruction of relative sea-level changes from mineralogical and petrographical evolution of a Quaternary sequence (borehole through Tenia Island, barrier-reef complex, New Caledonia) (From Coudray, 1976). Diss. = solution; Cem. = cementation; Recr. = recrystallization; - = absent or rare; + = present in low amount; ++ = common; FWV = fresh-water vadose; FWP = fresh-water phreatic; MP = marine phreatic.

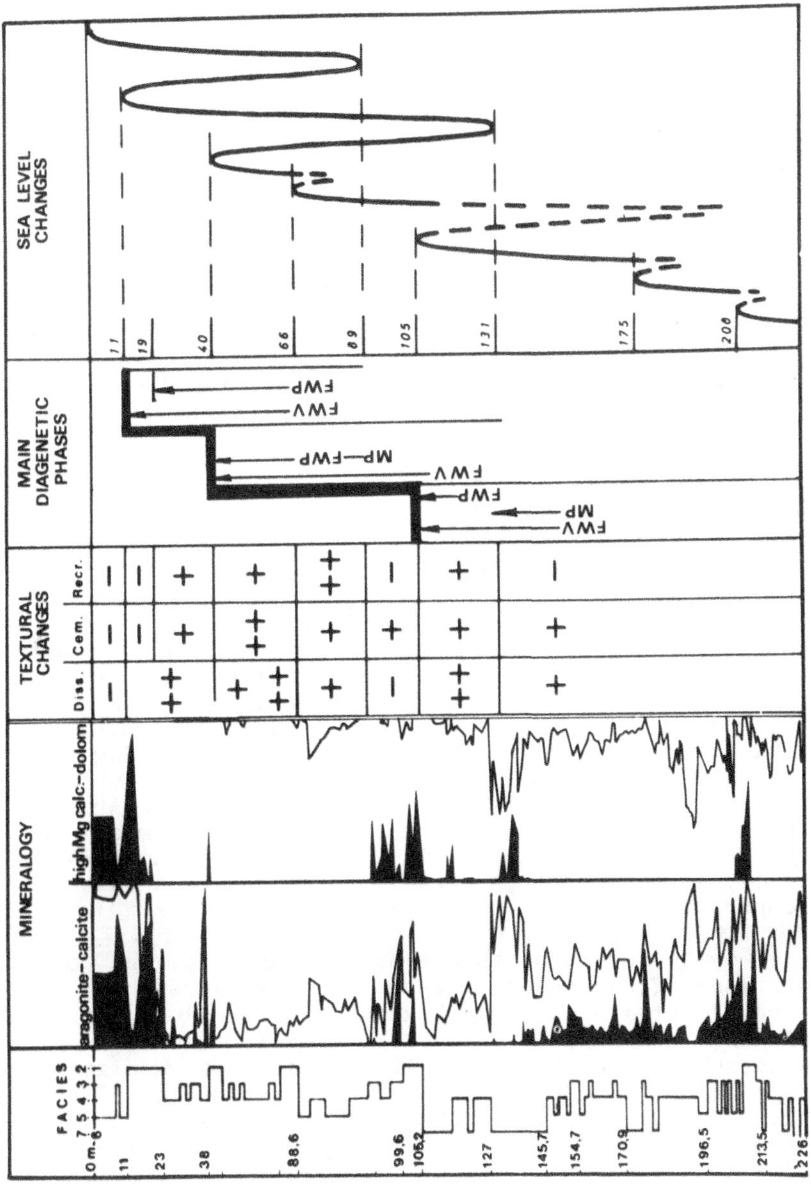
SEA LEVEL CHANGES
11
19
40
66
89
105
131
175
208
MAIN DIAGENETIC PHASES
FWP
FWV
MP—FWP
FWV
FWP
MP
FWV
TEXTURAL CHANGES
Diss.
Cem.
Recr.
MINERALOGY
aragonite–calcite
highMg calc.–dolom
FACIES
8 7 5 4 3 2 1
0 m.
11
23
38
88,6
99,6
105,2
127
145,7
154,7
170,9
196,5
213,5
226

generation of micritic calcites; priming of the same process at the expense of the second generation of high-Mg calcite micrites and the formation of meteoric sparry crystals.

In the light of these results, it appears that vertical opposite movements occurred under a relatively stable SL.

b) Pleistocene age

Mineralogical and petrographical analysis of a borehole which penetrated 226 m through the whole barrier-reef of New Caledonia has allowed the succession of depositional diagenetic environments of a Quaternary sequence to be reconstructed (Coudray, 1976). Concerning diagenesis, two kinds of diagnostic criteria have been used: 1) relative amount of the major carbonate mineral species (aragonite, high-Mg and low-Mg calcite, dolomite); 2) textural changes of originally marine sediments subsequent to their deposition, including solution, cementation and recrystallization.

Three major diagenetic sequences have been recognized as a result of three worldwide low stands of SL related to the three last glacials (Fig. 28).

The lower sequence (-226 to -105 m) exhibits two successive units: In the first, from 226 to 131 m deep, aragonite and very localized high-Mg calcite occur in skeletal grains only, associated with low-Mg calcite and syndepositional dolomite. Solution cavities are rare and affect aragonite (moldic porosity) or calcite (vuggy porosity in micritic and microsparitic dolomite matrices) materials. Slight cementation was manifested as dog-tooth shaped calcite. In the second phase from 131 to 105 m deep, dolomite is scarce, while calcite is widespread. Aragonite has been dissolved (moldic porosity) or transformed into microcrystalline calcite. Calcite occurs only as low-Mg type; micrite cements are locally recrystallized into microsparite.

It is obvious that this sequence has undergone post-depositional fresh-water diagenetic alterations which were more severe at its top. Such alterations reflect a fall of SL of about one hundred metres, related to the Kansan glacial (S stage of Ericson & Wollin, 1968). The intermediate sequence (-105 to -40 m) also includes two units: first, from 105 to 89 m deep, diagenetic alterations are slight; high-Mg calcite and aragonite are common. Early micrite cements generally are not recrystallized, and solution features are very scattered. Such a state of evolution can be considered to have been weakly controlled by meteoric processes. On the other hand, in the second unit, from 89 to 40 m deep, meteoric alterations have been intensive. Sediments consist entirely of low-Mg calcite and exhibit upward size-increasing solution cavities (moldic, vuggy and karstic). Other typical features are: well-developed late cements such as syntaxial overgrowths on echinoderm debris, drusy mosaics and blocky cements, a widespread recrystallization increasing downward and affecting both bioclasts and early cements. Such features are characteristic of a fresh-water phreatic environment; they are regarded as the result of lowered SL during the Illinoian glaciation (Ericson & Wollin, 1968).

The upper sequence (-40 to -11 m) typically shows vuggy and moldic solution voids, dog-tooth shaped calcite cements, and strongly recrystallized matrices and cements increasing in amount downward. The conservation of aragonite and low-Mg calcite in both skeletal detritus and matrices at the top of the sequence is confusing. However, changes in lithification and dating data obtained from coral samples (7000 ± 1000 BP at -8 m; 125 000 ± 20 000 BP at -11 m; 235 000 ± 15 000 BP at -18 3 m) provide convincing evidence of emergence ascribed to SL lowering during the last glacial period.

The overlying sediments of Holocene age are unlithified and very slightly altered.

4. Use of geochemical criteria

Allan & Matthews (1982) developed an isotopic model for stratigraphic variations in the $\delta^{13}C$ and $\delta^{18}O$ composition of carbonates, produced during early fresh-water diagenesis, and based largely upon empirical data from subaerially exposed coral-reef terraces of the late Pleistocene of Barbados (West Indies). Six potentially useful patterns have been defined with respect to isotope signatures and probable origins (Fig. 29).

1) The subaerial exposure surface is characterized by strongly ^{13}C - enriched carbonates, with respect to limestones which underwent mineralogical stabilization in the deeper vadose or phreatic zones. The $\delta^{13}C$ compositions of these underlying limestones are progressively heavier with increasing depth;
2) the subaerial exposure surface may also be marked by slight ^{18}O enrichment;
3) carbonates within one or two metres beneath an exposure surface show ^{18}O enrichment; thus abrupt shifts in ^{18}O values may differentiate sediments above the exposed zone from those below;
4) carbonates altered in the marine-meteoric mixing zone may be characterized by positive covariance between their $\delta^{18}O$ and $\delta^{13}C$ compositions;
5) the meteoric vadose-phreatic boundary (the 'water table') may be marked by a sharp increase in $\delta^{13}C$ values;
6) carbonates altered in meteoric water associated with subaerial exposure show wide variability in $\delta^{13}C$ values and narrow variability in $\delta^{18}O$ values.

Observations 1, 2 and 3, uniquely define the existence of a subaerial exposure surface. Observation 4 is sufficient to infer the position of an ancient marine-

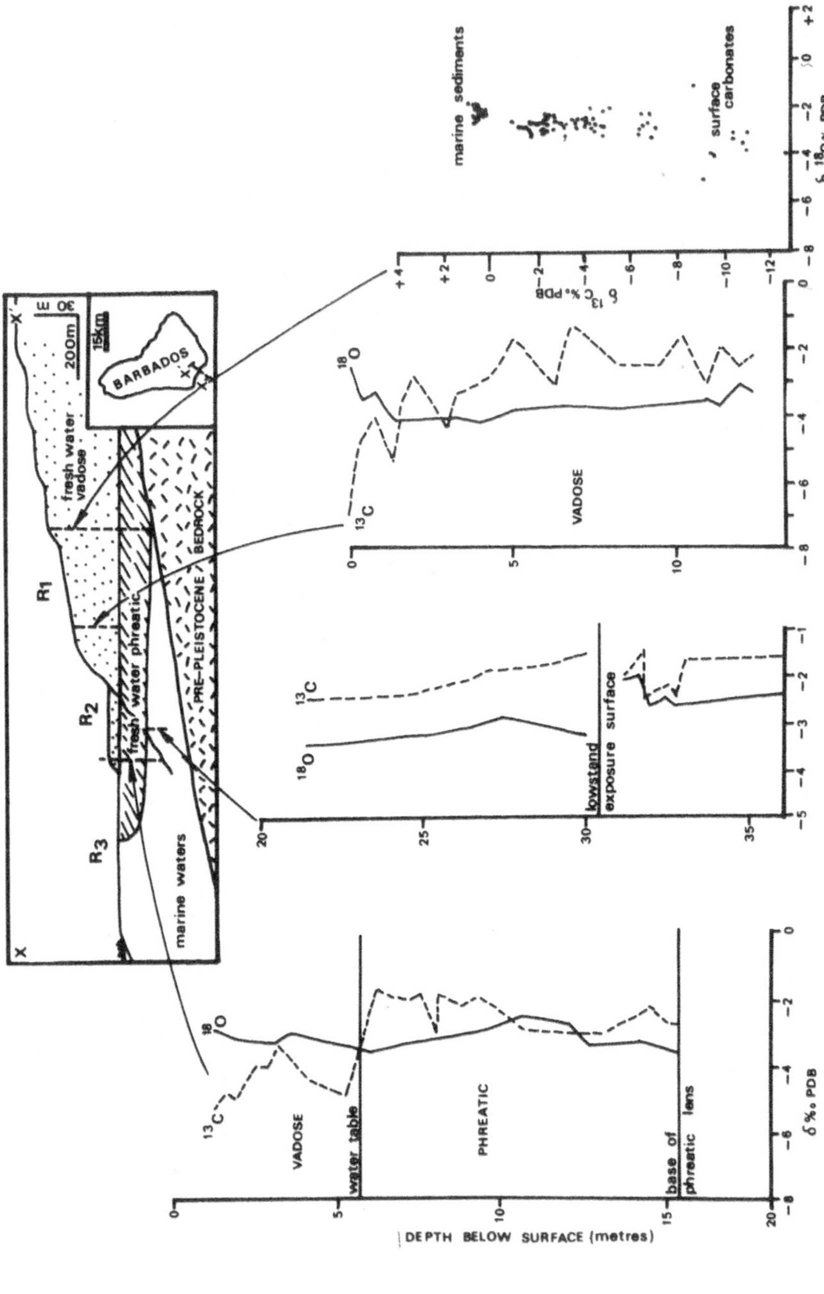
marine sediments
surface carbonates
δ 18O ‰ PDB
1
δ 13C ‰ PDB
VADOSE
2
lowstand exposure surface
3
fresh water vadose
fresh water phreatic
marine waters
PRE-PLEISTOCENE BEDROCK
BARBADOS
15km
200m
30 m
R1
R2
R3
VADOSE
water table
PHREATIC
base of phreatic lens
δ ‰ PDB
4
DEPTH BELOW SURFACE (metres)

phreatic mixing zone. Recognition of the top and base of a former fresh-water meteoric zone (observations 5 and 6) is somewhat less straight-forward.

PRACTICAL USE AND VALIDITY OF LABORATORY ANALYSIS

Studies of most diagenetic products in marine carbonates necessitate the use of various sophisticated analytical methods. In some cases, from observation of outcrops or hand-specimens, it is true that particular large-scale structures or textures such as laminations, desiccation cracks, birdeye and solution cavities, can easily be recognized. Likewise, after a sample has been polished and etched with dilute acid, use of a hand lens sometimes allows micritic and sparitic cements, internal sediments, boring and burrowing effects to be differentiated. Laboratory research includes petrographic study of thin sections and peels from which various categories of sedimentary components, crystal habit and assemblages can be identified. However, one generally needs to resort to more accurate techniques compiled in some comprehensive books (for example, Milner, 1962; Carver, 1971). Thus the crystallographical and optical characteristics of carbonate minerals are so close it is necessary to apply staining methods for identification of each species; the most commonly used stains are alizarine red S and Feigl's solution (Friedman, 1959) which lead to distinguish the four major carbonate minerals: aragonite, high-Mg and low-Mg calcite, and dolomite. Quantitative mineralogical analysis of a given component of the bulk rock is performed by X-ray diffraction. Determination of the concentration of chemical elements is obtained from X-fluorescence and atomic absorption methods

---

Figure 29 (see opposite). Isotope patterns associated with diagenetic environments in Pleistocene reef terraces, Barbados, West Indies (from Allan & Matthews, 1982, simplified). The upper scheme is a cross-section (XX') showing configuration of modern diagenetic environments through the reef terraces in question. Terraces R_1, R_2 and R_3 are related to sea level at 125 000, 105 000 and 82 000 BP respectively.

Graph 1. Shows carbon and oxygen isotopic compositions of stabilized Pleistocene carbonates and unaltered recent marine sediments.

Graph 2. $\delta^{13}C$ and $\delta^{18}O$ versus depth (in metres) for an entire analysed interval within fresh-water vadose environment. $\delta^{13}C$ compositions become heavier with depth, while $\delta^{18}O$ compositions are rather constant.

Graph 3. Shift in $\delta^{18}O$ values across buried exposure surface.

Graph 4. $\delta^{13}C$ and $\delta^{18}O$ versus depth (in metres) for an interval within both fresh-water vadose and phreatic environments.

(Volborth, 1969). Moreover the electron microscope is a tool specifically suitable for the determination of micrometer-sized constituents. This is particularly interesting for following distributional patterns of elements such as Sr (in aragonite-calcite inversion) or Mg (in high-Mg to low-Mg calcite transformation). Calculation of isotopic ratios is based on the use of mass spectrometer. Finally, examination of the finest particles and ultrastructures is at present commonly carried out by means of scanning electron microscopy; its wide range of magnifications (15X to 50 000), its ability to examine entire specimens rather than restricted portions of objects, and its great depth of field, make it a highly valuable tool for studying diagenesis of sediments. Moreover, Sippel (1965) and Long & Agrell (1965) used *cathodoluminescence* as a petrologic tool for thin section examinations. The luminescence (light emission effects) excited in thin sections by cathode rays is produced by electron bombardment of the specimen. Several results of cathodoluminescence investigations, using SEM, concerning marine sediments were published (see Krimsley & Hyde, 1971).

The combined application of these various techniques makes the identification of the various diagenetic products possible, but the most awkward problem is that of their dating. Absolute dating methods susceptible to be applied to recent carbonate sediments have been greatly improved during the last two decades, particularly those applied to shells and corals (Lalou & Nguyen, 1970; Chappell, Thom & Polach, 1978). The two most common methods are based upon determination of ^{14}C concentration and ^{290}Th - ^{234}U ratio.

At present great achievements are expected from the measure of racemization rate of amino-acids, mainly applied to date bones and marine sediments (Bada, Luyendyk & Maynard, 1970; Oustrière, 1979), reaching back in age as far as 700 000 years.

The main obvious disadvantage of the radiocarbon method is its unsuitability for dating material older than 40 000 BP. Its advantage is to date all carbon-rich deposits, including biologically or chemically precipitated carbonates, such as high-Mg calcite (Scoffin & McLean, 1978), aragonite (Ginsburg & James, 1976) or low-Mg calcite cements (Nguyen, 1973), on the condition that they should be abundant enough for sampling and free of diagenetic alterations. The uranium-thorium method which allows sediments younger than 300 000 years BP to be dated has been successfully applied to marine as well as subaerial carbonates (cave concretations: Nguyen & Lalou, 1969; Nguyen, 1973). An inventory of the extensive literature on this subject reveals that absolute dating literature on this subject reveals that absolute dating has been mainly performed from non-recrystallized deposits in order to date the moment of deposition. Dating of diagenetic products has claimed less of the attention of geochemists since cements only have been taken into account. Besides, the cements dated are generally large-sized (marine botryoids: Ginsburg & James, 1976; speleothems: Harmon *et al.*, 1977), making sampling easier. This

Table 3. An attempt to compile the main diagenetic products associated with nearshore marine and fresh-water environments

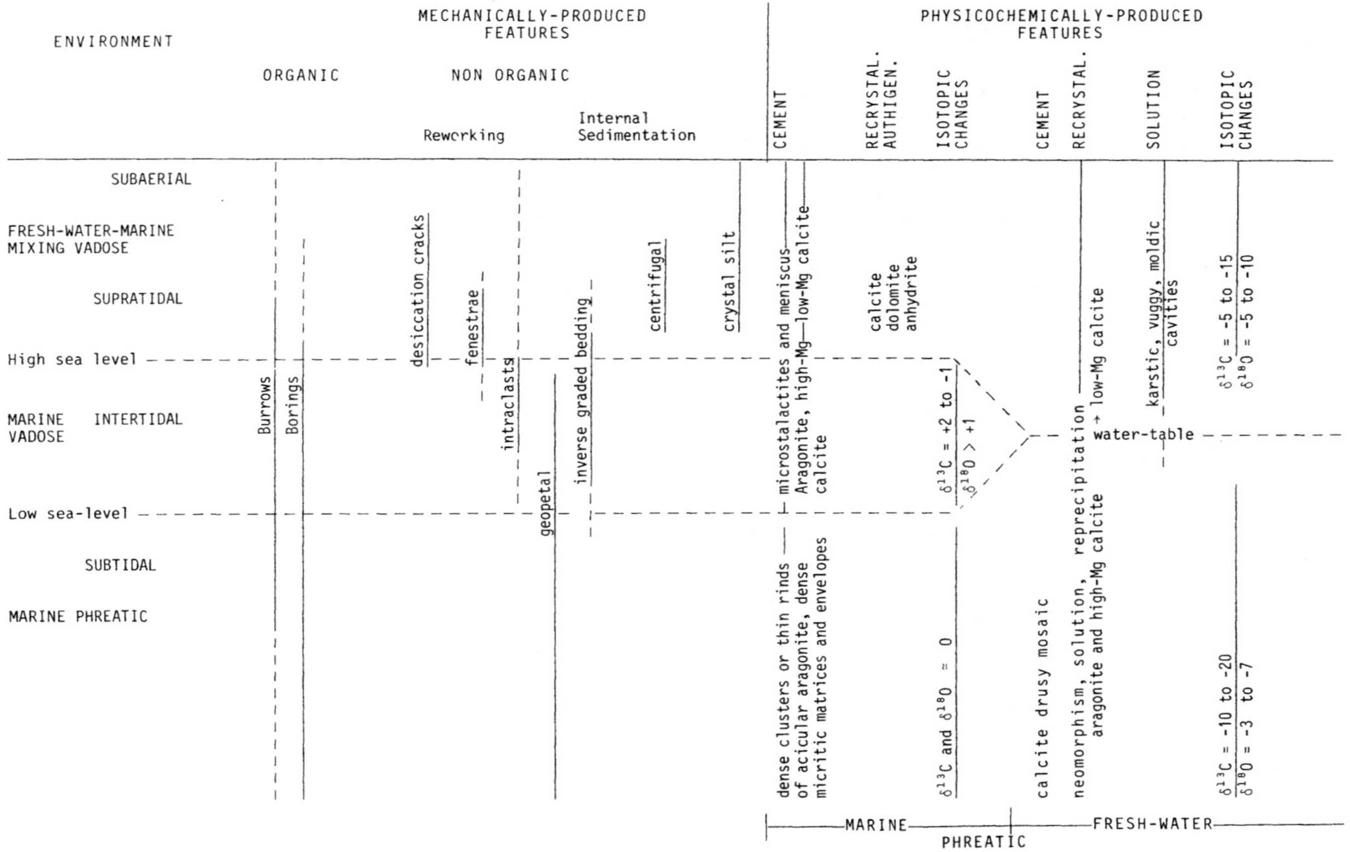

demonstrates that, once the problem of collecting has been solved, certain diagenetic products are suitable for dating provided that possible contamination is not neglected. As suggested by Oustrière (1979), since recrystallization of organic tests leads to introduction of newly formed organic matter, it is theoretically possible to date such a diagenetic phenomenon by means of the racemization of amino-acids. Likewise the relative age of solution cavities may be deduced, in some cases, from those of host rock and of the infilling internal sediments.

CONCLUSIONS

The diagenetic products of marine carbonates are believed to be rarely reliable indicators of paleo-SL except for a few products of biological origin and cements. Generally they provide information on the environment of deposition, this leading indirectly to localize SL just above or below the environment predefined (Table 3).

The main interest of use of the diagenetic features lies in reconstructing the various diagenetic environments, not only in space, but also in time, in order to obtain an evolutionary pattern of SL changes. Indeed as SL fluctuates, changes in water chemistry occur. Sediments initially deposited in marine environments will undergo the influence of fresh waters after emergence (regressive diagenesis); during their diagenetic history, the sequence would be reversed (transgressive diagenesis). Consequently since diagenetic processes will successively act at various scales, from that of a bed to that of a pore space or a particle, the corresponding diagenetic products typical of each depositional environment will be associated, superimposed or interbedded (diagenetic sequences in the sense of Purser, 1980). Recognition of these end products is based upon classical and recent criteria of sedimentary petrology. The principal difficulty is that of dating them.

However that may be, the diagenetic method of reconstruction of paleo-SL must be used as a complementary tool to other methods of SL research.

ACKNOWLEDGEMENT

We thank R.G.C. Bathurst for improving the English of our text.

REFERENCES

Aissaoui, D.M. & Purser, B.H., 1983. Nature and origins of internal sediments in Jurassic limestones of Burgundy (France) and Fnoud (Algeria). *Sedimentology*, 30, 273-284

Alexandersson, E. T. 1972. Micritization of cabonate particles: process of precipitation and dissolution in modern shallow marine sediments. *Bulletin Geology Institute University Uppsala,* N.S. 3, 201-236

Allan, J.R. & Matthews, R.K., 1982. Isotope signatures associated with early meteoric diagenesis. *Sedimentology,* 29, 797-818

Bada, J.L., Luyendyk, B.P. & Maynard, J.B., 1970. Marine sediments: dating by the racemization of amino acids. *Science,* 170, 730-738

Badiozamani, K., Mackenzie, F.T. & Thorstensen, D.C., 1977. Experimental carbonate cementation: salinity, temperature and vadose-phreatic effects. *Journal of Sedimentary Petrology,* 47, 529-542

Basan, P.B., (ed.), 1978. *Trace fossil concepts. Short Course Notes n°5,* Society of Economic Paleontologists and Mineralogists

Bathurst, R.G.C., 1966. Boring algae, micrite envelopes and lithification of molluscan biosparites. *Liverpool Manchester Geology Journal,* 5, 15-32

Bathurst, R.G.C., 1971. Carbonate sediments and their diagenesis. *Developments in Sedimentology,* (Elsevier, Amsterdam), 12

Bathurst, R.G.C., 1974. Marine diagenesis of shallow water calcium carbonate sediments. *Annual Review of Earth and Planetary Sciences,* 2, 257-274

Bathurst, R.G.C., 1980. Lithification of carbonate sediments. *Science Progress Oxford,* 66, 451-471

Bathurst, R.G.C., 1983. Early diagenesis of carbonate sediments. In: *Sediment Diagenesis,* eds. A. Parker & B.W. Sellwood, (D. Reidel Publishing Company, London), 349-377

Bourrouilh-Le Jan, F., 1973. Les dolomies et leurs genèses. *Bulletin Centre de Recherches de Pau-SNPA.* 7, 111-135

Bricker, O.P. (ed.), 1971. *Carbonate cements.* The Johns Hopkins University, Studies in Geology, 19

Bromley, R.G., 1970. Borings as trace fossils and *Entobia cretacea* Portlock, as an example. In: *Trace Fossils,* eds. T.P. Crimes & J.C. Harper, *Journal of Geology,* Special Issue, 3, 49-90

Bromley, R.G., 1978. Bioerosion of Bermuda reefs. *Palaeogeography, Palaeoclimatology, Palaeoecology,* 23, 169-197

Cabrol, P. & Coudray, J., 1982. Climatic fluctuations influence the genesis and diagenesis of carbonate speleothems in Southwestern France. *The National Speleological Society Bulletin,* U.S.A., 44, 112-117

Carver, E.R. (ed.), 1971. *Procedures in sedimentary petrology.* (Wiley, Interscience, London)

Chappell, J., Thom, B.G. & Polach, H.A., 1978. Radiometric dating of coral reefs. In: *Coral Reefs: Research Methods*, UNESCO, 81-91

Chave, K.E., 1954a. Aspects of the biogeochemistry of magnesium. 1. Calcareous marine organisms. *Journal of Geology*, 62, 266-283

Chave, K.E., 1954b. Aspects of the biogeochemistry of magnesium. 2. Calcareous sediments and rocks. *Journal of Geology*, 62, 587-599

Chilingar, G.V., Bissel, H.J. & Wolf, K.H., 1967. Diagenesis of carbonate rocks. In: *Diagenesis in Sediments*, eds. G. Larsen & G.V. Chilingar, *Developments in Sedimentology*, (Elsevier, Amsterdam), 8, 179-322

Choquette, P.W., 1968. Marine diagenesis of shallow-marine lime-mud sediments: insights from δO^{18} and δC^{13} data. *Science*, 161, 1130-1132

Coudray, J., 1976. Recherches sur le Neogène et le Quaternaire marins de la Nouvelle-Calédonie. *Editions Fondation Singer-Polignac*, Paris

Craig, H., 1953. The geochemistry of the stable carbon isotopes. *Geochimica et Cosmochimica Acta*, 3, 53-92

Curran, H.A., 1984. Ichnology of Pleistocene carbonates on San Salvadore, Bahamas. *Journal of Paleontology*

Dörges, J. & Herweck, Ö. 1975. Recent biocoenoses and ichnocoenoses in shallow-water marine environments. In: *The Study of Trace Fossils*, ed. R.W. Frey, (Springer-Verlag, New York), 459-491

Dunham, R.J., 1969. Early vadose silt in Townsend Mound (reef), New Mexico. In: *Depositional Environments in Carbonate Rocks*, ed. G.M. Friedman, Society of Economic Paleontologists and Mineralogists, Special Publication, 14, 182-191

Dunham, R.J., 1971. Meniscus cement. In: *Carbonate Cements*, ed. P.O. Bricker, The Johns Hopkins University, Studies in Geology, 19, 297-300

Ericson, D.B. & Wollin, G., 1968. Pleistocene climates and chronology in deep sea sediments. *Science*, 162, 1227-1234

Esteban, M., 1974. Caliche textures and microcodium. *Bulletin of the Geological Society of Italy*, 92, 105-125

Evamy, B.D. & Shearman, D.J., 1965. The development of overgrowth from echinoderm fragments. *Sedimentology*, 5, 211-233

Evans, G., Schmidt, V., Bush, P. & Nelson, H., 1969. Stratigraphy and geologic history of the sabkha, Abu Dhabi, Persian Gulf. *Sedimentology*, 12, 145-159

Fairbridge, R.W., 1967. Phases of diagenesis and authigenesis. In: *Diagenesis in Sediments,* eds. G. Larsen & G.V. Chilingar, *Developments in Sedimentology,* 8, (Elsevier, Amsterdam), 19-20

Farrow, G.E., 1971. Back-reef and lagoonal environments of Aldabra Atoll distinguished by their crustacean burrows. In: *Regional Variation in Indian Ocean Coral Reefs,* eds. D.R. Stoddart & M. Young, Symposium Zoological Society London n° 28, (Academic Press), 455-500

Folk, R.L., 1959. Practical petrographic classification of limestones. *Bulletin of the American Association of Petroleum Geologists,* 43, 1-38

Folk, R.L., 1965. Some aspects of recrystallization in ancient limestones. In: *Dolomitization and Limestones Diagenesis: a Symposium,* eds. L.C. Pray & M.C. Murray, Society of Economic Paleontologists and Mineralogists, Special Publication, 13, 14-48

Ford, D.C. & Ewers, R.O., 1978. The development of limestone cave systems in the dimensions of length and depth. *Canadian Journal of Earth Sciences,* 15, 1783-1798

Frey, R.W. (ed.), 1975. *The study of trace fossils.* (Springer-Verlag, New York)

Friedman, G.M., 1959. Identification of carbonate minerals by staining methods. *Journal of Sedimentary Petrology,* 29, 89-97

Friedman, G.M., 1964. Early diagenesis and lithification in carbonate sediments. *Journal of Sedimentary Petrology,* 34, 777-813

Friedman, G.F., 1975. The making and unmaking of limestones or the downs and ups of porosity. *Journal of Sedimentary Petrology,* 45, 379-398

Friedman, G.M. & Sanders, J.E., 1978. *Principles of Sedimentology,* (John Wiley & Sons, eds., New York), 144-179

Füchtbauer, H., 1975. *Sedimentary petrology. Part II: Sediments and sedimentary rocks,* 1, (John Wiley & Sons, eds., New York), 279-352

Gatrall, M. & Golubic, S., 1970. Comparative study on some Jurassic and Recent endolithic fungi using scanning electron microscope. In: *Trace Fossils,* eds. T.P. Crimes & J.C. Harper, *Journal of Geology,* Special Issue, 3, 167-178

Gavish, E. & Friedman, G.M., 1969. Progressive diagenesis in Quaternary to late Tertiary carbonate sediments: sequence and time scale. *Journal of Sedimentary Petrology,* 39, 980-1006

Gebelein, C.D., 1969. Distribution, morphology, and accretion rate of recent subtidal algal stromatolites, Bermuda. *Journal of Sedimentary Petrology,* 39, 32-49

Ginsburg, R.N., 1964. South Florida carbonate sediments. *Guidebook for Field Trip N°1, Geological Society of America, Convention 1964*, Geological Society of America, New York

Ginsburg, R.N. & James, N.P., 1976. Submarine botryoidal aragonite in the Holocene reef limestones, Belize. *Geology*, 4, 431-436

Goldring, R., 1964. Trace fossils and the sedimentary surface in shallow water marine sediments. In: *Deltaic and Shallow Marine Deposits*, ed. L.M.J.U. van Straaten, *Developments in Sedimentology*, 1, (Elsevier, Amsterdam), 136-143

Golubic, S., 1969. Distribution taxonomy and boring patterns of marine endolithic algae. *American Zoologist*, 9, 747-751

Golubic, S., Perkins, R.D. & Lukas, K.J., 1975. Boring micro-organisms and microborings in carbonate rocks. In: *The Study of Trace Fossils*, ed. R.W. Frey, (Springer-Verlag, New York), 229-259

Golubic, S., Campbell, S.E., Drobne, K., Cameron, B., Balsamm, W.L., Cinerman, F. & Dubois, L., 1984. Microbial endoliths: a benthic overprint in the sedimentary record and a paleobathymetric cross-reference with Foraminifera. *Journal of Paleontology*, 58, 351-361

Gross, M.G., 1964. Variations in the O^{18}/O^{16} and C^{13}/C^{12} ratios of diagenetically altered limestones in the Bermuda Islands. *Journal of Geology*, 72, 170-194

Hanor, J.S., 1978. Precipitation of beachrock cements: mixing of marine and meteoric waters vs. CO_2 - degassing. *Journal of Sedimentary Petrology*, 48, 489-502

Harmon, R.S., Ford, D.C. & Schwarz, H.P., 1977. Interglacial chronology of the Rocky and Mackenzie Mountains based upon Th^{230}/U^{234} dating of calcite speleothems. *Canadian Journal of Earth Sciences*, 14, 2543-2552

Hayasaka, I., 1935. The burrowing activities of certain crabs and their geologic significance. *American Midland Naturalist*, 16, 99-103

Harrison, R.S., 1974. *Near-surface subaerial diagenesis of Pleistocene carbonates, Barbados, West Indies*, Ph.D. thesis, Brown University

Harrison, R.S., 1975. Porosity in Pleistocene grainstones from Barbados: some preliminary observations. *Canadian Petroleum Geologists Bulletin*, 23, 383-392

Howard, J.D., 1972. Trace fossils as criteria for recognizing shorelines in the stratigraphic record. In: *Recognition of Ancient Sedimentary Environments*, eds. J.K. Rigby & W.K. Hamblin, Society of Economic Paleontologists and Mineralogists, Special Publication, 16, 215-225

Hudson, J.D., 1975. Carbon isotopes and limestone cement. *Geology*, 3, 19-22

Illing, L.V., Wells, A.J. & Taylor, J.C.M., 1965. Penecontemporary dolomite in the Persian Gulf. In: *Dolomitization and Limestone Diagenesis*, eds. L.C. Pray & R.C. Murray, Society of Economic Paleontologists and Mineralogists, Special Publication, 13, 89-111

Inden, R.F. & Moore, C. H., 1983. Beach environment. In: *Carbonate Depositional Environments*, eds. P.A. Scholle, D.G. Bebout & C.H. Moore, American Association of Petroleum Geologists, memoir 33, 211-266

James, N.P., 1972. Holocene and Pleistocene calcareous crust (caliche) profiles: criteria for subaerial exposure. *Journal of Sedimentary Petrology*, 42-44, 817-836

James, N.P., 1974. Diagenesis of scleractinian corals in the subaerial vadose environment. *Journal of Paleontology*, 48, 785-799

James, N.P., Ginsburg, R.N., Marszalek, D.S. & Choquette, P.W., 1976. Facies and fabric specificity of early subsea cements in shallow Belize (British Honduras) reefs. *Journal of Sedimentary Petrology*, 46, 523-544

Kinsman, D.J.J., 1964. *Recent carbonate sedimentation near Abu Dhabi, Trucial Coast, Persian Gulf*. Ph.D. thesis, Imperial College, London

Kinsman, D.J.J., 1969. Interpretation of Sr^{2+} concentrations in carbonate minerals and rocks. *Journal of Sedimentary Petrology*, 39, 486-508

Krinsley, D. & Hyde, P., 1971. Cathodoluminescence studies of sediments. *Proceedings of the 4th Annual Scanning Electron Microscope Symposium*, I.T.T. Research Institute of Chicago, Illinois, 409-416

Krumbein, W.C., 1942. Physical and chemical changes in sediments after deposition. *Journal of Sedimentary Petrology*, 12, 111-147

Lalou, C. & Nguyen, H.V., 1970. La méthode de datation par les déséquilibres de la famille de l'uranium. In: *Colloque sur les méthodes et les tendances de la stratigraphie*, Paris - Orsay, Reprint.

Land, L.S., 1967. Diagenesis of skeletal carbonates. *Journal of Sedimentary Petrology*, 37, 914-930

Land, L.S., 1970. Phreatic versus vadose meteoric diagenesis of limestones: evidence from a fossil water table. *Sedimentology*, 14, 175-185

Land, L.S. & Epstein, S., 1970. Late Pleistocene diagenesis and dolomitization, N. Jamaica. *Sedimentology*, 14, 187-200

Land, L.S. & Goreau, T.R., 1970. Submarine lithification of Jamaican reefs. *Journal of Sedimentary Petrology*, 40, 457-462

Land, L.S., Mackenzie, F.T. & Gould, S.J., 1967. Pleistocene history of Bermuda. *Bulletin of the Geological Society of America*, 78, 993-1006

Le Campion-Alzumard, Th., 1970. Cyanophycées marines endolithes colonisant les surfaces rocheuses dénudées (étages supralittoral et médiolittoral de la région de Marseille). *Schweizerische Zeitschrift für Hydrologie*, 32, 552-558

Le Campion-Alzumard, Th., 1979a. Les végétaux perforants en tant qu'indicateurs paléobathymétriques. In: *Les Indicateurs de Niveaux Marins, Oceanis*, 5, fasc. hors série, 259-264

Le Campion-Alzumard, Th., 1979b. Les Cyanophycées endolithes marines: systématique, ultrastructure, écologie et biodestruction. *Oceanologica Acta*, 2, 143-156

Lohmann, K.C. & Meyers, W.J., 1977. Microdolomite inclusions in cloudy prismatic calcites: a prospect criterion for former high-magnesium calcites. *Journal of Sedimentary Petrology*, 47, 1078-1088

Long, J.V.P. & Agrell, S.O., 1965. The cathodoluminescence of minerals in thin section. *Mineralogical Magazine*, 34, 318-326

Longman, M.W., 1980. Carbonate diagenetic textures from near-surface diagenetic environments. *Bulletin of the American Association of Petroleum Geologists*, 64, 4, 461-487

Loreau, J.P., 1970. Contribution à l'étude des calcarénites hétérogènes par emploi simultané de la microscopie photonique et de la microscopie électronique à balayage. Problème particulier de la micritisation. *Journal de Microscopie*, 9, 727-734

Loucks, R.G. & Folk, R.L., 1976. Fanlike rays of former aragonite in Permian Capitan reef pisolite. *Journal of Sedimentary Petrology*, 46, 483-485

Macintyre, I.G., 1977. Distribution of submarine cements in a modern Caribbean fringing reef, Galeta Point, Panama. *Journal of Sedimentary Petrology*, 47, 503-516

Magaritz, M., 1983. Carbon and oxygen isotope composition of recent and ancient coated grains. In: *Coated Grains*, ed. T.M. Peryt, (Springer-Verlag, Berlin), 27-37

McCall, P.L. & Tevesz, M.J.S., 1982. *Animal-sediment relations: the biogenic alteration of sediments*. (Plenum, New York)

McKenzie, J.A., Hsü, K.J. & Schneider, J.F., 1980. Movement of subsurface waters under the sabkha, Abu Dhabi, U.A.E., and its relation to evaporitive dolomite genesis. In: *Concepts and Models of Dolomitization*, eds. D.H. Zenger, J.B. Dunham & R.L. Ethington, Society of Economic Paleontologists and Mineralogists, Special Publication, 28, 11-30

Mercadier, H., Coudray, J. & Joseph, C., 1975. Dosages et corrélations aragonite-strontium dans les polypiers d'une série récifale quaternaire de Nouvelle-Calédonie. *Troisième réunion annuelle des Sciences de la Terre*, Montpellier, 261

Milliman, J.D., 1974. *Marine carbonates*. Part 1. (Springer-Verlag, Berlin)

Milner, H.B., 1962. *Sedimentary Petrography*. (George Allen & Unwin Ltd., London)

Montaggioni, L.F., 1978. *Recherches géologiques sur les complexes récifaux de l'archipel des Mascareignes (Océan Indien occidental)*. Unpublished D.Sc. thesis, University of Marseille, 2 volumes

Montaggioni, L.F. & Pirazzoli, P.A., 1984. The significance of exposed coral conglomerates from French Polynesia (Pacific Ocean) as indicators of recent relative sea-level changes. *Coral Reefs*, 3, 29-42

Moore, C.H. Jr., 1977. Beach rock origin: some geochemical, mineralogical, and petrographic considerations. *Geoscience and Man*, 18, 155-163

Müller, G., 1971. Gravitational cement: an indicator for the vadose zone of the subaerial diagenetic environments. In: *Carbonate Cements*, ed. P.O. Bricker, The Johns Hopkins University, Studies in Geology, 19, 301-302

Neumann, A.C., 1968. Biological erosion of limestones coasts. In: *Encyclopedia of Geomorphology*, ed. R.W. Fairbridge, (Reinhold Book Corporation, New York), 75-81

Newell, N.D., Fischer, A.G., Whiteman, A.J., Hickox, J.E. & Bradley, J.S., 1953. *The Permian Reef Complex of the Guadaloupe Mountains Region, Texas and New Mexico, a study in Paleoecology*. (Freeman, San Francisco)

Nguyen, H.V., 1973. *Contribution à l'étude du déséquilibre de la famille de l'uranium dans les eaux naturelles et les carbonates. Applications à la géochimie et à la géologie du Quaternaire*. Unpublished Ph.D. thesis, University of Paris

Nguyen, H.V. & Lalou, C., 1969. Comportement géochimique des isotopes de la famille de l'uranium et du thorium dans les concrétionnements des grottes: application à la datation des stalagmites. *Comptes Rendus Académie des Sciences*, Paris, 269, 560-563

Oustrière, P., 1979. *Mesure du taux de racémisation de l'isolencine dans les matériaux coralliens. Limitation des possibilités de datation due aux processus diagénétiques*. Unpublished thesis, University of Orléans

Perkins, R.D. & Halsey, S.D., 1971. Geologic significance of microboring fungi and algae in Carolina shelf sediments. *Journal of Sedimentary Petrology*, 41, 843-853

Pettijohn, F.J., 1957. *Sedimentary rocks,* 2nd Ed. (Harper & Row, New York)

Pingitore, N.E., 1970. Diagenesis and porosity modification in *Acropora palmata,* Pleistocene of Barbados, West Indies. *Journal of Sedimentary Petrology,* 40, 712-721

Plummer, P.S. & Gostin, V.A., 1981. Shrinkage cracks: desiccation or synaeresis? *Journal of Sedimentary Petrology,* 51, 1147-1156

Purdy, E.G., 1968. Carbonate diagenesis: an environment survey. *Geologica Romana,* 7, 183-228

Purser, B.H., 1969. Synsedimentary marine lithification of Middle Jurassic limestones in the Paris basin. *Sedimentology,* 12, 205-230

Purser, B.H. (ed.), 1973. *The Persian Gulf,* (Springer-Verlag, Berlin)

Purser, B.H., 1980. *Sédimentation et diagenèse des carbonates néritiques récents,* Editions Technip., 1, 159-344

Purser, B.H. & Loreau, J.P., 1973. Aragonitic, supratidal encrustations on the Trucial coast, Persian Gulf. In: *The Persian Gulf,* ed. B.H. Purser, (Springer-Verlag, Berlin), 343-376

Risk, M.J. & MacGeachy, J.K., 1978. Aspects of bioerosion of modern Caribbean reefs. *Revue de Biologie Tropicale,* 26, 85-105

Rooney, W.S. & Perkins, R.D., 1972. Distribution and geologic significance of microboring organisms within sediments of the Arlington Reef Complex, Australia. *Bulletin of the Geological Society of America,* 83, 1139-1150

Schroeder, J.H., 1973. Submarine and vadose cements in Pleistocene Bermuda reef rock. *Sedimentary Geology,* 10, 179-204

Scoffin, T.P. & McLean, R.F., 1978. Exposed limestones of the Northern Province of the Great Barrier reef. *Philosophical Transactions of the Royal Society,* London, 291, 119-138

Seilacher, A., 1964. Biogenic sedimentary structures. In: *Approaches to paleoecology,* eds. J. Imbrie & N.D. Newell, (Wiley & Sons, New York), 296-316

Seilacher, A., 1967. Bathymetry of trace fossils. *Marine Geology,* 5, 413-428

Shearman, D.J., Twyman, J. & Karimi, M.Z., 1970. The genesis and diagenesis of oolites. *Proceedings of the Geologists' Association (England),* 81, 561-575

Shinn, E.A., 1968a. Burrowing in recent lime sediments of Florida and the Bahamas. *Journal of Paleontology,* 42, 878-894

Shinn, E.A., 1968b. Practical significance of birdeye structures in carbonate rocks. *Journal of Sedimentary Petrology,* 38, 215-223

Shinn, E.A., 1968c. Selective dolomitization of recent sedimentary structures. *Journal of Sedimentary Petrology,* 38, 612-616

Shinn, E.A., 1973. Sedimentary accretion along the leeward, SE coast of Qatar Peninsula, Persian Gulf. In: *The Persian Gulf,* ed. B.H. Purser, (Springer-Verlag, Berlin), 199-209

Shinn, E.A., 1983a. Tidal flat environments. In: *Carbonate Depositional Environments,* eds. P.H. Scholle, D.G. Bebout & C.H. Moore, American Association of Petroleum Geologists, memoir 33, 171-210

Shinn, E.A., 1983b. Birdeyes, fenestrae, shrinkage pores and loferites: a reevaluation. *Journal of Sedimentary Petrology,* 53, 619-628

Shinn, E.A., Ginsburg, R.N. & Lloyd, R.M., 1965. Recent supratidal dolomite from Andros Island, Bahamas. In: *Dolomitization and Limestone Diagenesis,* eds. L.C. Pray & R.C. Murray, Society of Economic Paleontologists and Mineralogists, Special Publication, 13, 112-113

Shinn, E.A., Robbin, D.M. & Steiner, R.P., 1980. Experimental compaction of lime sediment. *Book of Abstracts, American Association of Petroleum Geologists - Society of Economic Paleontologists and Mineralogists Annual Meeting, Denver*, 120

Sippel, R.F., 1965. A simple device for luminescence petrography. *Review of Scientific Instrumentation,* 36, 1556-1558

Stehli, F.G. & Hower, J., 1961. Mineralogy and early diagenesis of carbonate sediments. *Journal of Sedimentary Petrology,* 31, 358-371

Steinen, R.P., 1974. Phreatic and vadose diagenetic modification of Pleistocene limestones: petrographic observations from the subsurface of Barbados, West Indies. *Bulletin of the American Association of Petroleum Geologists,* 58, 1008-1024

Stephenson, D.G., 1965. Fossil burrows on the coast of Kenya. *Nature,* 207, 850-851

Supko, P.R., 1971. 'Whisker' crystal cement in Bahamian rock. In: *Carbonate cements,* ed. P.O. Bricker, The Johns Hopkins University, Studies in Geology, 19, 143-148

Swinchatt, J.P., 1969. Algal boring: a possible depth indicator in carbonate rocks and sediments. *Bulletin of the Geological Society of America,* 80, 1391-1396

Takahashi, S., 1932. On the burrows of *Ocypoda ceratophthalma* Fabricius. *Kwagaku,* 2, 329-335

Taylor, J.M.C. & Illing, L.V., 1969. Holocene intertidal calcium carbonate cementation, Qatar, Persian Gulf. *Sedimentology,* 12, 69-107

Tebbutt, G.E., Conley, C.D. & Boyd, D.W., 1965. Lithogenesis of a distinctive carbonate rock fabric. *University of Wyoming Contribution to Geology*, 4, 1-13

Thomassin, B. & Coudray, J., 1981. Presence of wide hard-ground areas on lagoonal bottoms of the coral reef complex of Noumea (S.W. New Caledonia). *Proceedings of the fourth International Coral Reef Symposium*, Manila, 1, 511-521

Thommeret, Y. & J., Laborel, J., Montaggioni, L.F. & Pirazzoli, P.A., 1981. Late Holocene shoreline changes and seismotectonic displacements in western Crete (Greece). *Zeitschrift für Geomorphologie*, N.F., Supplement Band 49, 127-149

Thrailkill, J., 1968. Chemical and hydrologic factors in the excavation of limestone caves. *Bulletin of the Geological Society of America*, 79, 19-46

Thrailkill, J., 1971. Carbonates deposition in Carlsbad Caverns. *Journal of Geology*, 79, 683-695

Von Guembel, C.W., 1868. *Geognostishe Beschreibung des ostbayrishen Grenzgebirges*. (Gotha)

Warme, J.E., 1975. Borings as trace fossils and the processes of marine bioerosion. In: *The Study of Trace Fossils*, ed. R.F. Frey, (Springer-Verlag, New York), 181-227

Weimer, R.J. & Hoys, J.H., 1964. Burrows of *Callianassa major* Say, geologic indicators of littoral and shallow neritic environments. *Journal of Paleontology*, 38, 761-767

Winland, H.D., 1968. The role of the high-Mg calcite in the preservation of micrite envelopes and textural features of aragonite sediments. *Journal of Sedimentary Petrology*, 38, 1320-1325

Volborth, A., 1969. *Elemental analysis in Geochemistry, A-Major elements*. (Elsevier, Amsterdam)

Yonge, C.M., 1963. Rock-boring organisms. In: *Mechanisms of Hard Tissue Destruction*, ed. R.F. Sognnaes, American Association of Advanced Science, Publication 75, 1-24

12

Marine notches

P. A. PIRAZZOLI

ABSTRACT

The descriptive terms *infralittoral, midlittoral (tidal* and *surf-)* and *supralittoral* are used to classify sea-corrosion notches, according to their elevation relative to sea-level at the time of formation. On limestone coasts, *tidal* notches are the most common and most useful sea-level indicators.

Possible causes and rates of undercutting are summarized. The shape of the notch profile may give useful information on the rate of sea-level change and on tectonic movements.

A series of photographs with associated data provides examples of marine notches from various parts of the world.

INTRODUCTION

A marine notch (or *nip*) is an indentation or undercutting, a few centimetres to several metres deep, left by sea erosion in coastal rocks. Erosion notches vary greatly in size, shape, and external appearance and may be cut at various levels by diverse processes (Kelletat, 1942). Therefore, an acceptable classification should be based on objectively determinable characteristics that give an insight into the process of formation. In order to read and evaluate the records of past sea levels (SLs) from marine notches, the most relevant parameter probably is its elevation relative to SL at the time of formation. It seems therefore appropriate to organize notches into a continuous series of elevation ranges centered on SL.

Bionomic zonations are usually based either on tidal data (Stephenson & Stephenson, 1949, or, in microtidal areas, on site exposure, (Peres & Picard, 1963). Tide levels are everywhere easy to measure and tide predictions are known for most coastal areas in the world. The easiest classification would therefore refer to tidal levels and distinguish a *supratidal zone*, a *tidal zone* and a *subtidal zone*. However, a zonation system taking into account also site exposure would be closer to the actual zonation of biological activity, of marine erosion, and of notch development. This paper employs a multivariate approach and distinguishes three main zones.

- The *infralittoral* (or *sublittoral*, or *subtidal*) *zone* experiences continuous immersion.
- The *midlittoral zone* is characterized by intermittent immersion by tide or waves. In sheltered sites, this zone approximately covers the tidal range (= *intertidal zone*). In regularly exposed sites the vertical range of the surf (*surf zone* or swash zone) must be added to the tidal amplitude.
- The *supralittoral zone* covers the highest level of marine influence above high tide or above the surf zone at high tide, where spray effects predominate. Here live epilithic algae and rasping snails (*Littorina*, *Nerita*, etc). The vertical width of this zone is limited in very sheltered sites and increases to several metres in exposed sites.

The terms infralittoral, midlittoral, tidal, surf- and supralittoral are used in this paper to classify erosion notches. These terms are purely descriptive and have no genetic implications.

GENESIS OF NOTCHES

The development of a marine notch may be attributed to various processes, of chemical, physical, biological or mechanical origin. However, despite a relatively abundant literature on the subject, the mechanism of notch formation remains barely understood.

Where mechanical action such as abrasion by wave-borne sand or gravel predominates, notches cut in hard rock have a noticeably rounded and polished appearance which makes them easy to recognize. *Abrasion notches* will therefore be considered separately. *Structural notches* also are a special case.

In most other cases the processes of notch formation fall under the general heading of *'sea corrosion'*. This term, intentionally vague, was introduced by Guilcher & Pont (1957, p 48) because it 'offers the advantage of including all the physico-chemical and biological processes, and therefore does not need to change when theory changes.'

As a matter of fact, sea corrosion is especially active in calcareous and volcanic rocks (Ph. 4) (Guilcher *et al.*, 1962; Emery, 1962, p B66; Guilcher & Bodéré, 1975). The predominant agents of sea corrosion are usually assumed to be solution and bioerosion.

1. Solution

The view that dissolution processes act on limestone rocks in the intertidal zone is supported by several authors (Mac Fadyen, 1930; Umbgrove, 1939; Fairbridge, 1952; Guilcher, 1952, 1953; Kaye, 1957, 1959; Verstappen, 1960; Emery, 1962; Christiansen, 1963; Stoddart, 1969; Machida *et al.*, 1976; Pirazzoli, 1976; Flemming, 1978).

Dissolution can hardly be ascribed to rainfall and groundwater, as proposed by Wentworth (1939), because tidal notches are found not only in moist tropical regions but also in very arid coastal areas like the Red Sea (Guilcher, 1955) or on mushroom rocks on tidal flats around oceanic atolls (Ph. 3). At the very most, solution by effluent groundwater may occur only in very localized coastal sections near springs, as shown by Higgins (1980) at Varkiza and Itea, Greece.

The ability of sea water to dissolve calcium carbonate is still under discussion. Alkaline water of tropical seas, supersaturated with $CaCO_3$, is evidently unable to do this. Emery (1946) showed that periodic undersaturation of sea water may occur in tidal pools at night, but this phenomenon was not confirmed by Schneider (1976), and undersaturation does not seem to occur in open waters in the vicinity of notches (Emery, 1962). However, sea water is not in contact with the intertidal limestone but with an algal film on the rock (Hodgkin, 1970; Schneider, 1976).

Fairbridge (1948) suggested that undersaturation may result from nocturnal fall of temperature, but according to Davy de Virville (quoted by Guilcher, 1953), temperature would have no influence on pH.

According to other views undercutting would result from occasional undersaturation of open coastal waters (Kaye, 1959), either because of water turbulence (Kaye, 1957), by currents (Verstappen, 1960), or humic and soil acids washed in from the land (Fairbridge & Teichert, 1948, quoted by Higgins, 1980), or by surface film effects due to gaseous exchange with the air (Kuenen, 1933; Emery, 1962). In any case, these phenomena are difficult to reproduce in the laboratory (Revelle & Emery, 1957).

According to Trudgill (1976) selective solution of magnesium may occur even when the rock is saturated with respect to calcium, thus enhancing disintegration. In sandstones composed of non-carbonate sands with a calcareous cement, solution is believed to attack first the intergranular cement. Then, as shown by Guilcher & Pont (1957), alternate wetting and drying causes disintegration.

It therefore appears that solution by sea water is a possible cause of notch formation, especially on shores fringed with mangrove where undersaturation with respect to $CaCO_3$ has been shown to occur (Trudgill, 1976).

2. Bioerosion

Endolithic organisms penetrate limestone in littoral zones. Surface feeders grazing upon epi- and endolithic algae, simultaneously abrade the surface with their hard teeth and radulas, thereby directly eroding the rock. For example, Chiton *Acanthopleura* has teeth with a hardness of about 6 on the Mohs scale (Trudgill, 1976), and the cusps of the teeth of *Patella coerulea* are reinforced with iron and silica to give an effective hardness of 5-6.

By comparison, calcite has a hardness of about 4. Also the calcitic teeth of sea urchins are usually harder (3 to 4 on the Mohs scale) than a limestone substrate weakened by endolithic organisms. In addition, sea-urchins' teeth have a rapid growth rate, up to 20 mm in 3 months for *Paracentratus lividus* (Torunski, 1979).

In non-sandy sites, according to Trudgill (1976, Fig.4), the effect of grazing would contribute 64% to the process of notch formation.

Many borers (*Porolithon, Cliona* (?), *Lithophaga,* Cyanophyta, Chlorophyta, etc.) evidently use secretions of organic acids to cause dissolution and softening of the rock (Frémy, 1945; Ranson, 1955; Revelle & Emery, 1957; Guilcher & Pont, 1957; Emery, 1962; Nesteroff, 1965; Trudgill, 1976; Torunski, 1979). For *Lithophaga,* which penetrate as deep as one decimetre in the rock, removal of pre-softened particles is achieved by a rocking movement of the shells.

Bioerosion is therefore generally acknowledged to play an important role in the formation of notches.

Among other possible processes, spray impact and salt weathering are also active, especially in exposed areas.

CLASSIFICATION OF NOTCHES

1. Midlittoral notches

Generally speaking, the precision of a corrosion notch as a SL indicator (SLI) increases where the site is sheltered, where the tidal range is low and where the cliff face is vertical. Site exposure is the most important factor and two main types of midlittoral notches may be distinguished:

a) *tidal notches* in relatively sheltered sites and wholly or partly cut in the intertidal zone;

b) *surf notches* only in exposed sites and usually cut above high-tide level.

The two are differentiated by a characteristic morphologic discontinuity, the *surf bench,* which is created by increasing exposure in the range from one type to the other (Focke, 1978a) (Fig. 1).

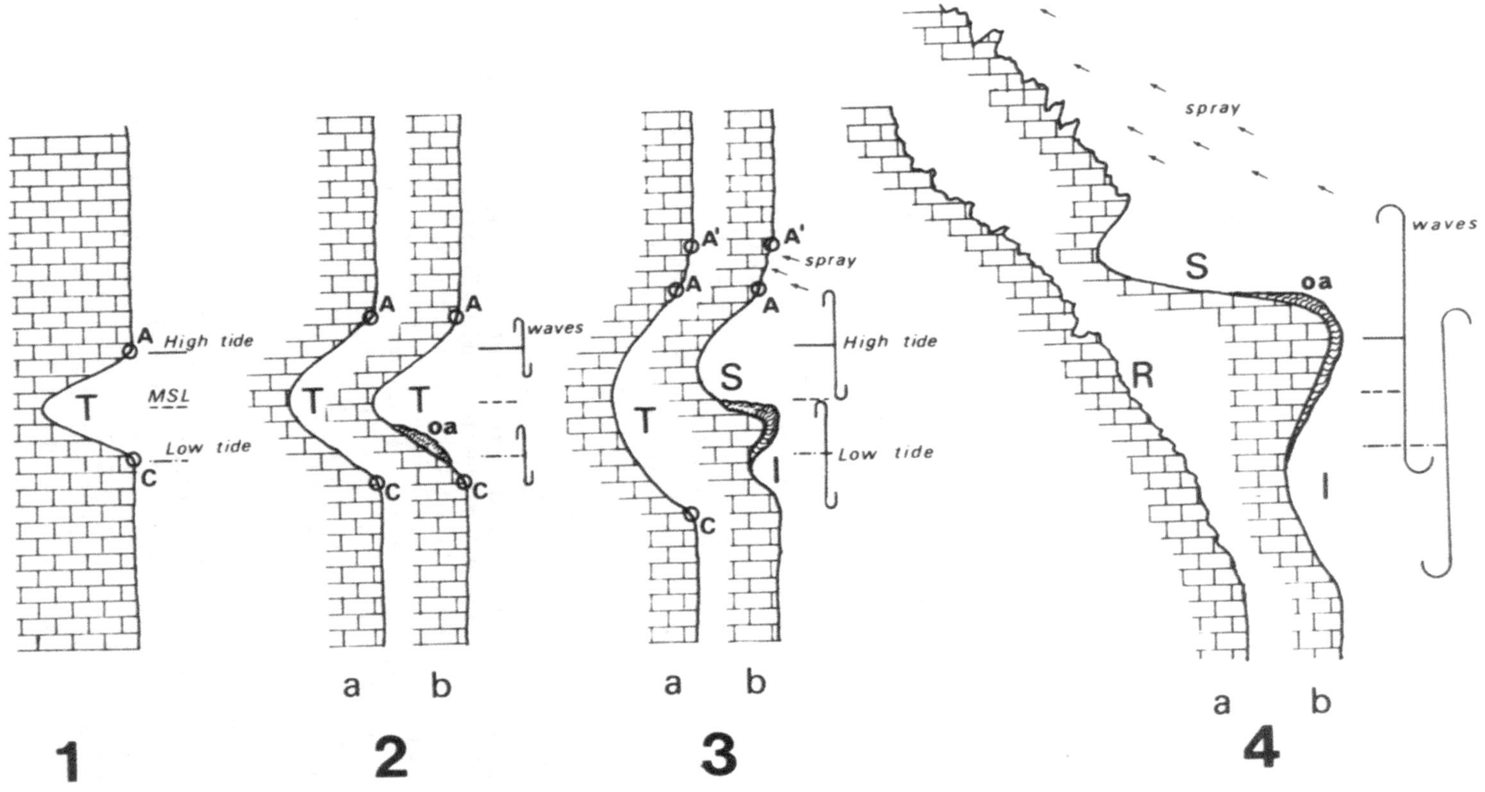

Figure 1. Morphological variations of sea-corrosion notch profiles as a function of site exposure
1 : very sheltered; 2 : moderately sheltered; 3 : moderately exposed; 4 : very exposed
a : profile without organic accretions (oa); b : profile with organic accretions.
T : tidal notch; S : surf notch (and bench); I : infralittoral notch

Figure 2. Model tidal notch profile

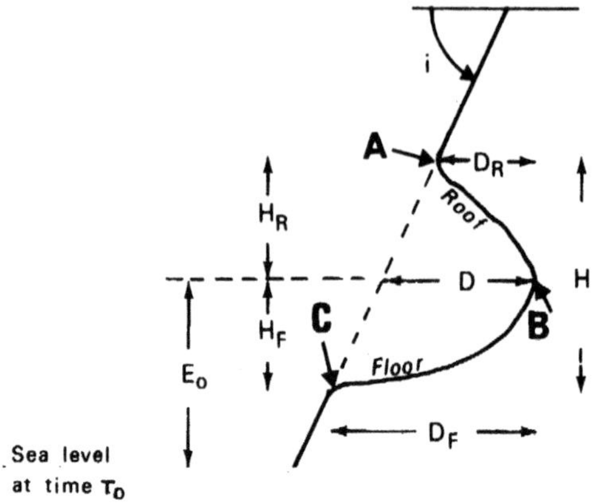

a) Tidal notches

This type is the most common and most useful SLI. On a vertical, sheltered cliff, the lower part of the *floor* or *base* (C in Fig. 2) usually extends to the limit of permanent immersion, ie. near the lowest tide level. Above C, the notch floor is flat if tidal range is small (Carobene, 1972). With a moderate tidal range, the most common profiles are recumbent V-shaped or U-shaped forms. The *depth* D of the notch increases gradually above C, to a segment with maximum convexity called the *retreat zone* or *reentrant,* which usually corresponds to neap tide range. The *retreat point,* or *vertex* B, is located near mean sea level (MSL). From that point the depth decreases gradually along the *roof* and becomes negligible near the highest tide level (the *top* A). Where the cliff ends slightly above the top of the notch, the upper part of the roof may form a projecting rock ledge, called the *visor.*

In a moderately exposed site, continuous wave action, though slight, may splash sea water onto the roof. Thus the top of the notch will then be shifted upwards, above the highest tide level. In this case the vertex becomes higher while the base remains at the lowest tide level or slightly below it. The *height* H of the notch therefore increases with exposure. In very exposed sites, however, tidal notches may be lacking altogether.

The vertical zonation of sea corrosion, described above, has been observed by the writer in many tidal notches in the Mediterranean, in Japan and in the South Pacific, and is well known (Mac Fadyen, 1930; Kuenen, 1933; Teichert, 1947; Fairbridge, 1948a, 1952; Kaye, 1959; Newell, 1961; Russell 1963; Hodgkin, 1964, 1970; Takenaga, 1968; Gill, 1973; Koba, 1974; Pirazzoli, 1976, 1979, 1980; Machida *et al.*, 1976; Montaggioni, 1979; Guilcher, 1979; Laborel, 1979b; Torunski, 1979; Kawana & Nishida, 1980) (see Ph. 1,2,3,8,15).

Sea corrosion therefore appears to be most active where the triple contact of rock, air, and sea is frequent and regular, and to decrease abruptly in the supralittoral and sublittoral zones. In very sheltered sites with high burrowing activity, however, the notch floor may be very irregular or even lacking (Kempf & Laborel, 1968; Focke, 1978a, b; Torunski, 1979). Under these conditions the notch extends into the infralittoral zone. Therefore, estimates of SL should be based only on the upper part of the notch profile (roof and vertex) (Ph. 6).

If SL fluctuates around two modes due, for example, to seasonal phenomena, the tidal notch would assume a recumbent w-shaped profile, with two distinct retreat points. Notch profiles of this type are reported by Flemming *et al.* (1978, p 42) from Israel, Gibraltar, and Cyprus.

Tidal notches are also found undercutting isolated stacks (relict from a higher reef surface) and isolated blocks emplaced by rock slides, cliff collapse and storm-wave throw. For wave-thrown coral blocks the term 'negro head', introduced by Flinders (1814) is considered unacceptable. The term 'mega block' (Battistini *et al.*, 1975) also should be avoided, because it has been pre-empted in a glacial context, and substituted by the more appropriate term 'reef block' (Stoddart, 1978). However, for the description of such remnant blocks with a tidal undercut, the term *mushroom rock* seems preferable. Moreover, it offers many possibilities in synthetic description if particular mushroom forms are specified (eg. puff-ball shape, chanterelle-shape, boletus-shape, etc.).

Small basins hollowed in calcareous or volcanic coastal platforms have often been described as typical intertidal features (Emery, 1946; Revelle & Emery, 1957; Guilcher, 1953, 1958, 1979; Battistini, 1957, 1964). The basins sometimes form a network of coalescent pools with overhanging edges. They are good indicators of the intertidal zone and can therefore be used with advantage to estimate former SLs (Dalongeville, 1979). On exposed sites, however, similar features may also form in the supralittoral zone (Guilcher, 1953; Debrat, 1974; Dalongeville, 1977) and confusion between supratidal and intertidal fossil basins should be avoided.

b) Surf notches

Limestone coasts exposed to persistent tradewinds and to strong surf and spray action commonly have a distinctive morphology. As a matter of fact, where water turbulence increases, the height of the notch also increases, and organic incrustations may develop on the floor of the notch (Fig. 1). They mostly comprise the calcareous algae (*Lithophyllum, Lithothamnium, Porolithon, Neogoniolithon*) and the vermetids (*Dendropoma, Petaloconchus, Spiroglyphus*) (Kempf & Laborel, 1968; Focke, 1978a, b; Laborel, 1979a). These organisms protect the substrate rock and thus locally inhibit erosion. While erosion proceeds above and below the accretion level, a surf bench protruding seaward will begin to form, and may become several metres wide. Surf benches may extend as high as 2 m above high-tide level. However, their level will decline with decreasing wave action (Focke, 1978b).

On the cliff behind a surf bench a corrosion notch often develops above high tide in the range regularly washed by waves (Ph. 7). The height of this notch is related to swash energy, rather than to tidal amplitude. A surf-notch profile is commonly asymmetric, with a flat floor extending to the vertex, which may be quite low in the notch profile.

The difference in elevation between a surf notch and a tidal notch may therefore exceed 2 m, but a surf notch is recognized by the presence of an adjacent surf bench, by its organic encrustations, and by the exposed situation. Still, care must be exercised to avoid confusing a fossil reef flat with a fossil surf bench. Both features are formed in the midlittoral zone, but while the former is intertidal or slightly subtidal, the latter is supratidal.

On the Eastern Mediterranean coasts surf benches called *trottoirs* are quite common and are rimmed by encrusting vermetids and algal accretions. As surf action there is relatively moderate, trottoirs are usually found not more than 0.2 to 0.4 m above MSL, ie. just above high tide (Sanlaville, 1977; Dalongeville, 1977a; Dalongeville & Sanlaville, 1978, 1979). Surf notches may therefore be considered as quite accurate SLIs in the Mediterranean.

2. Infralittoral notches

It has been mentioned above that biological processes may erode the floor of a tidal notch, thus extending the profile into the infralittoral zone. Other corrosion notches develop wholly underwater as illustrated by the notch below a surf bench in Curaçao (Focke, 1978a, Fig. 21), and the undercuts bored by *Cliona lampa* in Bermuda (Neumann, 1966). In both cases, however, the shape of the notch differs from those in the tidal zone, and confusion can be easily avoided.

Other underwater notches have been reported in aeolianite rocks, where an undercut edge sometimes delimits intertidal platforms at their seaward end (Guilcher, 1953; Gill, 1972). However the zonation of development of these notches is controversial and Fairbridge (1948a) interprets them as submerged fossil tidal notches.

3. Supralittoral notches

Some granular magmatic rocks and sandstones have a hard protective crust through which the interior is hollowed out by cavernous weathering to produce rounded or irregular notch-like profiles (Ph. 11). These erosional features (*wind blasting notches*) usually occur in or near the spray zone (Guilcher, 1958a) but may also be found farther inland, as in Corsica and Sardinia where they are known as *tafoni*.

Lines of weakness usually control their occurrence. Granular disintegration is probably due to alternate drying and wetting and to crystallization of salt spray (Mii, 1962a, b). Moreover, sand grains, driven by the wind, attack the inside of the hollow, and thus remove other sand grains. The surface of a wind-blasted notch commonly crumbles to the touch. In the spray zone, wind blasting features often grade laterally to honeycomb structures. Neither can be used as indicators of SL.

SEA-CORROSION RATES

Several estimates of erosion attributable to biologic activity are available. Rates of 0.8 cm/yr were measured for individuals of *Lithotrya*, of 0.9 cm/yr for *Lithophaga*, and of 4.9 cm/yr for *Echinometra* (Trudgill, 1976; Vita Finzi & Cornelius, 1973). Other boring rates refer to populations: 1.4 mm/yr for a dense field (ie. 100/m^2) of *Echinometra lucuntur* (Focke, 1978b), 0.51 to 0.75 mm/yr for the grazers *Patella coerulea*, and 1.1 mm/yr for sea urchins in the North Adriatic (Torunski, 1979).

Table 1. Estimates of intertidal erosion rates at various localities

Locality	Lat.	Long.	Lithology	Intertidal feature	Erosion rate (mm/yr)	References
N. Adriatic	45°N	14°E	limestone	notch	0.7-1.0	Torunski, 1979
S. France	43°N	5°E	limestone	notch	0.25-1.0	Blanc, 1979
Antikythira	36°N	23°E	limestone	notch	0.2-0.3	Pirazzoli *et al.*, 1982a (*cf.* ph.16)
La Jolla	33°N	117°W	sandstone	solution basins	0.3	Emery, 1946
Oman	26°N	56°E	limestone	notch	2.5	Vita Finzi & Cornelius, 1973
Puerto Rico	18°N	66°W	eolianite	notch	0.6-1.6	Kaye, 1959
Barbados	13°N	60°W	limestone	notches	0.2-2.0	Bird *et al.*, 1979
Sunda Str.	6°S	105°W	coral	notch	5.0	Verstappen, 1960
Aldabra	9°S	46°E	limestone	notch	0.6-1.0	Trudgill, 1976
Aldabra	9°S	46°E	limestone (exposed)	cliff or ramp	2.0-4.0	Trudgill, 1976
Rurutu	22°S	151°W	limestone	notch	1.0-1.6	This investigation (*cf.* ph.5)
Norfolk Isl.	29°S	168°E	dune limestone	steps cut by man	0.6-1.0	Hodgkin, 1964
Point Peron	32°S	116°E	eolianite	notch	1.0	Hodgkin, 1964
Kaikoura	42°S	174°E	limestone	platforms	0.4-2.5	Kirk, 1977
Kaikoura	42°S	174°E	mudstone	platforms	0.4-7.0	Kirk, 1977

Erosion due to the combined action of two organisms (such as fish browsing on clionid-weakened rock) are of the same order, 1 to 2 mm/yr (McLean, 1974; Bromley, 1978).

Generally, zones of erosion correlate well with zones of major eroding organisms (Emery, 1962; Hodgkin, 1970). In relatively sheltered sites, measured erosion rates vary with the density of grazing organisms and with the shape of the profile; a maximum value is found in the midlittoral zone, especially near MSL (Emery, 1962; Hodgkin, 1964, 1970; Trudgill, 1976).

Table 1 summarizes several reported rates of intertidal erosion on calcareous shores. Excluding the datum from Oman which seems to be an example of exceptionally rapid burrowing under favourable conditions and of the one from the Sunda Straits where an undercut coral boulder is assumed by Verstappen (1960) to date from the sea wave of the 1883 Krakatau eruption, the maximum erosion rates are usually less than 1.0 to 1.5 mm/yr. These values evidently set an upper limit consistent with notch formation and preservation in rocks varying from porous lithified dune sands (Norfolk, Point Peron) to porous Quaternary corals (Barbados, Aldabra), to hard dense limestones (North Adriatic, Rurutu). On the other hand, erosion rates evidently increase mainly with exposure and secondarily with temperature. Where surf and spray are the major factors, that is where erosion rates exceed 1.0 to 1.5 mm/yr on exposed coasts, the location of the maximum rate shifts from the midlittoral to the supra-littoral zone. Thus, the roof of the notch is attacked and the erosion profile expands from a notch, through a cliff, to a ramp as exposure increases (Trudgill, 1976; Focke, 1978b).

INFLUENCE OF THE CLIFF SLOPE

The slope of the cliff may hasten or retard the development of a notch profile. Assuming that on a sheltered limestone coast where tidal range is 1.0 m, the sea-corrosion rate is 1.0 mm/yr at MSL and decreases toward high- and low-tide levels, becoming negligible outside the tidal zone, then the most favourable conditions for notch formation occur where the cliff is vertical (Fig. 3). In this case the notch profile will correspond exactly to the zonation of sea corrosion and a symmetrical undercut will develop very rapidly on the cliff. Fifty years later the notch will be 5 cm deep, with the vertex situated at MSL.

If the cliff is steeper than 90°, overhangs will form rapidly, but the shape of the notch will be asymmetrical with a long roof and a short, steep floor (Ph. 8). On the other hand, if the cliff is much gentler that 90°, an undercut even 5 cm deep will require a long time to form: 300 years on a slope of 2:1 (about 27°). Again, the shape of the undercut will be asymmetrical, but the roof will be short and steep, and the floor long and flat.

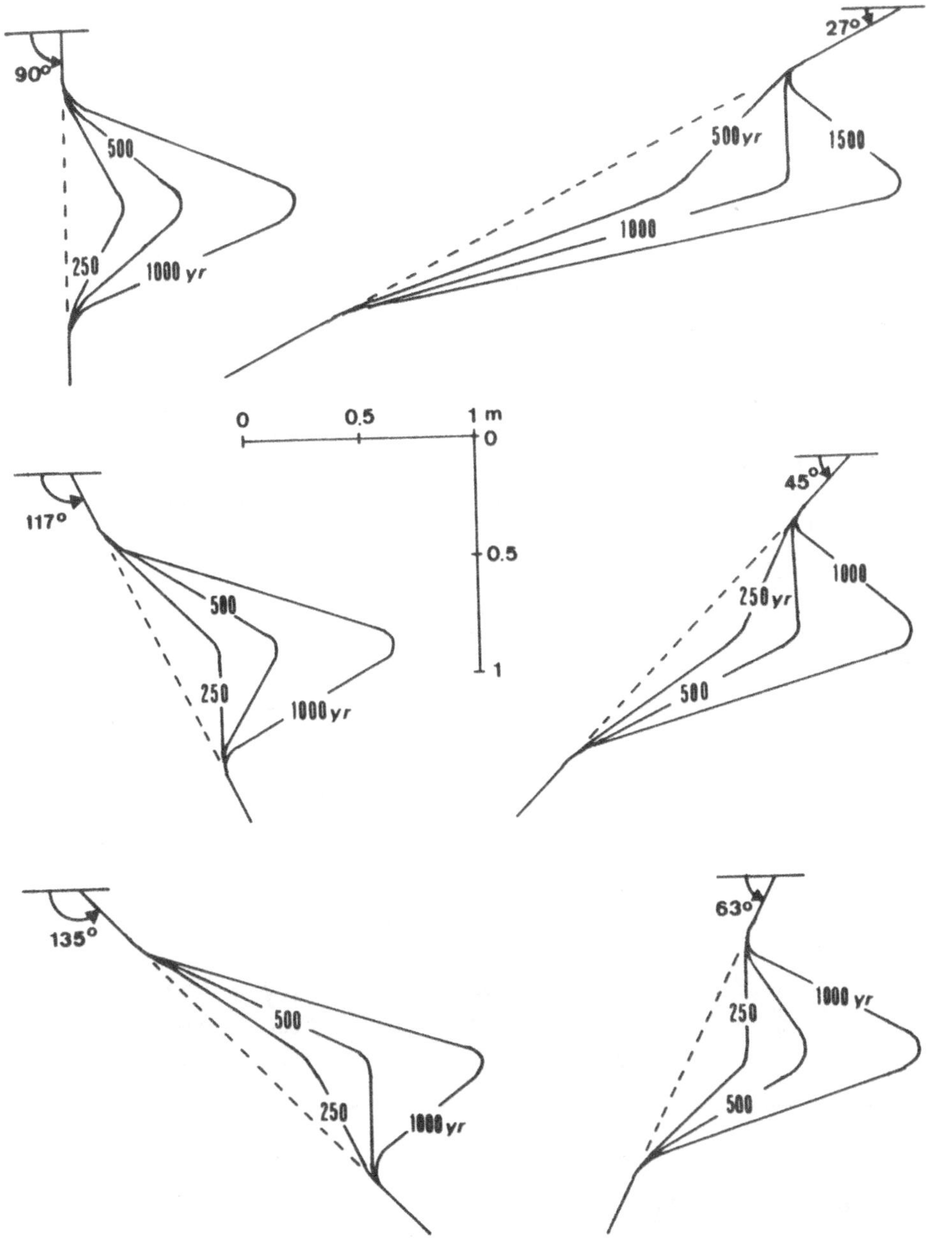

Figure 3. Influence of cliff slope on the development of corrosion notch profiles in a sheltered setting. Tidal range is assumed to be 1.0 m and erosion rate at mean sea level 1.0 mm/yr.

Limestone formations that are gently inclined toward the sea, such as along much of the Yugoslavian and Aegean coasts, are therefore unfavourable to sea-corrosion undercuts.

On cliff faces that slope less than 90°, if the tidal range decreases, then the period of time necessary for the development of a notch will be shorter; if the tidal range increases, it will be longer. This fact explains in part why corrosion notches are so rare in regions with high tidal ranges.

Lastly, if the maximum corrosion rate at MSL decreases, the period of time needed for the appearance of an undercut will necessarily increase. Table 2 summarizes the length of time required to develop a 5 cm deep undercut for various sea-corrosion rates, tidal ranges and cliff slopes.

Table 2. Periods of time (in years) necessary for appearance of an undercutting 5 cm deep on a sheltered limestone cliff.

Maximum corrosion rate at MSL (mm/yr)	Tidal range (m)	Cliff slope 90° or more	63°	45°	27°
	0.3	167	417	667	1167
0.3	1.0	167	1000	1834	3500
	2.0	167	1834	3500	6834
	0.3	84	209	334	584
0.6	1.0	84	500	917	1750
	2.0	84	917	1750	3417
	0.3	50	125	200	350
1.0	1.0	50	300	550	1050
	2.0	50	550	1050	2050

POSSIBILITIES FOR CONFUSION WITH NOTCHES OF OTHER ORIGIN

1. Abrasion notches

The mechanical action of sand and pebbles thrown by waves may erode notches and other features in a variety of rock types. Unlike corrosion surfaces, which appear bioeroded, rough, rugged and pitted, abrasion surfaces are generally smooth and polished. As a result, the profile of an abrasion notch is usually rounded, with obliteration of any angular surface.

Abrasion features develop in the surf zone and, generally, in any zone reached by waves or by hydrodynamic agents acting in combination with granular deposits. As wave energy lessens in shallow water, abrasion notches are found in the upper part of the midlittoral zone and their height is usually not related to the local tidal range.

Abrasion notches may exist in continuous or very short segments. Continuous undercuttings are commonly found at the angle between an abrasion platform and a cliff, near the level of the highest tides on a heavy sea. Shale or similar rock with vertical stratification or cleavage is particularly favourable for the development of abrasion platforms and notches. The vertical precision of this kind of indicator is generally weak (± several metres), depending on the site exposure.

Localized abrasion notches are usually found on rocks bordering a beach. Their height (generally less than 2 m) depends on the local wave energy. These notches rarely have the same profile for more than a few metres and a gentle longitudinal slope is not uncommon. Estimating the relation between fossil localized abrasion notches and the former SL is usually difficult, because an abrasion notch may have been shaped in the surf zone by waves as well as underwater by currents (Ph. 9).

2. Structural notches

In rock formations consisting of horizontal or gently inclined beds, differential erosion may excavate the weakest layers, along with joints and fissures and thus produce linear reentrants with notch-like profiles. Such features are therefore primarily of structural origin. Since differential erosion by subaerial weathering can produce the same effect as sea corrosion and abrasion, structural notches cannot be considered definitive SLIs.

The structural origin of a notch is usually easy to recognize in the field: the attitude does not necessarily remain horizontal but follows the original line of weakness. Whatever its slope, the profile may vary along its length; the height of the notch corresponds to certain layers in the rock, etc. (Ph. 10, 11).

MODIFICATION OF A TIDAL NOTCH DURING EMERGENCE OR SUBMERGENCE

Using again the type case of a sheltered limestone coast where tidal range is 1.0 m and the corrosion rate at MSL is 1.0 mm/yr. After a time, a tidal corrosion notch will form (Fig. 4a). If SL does not change the undercut deepens and assumes a recumbent V-shaped profile, the height of the notch being equal to the tidal range (Fig. 4b). The notch will deepen and undermine the cliff until the roof collapses. Then a new, similar notch will form farther inside the cliff.

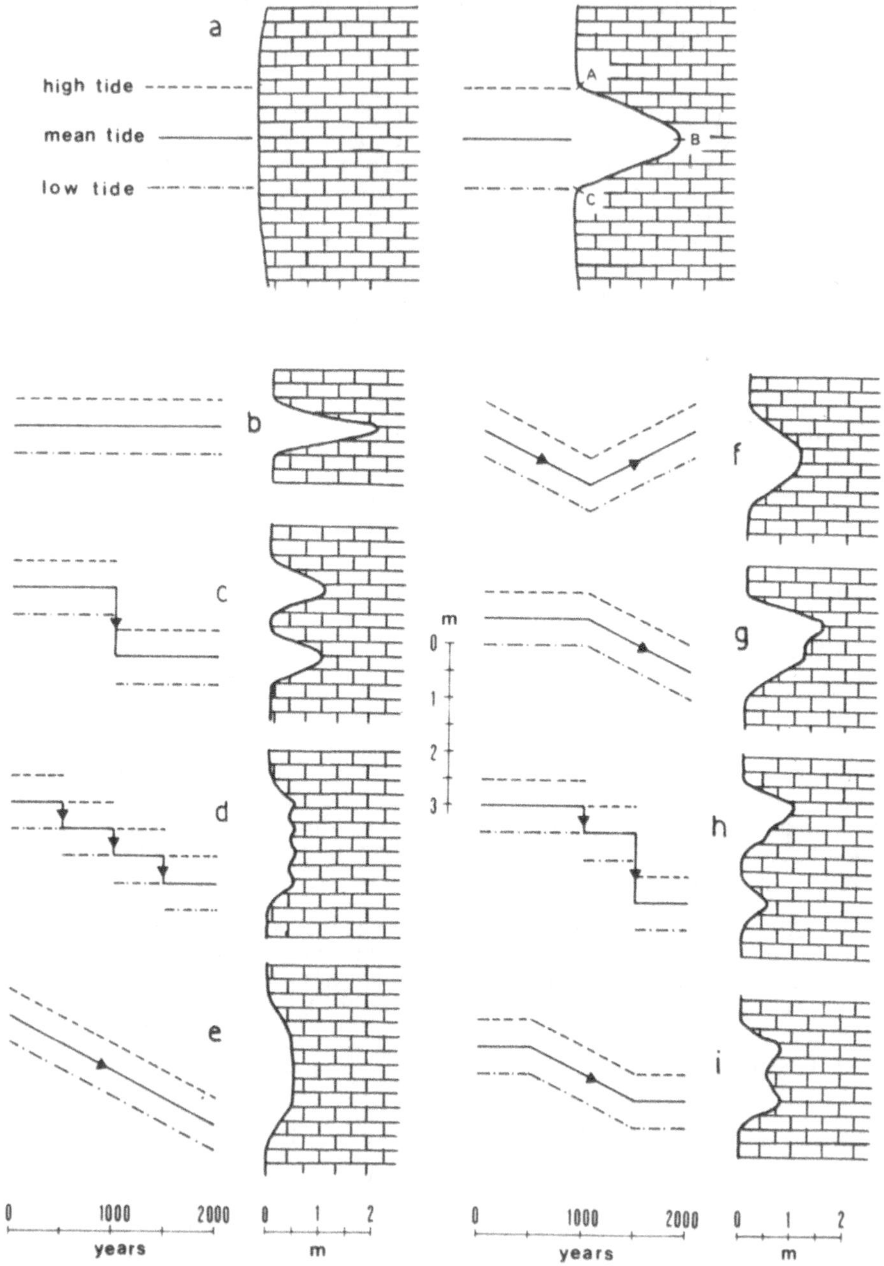

Figure 4. Notch profiles resulting from different combinations of relative sea-level change in sheltered places. The maximum sea-corrosion rate is assumed to be 1.0 mm/yr and to occur at mean sea level.

If, on the other hand, there is relative movement of SL, the notch profile will be modified. In the case of an emergence due to a relative fall of SL, several results are possible. A sudden emergence *greater* than the tidal range (Fig. 4c) is a unique case whereby the notch, completely emerged, will be preserved from further marine modification while a new notch will develop in the new lower tidal zone. A sudden emergence with vertical displacement *less* than the tidal range (Fig. 4d and h) allows only the roof of the notch to be preserved while the floor is attacked and so degraded to the new low-tide level. The height of the notch will thus increase. If later sudden vertical movements occur, successive overlapping indentations will produce so-called *ripple-notches* (Ph. 16). A slow rate (eg. 1 mm/yr) of gradual emergence (Fig. 4e) or a gradual SL oscillation (Fig. 4f) will result in a symmetrical profile, although the height will approximate the amplitude of the change. A gradual emergence following a period of stable SL (Fig. 4g) will cause only the floor of the notch to be modified. A gradual emergence between two periods of stable SL (Fig. 4i): the notch will assume a reclined w-shaped profile.

Many combinations of movements leading to submergence instead of emergence are possible, but even so notch profiles would be similar to those of Figure 4. In the case of submergence, borings and organic encrustations will appear on the roof (Ph. 1, 12). On the other hand very recent emergence may be indicated by limpet and chiton markings up to an abrupt limit below the roof.

Four main conclusions are:

1 - a tidal notch whose height corresponds to the local tidal range must have been formed during a period of stable SL;

2 - an emerged tidal notch with an unaltered floor and base indicates that emergence was rapid and greater than the tidal range (Ph. 2,8,14,15). Similarly, a submerged notch with an unaltered roof profile suggests that submergence was probably rapid and greater than the tidal range (Ph. 1);

3 - a tidal notch whose height is less than the present tidal range indicates that the tidal amplitude has changed. A possible example is mentioned by Hodgkin (1970, p 48);

4 - during vertical displacements less than the tidal range, either gradual (Fig. 4g) or sudden (Fig. 4h) SL changes can produce similar notch profiles. In order to determine whether the movement was fast or slow, other indicators (eg. biological) must be evaluated.

Wave and spray action have not been considered in the above cases. However, if the height of the midlittoral zone is compared to the tidal amplitude, diagrams similar to those of Figure 4 may aid in the interpretation of a wide range of field data. Finally it should be noted that not all erosion features at a given level are necessarily related to a single SL. Notches may be polycyclic and may contain morphological features relict from earlier SLs (Ph. 12,13). Notches varying in level from the mouth to the head of a bay may reflect differing exposure at the same SL.

DATING A NOTCH

It has been shown that if local erosion rates can be estimated, a fossil notch may indicate both the position and approximate duration of a SL stand as well as the speed of the SL change. On the other hand a notch, like any erosion feature, is inadequate to date a former SL. Related, dateable SLIs must therefore be sought such as:

a) *Marine organisms*. In the most favourable cases, vermetids (*Dendropoma, Petaloconchus*), calcareous algae (*Lithophyllum, Porolithon, Neogoniolithon*), barnacles, or even small bosses of coral, are found in living position encrusting the notch surface (Ph. 15) or other nearby rocks at the same level. Well-preserved samples often come from sea caves and hollows sheltered from rain and spray action. Infralittoral sessile shells, organic accretions, and corals are also suitable if in living position and the vertical zonation is narrow and well-known.

b) *Beach deposits* concordant in level with the notch or covering the floor may contain fossils suitable for radiometric or paleontologic dating. However, it is normally difficult to prove that the deposits were caused by the same, and not a later SL stand. However, where emerged beachrock and an intertidal notch occur at the same level they are likely to be coeval. In this case, the former SL may be dated by organic material included in the beachrock, except that fossils in beachrock may be re-cycled from older deposits (Pirazzoli, 1976, p 143; Dalongeville & Sanlaville, 1979; Montaggioni, 1979a; Hopley, this volume). In both cases, the inferred age of the beach deposits will be doubtful. Ideally, material in living position is always preferred.

c) *Archaeologic remains*. Dateable sherds may be fossilized in beach deposits and even in beachrock (Sanlaville, 1977). Tidal notches may be cut across man-made surfaces, as in the raised Roman harbour of Falasarna (Crete) and the submerged quarries of Karpathos, or in other archaeological remains (Flemming *et al.*, 1973, 1978; Dalongeville & Sanlaville, 1978; Pirazzoli, 1980; Flemming & Pirazzoli, 1981). In other cases, submerged archaeological remains may also provide useful chronometric control on the age of former SLs as well as on seismic subsidence evidenced by submerged tidal notches (Pirazzoli, 1980).

ACKNOWLEDGEMENTS

The writer is indebted to Richard G. Bromley, Edmund D. Gill, Jacques Laborel, Lucien F. Montaggioni and to the Editor for reviewing an earlier version of the paper and offering many useful comments, and to Douglas R. Grant for kindly revising the English manuscript.

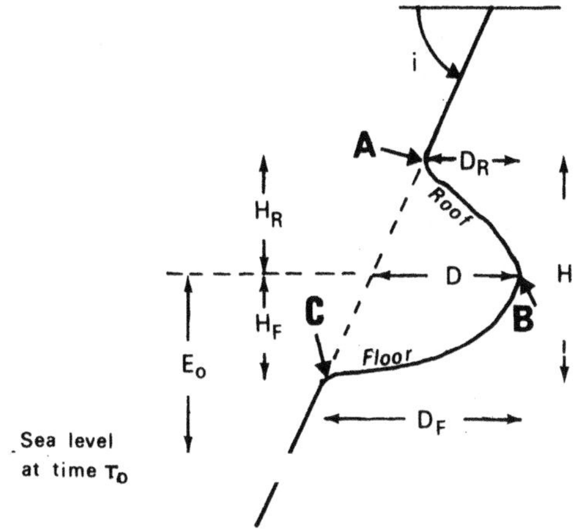

APPENDIX I: MEMORANDUM FOR FIELD WORK

1 - *Locality* (to be situated on a large-scale map). Local spring and neap tidal ranges, together with predictions of tide at time T_o should be given. Note the exposure of the site and meteorological conditions at time T_o (especially atmospheric pressure and wind).

2 - *Geometric characteristics of the notch profile:* i, D_R, D_F, H_R, H_F ($H = H_R + H_F$; if the notch is symmetrical: $D = (D_R + D_F)/2$). If the shape is not a simple one, the profile should be measured in detail. Note the accuracy of each measure.

3 - *Elevation:* to be measured if possible, with respect to a similar feature presently developing at the same site. Otherwise: E_o, to be measured with respect to sea level at time T_o. Note the accuracy of each measure.

4 - *Lithology* and dip of the bedding planes.

5 - *Appearance* of the notch surface (bioeroded (by what?), pitted, smooth, polished, etc.).

6 - *Continuity of the notch profile* (continuous, scattered, irregular, isolated).

7 - *Continuity of the elevation E_o* (constant, sloping, irregular).

8 - *Other* near-by *sea-level indicators*.

9 - *Other* ...

Photo 1. Kyra Panagia, Karpathos, Greece (35°35'N - 27°12'E). Spring tidal range (STR): 0.2 m (est.). Site is exposed to NE winds. Lithology: Paleogene hard limestone. Height (H) of the notch: 0.5 m. Elevation above MSL (E) of the notch: -1.1 m. *Interpretation:* According to the shape of this very well-preserved tidal notch, submergence was the consequence of a rapid movement, very likely of seismic origin, which evidently occurred during historical times (Pirazzoli, 1980) (Ph. P.A. Pirazzoli).

Photo 2. NW Kabara, Lau Group, Fiji (18°57'S - 178°57'W). STR: ~1.5 m. Site is protected from wave action by a fringing reef and a reef flat. Lithology: raised reef limestone. H: ~1.5 m. E: ~+1 m. A continuous regular notch, pitted by sea corrosion, encircles the vertical cliff. A smaller cliff, about 0.7 m high, separates the notch floor from a reef flat at present MLWS level. The photo was taken at low tide. At MHWS tide, sea level reaches the vertex of the notch. A similar notch, submerged at about -3 m, has been observed in the same area (Galzin *et al.*, 1979). *Interpretation:* This notch is typically tidal and has been formed during a period of stable sea level which, according to the depth of the notch, may have lasted 2000 or 3000 years. Emergence of about 1 m was probably rapid and recent, since the small cliff is still preserved despite its present position near MSL where sea corrosion reaches the highest rates (Ph. B. Salvat).

Photo 3. NW Kaukura, Tuamotu Islands, French Polynesia (15°42'S - 146°53'W). STR: 0.6 m. Water level is at about +0.1 m (0.1 m above MSL). Position in the middle of the reef flat is sheltered. Lithology: Quaternary coral limestone. H: at least 0.45 m (part of the notch floor is lacking). E: +0.55 m. *Interpretation:* Mushroom-rocks of this type are very common along the W side of the atoll. Locally called 'feo' or 'heo', they are remnants of a higher Pleistocene coral reef. When the notch was formed, the relative level of the ocean was 0.55 ± 0.1 m higher than at present (nearby, slightly emerged coral heads in living position were dated 2010 ± 110 BP and 2320 ± 95 BP [Pirazzoli & Montaggioni, in press]). Subsequent fall of sea level was slow enough to allow destruction by sea corrosion of most of the notch floor (Ph. P.A. Pirazzoli).

Photo 4. Kuroiwa, Asamushi, Aomori Pref., Japan (40°57'N - 140°51'E). STR: 0.6 m. Water level is at about MSL. Site is moderately exposed to the prevailing WNW winds. Lithology: Miocene dacite. H: >2 m. E: 1.5 m (est.). The notch is very high compared to tidal range. *Interpretation:* while corrosion features in volcanic rocks are less precise indicators of sea level than in limestone rocks, several benches at +0.2 to +0.5 m, together with emerged notches and sea caves at various levels, indicate recent emergence in this area (Ph. P.A. Pirazzoli).

3

4

Photo 5. Point Arei, Rurutu, Tubuai Islands, French Polynesia (22°27 S - 151°19'W). STR: ~0.6 m. Water level is at about MSL. Site is relatively sheltered from strong waves by a narrow reef flat. Lithology and geomorphology: this basaltic island is surrounded by a raised limestone reef, reaching up to 100 m in altitude. Several marine notches, at different levels, are cut in the hard limestone cliff. *Interpretation:* The two lowest notches, at +8 to +10 m and near the present sea level, are well visible. They have similar profiles, but the +8 to +10 notch is locally encumbered with calcareous columns. The lowest notch (H: 2 to 3 m; Depth: 4 to 8 m) was probably formed during Holocene times and is clearly polycyclic. In the photo, two small notches may be distinguished inside the Holocene notch: the lowest one, whose vertex stands at about +0.6 m (near the basis of the meter), is the less well marked. At +1.2 m another reentrant is deeper and more continuous. However, such a clear stepped profile is rare in Rurutu, and most of the times the notch has only one vertex, ranging from +1.7 to +0.5 m in elevation. Corrosion benches, hollowed by pools with overhanging edges, are also found at various elevations. Holocene sea level seems to have reached in this area a peak at +1.7 m. Then it slowly dropped to present sea level. The depth of the Holocene notch reaches, in places, 6 to 8 m. According to a possible duration of 5000 or 6000 years with the sea level 'near' to its present position, the intertidal erosion rate may be estimated to 1.0 to 1.6 mm per year in Rurutu (Ph. P.A. Pirazzoli).

Photo 6. Himi, Toyama, Japan (36°53'N - 137°00'E). STR: 0.2 m. Site is exposed to occasionally heavy storm waves in winter. Lithology: Pliocene soft mudstone. *Interpretation:* This tidal notch is formed at present by marine bioerosion. In many places, the notch floor is lacking, eroded by molluscs, and the notch profile extends into the infralittoral zone. Due to the softness of the rock, examples of raised notches are absent from this area, although sea level is known to have reached a higher level during Holocene (Fujii, 1967) (Ph. P.A. Pirazzoli).

5

6

Photo 7. Cape Ara, Okinawa, Japan (26°04'N - 127°41'E). STR: 1.7 m. Water level is at -0.7 m. Located at the S end of the island, the cape is exposed to E, S, and W. Lithology: Quaternary reef limestone. A wide corrosion platform is visible on the background. Nearer the cape, exposure increases and the platform becomes narrower and higher. Marine accretions develop near its outer edge. On the cape (in the foreground) the platform is a narrow bench situated ~0.5 m above high-tide level. At the back of the bench, a corrosion notch develops in the cliff situated between the bench and a higher platform. *Interpretation:* This notch is an example of a surf notch developing above a surf bench. The position of the notch vertex depends on swash energy and is therefore not related to a former MSL (Ph. P.A. Pirazzoli).

Photo 8. Cape Melissa, Crete, Greece (35°06'N - 24°34'E). STR: 0.2 m (est.). Site is exposed to occasional storms from S in winter. Lithology: pre-Cenozoic hard limestone. The notch profile is regular and continuous along the cliffs. *Interpretation:* Due to the cliff slope (about 135°) the profile of this tidal notch is typically asymmetric, with a long flat roof and a short steep floor. The notch was raised to the present elevation by the same earthquake *c.*1530 ± 40 BP (Thommeret *et al.*, 1981), which uplifted the ripple notches of Antikythira (*cf* h. 16) (Ph. P.A. Pirazzoli).

Photo 9. Oorio (2.5 km N of-), Tottori, Japan (35°35'N - 134°18'E). STR: 0.15 m. Site is moderately exposed to N winds. Lithology: Late Cretaceous biotite granite. H: about 0.8 m. E: from +0.5 to -1.3 m. *Interpretation:* Already described by Toyoshima (1965), this notch is an excellent example of a sloping abrasion notch on rocks bordering on a beach. Well rounded and polished by sand abrasion, the notch begins in the surf zone (at about +0.5 m), then slopes down, continues underwater, and disappears more than ten metres farther on, 1.5 to 2.0 m deep. In the foreground: S. Akagi and Y. Toyoshima, September 1974 (Ph. P.A. Pirazzoli).

Photo 10. Punta Cancun, N.E. Quintana Roo, Mexico (21°08'N - 86°46'W). STR: about 0.5 m. Site is exposed to the swell coming from E. Lithology: Holocene eolianite (Ward, 1975, p. 512). H: 0.6 m. E: +2.5 m. The height of this V-shaped notch approximately corresponds to the interval of the tidal range, however in such an exposed site the height of a tidal notch should be larger than the tidal range. Furthermore, several bedding planes are parallel to the notch axis and the notch profile is irregular and discontinuous. According to W.C. Ward (1979, pers. comm.) similar bedding planes are found in other parts of the island, but these are at various elevations and there is no notch that is at a constant elevation around the island. *Interpretation:* The notch is therefore considered to be structural and its vertex is not related to a former MSL (Ph. O. van de Plassche).

9

10

Photo 11. Shirahama, Wakayama Pref., Japan (33°41'N - 135°20'E). STR: 1.4 m. Water level is 5 cm above MSL. Site is exposed to prevailing E winds. Lithology: Miocene pebbly conglomerate with intercalated lens of sandstone. Several notches, commonly related to bedding planes, are visible in the cliff. Notch profiles are irregular and discontinuous. The surface of the notches do not crumble to the touch but sand grains and some pebbles may be easily dislodged. *Interpretation:* Already described by Mii (1962a, Fig. 28), these notches are typically structural. Their development is mainly ascribed to the combined action of spray and of wind blasting (Ph. P.A. Pirazzoli).

Photo 12. Nauplia, Greece (37°33'N - 22°47'E). STR: 0.3 m (est.). Site is very sheltered. Lithology: hard limestone. H: 5 m. E: ~5 m. A narrow bench extends the notch floor. Although the notch surface appears to be rather smooth, niches hollowed by sea urchins are visible on the floor and holes burrowed by molluscs can be found on the notch roof, as high as +6.1 m. In the background: a last-glacial age breccia laterally overlies marine deposits containing *Strombus bubonius*. According to Keraudren (1971, pl. X and Fig. 40 III), the notch is Neotyrrhenian. *Interpretation:* As the height is disproportionate to local tidal range and exposure, the notch is clearly polycyclic. This profile suggests several sea levels; the last, of probable Neotyrrhenian age, rejuvenated an existing shape (Ph. P.A. Pirazzoli).

11

12

Photo 13. Ay. Roumeli, Crete, Greece (35°14'N - 23°57'E). STR: 0.2 m (est.). Site is exposed to S winds (occasional heavy storms in winter). Lithology: Carboniferous hard limestone. H: about 3 m (the base of the notch is difficult to situate exactly). E: +2.5 m. The notch surface is smooth, although the present environment is non-abrasional. Two corrosion lines may be seen at the level of the vertex (arrow A) and just below the top of the notch (arrow B). *Interpretation:* Notwithstanding a previous interpretation by the writer, this notch is not Holocene, but clearly polycyclic, and was probably formed by several Pleistocene sea levels. During the Holocene the notch was rejuvenated and the two corrosion lines were produced. The lower line (A) coincides, at +2.5 m, with the apparent retreat point of the notch: it was tentatively dated at 2400 ± 70 BP (MC-1378, -1379) (Pirazzoli & Thommeret, 1977). However, as the dated fauna is infralittoral, the 2400 BP sea level was probably slightly above A. The upper line (B), at +4.25 m, seems to correspond to the highest level reached by the sea in this area during Holocene. On it some encrusting algal accretions (*Neogoniolithon*) were dated 1860 ± 70 BP (Laborel *et al.*, 1979). Subsequently the notch was raised by an earthquake around 1530 ± 40 BP, at the same time as Cape Melissa (Ph. 8) and Antikythira (Ph. 16) (Ph. P.A. Pirazzoli).

Photo 14. Araki, Kikai-jima, Japan (28°18'N - 129°55'E). STR: 1.3 m. Lithology: Late Quaternary reef limestone. Situated near the Ryukyu trench, this island is completely surrounded by raised reefs. The highest Holocene marine marks reach 12-13 m in elevation. Three superimposed notches, between +9.5 and +13 m, are visible (Pirazzoli, 1974, 1978). *Interpretation:* The height of the upper two notches approximately corresponds to the tidal range. The floor of the third notch is eroded but in its vicinity are remnants of a corrosion platform at +9.4 m which probably corresponds to the base of the third notch. Each of the three notches indicates a separate sea-level stand probably more than 6000 years old (Konishi *et al.*, 1974; Nakata *et al.*, 1978; Ota *et al.*, 1978). According to the shape of the profiles, uplift was spasmodic rather than continuous (Ph. P.A. Pirazzoli).

13

14

Photo 15. Cape Teodoco, Rhodes, Greece (36°15'N - 28°13'E). STR: 0.2 m. Site is relatively sheltered. Lithology: Mesozoic hard limestone. Several superposed corrosion notches may be distinguished. For the highest and best preserved notch H: 0.5 m; E: +2.95 m; Depth: 0.7 m. *Interpretation:* This U-shaped notch is clearly tidal. An accretion of the vermetid gastropod *Dendropoma* in living position is visible under the hammer. It was dated 4270 ± 100 BP (MC-2347) and 4340 ± 100 BP (MC-2201) (Pirazzoli *et al.*, 1982b). A sudden uplift must have occurred since that time (Ph. P.A. Pirazzoli, reproduced with the courtesy of the 'Annales de l'Institut Océanographique').

Photo 16. Antikythira (Cerigotto), Greece (35°53'N - 23°18'E). STR: 0.2 m (est.). Site is moderately exposed to winds from NW. Lithology: hard limestone. Nine ripple notches may be distinguished, with E = +2.7, +2.5, +2.35, +2.15, +2.0, +1.75, +1.5, +1.3, and +1.1 m. Scale is given by a meter placed on the +1.5 m level. The depth of the notches varies from 3 to 7 cm. *Interpretation:* This is an exceptional series of emerged ripple notches. Five were dated by ^{14}C and the age of three others was inferred from correlations (Pirazzoli *et al.*, 1982a). These result from a series of rapid but small sinkings, at intervals of about 200-250 years, which occurred between 4000 and 1700 BP. A sudden upheaval of 2.7 m took place around 1530 ± 40 BP, raising at one time the whole sequence of shorelines. Intertidal corrosion rate was here of 0.2 to 0.3 mm/yr. (Ph. P.A. Pirazzoli).

15

16

REFERENCES

Battistini, R., 1957. La zonation de l'estran dans les grès littoraux au Cap Mandevitse près de Morombé (côte sud-ouest de Madagascar). *Mémoires de l'Institut Scientifique de Madagascar*, D, 8, 183-193

Battistini, R., 1964. L'âge absolu de la plature de corail mort flandrienne de 50-60 cm d'Itampolo (côte sud-ouest de Madagascar). *Madagascar, Revue de Géographie*, 4, 109-112

Battistini, R. *et al.*, 1975. Eléments de terminologie récifale indopacifique. *Téthys*, 7, 1-111

Bird, J.B., Richards, A. & Wong, P.P., 1979. Coastal subsystems of western Barbados, West Indies. *Geografiska Annaler*, 61A, 221-236

Blanc, J., 1979. Vitesse de formation des encoches littorales. In: *Les Indicateurs de Niveaux Marins*, ed. Groupe Nivmer, *Oceanis*, 5, Fascicule hors-série, 325-326

Bromley, R.G., 1978. Bioerosion of Bermuda reefs. *Palaeogeography, Palaeoclimatology, Palaeoecology*, 23, 169-197

Carobene, L., 1972. Osservazioni sui solchi di battente attualdi ed antichi nel golfo di Orosei in Sardegna. *Bolletino della Società Geologica Italiana*, 91, 503-601

Christiansen, S., 1963. Morphology of some coral cliffs, Bismark Archipelago. *Norsk Geografisk Tidsskrift*, 62, 1-23

Dalongeville, R., 1977. Formes littorales de corrosion dans les roches carbonatées au Liban. Etude morphologique. *Mediterraneé*, 3, 21-33

Dalongeville, R., 1977a. Trottoir-encorbellement-encoche: aspects du littoral actuel libanais. *Bulletin du Laboratoire Rhodanien de Géomorphologie*, 1, 53-58

Dalongeville, R., 1979. Morphologie littorale des roches carbonatées à Jerba (Tunisie). *Revue Tunisienne de Géographie*, 3, 9-30

Dalongeville, R. & Sanlaville, P., 1978. Les changements de la ligne de rivage en Méditerranée orientale, à l'époque historique. Exemple de la côte levantine. *Colloques Internationaux du C.N.R.S.*, 578, 19-32

Dalongeville, R. & Sanlaville, P., 1979. Rivages holocènes de Turquie méridionale. *Bulletin Rhodanien de Géomorphologie*, 4-5, 5-15

Debrat, J.M., 1974. Etude d'un karst calcaire littoral méditerranéen. Exemple du littoral de Nice à Menton. *Méditerranée*, 2, 63-85

Emery, K.O., 1946. Marine solution basins. *The Journal of Geology*, 54, 209-228

Emery, K.O., 1962. Marine geology of Guam. *United States Geological Survey Professional Paper*, 403-B, B1-B76

Fairbridge, R.W., 1948. Notes on the geomorphology of the Pelsart Group of the Houtman's Abrolhos Islands. *Journal of the Royal Society of Western Australia*, 33, 1-43

Fairbridge, R.W., 1948a. The geology and geomorphology of Point Peron, Western Australia. *Journal of the Royal Society of Western Australia*, 34, 35-72

Fairbridge, R.W., 1952. Marine erosion. *Proceedings of the Seventh Pacific Science Congress*, 3, 347-359

Flemming, N.C., 1978. Holocene eustatic changes and coastal tectonics in the northeast Mediterranean: implications for models of crustal consumption. *Philosophical Transactions of the Royal Society, London*, A, 289, 1362, 405-445, + Appendix 1

Flemming, N.C. & Pirazzoli, P.A., 1981. Archéologie des côtes de la Crète. In: *Niveau de la Mediterranée depuis l'Antiquité. Dossiers de l'Archéologie*, 50, 68-81

Flemming, N.C., Czartoryska, N.M.G. & Hunter, P.M., 1973. Archaeological evidence for vertical earth movements in the region of the Aegean island arc. In: *Science Diving International*, ed. N.C. Flemming, *Proceedings of the 3rd Symposium on the Scientific Committee of the Confédération mondiale des activités subaquatiques*, 47-65

Flemming, N.C., Raban, A. & Goetschel, C., 1978. Tectonic and eustatic changes on the Mediterranean coast of Israel in the last 9000 years. *Progress in Underwater Science*, 3, 33-93

Flinders, M., 1814. *A voyage to Terra Australis*. (Nicol, London), 2 volumes

Focke, J.W., 1978a. Limestone cliff morphology and organism distribution on Curaçao (Netherlands Antilles). *Leidse Geologische Mededelingen*, 51, 131-150

Focke, J.W., 1978b. Limestone cliff morphology on Curaçao (Netherlands Antilles), with special attention to the origin of notches and vermetid/coralline algal surf benches. *Zeitschrift für Geomorphologie*, 22, 329-349

Frémy, P., 1945. Contribution à la physiologie des thallophytes marins perforant et cariant les roches calcaires et les coquilles. *Annales de l'Institut Océanographique*, 22, 107-144

Fujii, S., 1967. Postglacial deposits and their carbon-14 datings in the Japanese Islands (in Japanese, with English abstract). *Daionki Kenkyu* (The Quaternary Research), 6, 192-199

Galzin, R., Ricard, M., Richard, G., Salvat, B. & Toffart, J.L., 1979. Le complexe récifal de Kabara (Lau, Fiji). Géomorphologie, biologie et socio-ecologie. *Annales de l'Institut Océanographique*, 55, 113-134

Gill, E.D., 1972. The relationship of present shore platforms to past sea levels. *Boreas*, 1, 1-25

Gill, E.D., 1973. Rate and mode of retrogradation on rocky coasts in Victoria, Australia, and their relationship to sea-level changes. *Boreas*, 2, 143-171

Guilcher, A., 1952. Formes et processus d'érosion sur les récifs coralliens du nord du Banc Farsan (Mer Rouge). *Revue de Géomorphologie dynamique*, 3, 261-274

Guilcher, A., 1953. Essai sur la zonation et la distribution des formes littorales de dissolution du calcaire. *Annales de Géographie*, 331, 161-179

Guilcher, A., 1955. Géomorphologie de l'extrémité septentrionale du Banc Farsan (Mer Rouge). *Annales de l'Institut Océanographique*, 30, 55-100

Guilcher, A., 1958. Coastal corrosion forms in limestones around the Bay of Biscay. *Scottish Geographical Magazine*, 74, 137-149

Guilcher, A., 1958a. *Coastal and submarine morphology*. (Methuen, London; John Wiley & Sons, New York)

Guilcher, A., 1979. Indicateurs résultant de processus d'érosion. In: *Les Indicateurs de Niveaux Marins*, ed. Groupe Nivmer, *Oceanis*, 5, Fascicule hors-série, 319-324

Guilcher, A., Berthois, L. & Battistini, R., 1962. Formes de corrosion littorale dans les roches volcaniques, particulièrement à Madagascar et au Cap Vert (Sénégal). *Cahiers Océanographiques*, 14, 208-240

Guilcher, A. & Bodéré, J.C., 1975. Formes de corrosion littorale dans les roches volcaniques aux moyennes et hautes latitudes dans l'Atlantique. *Bulletin de l'Association de Géographes français*, 426, 179-185

Guilcher, A. & Pont, P., 1957. Etude expérimentale de la corrosion littorale des calcaires. *Bulletin de l'Association de Géographes Français*, 265-266, 48-62

Higgins, C.G., 1980. Nips, notches, and the solution of coastal limestone: an overview of the problem with examples from Greece. *Estuarine and oastal arine cience*, 10, 15-30

Hodgkin, E.P., 1964. Rate of erosion of intertidal limestone. *Zeitschrift fur Geomorphologie*, 8, 385-392

Hodgkin, E.P., 1970. Geomorphology and biological erosion of limestone coasts in Malaysia. *Geological Society of Malaysia*, 3, 27-51

Hopley, D., (this volume). Beachrock as a sea-level indicator

Kawana, T. & Nishida, H., 1980. Preliminary report of the formative periods of the notches on Yoron, Okinawa, Miyako and Ishigaki islands, the Ryukyu Islands, South Japan (in Japanese, with English abstract). *Geological Studies of the Ryukyu Islands*, 5, 103-123

Kaye, C.A., 1957. The effect of solvent motion on limestone solution. *The Journal of Geology*, 65, 35-46

Kaye, C.A., 1959. Shorelines features and Quaternary shoreline changes, Puerto Rico. *United States Geological Survey Professional Paper*, 317B, 49-140

Kelletat, D., 1982. Hohlkehlen sowie rezente organische Gesteinsbildungen an den Kusten un ihre Beziehungen zum Meeresniveau. *Essener Geographischen Arbeit.*, Band 1, 1-27

Kempf, M. & Laborel, J., 1968. Formations de vermets et d'algues calcaires sur les côtes du Brésil. *Recueil des Travaux de la Station marine d'Endoume, Bull.*, 43, fasc. 59, 9-23

Kéraudren, B., 1970/1972. Les formations quaternaires marines de la Grèce. *Bulletin du Musée d'Anthropologie Préhistorique de Monaco*, 16, 5-154 (1970), 17, 87-169 (1971), 18, 245-282 (1972)

Kirk, R.M., 1977. Rates and forms of erosion on intertidal platforms at Kaikoura Peninsula, South Island, New Zealand. *New Zealand Journal of Geology and Geophysics*, 20, 571-613

Koba, M., 1974. Coastal geomorphology of Okino-erabu Island and postglacial sea levels (in Japanese, with English abstract). *Tôhoku Chiri* (Annals of the Tohoku Geographical Association), 26(1), 37-44

Konishi, K., Omura, A. & Nakamichi, O., 1974. Radiometric coral ages and sea-level records from the late Quaternary reef complexes of the Ryukyu Islands. *Proceedings of the Second International Coral Reef Symposium*, 2, Brisbane, 595-613

Kuenen, P.H., 1933. Geology of coral reefs. *The Snellius Expedition in the eastern part of the Netherlands East Indies*, Utrecht, 5, 2, 126 pp

Laborel, J., 1979a. Les Gastéropodes Vermetidés: leur utilisation comme marqueurs biologiques des rivages fossiles. In: *Les Indicateurs de Niveaux Marins*, ed. Groupe Nivmer, *Oceanis*, 5, Fascicule hors-série, 221-239

Laborel, J., 1979b. Discussion. In: *Les Indicateurs de Niveaux Marins*. ed. Groupe Nivmer, *Oceanis*, 5, Fascicule hors-série, 324-325

Laborel, J., Pirazzoli, P.A., Thommeret, J. & Thommeret, Y., 1979. Holocene raised shorelines in Western Crete, Greece. *Proceedings of the '1978 International Symposium on Coastal Evolution in the Quaternary'*, São Paulo, Brasil, 475-501

Mac Fadyen, W.A., 1930. The undercutting of coral reef limestone on the coast of some islands in the Red Sea. *The Geographical Journal*, 75, 26-34

Machida, H., Nakagawa, H. & Pirazzoli, P.A., 1976. Preliminary study on the Holocene sea levels in the central Ryukyu Islands. *Revue de Géomorphologie Dynamique*, 25, 49-62

McLean, R.F., 1974. Geologic significance of bioerosion of beachrock. *Proceedings of the Second International Coral Reef Symposium*, 2, Brisbane, 401-408

Mii, H., 1962a. Coastal geology of Tanabe Bay. *Science Reports, Tohoku University, Geology*, 34, 1-93

Mii, H., 1962b. Some ancient shore features. *Science Reports, Tohoku University, Geology*, Special Volume 5, 361-371

Montaggioni, L.F., 1979. Le problème de l'absence de hauts stationnements marins d'âge holocène dans l'archipel des Mascareignes, Océan Indien occidental. *Comptes Rendus de l'Académie des Sciences, Paris*, D, 288, 1591-1594

Montaggioni, L.F., 1979a. Application du principe de Bruun à la détermination des variations du niveau marin au course de l'Holocène: cas des îles Maurice et La Réunion, Océan Indien. *Marine Geology*, 31, M29-M38

Nakata, T., Takahashi, T. & Koba, M., 1978. Holocene emerged coral reefs and sea-level changes in the Ryukyu Islands (in Japanese, with English abstract). *Chirigaku Hyôron*, (Geographical Review of Japan), 51, 87-108

Nesteroff, W., 1965. Recherches sur les sédiments marins actuels de la région d'Antibes. *Annales de l'Institut Océanographique*, 43, 1-136

Neumann, A.C., 1966. Observations on coastal erosion in Bermuda and measurements on the boring rate of the sponge *Cliona lampa*. *Limnology and Oceanography*, 11, 92-108

Newell, N.D., 1961. Recent terraces of tropical limestone shores. *Zeitschrift für Geomorphologie*, Supplement Band, 3, 87-106

Ota, Y., Machida, H., Hori, N., Konishi, K. & Omura, A., 1978. Holocene raised coral reefs of Kikai-jima (Ryukyu Islands). An approach to Holocene sea-level study (in Japanese, with English abstract). *Chirigaku Hyôron* (Geographical Review of Japan), 51, 109-130

Pérès, J.M. & Picard, J., 1963. Nouveau manuel de bionomie benthique en Mer Mediterranée. *Recueil des Travaux de la Station Marine d'Endoume, Bull.* 31, fasc. 47, 1-137

Pirazzoli, P.A., 1974. Terrasses marines à Kikai-kima (Japon). *Photo Interprétation*, 6, 14-22

Pirazzoli, P.A., 1976. Les variations du niveau marin depuis 2000 ans. *Mémoires du Laboratoire de Géomorphologie de l'Ecole Pratique des Hautes Etudes*, 30, 1-421

Pirazzoli, P.A., 1978. High stands of Holocene sea levels in the northwest Pacific. *Quaternary Research*, 10, 1-29

Pirazzoli, P.A., 1979. Encoches de corrosion marine dans l'arc héllénique. In: *Les Indicateurs de Niveaux Marins*, ed. Groupe Nivmer, *Oceanis*, 5, Fascicule hors-série, 327-334

Pirazzoli, P.A., 1980. Formes de corrosion marine et vestiges archeologiques submergés: interprétation néotectonique de quelques exemples en Grèce et en Yougoslavie. *Annales de l'Institut Oceanographique*, 56, 101-111

Pirazzoli, P.A. & Montaggioni, L.F., in press. Late Holocene sea-level changes in the north-west Tuamotu Islands, French Polynesia. Pacific Ocean. *Quaternary Research.*

Pirazzoli, P.A. & Thommeret, J., 1977. Datation radiométrique d'une ligne de rivage a +2.5 m près de Aghia Roumeli, Crète, Grèce. *Comptes Rendus de l'Académie des Sciences, Paris*, D, 284, 1255-1258

Pirazzoli, P.A., Thommeret, J. & Y., Laborel, J. & Montaggioni, L.F., 1982a. Crustal block movements from Holocene shorelines: Crete and Antikythira (Greece). *Tectonophysics*, 86, 27-43

Pirazzoli, P.A., Montaggioni, L.F., Thommeret, J. & Y. & Laborel, J., 1982b. Sur les lignes de rivage et la néotectonique à Rhodes (Grèce) à l'Holocene. *Annales de l'Institut Océanographique*, 58(1), 89-102

Ranson, G., 1955. Observations sur les principaux agents de la dissolution du calcaire sous-marin, dans la zone côtière des îles coralliennes de l'Archipel des Tuamotu. *Comptes Rendus du l'Académie des Sciences, Paris*, 240, 806-808

Revelle, R. & Emery, K.O., 1957. Chemical erosion of beach-rock and exposed reef rock. *United States Geological Survey Professional Paper*, 260-T, 699-707

Russell, R.J., 1963. Recent recession of tropical cliffy coasts. *Science*, 139, 9-15

Sanlaville, P., 1977. Etude géomorphologique de la région littorale du Liban. *Publication de l'Université Libanaise, Section Etudes Géographiques*, I, Tome 1, 1-402, 43 pl.; Tome 2, 403-859; 9 maps

Schneider, J., 1976. Biological and inorganic factors in the destruction of limestone coasts. *Contributions to Sedimentology*, 6, 1-112

Stephenson, T.A. & Stephenson, A., 1949. The universal features of zonation between tide-marks on rocky coasts. *Journal of Ecology*, 37, 289-305

Stoddart, D.R., 1969. Geomorphology of the Marovo elevated barrier reef, New Georgia. *Philosophical Transactions of the Royal Society, London*, B 255, 383-402

Stoddart, D.R., 1978. Descriptive reef terminology. In: *Coral Reefs: Research Methods*, Unesco, 5-15

Takenaga, K., 1968. The classification of notch profiles and the origin of notches (in Japanese, with English abstract). *Chigaku Zashi* (Journal of Geography, Hiroshima), 77, 329-341

Teichert, C., 1947. Contributions to the geology of Houtman's Abrolhos, Western Australia. *Proceedings of the Limnological Society of New South Wales*, 71, 145-196

Thommeret, Y. & J., Laborel, J., Montaggioni, L.F. & Pirazzoli, P.A., 1981. Late Holocene shoreline changes and seismo-tectonic displacements in western Crete (Greece). *Zeitschrift für Geomorphologie*, Supplement Band 40, 127- 169

Torunski, H., 1979. Biological erosion and its significance for the morphogenesis of limestone coasts and for nearshore sedimentation (Northern Adriatic). *Senckenbergiana maritima*, 11, 193-265

Toyoshima, Y., 1965. Some wave-cut features in granitic regions, Uradome coast, Tottori Prefecture (in Japanese, with English abstract). *Tottori daigaku gakugeigakubu Kenkyû hôkoku* (Liberal Arts Journal, Tottori University), 16, 46-59

Trudgill, S.T., 1976. The marine erosion of limestones on Aldabra Atoll, Indian Ocean. *Zeitschrift für Geomorphologie*, Supplement Band, 26, 164-200

Umbgrove, J.H.F., 1939. De atollen en barrière-riffen der Togian-Eilanden. *Leidsche Geologische Mededelingen*, 1, 132-187

Verstappen, H.T., 1960. On the geomorphology of raised coral reefs and its tectonic significance. *Zeitschrift für Geomorphologie*, 4, 1-28

Vita Finzi, C. & Cornelius, P.F.S., 1973. Cliff sapping by molluscs in Oman. *Journal of Sedimentary Petrology*, 43, 31-32

Ward, W.C., 1975. Petrology and diagenesis of carbonate eolianites of northeastern Yukatan Peninsula, Mexico. In: *Belize Shelf - Carbonate Sediments, Clastic Sediments, and Ecology*, eds.K. F. Wantland & W.C. Pusey, *American Association of Petroleum Geologists, Studies in Geology*, 2, 500-571

Wentworth, C.K., 1939. Marine bench-forming processes. II, Solution benching. *Journal of Geomorphology*, 2, 3-25

13

Submerged forests as sea-level indicators

A. HEYWORTH

ABSTRACT

Submerged forests are defined, and their origin is discussed. Subjects considered are: a) the present-day relationship of trees (*Quercus* and *Pinus*) to tidal levels, b) the precision with which these relationships can be applied to submerged forest trees, c) methods of sampling the associated forest-bed peats for pollen analysis, and the correlation of pollen evidence with the trees themselves, d) identification of trees which may have been fortuitously preserved in submerged forests, although, when living, they had little relationship to sea level, and e) response of forest growth to transgressions and regressions. Conclusions reached are 1) that the lowest levels of tree growth can be closely related to a point between the local mean high water spring tide and highest astronomical tide and 2) that the preservation of trees in submerged forests is an accurate reflection of past sea levels.

INTRODUCTION

Submerged forests are here defined as assemblages of tree remains, clearly in their growth position, and so situated that they are regularly covered by the tide. These forests are usually found as peat beds, containing the stumps and trunks of trees, which outcrop on the foreshore. They are, in many cases, visible only when beach levels are low, being covered, for much of the year, by sand or shingle. The above definition can be widened to include the buried inland extensions of such forests, and also those beds which, whilst clearly originating in the same way as submerged forests, are now, as a result of uplift, above the reach of the highest tides.

This chapter is concerned specifically with submerged forests in the British Isles, where the trees present are the normal terrestrial species, such as oak, pine, and alder.

Similar submerged forests occur elsewhere in the temperate zone, and have a similar relationship to past sea levels (SLs). This relationship depends on the fact that the trees could not survive if subject to more than occasional tidal cover. The same considerations do not apply to species showing adaptation to an intertidal environment, so that submerged peats derived from coastal mangroves cannot be regarded as strictly analogous features.

OCCURRENCE

Submerged forests of various ages are of widespread occurrence in Britain. They are *absent* from three main types of coastline: 1) where changes in estuarine channels have removed the beds; 2) on a receding coastline where a coastal alluvial belt is now, or in the past, absent; and 3) where accretion has caused the coastline to advance across the earlier intertidal area, so that any submerged forest beds which might have been present are now well inland.

Two main types of submerged forest may be recognised. Firstly, there are those sites where trees growing on pre-Holocene deposits were drowned by the rising sea, so that a continuous basal submerged forest bed rests on an undulating surface, the age of the bed decreasing with increasing altitude. The second type is seen where trees grew out across a more or less horizontal Holocene alluvial surface, on an accreting coastline. A very slight rise in SL was sufficient, in this instance, to drown trees over a wide area, and produce extensive horizontal submerged forest beds. A forest of this type, at Borth, in Cardigan Bay, Wales, is shown in Figure 1. The possibility of subsequent compaction of the underlying clays must be borne in mind at such sites.

It is clear, in many instances, that a coastal barrier, which was originally between the growing trees and the sea, has subsequently been pushed up and back by wave action, across the forest bed, which now appears on the foreshore. In other cases, shingle barriers are seen to have been overtaken by the rising sea, and largely left behind by it, so that the submerged forest is found behind the 'fossil' shingle ridge on the beach, at a lower level than the present storm beach (Kidson & Heyworth, 1973).

ORIGIN

Submerged forest trees were clearly overtaken by a rising water table shortly after their death, and thereby preserved. Often the trees are embedded in peat of carr or fen origin, showing a constant relation to a rising water table. Sometimes the trees are rooted in peat, but more often in a soil, or in estuarine silts and clays. The relationship of the rooting horizon to the water table is the same in each case. It is the almost perfect state of preservation of most submerged forest trees which is the immediately apparent indication of a rising water table. Whether or not the

Figure 1. The submerged forest at Borth, Cardigan Bay, Wales. Most of the trees in the photograph are pine. The forest bed lies on an estuarine clay surface, which is very nearly horizontal over a distance of *c.* 3 km. The roots of the trees are confined within a vertical range of *c.* 15 cm. The clay surface was first colonised by marsh species, followed by the pines, and then by alder carr and fen, which buried the tree stumps. Much of the upper peat has subsequently been removed by the sea. These trees give radiocarbon dates of *c.* 5000 BP.

trees were killed as a result of such a rise, and whether this rise was directly related to SL rise, are more complex problems.

The first point which needs to be established, however, is the present-day relationship of similar species to SL. As the trees concerned are not specifically coastal species, the important datum is clearly the lowest point, relative to SL, at which the trees are able to grow. There appears to be no possibility that the trees could have changed their environmental requirements during the period concerned. No opportunity has arisen for adaptive changes, such as the development of salt-tolerant strains, since the species have not been confined to a coastal belt. Regeneration and recolonisation from an inland reservoir have been continual throughout the Holocene.

Many authors have apparently assumed that the presence of mature trees in a deposit implies that the sea must have been appreciably below the level at which they grew, or at a considerable distance. Reid (1913), in discussing the submerged forest beds of South Wales, refers to 'a true forest growth, such as could only live at an elevation clear of the highest tides', and suggests that the trees must have grown 'a few feet' above the highest level of the salt marshes 'in order to keep the oak roots well clear of the highest tides during a westerly gale. An exceptional gale occurring only once during the lifetime of an oak might bank up the sea water sufficiently to kill a tree, if it grew at a lower elevation'.

Subsequent authors have largely followed this reasoning. However, work on present-day coastal woodlands shows that oak trees (*Quercus robur*) grow, and are perfectly healthy, at the upper limit of the salt marsh. Their roots are below the level reached by several high tides each year, and their trunks are submerged to a depth of up to 1 m by exceptionally high storm levels, many times in the life of a tree. In Morecambe Bay, NW England, where mean high water spring tide (MHWST) is at *c*.5.3 m above OD (Ordnance Datum), the lowest oaks are found where the surface is at a height mid-way between this level and that of highest astronomical tide (HAT) (Kidson & Heyworth, 1979). In the estuary of the Dyfi (Dovey) in Cardigan Bay, the lowest oaks again occur at a level mid-way between MHWST and HAT, although MHWS at this site is at *c*.3 m above OD. The roots of these trees are, of course, lower than this, and below tide level for longer periods. Other sites investigated confirm that, regardless of the tidal range, the lowest trees are found at a height mid-way between MHWST and HAT. The significance of this particular height is that it is the lowest level at which there are periods of three or four months free of tidal cover in most years, enabling tree seeds to germinate and become established (*op. cit.*). A level as little as 10 cm above the same datum is covered by the tide at only very infrequent intervals. It appears that the lowest level at which oak trees will grow can be predicted to ± 10 cm. It should be noted that the lowest full-grown trees in these coastal woodlands are oaks, although small specimens of alder (*Alnus glutinosa*) and willow (*Salix* spp.) are sometimes found a few metres seaward and a few cm lower than the oaks.

The ability of oaks to withstand prolonged immersion in sea water appears to be due to the ability of the root hairs to control the entry of ions to a greater extent than other species. The soil water surrounding the roots is clearly not sea water, but must be more or less brackish, depending on the fresh-water table, and the permeability of the soil. At the top of the tidal range neither the periods of tidal cover nor the head of sea water are sufficient to allow the salt water to completely displace the fresh water already in the soil. Germinating acorns are, therefore, most susceptible to killing by salt water before they have taken root.

It should be remembered that in winter, which is usually the period of most severe storms, the soil will be saturated with fresh water. Flooding by sea water during the summer would be much more likely to be fatal.

Figure 2a. Oak tree in the Dyfi estuary, Wales. The tide level shown is just above the mid-point of mean high water spring tide and highest astronomical tide. This tree shows no ill-effects from the regular submergence of its roots by sea water.

The same considerations are also important in deciding the way in which the submerged forest trees were killed. Figure 2a shows an oak tree on the shore of the Dyfi estuary. The level of the tide shown in the photograph was at *c*.10 cm above the mid-point of MHWST and HAT. The tree shows no sign of any ill-effects from the salt water.

Figure 2b shows oaks some 100 m from that in Figure 2a. In 1977, a sea wall was built across the mouth of an inlet, leaving the tree in Figure 2a on the seaward side, and the trees in Figure 2b on the landward side. This sea wall, of clay and shingle, is seen in the background of Figure 2b. The water in the drainage channel shown is only very slightly brackish, and is maintained at a more or less constant level, which is very close to that of the tide in Figure 2a. This level is that of the surface of the silt in the embayment, which was, prior to the construction of the sea wall, built up to the level of high spring tides. Areas of sward are maintained at a level a few cm above the water table.

Figure 2b. Oak trees at the same altitude as that shown in Figure 2a. and some 100 m distant from it. An embankment has been constructed across the mouth of the inlet, and the resultant ponding up of water has, in less than two years, killed the trees shown in the photograph, although the tide no longer has access to them.

Within two years of the completion of the sea wall, the trees adjacent to the channel were dead. This confirms the deduction, based on studies of submerged forests, that oaks in this situation were killed not by tidal immersion in salt water, but by a rise in the water table, which caused the roots to be constantly immersed in stagnant acid water. Other species, notably alder, which are adapted to waterlogged conditions, would probably be able to colonise areas previously occupied by oak woodland. It is also possible that, at some sites, oaks could recolonise the area to a limited extent, the new trees producing only very shallow roots, and avoiding the anaerobic conditions which killed the deeper roots of their predecessors. Much would depend on the rate of SL rise at the time.

During periods of appreciable rise, the cause of submerged forest formation can be suggested, following the pattern suggested by the modern equivalent.

As the sea rose, the coastal barrier would be pushed back, and up the slope of the previous land surface. This slope might be quite steep, or, where trees had colonised a marsh surface, very slight. On the landward side of the coastal barrier, a belt of ponded-up water would move ahead of it, killing the trees as it advanced, and producing conditions favourable to the preservation of their remains. This belt would be narrow on a steeply sloping surface, and very wide on an almost horizontal area.

Trees which did not fall, and were thus left protruding above the waterlogged horizon, would quickly disappear as a result of insect, fungal and similar biological activity.

Oaks would be succeeded by alder carr and fen species, which might persist for hundreds or thousands of years, as peat accumulation kept pace with the rise in the water table. This would explain the commonly encountered stratigraphy of submerged forest beds, where the remains of large trees are buried in peat consisting largely of the remains of *Alnus, Salix, Phragmites,* etc. In some cases, particularly when SL rise was rapid, or during severe storms, the landward movement of the coastal barrier would not suffice to prevent it from being overtopped or breached, and the fen area would then be inundated, and brackish silts and clays would cover the submerged forest.

Where no coastal barrier existed, submerged forests would be formed only on low-energy coasts, where the trees grew at the upper limit of extensive salt marshes. In these areas, the surface of the marsh would rise at the same rate as SL, and trees would be preserved by burial in salt-marsh peaty clays or silts.

Not all submerged forests are of oak. Many consist largely of the remains of pine (*Pinus sylvestris*). This species is more susceptible to salt water even at low concentrations than is oak. Periods when coastal pine forests were able to flourish, and to colonise alluvial surfaces were, therefore, relatively brief. The seaward limit of pines is, in general, farther inland than that of oaks, but the vertical difference between the species, as SL indicators, is very small, probably a few cm, since the conditions necessary for preservation of the tree remains are the same for both species.

SAMPLE SELECTION AND IDENTIFICATION OF REPRESENTATIVE ALTITUDE

Submerged forests are probably the most easily recognised indicator of SL change. There are occasional intertidal outcrops of buried Pleistocene forest beds, but there is no real danger that these could be confused with Holocene deposits.

Tree remains in submerged forests are usually in a remarkably good state of preservation, and the species are easily identified.

Most submerged forest trees are quite clearly in their growth position, the stumps being firmly rooted in their original substrate. Many fallen trunks do occur, however, and if these are isolated specimens, they should be avoided in sampling, since they may have been drifted to their present position, either shortly after their death, or by later erosion of the forest bed. On the other hand it is rare for an *in situ* stump to be more than *c*.1 m high. Large trunks are found only in a prostrate position. These usually have their roots attached, having been uprooted from the soft underlying layer, either before, or shortly after they died. The original rooting horizon of these trunks can usually be identified from adjacent stumps, and they can then be safely used for dating purposes. These fallen trunks often provide a better source of dateable material than do the stumps, which may be considerably eroded, so that the bark and outer rings may be missing as seen in Figure 1. This is important if ring-width measurements are to be made (see Heyworth, this volume).

Samples of wood can be taken from any level of a submerged forest trunk or stump, so that the altitude of the sample itself may not have any particular significance. The altitude which is important is that of the contemporaneous water table. In many cases submerged forest trees have a horizontal spread of roots, with a vertical range of perhaps 10-15 cm. The trees were killed only when these roots were submerged, so that the significant altitude is that of the horizontal plane of their upper surface. If an extensive horizontal spread of roots is not visible, then the apex of the angle formed between the trunk and the diverging roots is, in most cases, a sufficiently accurate altitude to use. This makes some allowance for the fact that, at a time of rapid SL rise, trees would probably have survived until they were brought to a point slightly lower in the tidal range than is the case today. It might be argued that deep-rooted trees could be killed by a rise in the water table well below ground level, but in this case the trunks would not have become waterlogged, and so would not have been preserved. Such trees might, in any case, produce new roots at a higher level, with the same relationship to the new water-table.

If the tree roots are embedded in, or overlain by peat, a vertical section should be dug through the peat, exposing the rooting horizon, and peat samples should be taken, at not more than 1 cm vertical intervals, from the whole of the section, for pollen analysis. It is often possible to identify the layer within the peat which contains the pollen from the preserved trees. This may be from 0 to perhaps 50 cm above the roots themselves. It must be borne in mind that, for example, an oak pollen maximum, may not be derived from the immediately adjacent trees, but from a surrounding woodland. However, in a submerged forest site it is almost always the case that, as the growth was controlled by a horizontal water table, the pollen peaks resulting from individual trees are very nearly synchronous. Where the submerged forest bed rests on a steeply sloping surface, greater caution must be exercised in the interpretation. It must also be remembered that trees may shed little pollen in their first 50 or so years, and may then be prolific for a further 250 or more years.

Pollen analysis of the peat matrix will also be of great help in clarifying the relationship of the water table to vegetational change and to peat formation.

Samples for dendrochronological investigation (see Heyworth, this volume) should be taken from several trees in the same area, so that small differences in altitude can be related to differences in the dates of individual trees. The lowest level at which trees were growing at a particular date is the altitude which should be used for the deduction of SL. Trees can, of course, die for reasons unconnected with SL or water-table movements, and some individuals may be more susceptible than others to such changes. There seems no possibility, however, that any could survive at a level appreciably below that of the mid-point between local MHWST and HAT.

Unrepresentative samples

Submerged forest beds which are buried conformably by fen or carr peat, or by estuarine silts and clays, were undoubtedly formed as a result of a rise in the SL-determined water table. Some submerged forests, however, are overlain by *Sphagnum* peat indicative of a raised bog. It is possible in these cases, that the trees were killed, not by a SL-determined rise in the water table, but by the accumulation around them of the *Sphagnum* peat. Particular care should be taken, in such situations, to determine, from the evidence of pollen and macroscopic remains, whether the growth of *Sphagnum* peat was initiated by a rise in the water table, which also killed the trees, or whether it was brought about by climatic and edaphic factors.

In coastal locations it is likely that some trees would be overtaken and killed by encroaching dunes. Such trees would probably be less well-preserved than those killed by water logging, but it might be possible to confuse the two. Trees associated with aeolian deposits should, in general, not be used, as their relationship to the water table is very uncertain.

EVIDENCE OF TRANSGRESSIONS AND REGRESSIONS

During a regressive phase, whether or not this was due to a fall in SL, the woodland margin would migrate seawards, and unless there was renewed transgression, the trees which had colonised marsh surfaces would not be preserved by inundation or by burial in fen peat. No evidence of such an episode would be left behind, unless the trees were overtaken by the seaward extension of a raised bog. Some areas of the inland equivalent of the submerged forest at Borth (Fig. 1) are buried by *Sphagnum* peat in this way. (Godwin, 1956; Wilks, 1977).

In general, if mature trees are found to be overlain by peat of alder carr, fen, or marsh origin, a rising water table, possibly representing a transgressive phase, is indicated. Trees rooted in estuarine sediments, and overlain by *Sphagnum* peat, imply a regressive phase.

It should be borne in mind that a marine transgression resulting from a local coastal change, such as the destruction of a spit, rather than from a rise in SL, is unlikely to leave evidence which could be confused with that produced by such a rise. On the other hand, a local regression may leave traces very similar to those which would result from a fall in SL. This distinction is due to the fact that a regression is associated with, and may, indeed, be the consequence of, the accumulation of sediment, whereas a local transgression merely results in salt water gaining access to sites which were previously fresh water, whilst leaving the water table within this zone unchanged.

A local flooding episode cannot, other than briefly, raise the water table above that determined by SL at the site; the water will simply drain away. Sediment, on the other hand, can not drain away. Sands, silts and clays, deposited by normal coastal processes may remain in their new position indefinitely, and areas previously at, or below the water table may be raised above it, either gradually or rapidly.

There are three principal ways in which a local regression may be recorded in the sediments:

1) Sediments laid down in salt or brackish conditions may be replaced by fresh-water sediments as the coastline advances. Clear evidence of the water table should be evident in the sediments.

2) Areas previously subject to occasional tidal inundation, but above the general water table, may be colonised by trees. Conditions may be dry enough to prevent further peat formation, resulting in a gap in the water-table record, during which an unrecorded fall and recovery of the level might have occurred.

3) On an advancing coastline, storm-beach material or blown sand may accumulate in areas previously occupied by marsh or intertidal zones. However, on a retreating coastline, marshes behind a coastal barrier may be infilled by such material, and a transgression may thus be associated with regressive sequences in the sediment.

The confusion between a local regression and a fall in SL arises only if studies are confined to individual sedimentary sequences, where a lessening of marine influence is interpreted as evidence of a fall in SL.

If deductions are based only on dated horizons, with the same or comparable relationships to a water table then many misleading conclusions, based on the imperfect interpretation of sedimentary sequences, will be avoided.

When attempting to distinguish regression phases, the lowest trees in the assemblage should again be dated, these being the first to colonise areas which were previously fen or marsh. In such instances the important date is that of the establishment of the seedling tree. The date of the death of the tree is largely irrelevant, if preservation was the result of burial by ombrogenous peat. The central parts of submerged forest trees are almost always well-preserved, with wide rings, so that ring-width matching of the early years of growth, is often relatively easy.

If the presence of long-lived trees in a submerged forest shows a long period free from transgression, although the trees were eventually preserved by a rise in the water table, then it is possible that SL fell and rose again during the lifetime of a tree. The death of the tree and its subsequent preservation are the most reliable indicators of the water table. As many trees as possible, therefore, should be examined, to determine whether, during the life-time of long-lived trees, other nearby specimens at the same altitude died and were preserved. If this proves to be the case, then an appreciable fall in SL can be ruled out.

REFERENCES

Godwin, H., 1956. *History of the British Flora.* (Cambridge University Press)

Heyworth, A., (this volume). Dendrochronological dating

Kidson, C. & Heyworth, A., 1973. The Flandrian sea-level rise in the Bristol Channel. *Proceedings of the Ussher Society,* 2, 6, 565-584

Kidson, C. & Heyworth, A., 1979. Sea 'level'. *Proceedings of the 1978 International Symposium on coastal evolution in the Quaternary,* eds. K. Suguio, Th.R. Fairchild, L. Martin & J.-M. Flexor, (Sao Paolo, 1979), 1-28

Reid, C., 1913. *Submerged forests.* (Cambridge University Press)

Wilks, P., 1977. *Holocene sea-level change in Cardigan Bay.* Upublished PhD thesis, University of Wales

14

Analysis of botanical macro-remains

K.-E. BEHRE

ABSTRACT

Analysis of botanical macro-remains can contribute in various ways to the study of vertical and horizontal sea-level movements and their ecological consequences. Sources for investigation are mainly peat, wood remnants from river banks, and the well-preserved material from archaeological excavations in coastal areas. Fruits, seeds, and different kinds of vegetative plant remains can be identified mostly to species level.

By the use of certain indicator species, or even more reliably, by the reconstruction of the fossil plant communities, it is possible to determine the degree of salinity and the position relative to local mean high water, both of which are reflected by the vegetation.

The plant sociologic evaluation leads to the reconstruction of the paleo-environment and its changes in the course of trans- and regressive phases. These cycles can be traced close to the coast as well as in the perimarine area of the hinterland. In each case, however, interpretation has to be done very carefully to separate the real changes of sea level from other possible factors such as rate of sedimentation, compaction, influence from the hinterland, and the inflow of fresh water from the rivers.

INTRODUCTION

On flat coasts, the easily recognized boundary of the closed cover of higher plants is a good indicator of the local mean high water (MHW) level. Below this line only very few higher plants grow, advancing as pioneers into the regularly submerged tidal flat area. Above local MHW there is usually a well-defined succession of halophytic communities, ranging from the lowest, which are subjected to frequent flooding by salt water, to those on higher levels where the interval between the floodings becomes progressively greater. The vegetation of vascular plants, therefore, reflects the position of a certain place in relation to local MHW, up to the highest areas that are reached by salt water only during storm floods.

Under natural conditions the elevated levees of the river banks further inland are covered with a succession of forest communities, the species composition varying according to the frequency of submergence.

In protected areas and under natural, ie. undyked conditions, there is a brackish transitional zone which extends up to the areas with a fresh-water regime. Salinity, water level and other factors can be recognized there by the changing composition of the vegetation.

The establishment of plant cover and the initiation of soil formation in a coastal area normally suggest that sea level (SL) has ceased to rise and a regressive development is taking place. Exceptions are the outer rim of the salt marshes, and river banks, where the level of deposition exceeds MHW as a result of special depositional conditions. Along the seaward rim of the salt marshes sedimentation takes place under short-lasting storm-flood conditions with extremely high water levels, whereas soil formation is going on during most of the time and under normal coastal processes. On the river banks sedimentation takes place only during phases of extreme fluvial discharge.

It is easy to understand that fossil plant remains, as for example peat, may be primary indicators of former SL changes and are essential to the reconstruction of the paleo-environments in coastal areas.

The advantage of macro-fossils compared with pollen analysis is the greatly increased possibility of determination, in most instances to genus and in many instances to species level. Another important advantage is that macro-fossil remains as, for example, stems, seeds, and wood were deposited at the very place they were produced or at least not too far from it, while on the contrary, pollen is normally dispersed over a large area.

The main sources of botanical macro-remains are peats, particularly intercalated peats in clay or sand profiles, wood remains from borings or exposures in river banks, and organic debris from archaeological excavations in the coastal area. Not included here is the investigation of tropical mangrove forests (between 30° North and 30° South).

In the context of SL changes, plant macro-fossil analysis is expected to answer two main questions: firstly, the degree of salinity in the area and secondly, the position of the respective layers with respect to groundwater, MHW and storm-flood levels. In both cases the presence or absence of indicator species and the reconstruction of the paleo-environment enable us to draw the appropriate conclusions.

In detail, all data have to be checked critically with regard to the specific local situation concerning the relationship to the sea or to rivers, possible compaction of the sediment and other factors, as will be shown later.

TECHNIQUES OF SAMPLING, PREPARATION, AND IDENTIFICATION OF PLANT MACRO-FOSSILS

1. Peat beds

Peat formation takes place under conditions ranging from fresh water to moderately brackish water, but it does not occur in places where the tidal range is too great. The kind of bog vegetation community is dependent on different factors: such as salinity, the general nutrient content, and particularly the hydrography, ie. the relative situation to the local mean water level. The communities succeed each other according to the so-called 'verlandungs' series, beginning with reed growth, followed by sedge and carr peat and, in places later on, by raised bog peat.

The composition of peat beds, therefore, enables us to reconstruct the paleo-environment with regard to those factors which depend on SL changes. In order to do so, we need an exact knowledge of the peat components, based on the identification of the plant macro-remains.

Two main sequences containing peat beds can be distinguished:

a) Intercalated peats

So-called 'schwimmende Torfe', ie. one or several peat layers bounded above and below by minerogenic sediments, often clay. This sequence frequently occurs in areas near the modern coast line; the macro-fossil analysis gives evidence of the degree to which a fresh-water environment developed at a certain time, the latter being provided by pollen analysis or radiocarbon dating.

Sampling of these peats is usually done by coring, and also under favourable circumstances from exposures. The majority of peats, eg. reed, sedge, carr and raised bog peats, can be recognized and classified in the field. Normally, therefore, it is possible to determine, by extensive borings, the horizontal distribution of a particular peat type, and the position of transitions from one peat type to another. Some of the cores have to be retained and from these a number may later be selected for further investigation in the laboratory.

It is important that the position of the rooting zones in the clays above and below the peat should be carefully noted in the field. They can be found by frequently breaking apart the core segments. A sharp distinction has to be made, however, between vertical roots from different plant species and the horizontal rhizomes of *Phragmites*. The latter, which are frequently encountered in clay layers below reed peat, belong to plants that lived several decimetres higher and they were driven into the clay (long) after its deposition. As confirmed by subsequent diatom analysis, these organic constituents, often together with a certain humus content, may show transitional brackish conditions at the end of the minerogenic deposition, before peat formation started. Within the peat, roots from brushwood horizons may run into the underlying peat layers.

b) Continuous peat profiles from the underlying pre-Holocene deposits to the surface.

Continuous peat sections are developed in the perimarine backswamp area, and there is a gradual transition from the (seaward) zone of the intercalated peats. These backswamp profiles reflect a paleo-environment in which water-level changes are mainly governed by SL fluctuations.

Sampling of peat by coring may sometimes be difficult if the peat is too wet, so it may be necessary to use a Livingstone piston corer.

All peat samples can easily be conserved by an addition of a 5% solution of formaldehyde and by subsequent storing in air-tight sealed plastic bags which prevent them from drying out. If possible, storage in a cold-room at a temperature of +2 to +4°C is recommended.

Before investigation, the material is broken up and softened for a number of days in a 10% KOH, until it readily falls apart. If necessary, it may be heated or boiled. It is then sieved, using soil sieves with mesh sizes 3 mm, 0.75 mm and 0.25 mm. To ensure that the small seeds of *Juncus*, *Calluna*, etc. are not lost during sieving, the smallest mesh size should not be greater than 0.25 mm; indeed mesh sizes of 0.2 mm and 0.1 mm are to be recommended to ensure retention of the smallest seeds, but fine mesh sieves tend to clog. For further details of sampling and investigation procedures the reader is referred to the papers of Wasylikowa (1979) and Grosse-Brauckmann (1982).

Peat samples may sometimes be difficult to work with, as the main components are often fine stems, mosses, leaf remnants, and other soft tissues which have to be separated very carefully under the stereomicroscope. For the identification of these peat remnants there exist only a few suitable works. Particularly helpful are the papers of Grosse-Brauckmann (1972, 1974) and the book by Katz, Katz & Skobeyena (1977), which contain many descriptions and photographs of the vegetative parts of higher plants as well as mosses.

2. River banks

Under natural conditions the elevated levees along the lower course of the tidal rivers are covered with river bank forests ('Auenwälder'), the zonation of which is adapted to the frequency of submergence, that is, to the relative height above MHW. These river-bank forests comprise the low-lying willow belt and the more elevated tidal forest.

Individual pieces of wood, as may sometimes be found during coring in the clays which constitute the banks of the larger rivers, are frequently not given due attention. These pieces provide information as to the former woodland cover of the river banks and, as will later be illustrated, enable us to draw conclusions as to the local MHW level. Of course, one cannot be certain, from coring alone, that the wood samples really grew where they were found, or whether they were redeposited. But the experience gained from large exposures, or from archaeological excavations in the elevated levees, shows that the great majority of those pieces belong to tree trunks with roots or to the roots themselves.

Occasionally, well-developed river bank woodland facies may be encountered, from which not only wood, but also much leaf material and fruits may be recovered, thus supplementing considerably the determinable macrofossil material.

Sampling of wood for species identification is easy. Only small pieces of a few cubic centimeters are necessary. They must be kept moist until investigation. For identification, tangential, radial, and cross sections are readily made, using a simple razor blade rather than a microtome. Sections of rare or important species should be mounted in glycerine and retained. Several excellent atlases with keys are available now. For Europe I mention only the works of Greguss (1959) and Grosser (1977).

Wood, as macro-fossil material, has the disadvantage that it can seldom be determined to species level. In limited areas, as for instance northern and central Europe, however, most genera are represented by few species only, and then, in many instances, the ecological requirements of the individual species are such that a determination to genus level suffices.

3. Plant remains from archaeological sites

Archaeological excavations in coastal areas contribute a variety of important kinds of data to the question of SL changes, but only the value of macro-fossil analyses from these sites will be discussed here. Around the southern North Sea in particular, many archaeological excavations in water-logged sites have recently been carried out. The generally excellent conditions of preservation in both the settlements on flat ground ('Flachsiedlungen') and those on artificially elevated mounds ('Wurten', 'Terpen') have resulted in the conservation of large amounts of macroscopic plant material from various prehistoric times. For the problems connected with SL changes, the botanical remains from these excavations are of particular significance for two reasons. Firstly, they consist predominantly of plant remains from mineral ground which, of course, is never represented in peat profiles; indeed in areas of saline conditions peat accumulation cannot take place. Secondly, the greater part of the material recovered from archaeological sites dates to relatively recent times, during which peat growth in coastal areas was insignificant. The archaeological sources, therefore, complement the material from geological strata both horizontally and vertically, that is, spatially and chronologically.

The problems associated with and the procedures to be followed in sampling at archaeological sites are quite different from those described above. The plant remains here have generally not grown on the site itself but rather have originated in the surrounding pastures, fields, and natural communities. They therefore represent different biotopes and may well have been mixed at the sampling site. The excavation should be carefully examined by the researcher him-/herself with a view to finding locations with suitable homogenous material, for example hay or litter layers. Strictly regular or random sampling of the excavated surface or sampling by vertical monoliths encompassing a number of habitation layers cannot be recommended from both the point of view of purity of sample and of dating. In prehistoric settlements spanning several habitation phases, where older strata are frequently disturbed during later occupations, particular attention should be paid to ensuring correct dating of the exact position of the find. This may be done by pottery, metal finds, dendrochronology or radiocarbon assay, depending on the particular circumstances.

Due to the desirability of obtaining homogenous material, the size of the samples may differ. Normally their volume ranges from 100 to 1000 ml. The samples are preserved in the same way as peat samples with 5% formaldehyde, and sealed in plastic bags. Cool storage is recommended, and the further treatment is also similar to that for the peat samples (see page 416).

The most time-consuming part of the investigation is the recognition and picking out of the fruits, seeds etc. from the separate fractions by means of a stereomicroscope. As to the finest fraction, it normally suffices to investigate part of it (1/4-1/100). This fine fraction usually contains many seeds which, however, represent only a small number of species (1/100 of this fraction may yield up to 2000 seeds of *Juncus gerardii* for instance). After determination, the fruits and seeds are preserved in a solution of alcohol, glycerine and water in the ratio 2:1:1.

There are several guides to the identification of

fruits and seeds, the most important being those by Beijerinck (1947), Berggren (1969, 1981), Bertsch (1941), Katz, Katz & Kipiani (1965), and Martin & Barkley (1961). In recent years, the botanical investigation of archaeological sites in coastal areas has resulted in extensive publications which include many illustrations and descriptions of halophyte and wetland species, eg. Körber-Grohne (1967), Behre (1970, 1976) and Van Zeist (1974); consultation of these might be useful. In addition to such literature a comprehensive reference collection of fruits and seeds must be regarded as essential.

EVALUATION AND PRESENTATION OF RESULTS

1. Salinity and height relationship to water level

Plant macro-fossil analysis yields results relevant to many different aspects of the history of the vegetation and paleo-environment. With respect to SL changes we have to concentrate on the main facts which enable us to draw conclusions on this subject. This method should provide answers to two important questions:

a) the degree of salinity in the area (ie. in soil and water); and

b) the relative height of the respective layers with regard to mean water level (*c.* groundwater), local MHW and storm-flood levels.

2. Indicator species

The presence or absence of indicator species with different relationships to the single factors yield fixed points on the continuous ecological scale. For a careful classification of individual species with respect to salt concentrations and groundwater levels the publications by Iversen (1936), Gillner (1960), and Ellenberg (1979) should be consulted.

With reference to salinity, the plant species can be divided into halophytes and glycophytes (fresh-water plants). According to their different salt tolerance, the halophytes can be further differentiated into euhalobic, mesohalobic and oligohalobic species.

Table 1 gives a small selection of species the remains of which frequently occur in macro-fossil analyses. They are arranged according to increasing salt tolerance from top to bottom in the list. As salinity is caused by flooding, this order also reflects the position relative to MHW.

A vertical range of the occurrence of the indicator species in relation to MHW can be given only on a local basis, because it depends on the frequency of submergence (ie. tidal and stormflood range, weather effects, *cf.* Ranwell, 1975) and the salt concentration of soil and water in the respective areas.

Table 1. Some indicator species for fresh-water and saline conditions. The order follows the increasing salt tolerance from top to bottom.

Species		
Glyceria fluitans		glycophytes
Filipendula ulmaria		glycophytes
Alisma plantago-aquatica		glycophytes
Lythrum salicaria		glycophytes
Lycopus europaeus		glycophytes
Lychnis flos-cuculi		glycophytes
Juncus gerardii	mesohalobic	halophytes
Triglochin maritimum	mesohalobic	halophytes
Aster tripolium	mesohalobic	halophytes
Puccinellia maritima	euhalobic	halophytes
Limonium vulgare	euhalobic	halophytes
Salicornia europaea	euhalobic	halophytes

In Figure 1 an application of the indicator species method is given as an example of how to present the results in the form of pie graphs. Analysis of macroscopic plant material from suitably located sites on the lower Ems river and its estuary, all dated to the 1st to 3rd century AD, allowed the determination of the configuration of the former coastline, by reconstructing the degree of salinity. To indicate the degree of salinity prevailing at each site, the presence of diagnostic halophyte/glycophyte species is shown (altogether 24 indicator species were used, *cf.* Behre, 1979). It is clear or evident that there was a gradual change in vegetation and, therefore, in salinity. At Bentumersiel, Jemgumkloster, and Boomborg, for example, the Ems waters were only slightly brackish, while at Pilsum strongly saline conditions prevailed. It can therefore be concluded that the zonation in vegetation here during the Roman period was basically similar to that prevailing today, even though the Dollart bay had not yet been formed.

The same method can be applied to SL movements not only horizontally but also vertically, as far as suitable samples are available. One example can be given here. The site of Boomborg (Fig. 1) shows 22% of halophyte indicator species for the Roman period at a surface level around the settlement of about +0.40 m above ordnance datum. Below this site there is a settlement of Older Iron Age (ie.about 800 years earlier) beginning at a level of -0.70 m below ordnance datum with no halophytes at all. This means, that during the interval there must have been a considerable rise of MHW and, in consequence, a pronounced increase in water salinity.

As is generally known, there are also species indicating the relation of their altitude to mean water level

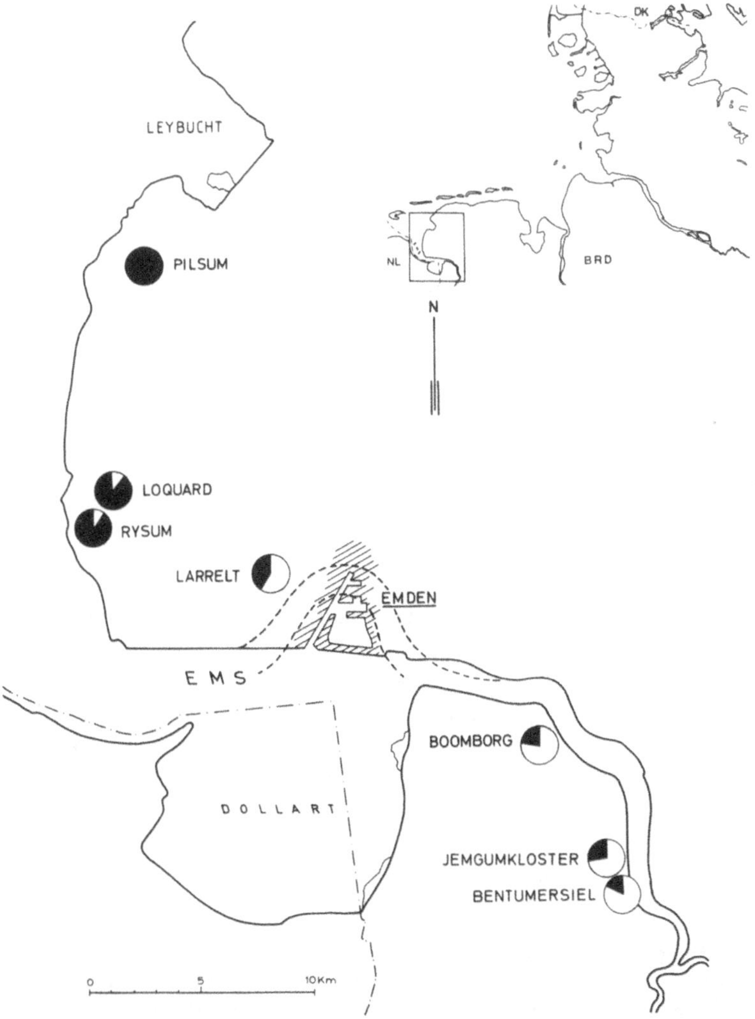

Figure 1. Ratio of characteristic halophytes (salt-marsh species, black) to specific glycophytes (fresh-water species, white) from several archaeological sites of the Roman period.

in fresh-water dominated areas. They are constituents of the peat forming plant communities or of the peat and wood communities above local mean water level.

3. Plant communities

If possible, changes of SL, or the related water level in perimarine areas, should be determined by the reconstruction of plant communities. In fossil material good indicator species may not always be represented or the species

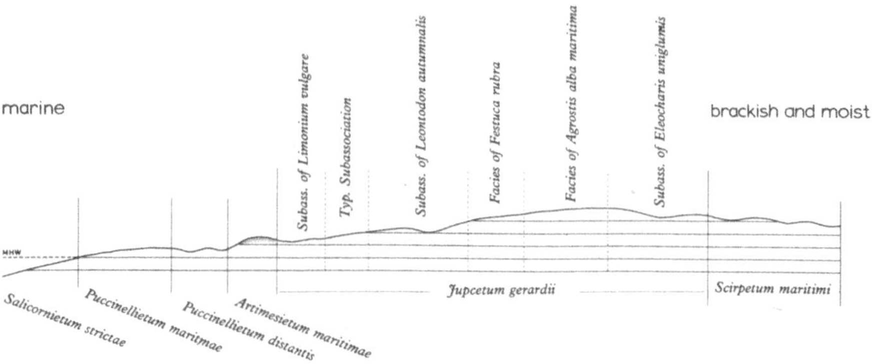

Figure 2. Schematic representation of the salt-marsh community series (*Armerion maritimae*) on the coast of the southern North Sea.

recorded may have reliable indicator values at one end of the ecological scale only. In order to find out about SL movements it is therefore recommended that reconstructions of the *plant communities* are attempted and that these are used rather than individual species to reconstruct changes in salinity conditions and groundwater level.

So far, fossil salt-marsh communities have been reconstructed only on the basis of material recovered from prehistoric settlement sites. Fossil communities corresponding to the present-day *Puccinellietum maritimi* or *Juncetum gerardii* enable the determination of the position above MHW of the farmed area around the dwellings. In the *Armerion maritimae* or salt-marsh communities series (*cf.* Tüxen, 1957; Westhoff & Den Held, 1975), the species composition responds to changes in ecological conditions effected by differences in elevation of a few decimetres.

In this case also, tidal range and other factors which have been mentioned already in connection with the indicator species (see page 419) allow only local or regional fixation of the exact height of certain plant communities to MHW. The succession of salt-marsh communities for the German North Sea coast is shown in Figure 2; others are published by Chapman (1974) for Norfolk, Gillham (1957) for NW Scotland, Johnson & York (1915) for the Northeast of the USA, or summarized by Beeftink (1977) for West and North Europe or MacDonald (1977) for Pacific North America.

To reconstruct former salt-marsh communities from botanical macro-remains, quantitative analyses are necessary, and the samples have to be large enough to contain a representative number of fruits, seeds etc. from the respective association. Furthermore, they must be homogenous and not too much contaminated by species of other plant communities.

For the presentation of results, the tabular system of the Braun-Blanquet school of phytosociology, with its diagnostic species at various classificatory levels, is recommended (*cf.* Körber-Grohne, 1967; Van Zeist, 1974; Behre, 1976). An example for the reconstruction of the association of a former *Juncetum gerardii* is given in Table 2. Twenty-eight samples from this community were put together and arranged according to the characteristic species of different levels of plant sociology. In this (exceptional) case it was even possible to distinguish several sub-associations, namely from the low lying initial sub-association of *Limonium vulgare* close to the MHW line to the more elevated sub-association of *Leontodon autumnalis* (from Behre, 1976).

A basic assumption in the evaluation of the results of peat analysis is the existence of a 'verlandungs'- or hydroseral successional sequence, which in its classical form includes the following stages: organic mud arising from open-water communities, reedswamp, tall sedge and low sedge (*Magnocaricion* and *Parvocaricion*) peats, carr peat and, when the growing surface exceeds water-table level, raised bog peat (autogenous series, *cf.* Menke, 1968). As a result of SL changes, but sometimes due also to local causes, the sequence can be progressive or retrogressive, depending on whether the water table was rising or falling.

It is important to distinguish between the *autogenic* and *allogenic* series (*cf.* Menke, *op.cit.*). While the former may be regulated by external (eg. falling water-table level) as well as by internal factors (eg. change in habitat conditions as a result of the 'verlandungs' process), the latter, which usually indicates increase in wetness and the onset of brackish conditions, is always governed by external factors alone. As the recognition of peat types by examination of the macro-remains is usually possible even in the field, the former hydrological conditions can normally be established with a high degree of certainty.

Not all peats are suitable for investigations with respect to SL movements. In peat profiles of perimarine back swamp areas, ingression cycles may be registered in various ways, eg. by the sequence: carr, fern swamp, reed, transitional fen and raised bog or by the sequence raised bog - reed and again raised bog. These peats are often located at a considerable distance from mineral ground, so that the peat forming communities and the hydrological and trophic conditions which they reflect are primarily determined by SL movements. Such a conclusion is supported by the synchronism of the changes in peat type, eg. the initiation and widespread extension of raised bog on top of fen peat in certain NW German perimarine areas at about the birth of Christ. This in return permits the recognition and dating of SL changes in perimarine areas. Examples of macro-fossil analyses of long peat profiles exist from the lower Weser (Grosse-Brauckmann, 1963), from the lower Ems (Behre, 1970) and from the Rhine in the Netherlands (Van der Woude, 1981). One has, however, to be careful in the

Table 2. Reconstruction of the salt-marsh community *Juncetum gerardii* and its subassociations from botanical macro-remains, gained in the Viking Age excavation Elisenhof/Schleswig-Holstein (excerpt, after Behre, 1976).

	Salzbinsenwiese *(Juncetum gerardii)*																												
	Initial-Subass. of *Limonium vulgare*							Typical Subass.						Subass. of *Leontodon autumnalis* without *Bromus mollis*						Subass. of *Leontodon autumnalis* with *Bromus mollis*									
Sample number	111	192	512	959	415	956	181	368	784	831	274	124	110	149	158	390	2	3	157	825	109	1	121	191	204	431	266	432	
No. of species	34	41	22	31	22	38	35	29	24	23	68	31	32	41	41	33	22	31	21	58	30	40	36	44	37	36	32	44	
Phytosociological order																													
Juncetum gerardii (Salzbinsenwiese)																													
Juncus gerardii Sa+Kr	3	4	3	5	5	5	5	5	5	5	5	5	4	5	4	5	2	5	5	5	5	4	5	4	3	4	3	3	28
Triglochin maritimum	3	2	2	1	+	1	1	1	+	1	+	+	2	2	2	1	4	+	2	1	2	1	1	2	1	2	1	2	28
Glaux maritima Sa+Kr+Bl	1	+	1	+	+	+	+	1	+	+	+	+	1	+	2	1	+	+	.	+	+	+	+	2	+	+	1	1	27
Aster tripolium	1	+	2	+	+	+	+	+	+	r	+	+	+	+	1	+	1	.	+	1	+	+	+	+	+	+	+	+	27
Plantago maritima Sa+Kr	2	+	r	r	+	r	1	+	1	+	.	+	1	.	2	+	1	+	r	+	1	1	+	2	+	1	1	+	26
Carex distans	.	+	+	r	+	+	.	+	r	+	+	+	+	+	+	.	+	+	+	+	+	+	+	+	r	.	.	+	23
Agrostis (stolonifera)	+	1	1	1	+	+	+	+	.	+	+	+	+	1	1	+	.	.	.	2	+	+	+	+	2	1	.	1	23
Armeria maritima K	+	+	.	.	.	.	+	+	r	.	.	+	+	+	+	+	1	+	.	+	+	+	+	1	+	+	r	+	21
Puccinellia maritima	+	+	r	+	+	.	+	+	+	+	+	.	.	r	+	+	.	.	.	+	+	+	+	+	+	+	.	.	20
Spergularia salina	+	+	.	.	+	.	+	.	+	.	+	+	.	+	+	.	.	+	.	.	+	+	+	+	.	+	1	1	17
Centaurium (vulgare)	+	+	2	+	.	.	.	.	+	.	+	+	.	.	+	r	.	+	.	+	.	+	.	+	1	.	.	.	14
Cochlearia officinalis	.	.	.	r	.	.	.	.	.	.	.	.	.	.	.	.	.	.	.	+	.	.	.	.	.	.	.	.	2
Althaea officinalis	.	.	.	.	.	+	.	.	.	.	+	.	.	.	.	.	.	.	.	.	.	.	.	.	.	.	.	.	2
Subass. of *Limonium vulgare*																													
Limonium vulgare K+Blbl	+	r	+	1	+	+	r	.	.	.	.	.	.	.	.	.	.	.	.	+	+	.	.	.	.	.	.	.	9
Spergularia marginata	+	+	1	+	+	+	+	+	+	+	+	+	+	+	+	+	+	.	.	+	.	+	+	+	1	+	+	+	25
Subass. of *Leontodon autumnalis*																													
Leontodon autumnalis	.	.	.	.	.	.	1	.	.	.	.	.	.	+	1	+	+	+	1	+	r	2	+	r	+	1	r	+	16
Odontites rubra	+	+	2	.	+	.	.	+	+	.	+	.	.	+	+	r	+	+	+	+	r	2	+	1	1	+	+	+	22
Trifolium repens Blbl	+	+	+	r	+	r	.	+	+	.	+	+	+	+	+	+	.	.	.	+	+	+	+	+	1	1	r	2	23
Poa pratensis/trivialis	+	1	+	+	.	+	1	+	r	+	+	+	+	+	1	r	.	.	1	+	+	+	+	1	2	1	.	2	24

No. of species	34	41	22	31	22	38	35	29	24	23	68	31	33	41	41	32	22	31	21	58	30	40	36	44	37	36	32	44	
Potentilla anserina	r	r	.	r	.	r	r	.	.	.	+	.	.	r	+	.	r	+	r	+	.	+	r	.	r	r	.	+	17
Trifolium fragiferum K	+	.	.	.	.	.	+	+	.	.	.	.	+	.	.	r	.	.	.	r	.	.	+	+	+	.	r	+	11
Bromus mollis	.	+	.	.	r	.	.	.	.	.	r	+	+	.	.	.	.	.	.	+	+	+	+	+	+	r	+	+	14
Festuca rubra	+	+	.	+	+	+	r	+	.	.	+	+	+	.	+	+	.	.	.	+	+	+	+	+	+	+	r	+	21
Sagina (maritima)	+	+	1	.	.	+	+	.	.	.	+	.	r	+	.	.	+	2	.	+	+	2	+	+	1	1	+	2	19
Juncetalia maritimi (Übr. Salzwiesen)																													
Puccinellia distans	+	+	+	+	+	+	+	+	.	+	+	+	+	+	+	+	.	.	.	+	+	+	+	+	+	+	.	.	22
Oenanthe lachenalii	+	+	.	+	.	+	.	.	.	.	+	.	.	.	.	.	.	.	.	.	.	.	.	.	+	.	.	r	7
Plantago coronopus	.	.	.	.	.	.	.	.	.	.	.	+	.	+	r	.	.	.	.	.	.	+	.	.	.	.	.	+	5
Apium graveolens	.	.	.	.	.	.	.	.	.	.	+	.	.	.	.	.	.	.	.	+	.	.	.	.	.	.	.	.	2
Salicornietum strictae (Quellerwatt)																													
Salicornia europaea	+	+	.	+	.	.	+	.	+	.	+	+	+	+	+	.	1	+	.	+	+	r	r	+	+	.	1	.	19
Cakiletalia maritimae (Meersenf-Spülsäume)																													
Atriplex littoralis/ hastata	+	+	+	.	.	+	1	+	+	+	+	+	.	1	+	+	r	+	r	+	+	+	+	+	+	+	+	+	25
Atriplex patula-Typ	+	.	r	+	+	+	1	+	+	+	1	+	+	+	1	+	+	+	.	+	+	.	+	+	.	+	+	+	24
Suaeda maritima	+	+	+	+	.	+	+	.	+	.	+	+	r	+	1	.	+	+	.	+	+	+	r	+	+	.	3	.	21
Matricaria maritima	.	+	.	+	r	+	+	+	.	r	+	+	r	+	+	.	.	.	.	+	.	.	.	+	+	+	+	+	18
Brassica campestris	.	.	.	.	.	.	.	.	.	.	+	.	.	.	.	.	.	.	+	r	.	r	.	.	.	.	.	.	4
Plantaginetalia majoris (Tritt-u. Flutrasen)																													
Plantago major Sa+Kr	+	2	2	+	+	+	+	+	+	+	+	+	+	+	1	+	+	+	1	1	+	+	+	1	2	+	+	2	28
Polygonum aviculare	r	+	.	.	.	+	+	r	.	+	+	+	+	+	+	.	.	+	.	+	.	r	+	+	.	+	+	+	19
Agropyron repens	r	.	r	.	.	+	.	.	.	.	+	.	+	+	r	+	.	.	r	+	.	+	+	+	.	+	+	.	15
Agropyron littorale	.	+	.	.	.	+	.	.	r	.	+	.	+	+	.	r	.	.	.	+	.	+	.	.	+	r	r	.	12
Ranunculus repens	.	+	.	.	.	.	r	.	.	.	+	.	.	.	+	.	.	+	r	+	.	+	.	+	r	+	.	+	12
Lolium perenne	.	+	.	.	.	.	.	.	.	.	.	.	.	.	r	.	.	.	.	+	.	.	.	+	.	.	.	r	5
Alopecurus geniculatus	.	.	.	.	.	.	.	.	.	.	.	.	.	.	.	.	.	.	.	.	.	.	.	r	.	.	.	.	1

The vertical columns represent samples; the numbers indicate abundancy classes of fruits, seeds etc. in the following way:
r = 1 specimen + = <1% 1 = 1 - 5% 2 = 5 - 25% 3 = 25 - 50% 4 = 50 - 75% 5 = >75%

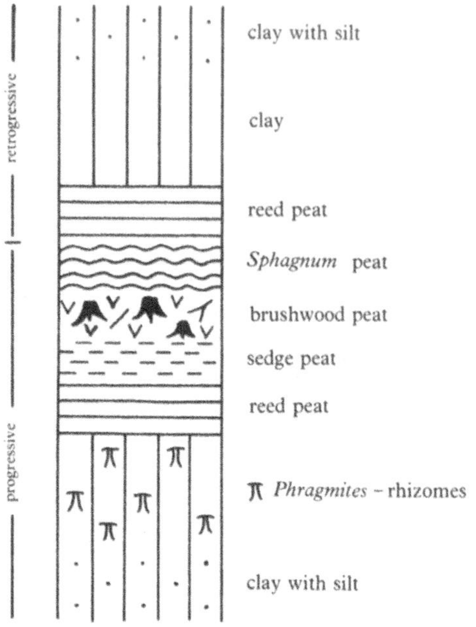

Figure 3.

Schematic section through an intercalated peat, comprising a complete 'verlandungs' series (progressive), followed by retrogressive development.

catchment area of great rivers where other factors may dominate.

With regard to SL changes particular interest has always been taken in the peats intercalated in minerogenic Holocene sequences. These intercalated peats are widespread and witness fresh-water environment preceded and followed by marine or brackish conditions. The peat-forming communities recognized by macro-fossil analysis do not only show the degree of fresh-water influence but also the tendency in paleoecological development. There often occurs a complete 'verlandungs' series, beginning with clay which includes reed remains, followed by reed, sedge, and carr peat and, in places, raised bog peat (progressive development), the whole series being covered by clay again. Sometimes even the change in tendency from progressive to retrogressive development can be traced (*cf.* Fig. 3), thus showing a complete cycle.

Of special importance is the basal peat layer which often overlies the pre-Holocene surface. The analysis of macro-remains ascertains the paleoecological conditions and makes sure whether such a peat is a real basis peat in the sense of Lange & Menke (1967), formed under the influence of rising SL, or whether it is merely an older peat in this basal position without direct context to SL rise.

The evaluation of the samples in order to reconstruct the former peat-forming communities has to be done in the same way as for the salt-marsh associations (see pages 422-3).

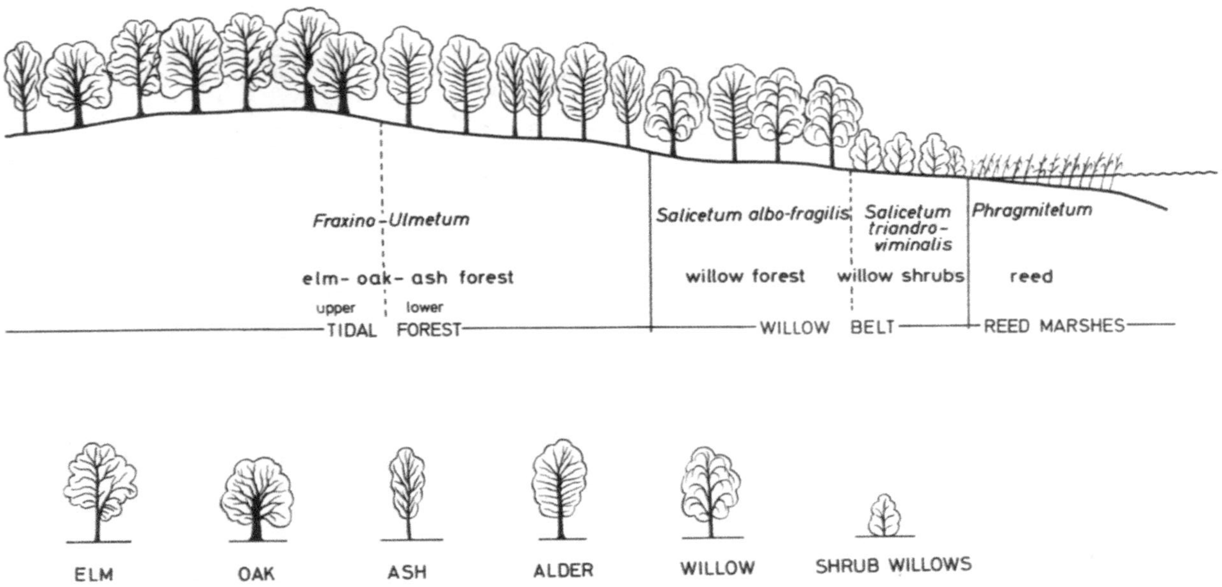

Figure 4. The zonation of river-bank forests ('Äuenwalder') according to their elevation above water level. Reconstruction for the lower Ems area (West Germany) for early Subatlantic times.

Some difficulties arise, because the distinguishable peat components are not mainly fruits and seeds, as in salt-marsh samples, but consist to a great extent of vegetative plant remains, and the estimation of their quantity may of course imply some subjectivity. Beside the already mentioned system which follows Braun-Blanquet, there are several different abundance scales which have been used for the quantification of mosses and other peat remains (reviewed by Dickson, 1982).

The reconstruction of former river-bank forests can only be achieved by the help of wood remnants, since leaves, fruits etc. are only rarely found in this context. There should be a minimum number of at least a hundred wood samples to guarantee the necessary statistical security. It is an important advantage that the two main communities of river-bank forests are characterized and dominated by different trees: the low lying willow coppice and willow forest = willow belt ('Weichholzaue', composed of soft wood species), starting from about MHW level with willow species (*Salix*) and followed on the more elevated natural levees by the tidal forest or 'Hartholzaue' (composed of hard wood species) formed by ash (*Fraxinus*), and in the highest parts by oak (*Quercus*) and elm (*Ulmus*). The only important tree which frequently occurs in both forest types is the alder (*Alnus glutinosa*).

In this way the forest communities reflect the position of different zones of the natural levees in relation to MHW. In Europe most of the original river-bank forests have been destroyed by cultivation. One of the last examples in the tidal fresh-water area was the Brabantse Biesbos in the Netherlands which was thoroughly investigated by Zonneveld (1960). The reconstruction of river-bank forests along the mouth of the Ems river (Germany) in the early Subatlantic period is shown in Figure 4 (*cf.* Behre, 1970, 1979).

INTERPRETATION OF DATA - LIMITATION OF THE METHOD

Every interpretation depends on the material that was investigated. Therefore, particular attention has to be paid to the conditions under which the fossil-bearing material was deposited. Peats are sedentary deposits and generally autochthonous, ie. they are formed *in situ* and represent, in terms of palaeontology, thanatocoenoses. This is also the case with wood and leaf layers originating from woodland river-bank communities ('Auenwaldfazies'). It should be carefully ascertained during the actual sampling that one is not dealing with secondary material such as redeposited peat or drift wood, although these possibilities cannot always be excluded. On the other hand, even primary material, for instance *Phragmites* rhizomes, has to be considered carefully.

Macro-fossil material from prehistoric sites, on the other hand, is allochthonous and represents taphocoenoses. Though it may come from a single source and hence be particularly suitable for investigations, the possibility must always be taken into consideration that it may have been collected by human effort for a particular purpose, eg. as food, fodder, or bedding for cattle, or as building material. This applies particularly to wood, which may have been transported over a considerable distance to a particular settlement.

The *quantitative evaluation* of the results of macro-fossil analysis is especially difficult. The actual results are usually presented as number of fruits, seeds etc. pertaining to each taxon. This may then provide a basis for classification according to phytosociological principles which are not based mainly on frequencies but on characteristic species. It must be emphasized that results presented in this way reflect only the frequency of fossilized parts and provide limited indication as to the actual composition of the vegetation.

The frequency of fossilized remains mainly depends on the following factors:

1. the former frequency of the particular species;
2. its seed-production capacity; this may vary considerably according to species;
3. the distance between the location in which the species grew and the actual sampling site;
4. the manner and effectiveness of natural seed dispersal;
5. the preservation characteristics of the species. Some species are never represented in the macro-fossil record; that they were nevertheless present may be deduced from other evidence such as pollen;
6. in the case of archaeological sites, the whole or part of a particular species may intentionally have been collected by man, while other species only accidentally occurred at the sampling site.

The results of macro-remain analysis always have to be used critically if applied to indicate the absolute or relative height of MHW. Interpretation has to be based on the local regional situation, which today may be quite different from what it used to be. Flat coasts in clay areas with a tidal range up to about 3 metres are the most suitable places for reliable interpretation. In macro-tidal areas there are only limited possibilities for the formation and preservation of macro-fossil-bearing beds and there may arise difficulties with regard to interpretation. For example the vertical range of salt-marsh communities does not only depend on the frequency but also on the duration of submergence, and therefore is connected with the tidal (and storm surge) range.

Another serious problem is that of compaction of sediments (see Chapter 22). Strongest compaction is found in peat, particularly if overlain by minerogenic deposits or drained by modern cultivation. Compaction in river-bank sediments on the other hand is less. Thus, all interpretations concerning absolute height have to be done with great care, considering all possible sources of error. *In many cases the results are useful only in chronological sense.*

The advantage of macro-fossil analysis is the possibility of reconstructing the former plant communities in coastal areas, ie. the paleo-environment of different times. One has to keep in mind, however, that these former ecosystems may have differed strongly according to natural factors other than changes of MHW, as for instance coast configuration (open, sheltered), kind of substratum (clay, sand, or rocks, which are not considered here), rates of sedimentation and others. Special problems may arise in the backswamps of the hinterland, the perimarine area. The water regime and also nutrient supply may be dominated either by the sea or by the water coming from the immediate hinterland or sources at a greater distance. Therefore the distance from the sea and from the nearest river as well as the size of the catchment area of the latter are important for the interpretation.

Under natural conditions there exists along flat coasts and estuaries a wide brackish transitional zone with a mosaic of ecosystems, according to the above-mentioned factors. In most parts of Europe this transition zone was cut off by dyking within the last 1000 years.

For the interpretation of the sometimes very complicated conditions in the backswamp area it is not only necessary to thoroughly investigate isolated sites, but to follow long sections, the lithological changes (clay, sand, different kinds of peat) of which can be observed in the field with the naked eye.

In spite of the obvious limitations of hinterland investigations with respect to SL changes, these areas offer possibilities of obtaining evidence that cannot be acquired from near-coastal deposits.

CONCLUSIONS

The analysis of botanical macro-remains provides, in various ways, results which can be used as information on former SL changes. Although this method is sometimes very time-consuming, it yields knowledge that cannot be gained from other sources.

According to the ecological requirements of the plant species from which the macro-remains derive, these fossils give information mainly on the level around and above MHW and are thus complementary to molluscs, foraminifera and other indicators which occur in the environment below MHW.

Botanical macro-remains enable the reconstruction of the former plant communities and the paleo-environment in the investigated areas. In this way the local MHW, groundwater, salinity and other ecological factors can be estimated. The results from the backswamp areas cannot normally be used for the determination of absolute SL height, but after critical consideration of other factors involved, changes in strength and directions of SL movements can be traced, as they are reflected by the pro- or retrogressive succession of bog vegetation in the hinterland. Necessary for those comparisons are, of course, datings which have to be obtained by pollen analysis or ^{14}C-dates of the investigated profiles. Macro-fossil analysis itself cannot be used for dating.

Plant macro-remains from archaeological excavations as well as wood remains from river banks, both from locations on mineral soil, give the opportunity to get information on the interval between MHW and maximum storm-flood level and on the height of the latter.

Up to now thorough analysis of botanical macro-remains with respect to SL movements has been employed mainly in the area around the southern North Sea. The possibilities, demonstrated in this chapter, may encourage further application of this method.

REFERENCES

Beeftink, W.G., 1977. The coastal salt marshes of western and northern Europe: an ecological and phytosociological approach. *Ecosystems of the World 1,* 109-155, (Elsevier, Amsterdam)

Behre, K.-E., 1970. Die Entwicklungsgeschichte der natürlichen Vegetation im Gebiet der unteren Ems und ihre Abhängigkeit von den Bewegungen des Meeresspiegels. *Probleme der Küstenforschung im südlichen Nordseegebiet,* 9, 13-48

Behre, K.-E., 1976. Die Pflanzenreste aus der frühgeschichtlichen Wurt Elisenhof. *Studien zur Küstenarchäologie Schleswig-Holsteins, Ser. A.,* 2, 183 pp

Behre, K.-E., 1979. Zur Rekonstruktion ehemaliger Pflanzengesellschaften an der deutschen Nordseeküste. Symposiumsband '*Werden und Vergehen von Pflanzengesellschaften*', ed. R. Tüxen, (Vaduz), 181-214

Beijerinck, W., 1947. *Zadenatlas der nederlandsche flora.* (Veenman & Zonen, Wageningen)

Berggren, G., 1969. *Atlas of seeds, 2. Cyperaceae.* (Swedish Natural Science Research Council, Stockholm)

Berggren, G., 1981. *Atlas of seeds, 3. Salicaceae-Cruciferae.* (Swedish Natural Science Research Council, Stockholm)

Bertsch, K., 1941. Früchte und Samen. *Handbücher der praktischen Vorgeschichtsforschung, 1,* (F. Enke, Stuttgart)

Chapman, V.J., 1974. *Salt marshes and salt deserts of the world.* (Cramer, Lehre), 2nd edition

Chapman, V.J., 1976. *Coastal vegetation.* (Pergamon Press, Oxford), 2nd edition

Dickson, J.H., 1982. Bryophyte remains. In: *Project Guide to IGCP - Project 158 B: Palaeohydrological changes in the temperate zone in the last 15000 years,* ed. B.E. Berglund, 131-147

Ellenberg, H., 1979. Zeigerwerte der Gefässpflanzen Mitteleuropas. *Scripta Geobotanica,* 9, (2nd edition)

Gillham, M.E., 1957. Coastal vegetation of Mull and Iona in relation to salinity and soil reaction. *Journal of Ecology,* 45, 757-778

Gillner, V., 1960. Vegetations- und Standortuntersuchungen in den Strandwiesen der Schwedischen Westküste. *Acta Phytogeographica Suecica,* 43, 198 pp

Greguss, P., 1959. *Holzanatomie der europäischen Laubhölzer und Sträucher.* (Akademiai Kiado, Budapest)

Grosse-Brauckmann, G., 1963. Uber die Artenzusammensetzung von Torfen aus dem nordwestdeutschen Marschenrandgebiet (Eine pflanzensoziologische Auswertung von Grossrestuntersuchungen). *Vegetatio,* 11, 325-341

Grosse-Brauckmann, G., 1972. Uber pflanzliche Makrofossilen mit teleuropäischer Torfe. I. Gewebereste krautiger Pflanzen und ihre Merkmale. *Telma,* 2, 19-55

Grosse-Brauckmann, G., 1974. Uber pflanzliche Makrofossilien mitteleuropäischer Torfe. II. Weitere Reste (Früchte und Samen, Moose u.a.) und ihre Bestimmungsmöglichkeiten. *Telma,* 4, 51-117

Grosse-Brauckmann, G., 1982. Vegetative Macrofossils. In: *Project Guide to IGCP - Project 158B: Palaeohydrological changes in the temperate zone in the last 15000 years,* ed. B.E. Berglund, 111-130

Grosser, D., 1977. *Die Hölzer Mitteleuropas. Ein mikrophotographischer Lehratlas.* (Springer, Berlin-Heidelberg-New York)

Iversen, J., 1936. Biologische Pflanzentypen als Hilfsmittel in der Vegetationsforschung. *Meddelelser fra Skalling Laboratoriet,* 4. 244 pp

Johnson, D.S. & York, H.H., 1915. The relation of plants to tide-levels. *Carnegie Institution Washington,* Publication, 206, 162 pp

Katz, N.Y., Katz, S.V. & Kipiani, M.G., 1965. *Atlas and keys of fruits and seeds occurring in the Quaternary deposits of the USSR.* (In Russian), (NAUK, Moscow)

Katz, N.Y., Katz, S.V. & Skobeyeva, E.I., 1977. *Atlas of the plant remains in peats.* (In Russian), (NAUK, Moscow)

Körber-Grohne, U., 1967. Geobotanische Untersuchungen auf der Feddersen Wierde. *Feddersen Wierde 1,* 357 pp (Steiner, Weisbaden)

Lange, W. & Menke, B., 1967. Beiträge zur frühpostglazialen erd- und vegetationsgeschichtlichen Entwicklung im Eidergebiet, insbesondere zur Flussgeschichte und zur Genese des sogenannten Basistorfs. *Meyniana*, 17, 29-44

MacDonald, K., 1977. Plant and animal communities of Pacific North American salt marshes. *Ecosystems of the World 1*, (Elsevier, Amsterdam), 167-191

Martin, A.W. & Barkley, W.D., 1961. *Seed identification manual*. (University of California Press, Berkeley, Los Angeles)

Menke, B., 1968. Ein Beitrag zur pflanzensoziologischen Auswertung von Pollendiagrammen, zur Kenntnis früherer Pflanzengesellschaften in den Marschenrandgebieten der schleswig-holsteinischen Westküste und zur Anwendung auf die Frage der Küstenentwicklung. *Mitteilungen der Floristisch-soziologischen Arbeitsgemeinschaft*, N.F. 13, 195-224

Ranwell, D.S., 1972. *Ecology of salt marshes and sand dunes*. (Chapman & Hall, London)

Tüxen, R., 1957. Die Pflanzengesellschaften des Aussendeichslandes von Neuwerk. *Mitteilungen der Floristisch-soziologischen Arbeitsgemeinschaft*, N.F. 8, 227-247

Wasylikowa, K., 1979. Plant macrofossil analysis. In: *Palaeohydrological Changes in the Temperate Zone in the Last 15000 Years. IGCP Subproject 158B. Lake and mire environments*, ed. B.E. Berglund, 291-313

Westhoff, V. & Held, A.J. Den, 1975. *Plantengemeenschappen in Nederland*. (Thieme, Zutphen), 2nd edition

Woude, J.D., van der, 1981. *Holocene paleoenvironmental evolution of a perimarine fluviatile area*. Ph.D. thesis, Free University, Amsterdam

Zeist, W., van, 1974. Palaeobotanical studies of settlement sites in the coastal area of the Netherlands. *Palaeohistoria*, 16, 223-371

Zonneveld, I.S., 1960. De Brabantse Biesbosch. *Mededelingen van de Stichting voor Bodemkartering, Bodemkundige Studies*, 4 (3 parts)

15

Foraminifera as sea-level indicators

D. B. SCOTT and F. S. MEDIOLI

ABSTRACT

There are over 1000 foraminiferal species presently living in marine environments, but relatively few of these can be used as direct indicators of sea level. We detail here certain assemblages occupying marsh environments that can be used for that purpose. Marsh foraminiferal assemblages have been shown to occur worldwide within narrowly defined vertical zones some of which extend less than 10 cm in total vertical range. These assemblage zones, once known, can be relocated in paleo-marsh deposits and accurately related to paleo-sea level. The key to success of these organisms as sea-level indicators is that their primary controlling factor appears to be elevation above mean sea level.

Other foraminiferal assemblages are controlled by a series of parameters (eg. salinity, temperature, dissolved O_2, etc.) which often have no direct relationship with actual water depth. Under appropriate circumstances, some information concerning sea level can be derived from other foraminiferal assemblages using indirect means. For instance areas which have undergone substantial emergence often contain small basins which, prior to emergence, were marine. The marine sediments at the bottom can be identified using foraminifera and the sill elevation of the basin can then be related to some prior sea-level stand.

Both these methods have imperfections, but in areas where both emergence and submergence have taken place they can be combined to obtain a relatively complete sea-level history.

INTRODUCTION

Foraminifera are uni-cellular testate marine organisms with a long stratigraphic history (for complete details see Loeblich & Tappan, 1964). They produce four basic test (shell) types: 1) calcareous (secreted by the organism), 2) agglutinated or arenaceous (made of foreign particles, usually sand, cemented together by the organism), 3) porcelaneous (combination of organic and calcareous material) and 4) organic walled forms (allogromiids). The first three types fossilize readily whereas the last one rarely does.

Foraminifera have two basic modes of life: planktonic and benthonic. The planktonic forms, of which there are only about 30 living species, all secrete tests of $CaCO_3$ while the benthonics, of which there are almost 1000 reported species (Boltovosky & Wright, 1976) may have any one of the four test types.

Although the planktonic forms have relatively few species they may occur in such high concentrations that their empty tests are often an important constituent of the deep sea floor sediment (ie. '*Globigerina*' ooze). Planktonic species, however, are restricted to open ocean conditions and live only in the water column; hence they are of little value as sea-level indicator (SLI).

Benthonic foraminiferal assemblages, as one might suspect from their diversity, occupy almost every marine benthic environment. As with other marine organisms, foraminiferal assemblages are controlled by any number of parameters whose variability and importance change depending on the environment. In marginal marine environments, where environmental variability is high, physical parameters such as salinity, temperature, and elevation (or perhaps more importantly temporal variation of these factors) appear to be the strongest influence on assemblage composition. In offshore areas (continental shelf and deep-sea) the elements influencing assemblage distribution become much less distinct with subtle 'water-mass' characteristics being used to explain distributions (eg. Streeter, 1973; Lohman, 1978). Additionally in the environmentally stable offshore areas bioecological factors (eg. predation, competition) may obscure the effects of physical parameters. For a more complete treatment of foraminiferal ecology see: Phleger (1960), Murray (1973), Boltovosky & Wright (1976), and Haynes (1981).

Of specific interest to readers of this manual are foraminiferal assemblages that can reliably and accurately be related to a certain elevation or water depth. No open sea assemblage falls into this category since their distribution limits depend on a particular water mass whose position is generally not depth dependent. Hence we are left only with marginal marine (ie. marsh-estuarine-lagoonal) assemblages to use as accurate SLIs.

Estuarine-lagoonal areas have been investigated extensively from a number of regions (eg. Scott *et al.*, 1976, 1977, 1980; Ellison & Nichols, 1976) and these assemblages do not appear to respond specifically to depth changes but rather to salinity changes. It appears that the use of estuarine foraminiferal assemblages may be helpful in a relative sense (ie. to establish that one area was shallower than another) but these assemblages are not sufficiently accurate as absolute indicators of SL.

By contrast, marshes, those vegetated areas between higher high water (HHW)[1] and mean sea level (MSL) contain an assemblage of foraminifera that appears to be so sensitive to changes in elevation that a former SL can be relocated by means of their fossil remains with an accuracy of ±10 cm (Scott & Medioli, 1978, 1980a).

In the following sections we discuss the method of collecting and using marsh foraminiferal assemblages as SLIs. The description of this method is drawn from an extensive data base covering almost all known types of marsh environment: semi-arid, hypersaline (Scott, 1976a, b); cold-temperate (Scott & Medioli, 1978, 1980a, b; Scott *et al.*, 1981; Deonarine, 1979); Mediterranean (Scott *et al.*, 1979); and sub-arctic (Scott & Martini, 1982). These studies and less detailed ones (eg. Phleger, 1966, 1967, 1970; Ellison & Nichols, 1976; Lutze, 1965, 1968, 1974) from many other areas indicate that the same relationships remain true in a worldwide sense.

Following discussion of the marsh foraminiferal assemblages we discuss another method based on the use of other foraminiferal assemblages as indirect indicators of former SLs.

GENERAL ASPECTS OF MARSHES

The conspicuous visual aspect of a marsh environment is the plant zonation. Most tidal marshes are in fact divided vertically into 2 to 3 plant zones depending on climate, fresh-water input, and tidal regime (see Chapman, 1960, 1976 for complete treatment of vegetation). The marsh itself is confined vertically between HHW and MSL with elevation above MSL apparently being the primary limiting factor of marsh-zone development. In more temperate climates with moderate rainfall, *Spartina* spp. dominate the entire plant spectrum of the marsh, while in arid coastal areas *Salicornia* spp. dominate. In the fossil state *Spartina* marshes tend to form peat while the *Salicornia* marshes usually develop into an organic mud, much less compactable than peat.

The vertical range of a marsh (MSL to HHW) appears to apply worldwide except in greatly expanded tidal ranges

---

[1]By HHW we mean the high water mark reached biweekly by the spring tides, not necessarily the daily high tide mark.

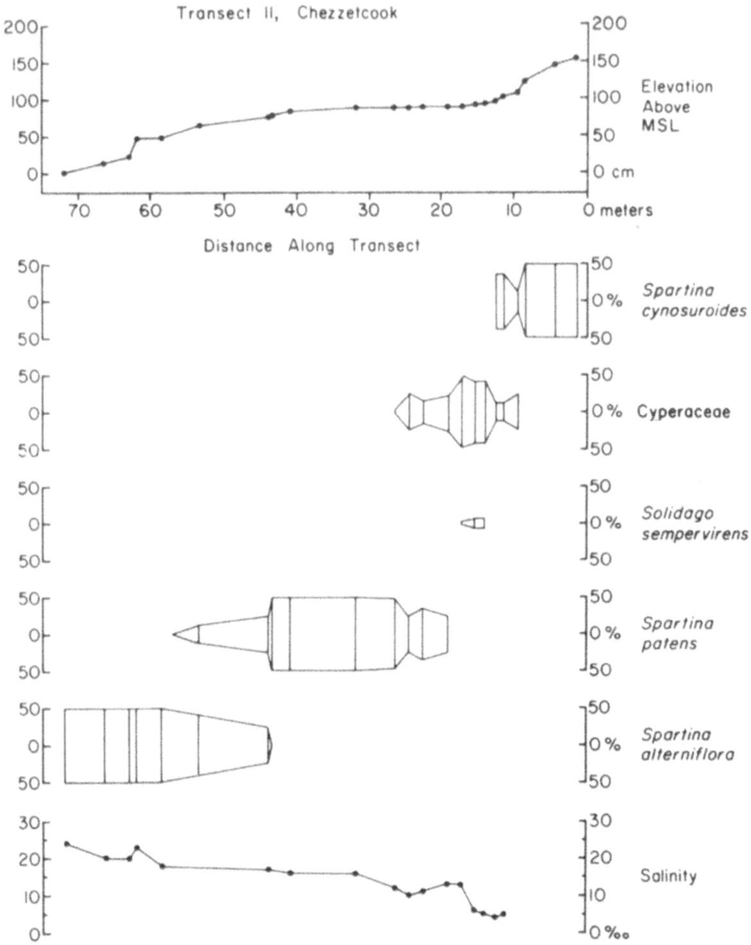

Figure 1. Plant and salinity vertical distributions along one of the transects (from Scott & Medioli, 1980a, transect II). Dots in upper parts of diagram represent sampling localities while vertical bars next to plant names are percent occurrences at each location. Salinity is shown at the bottom; note the trend with decreasing elevation.

such as those found in the Bay of Fundy, Nova Scotia. In such cases the marsh will expand vertically to accommodate some of the tidal increase, but it appears that if the total tidal range exceeds 7 m the base of the marsh will no longer reach down to MSL (Scott & Medioli, 1980a). In the expanding vertical marsh sequences high-marsh zones appear to retain their absolute vertical ranges of 30-80 cm while the low-marsh zones absorb most of the increase in vertical range.

Because of the low pH conditions common to most marshes (Phleger & Bradshaw, 1966), the most widespread marsh foraminiferal species are the agglutinated forms which are resistant to dissolution in the low pH sediments. These forms are preserved in subsurface sediments in abundance and form the basis of the method to be described below.

MARSH FORAMINIFERAL ZONATION

Certain well-defined foraminiferal assemblages characterize specific elevation intervals within the salt-marsh vertical range. The composition of the assemblage depends on a series of factors, most important of which appears to be elevation (at least for high-marsh faunas). One example of a detailed transect is shown here from Nova Scotia (Figs. 1 and 2). Figure 1 illustrates the plant-salinity vertical relationships while Figure 2 demonstrates that the same type of relationship is true for marsh foraminifera. This transect was the easiest to present visually since there is a steady slope, but all other transects indicate the same vertical boundaries (Scott & Medioli, 1980a). Of particular note on Figure 2 is that at +100 cm there is a marked decrease in the total numbers of foraminifera from more than 500/10 cm^3 just below +100 cm to less than 100/10 cm^3 above this level. This break signifies the effective upper limit of tidal activity, above which no foraminifera are found.

Data from southern California and Nova Scotia are summarized together in Figure 3. The same types of zones are recognizable from both areas but none of the plant species is common to both. In contrast, the foraminiferal zone species are almost the same so that all the zone I species found in California are also found in Nova Scotia although more agglutinated species are found in temperate marshes. In areas with higher salinities a few calcareous species replace some of the more primitive agglutinated forms but the former are seldom preserved in subsurface sediments (Scott, 1976a). The similarity disappears in the low-marsh zone IIB, indicating that low-marsh faunas tend to mirror local estuarine faunas; this is true even within the same estuary as salinities increase seaward (see Scott & Medioli, 1980a; Scott *et al.*, 1980). Also on Figure 1 salinity is shown as increasing with decreasing elevation (ie. more tidal influence) and this is a consistent relationship in all marshes where there is sufficient fresh-water runoff (the relationship is reversed in arid areas). But while high-marsh salinities are not the same at the head of an estuary as at the mouth, zone I or high-marsh faunas, remain the same. This demonstrates that these zone I species are elevation, rather than salinity, controlled.

The zones and subzones outlined in Figure 3 can be assigned definite elevation ranges. If the assemblages that characterize these zones are found in buried or raised sediments an accurate former SL can be assigned and a SL point determined. For example in Nova Scotia if a

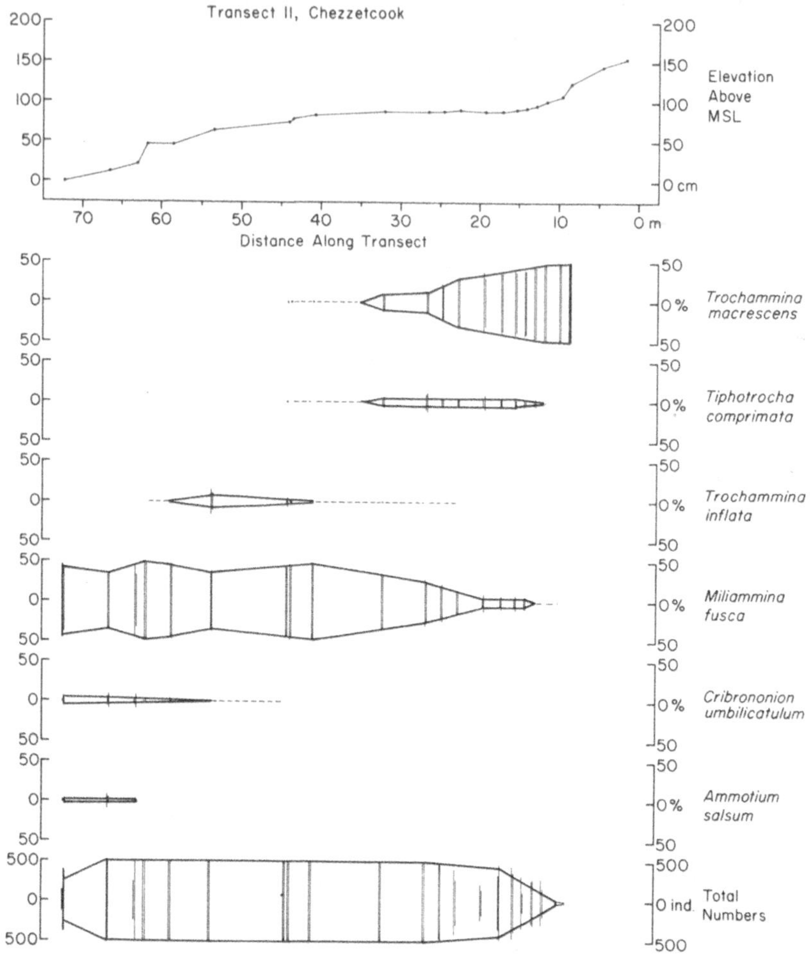

Figure 2. Foraminiferal vertical distributions from the same transect shown in Figure 1. The double vertical bars represent the two (replicate) samples obtained at each locality. Horizontal connecting lines represent subjective averaging, hence not all vertical bars connect with the lines. Also notice that there is substantial variance even at one locality between replicates, especially in the total numbers/10 cm^3. Note the decrease in total numbers per sample and the monospecific fauna (*T. macrescens*) occurring at the HHW mark (here at +100). Although total numbers are shown as only reaching values as high as 1000/10 cm^3 they do attain much higher numbers, but variations in the over 1000 mark are random and not shown here.

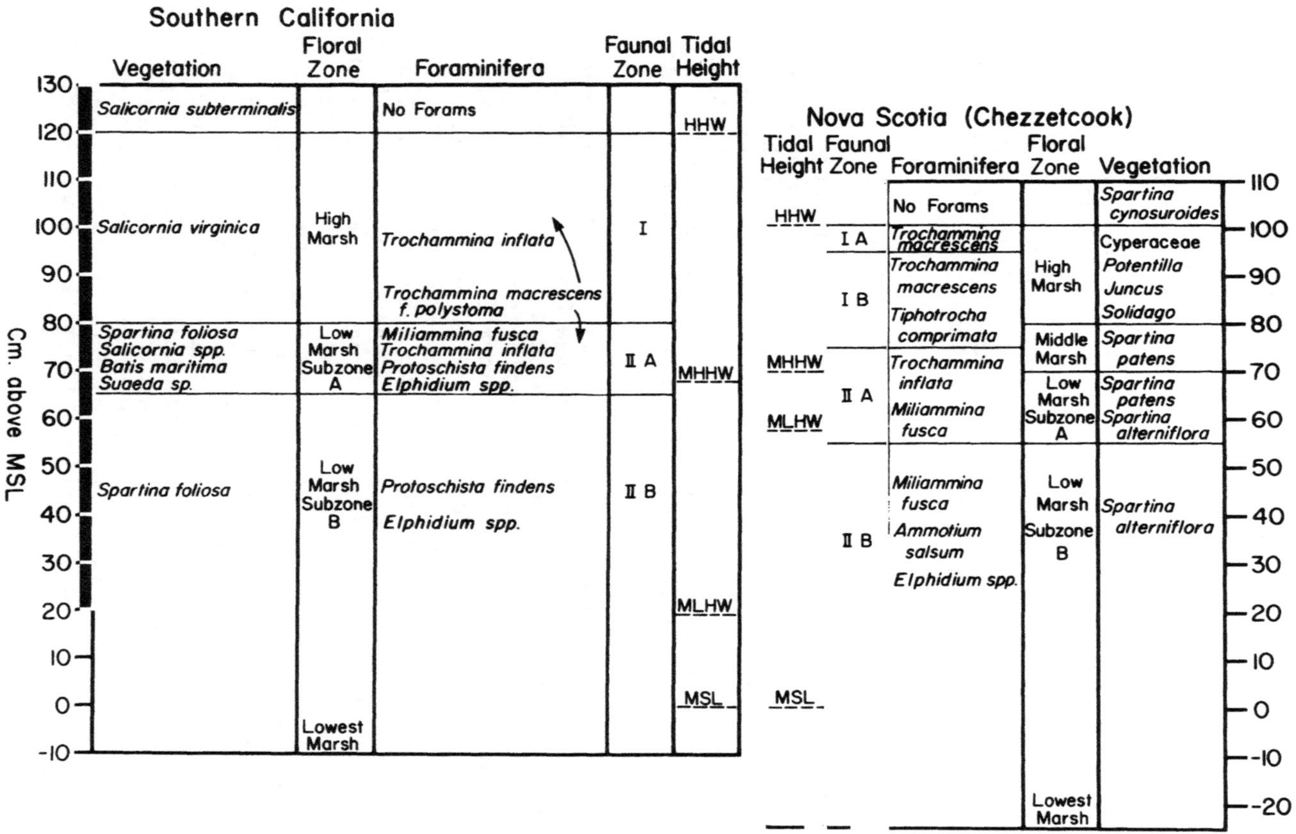

Figure 3.

The diagram shows a direct comparison of the southern California marsh data (from Scott, 1976a) and the Nova Scotia situation (from Scott & Medioli, 1980a). The comparison is interesting because of the large difference in climate between the two areas. Note that although plant species show no similarities, marsh foraminiferal species, especially in zone I and IIA, are similar.

monospecific assemblage with *T. macrescens* was observed in buried sediments,it would be safe to assume that this deposit formed within the depth range of zone IA (Fig. 3). Since the total vertical range of this zone is 10 cm the position of the former SL can be established with an accuracy of ±10 cm (other sources of error not withstanding). This of course is the optimum situation since lower zones occupy wider vertical ranges. However, few zone I faunas would produce less than ±30 cm accuracy. The zone II faunas (with the exception of some zone IIA assemblages) have wide vertical ranges, in addition to mirroring adjacent estuarine faunas, consequently they do not appear to make suitable SLIs.

Zone I faunas always appear to be characterized by the same agglutinated forms: usually *Trochammina inflata*, *T. macrescens* (formae *macrescens* and *polystoma*), and *Tiphotrocha comprimata*. Hence when an assemblage dominated by these species is observed, it is fairly safe to assume that the deposit formed at least in the upper quarter of the tidal range. These species and some other common forms are illustrated in the color plate that goes with this chapter. Complete taxonomy and illustrations of all marsh species can be found in Scott & Medioli (1980a).

APPLICATION OF METHOD

To show how we have applied the method we present three examples, two from Nova Scotia and one from Prince Edward Island. In all areas our first step has always been to establish the local foraminiferal zonation. We strongly recommend that SL studies using foraminifera as indicators be preceded by some study of present-day surface distribution of marsh foraminifera in areas where no such study has been done previously.

The *first example* is from Wallace Basin, Nova Scotia (Fig. 4; for details on location see Scott & Medioli, 1980a). Although these drill holes are relatively shallow, it is possible to see marsh foraminiferal zone changes in the diagrams and to determine a former SL. In the two sequences illustrated foraminiferal numbers decrease markedly at -75 cm in D.H. I and at -147 cm in D.H. VIII. These are the levels which must be used as the SLIs, even though zone IB does not represent HHW, since levels below these in the sequence have few foraminifera and could have formed anywhere above HHW. This is one of the better examples we can illustrate showing that the zone IA is not always preserved in subsurface sediments and demonstrates how error limits can be placed on marsh deposits. Since locally zone IB has a relatively wide vertical range (60 cm total range), the best accuracy that can be obtained is ±30 cm. The center of this zone is at local mean higher high water (MHHW) so that the indicator levels formed at local MHHW ±30 cm. The surface of the drill holes is in zone IIA which forms 40 cm below local MHHW (and only has a 10 cm range). Hence to obtain the total amount of SL rise

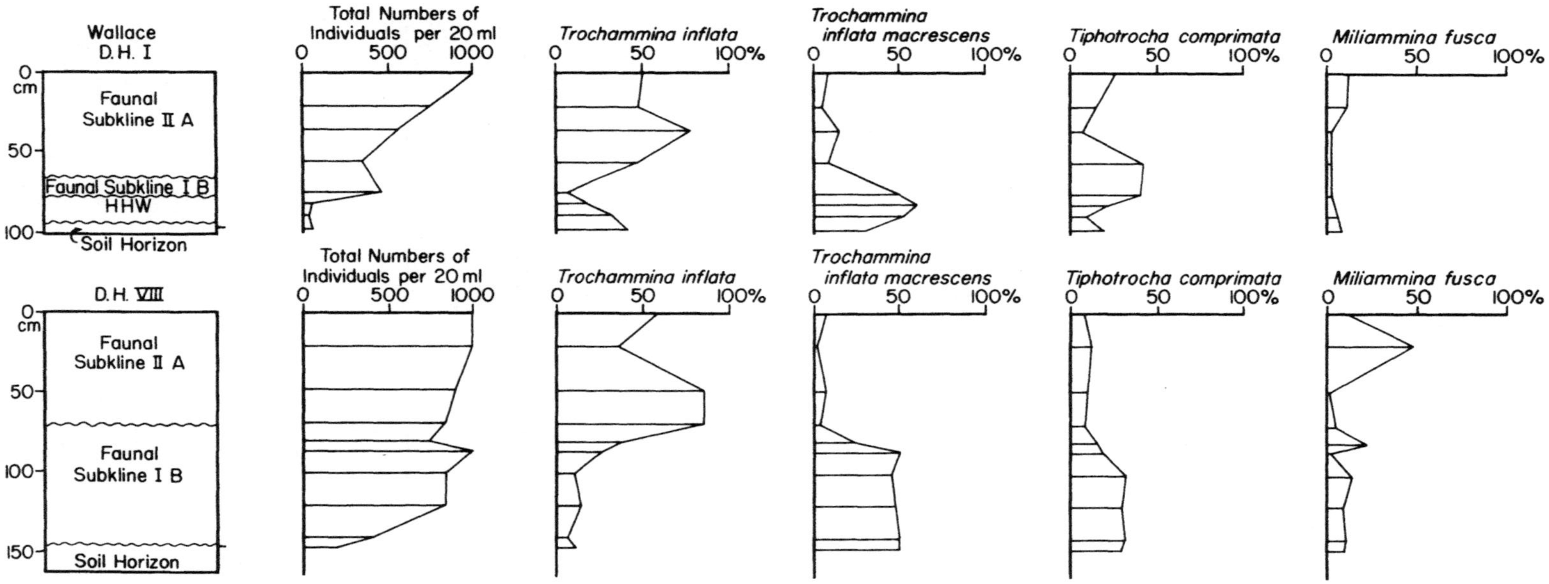

Figure 4. Set of drill holes from Wallace Basin, Nova Scotia (for details on location see the map in Scott & Medioli, 1980a). At the surface of the drill hole a fauna characteristic of zone IIA in this area is observed and this gradually changes to a zone IB at the base. Hence the sequence shows a gradual sea-level rise. The ^{14}C dates were taken from last interval where substantial foraminiferal numbers occurred since below this the absence of foraminifera could not provide a reliable sea level.

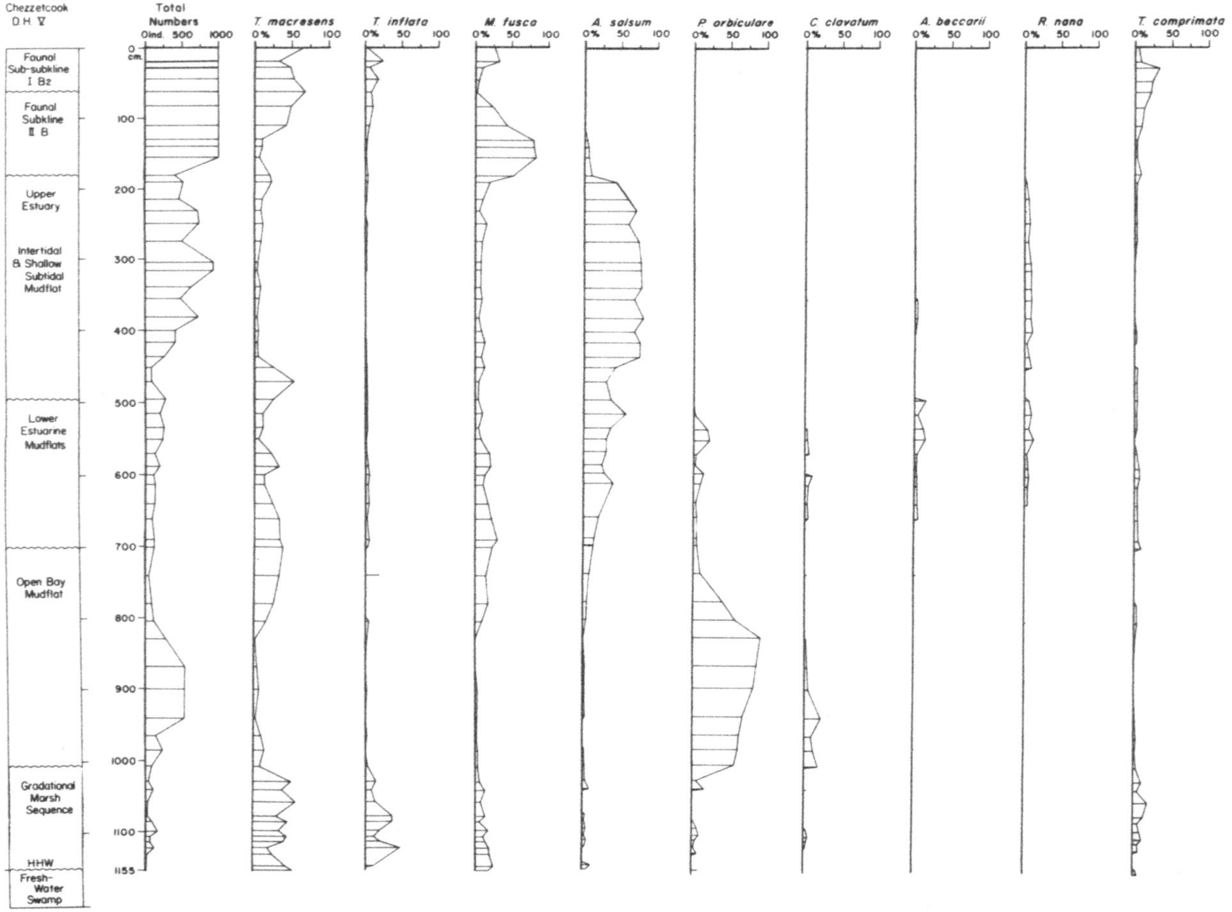

Figure 5.

This drill hole from Chezzetcook Inlet, Nova Scotia is the longest marsh-estuarine sequence we have encountered in eastern North America (see Scott, 1980 for location). It not only indicates a sea level 12 m below present, but several interesting paleo-environmental features as well. Once more the only point suitable for dating is at the 1155 cm mark: the fresh-water/salt-marsh peat contact.

recorded at these two drill holes, one must add 40 cm to the drill-hole depth to arrive at 115 ± 30 cm at D.H. I and 187 ± 30 cm at D.H. VIII. In most areas we attempt to drill or core in a vegetation zone with an established small vertical range; this avoids the time consuming and sometimes impossible task of tying in the surface of the drill hole to a bench mark.

The *second example* comes from the longest estuarine sequence we have found in the area (12 m, Fig. 5). For a complete stratigraphic discussion of this area we refer to Scott (1980). This drill hole records a large number of stratigraphic events but only the contact between the fresh water and marsh peat is useful for a SL determination. This contact point represents the first transgression into the area and the marsh foraminifera allow differentiation of the fresh water from marsh peat as well as yielding an accurate SL. In this case the fresh-water peat was much older than the overlying salt-marsh peat formation and the fresh-water peat was already compacted, otherwise there could have been some autocompaction effects (Kaye & Barghoorn, 1964). The point dated here was the contact point at 1155 cm which represented a foraminiferal zone IA assemblage (HHW ± 10 cm) and represented 1185 cm ± 10 cm of SL rise (top of drill hole was 30 cm below HHW).

The *third example* is from Prince Edward Island (Fig.6) and is a classic example of marsh-zone development. For all details on this area see Scott *et al.* (1981). The foraminiferal stratigraphy is shown only for the longest section and shows the well-developed zone IA at the base of the hole with the total foraminiferal numbers dropping just below it. The HHW line is shown at 25 cm above the surface of the drill hole so 25 cm is added to the total depth of the section to obtain a total SL rise at this location of 250 cm ± 10 cm. At this site the base of each of these drill holes was used as a SL point avoiding the problem of compaction within the section.

COLLECTION, PRESERVATION AND DATING OF SAMPLES

Preservation of marsh foraminifera. Very few natural phenomena can destroy agglutinated marsh foraminiferal remains, particularly in the reducing environment characteristic of a marsh. In the laboratory some specimens have actually survived pollen processing (with hot HF). However, exposure of marsh sediments to oxidation, or perhaps more accurately to the action of oxidizing bacteria, will destroy the organic wall of the foraminiferal tests and cause them to collapse. This was first noticed when we examined marsh sediments exposed on an erosional bank and found few or no foraminifera where they should have been abundant. Subsequent drilling in the same marsh deposit, but away from the exposed bank, demonstrated that there were abundant foraminifera in the fresh unexposed sediments. The same destruction of foraminiferal tests will take place if collected samples are left to stand at room temperature for any length of time. In light of this we suggest that

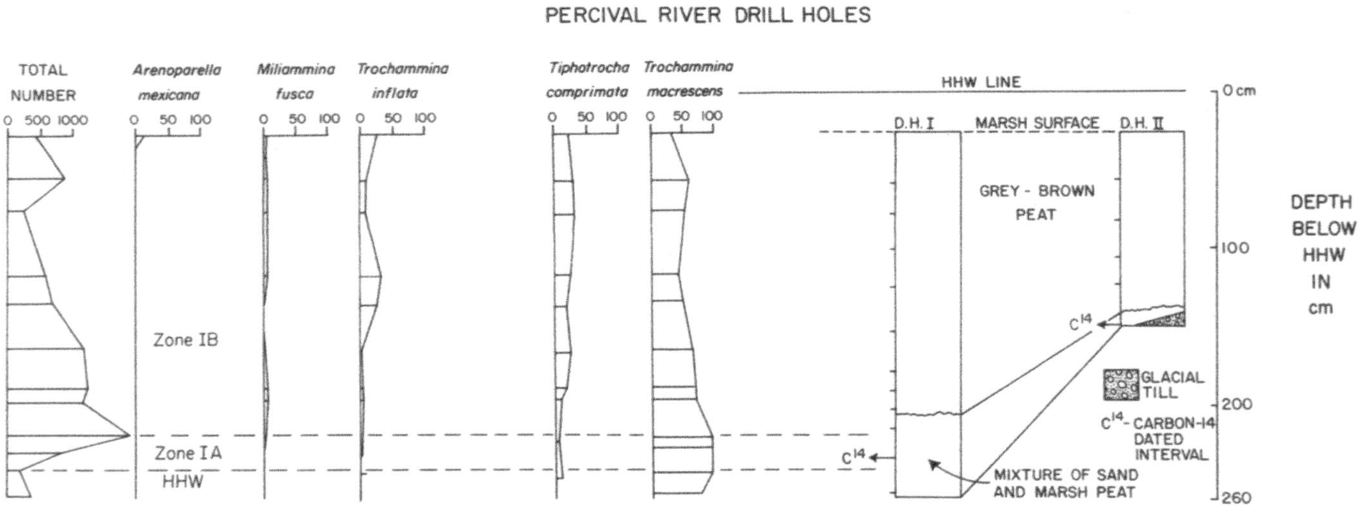

Figure 6. This short sequence from Percival River, Prince Edward Island, illustrates the same type of interval as shown in Figure 4 (see Scott *et al.*, 1981 for location). Again note here that the number of foraminifera/sample decreases toward the base of the sequence and at the interval which was chosen as the ^{14}C dating interval, there is a zone IA fauna indicative of that 5 cm interval just below HHW.

either the samples be processed immediately after collection (the safest way) or if this is impossible, that preservative be added immediately to stop any bacterial growth. When adding preservative to unprocessed marsh samples care must be taken so that the preservative gets thoroughly mixed into the sometimes tightly bound marsh sediments.

Collection of surface data There is a good basic knowledge of most types of marsh environments and their associated vertical foraminiferal zones but, as mentioned earlier, we do suggest detailed studies in new areas of investigation. Although, as we have noted earlier, the same species do re-occur in almost all zone I faunas local subzones may develop that can enhance the accuracy of the method (such as subzone IB_1 in Nova Scotia) and also provide information on salinity as well as elevation. On Prince Edward Island we noticed local effects from man-made causeways that would have been overlooked (Scott *et al.*, 1981) had we not performed preliminary surface studies. *Methods of collection, preparation, and examination of surface material are detailed in* Scott & Medioli (1980a).

Collection of subsurface material We have briefly mentioned the problem of compaction of peat in relation to Figure 5. In areas where marsh formation is predominantly peat, only the basal sections resting on a hard substrate can be used as SL points. Any peat above the base can be compacted down to as little as 1/5 of its original thickness (Kaye & Barghoorn, 1964).

In other areas where the marsh sediments are relatively non-compactable muds (eg. California and Greece) compaction may be less of a problem and one long marsh section may be suitable for several SL points. However, it is always preferable to use only a basal section since it will usually contain the most precise indicator and there will be no doubts about compaction.

Finally it is suggested that a complete stratigraphic section similar to that in Figure 5 be obtained from each study area. Although a reliable SL curve can be obtained from basal sections only, the complete environmental history of an area can help to explain the effects of SL change. Also, the effects of SL rise may be masked by other effects such as increased sedimentation rates.

Problems with radiocarbon dating In our earlier work we always attempted to use the small pieces of detrital wood that collected in paleo-strandlines. The reasoning here was these small pieces of wood could not have survived much transport, therefore they had to be contemporaneous with the time of deposition. However, even though this is a valid assumption the possiblity cannot be excluded that pieces of wood were reworked. Moreover, we often do not find sufficient detrital wood to obtain a reliable date so we have begun to use whole peat samples. Whole peat always provides sufficient organic material but it has the inherent problem of root contamination from above. However, this appears to be insignificant where we have been able to obtain both peat and wood dates from the same level. Therefore, we have now switched to using whole peat samples exclusively for dating. For checking on dates, particularly younger dates, local pollen profiles can be used. Most areas have a history of introduced pollen and identification of first occurrences of the introduced pollen can often provide an age within ± 10 years (Mudie & Byrne, 1980). Lead 210 dating may also be used for dating material less than 100 years old.

OTHER FORAMINIFERAL ASSEMBLAGES AS SEA-LEVEL INDICATORS

Although marsh foraminiferal assemblages provide highly reliable paleo-SL determinations, marsh deposits are not always present or preserved at critical locations, particularly in areas that have undergone emergence and subsequent subaerial erosion. For these cases an indirect method for using other foraminiferal assemblages as SLIs was developed. This method is substantially less accurate than the previous one but it does provide a tool that can be used to obtain a continuous emergence curve (Scott & Medioli, 1980c).

In areas with many small basins (as most formerly glaciated ones are) sediment is continuously deposited either under fresh-water or marine conditions. If the area has experienced either regression or transgression, there is an excellent possibility that both marine and fresh-water sediments will have been deposited in the same basin. These sediments can then be differentiated using foraminiferal and fresh-water rhizopod ('thecamoebian') remains, the transition point can be ^{14}C dated, and the SL determined by equating it to the elevation of the basin sill at the moment of cut-off (Fig. 7). This method has been used successfully with diatoms (eg. Berglund, 1971; Alhonen, 1978; see also Palmer & Abbott, this volume), however, diatoms are sometimes extremely rare in some periglacial environments (few were found in the marine sediments from New Brunswick, A. Palmer, pers. comm.).

a) Sample site selection

Selection of the proper type basin (either submerged or emerged) is often the most difficult part of the operation. Not all areas are suitable, especially those areas which have relatively flat terrain. The process of selecting emerged basins (present day fresh-water lakes) is much easier since sites can be investigated directly; submerged basins often must be located using bathymetric charts and/or seismic profiling and the nature of the basin sills cannot be examined directly.

The ideal type of basin is one where there is an easily definable sill, preferably in bedrock so that intervening erosion is less of a problem. Experience has shown that small and relatively shallow (3-19 m water depth below sill) basins provide the most reliable SL results since marine influence is almost instantaneously affected as the

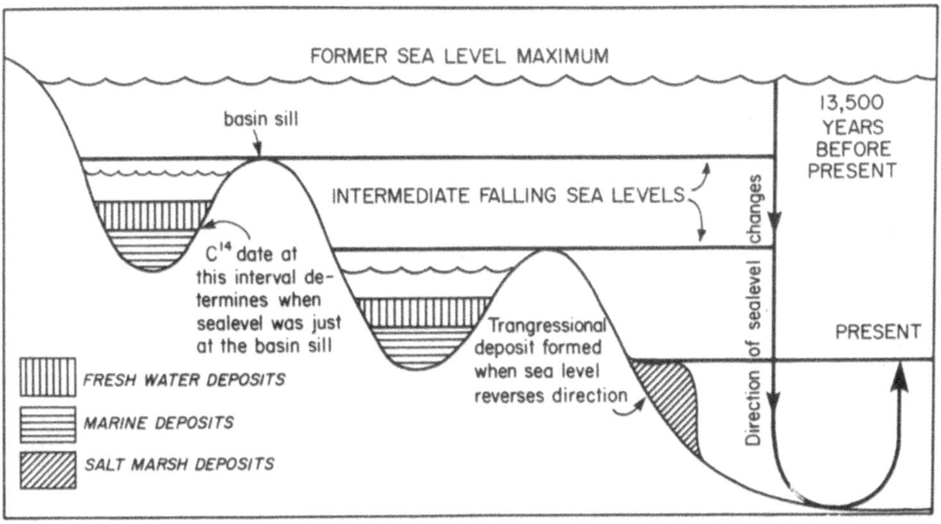

Figure 7. This diagram is an idealized conception of how to use formerly submerged (or emerged) basins as sea-level indicators. The foraminifera and thecamoebians are used to differentiate the marine from fresh-water sediments and to identify the exact transition point from where the ^{14}C date should be obtained. The actual sea-level curve shown is of course illustrative but it does demonstrate how the marsh foraminiferal method and this one can complement each other.

sill becomes emerged or submerged; this means that the marine/fresh-water transition relates directly to when SL (here we assume MSL) was at the elevation of the sill. The term marine here refers to any tidal influence and may represent any range of salinity from brackish to normal as long as there was some tidal flow over the sill.

When suitable basins have been located cores must be obtained that penetrate the entire sequence of the fresh-water/marine transition sediments. Preliminary surveys of prospective basins must be carried out to locate the deepest sediment thickness. In the case of marine basins it may be possible to locate the best coring site with seismic profiling. In small fresh-water basins usually locating the deepest water is sufficient. In offshore basins a large oceanographic vessel capable of handling a heavy coring head will be required for coring; in some cases a vibra-core may be necessary to penetrate a sand or gravel lag deposit. In small lakes a much less expensive method can be employed with a small portable platform and a manually operated Livingstone corer (see Wright, 1967, for details of this corer).

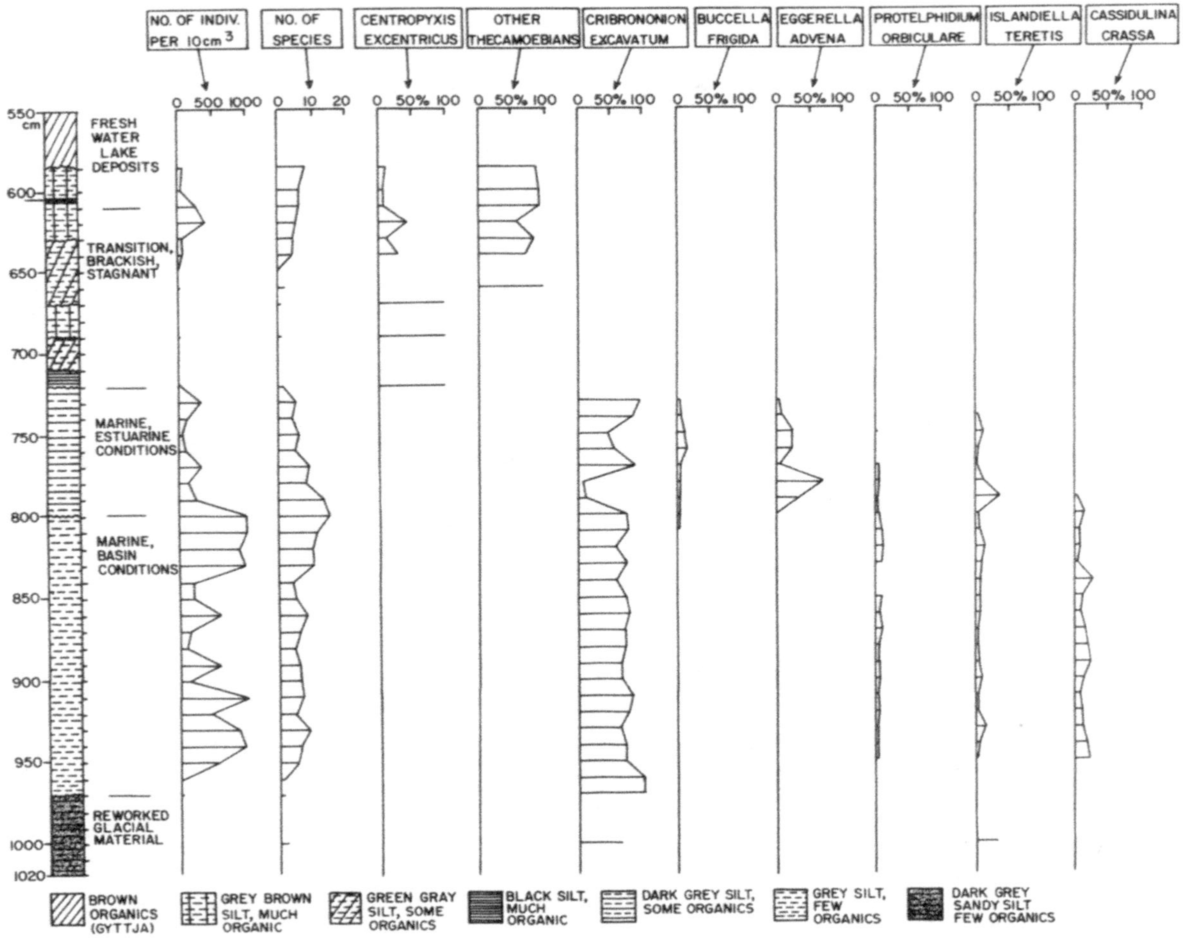

Figure 8.

Diagram illustrating one of the down-core sequences encountered in S.W. New Brunswick (see Scott & Medioli, 1980c for details). The horizontal bars represent percentages at each level. The particular fauna seen here is characteristic of early post-glacial sediments both in Canada and Europe. The most important part of this sequence in reference to sea level is the 710-720 cm segment which represents the time when sea level was just dropping below the lake sill.

Once the cores have been obtained samples can be processed and examined in the same manner as other micropaleontological samples (see Scott *et al.*, 1980).

Following collection of cores, the elevation of the basin sill must be determined so that the marine/fresh-water transition can be related to a specific level.

b) Application of method

The ideal type of result is illustrated in Figure 8. Here a clear break occurs in foraminiferal numbers at 720 cm indicating the cessation of marine conditions. The transition takes place very rapidly since there appears to be no progression from totally marine to estuarine conditions where one might expect to find foraminifera indicating lowered salinities. In situations of slow emergence a progression of marine faunas might be found. The interval directly above this (710-720 cm), which is barren of foraminifera, probably indicates a stagnant brackish period when sea water only reached the area during storms (which is sufficient to prevent fresh-water colonization). The stagnant sequence forms a rich organic zone since lack of circulation creates low oxygen, reducing conditions favourable to the preservation of organic carbon and this organic zone provides sufficient carbon for a reliable Carbon-14 date (Scott & Medioli, 1980c). Directly above this short, organic-rich horizon there is, what we call here, a brackish sequence. This term is used informally since we do not know salinity values; however, the presence of the thecamoebian species *Centropyxis excentricus,* even in such low numbers, indicates that salinity could not have been much higher than 1-2 ‰ (eg. DeCloitre, 1953; Medioli & Scott, 1983). The residual salinity probably comes largely from marine salt trapped in the sediments below, which is slowly released into the sediments above, and the duration of this sequence is a measure of how long it takes to completely flush one of these small basins (sometimes up to 6000 years, see Fig. 9).

To obtain a SL curve for a certain area several basins must be located to provide a sufficient number of points (as schematically shown in Fig.7). This is how the curve in Figure 9 from S.W. New Brunswick, Canada was derived.

c) Notes on method

Most errors associated with this method derive from poor selection of sites and this can be minimized if the above procedures are followed. Dating of total carbon content may also be a problem if there is an 'old' carbon source nearby. In this case it may be necessary to locate shells or other discrete carbon sources for a date. Pollen stratigraphy may also be useful.

The largest source of vertical error may be the assumption that tidal activity behind the sill ceases when the sill reaches the elevation of MSL. It is possible that

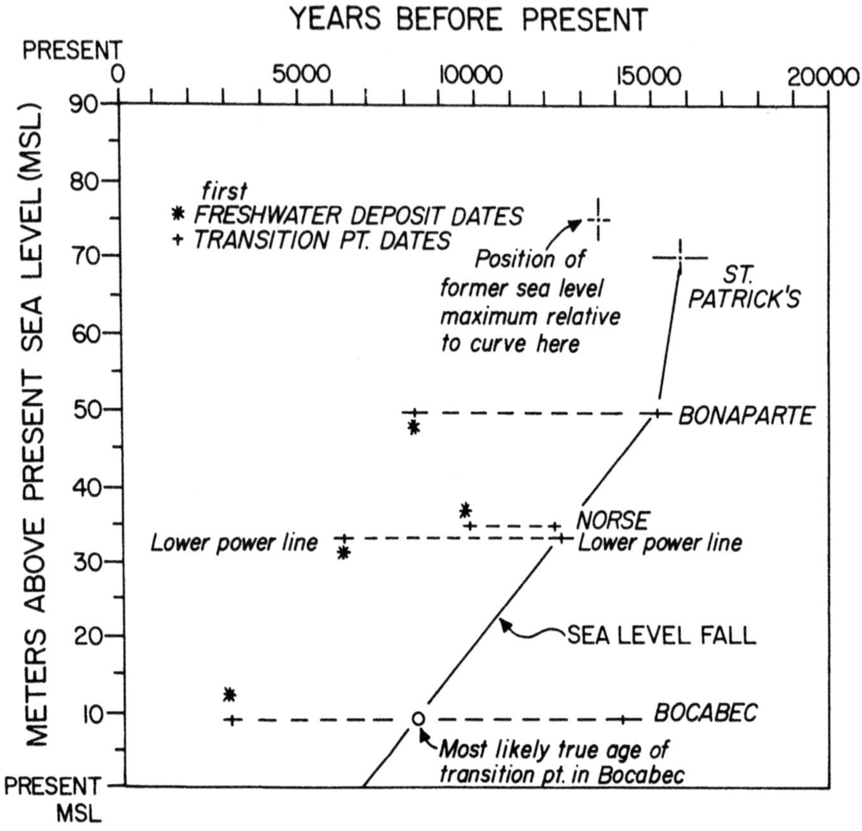

Figure 9. Emergence curve constructed in S.W. New Brunswick (see Scott & Medioli, 1980c for details) using the basin method. Lower Power Line is the lake illustrated in Figure 8. Note the wide time interval between the first fresh water and the actual last marine; this is why it is so important to identify exactly the transition point where the foraminifera abruptly disappear.

a small basin may remain actively tidal after its sill has risen above MSL or that in larger basin, tidal activity may cease before the water depth on the sill has shallowed to MSL. There would be a possibility of refining this technique if we were able to study a series of basins through their transition from marine to fresh water. However, even if tidal influence ended in a basin at either HHW or LLW instead of MSL, the total error would be 'only' ± 1 m if average tidal ranges are assumed.

The foraminifera should present little problem; different assemblages may occur in the core as the basin shallowed with lowering SL, but all foraminifera will disappear at the transition (as they do at the HHW mark in marsh foraminiferal zonations). The only problem we have encountered with the use of foraminifera in this method was in northern Maine (USA) where the sedimentation rates were so high during early post-glacial times that the fauna was diluted to the extent that there were only one or two specimens per sample (this was noted in the same deposit by Buzas, 1965). The low abundances made it impossible to delineate exactly where the transition had taken place.

One rather important advantage of this method is that in areas that have experienced both fall and rise of relative SL, a strategically located basin may record two SL points: one emergent time and one submergent time. This type of situation was located in Bedford Basin, Nova Scotia and delineated using foraminifera and thecamoebians (Miller *et al.*, 1982).

REFERENCES

Alhonen, P., Eronen, M., Salomaa, R. & Uusinoka, R., 1978. A contribution to Holocene shore displacement and environmental development in Vantaa, south Finland: the stratigraphy of Lake Lammaslampi. *Bulletin of the Geological Society of Finland*, 50, 69-79

Boltovosky, E. & Wright, R., 1976. *Recent Foraminifera*. (Published by Dr. W. Junk, The Hague)

Berglund, B.E., 1971. Littorina transgressions in Blekinge, south Sweden. A preliminary survey. *Geologiska Föreningens i Stockholm Förhandlingar*, 93, 3, 625-652

Buzas, M.A., 1965. Foraminifera from the late Pleistocene clay near Waterville, Maine. *Smithsonian Miscellaneous Collections*, 145(8), 30 pp

Chapman, V.J., 1960. *Salt marshes and salt deserts of the world*. (Leonard Hill LTP, London)

Chapman, V.J., 1976. *Coastal vegetation*. (second edition; Pergamon Press, Toronto)

DeCloitre, L., 1953. Recherches sur les Rhizopodes thecamoebiens D'A.O.F. *Memoires de l'Institut Francais d'Afrique Noire*, 31, Ifan Dakar, 59-74

Deonarine, B., 1979. Foraminiferal distribution in two Nova Scotia marshes. *Maritime Sediments*, 15(2), 35-46

Ellison, R.L. & Nichols, M.M., 1976. Modern and Holocene foraminifera in the Chesapeake Bay Region: 1st Int. Symposium on Benthonic Foraminifera of the Continental Margins. Part A. Ecology and Biology. *Marine Sediments, Special Publication*, 1, 131-151

Haynes, J.R., 1981. *Foraminifera*. (John Wiley and Sons, New York)

Kaye, C.A. & Barghoorn, E.S., 1964. Late Quaternary sea-level change and crustal rise at Boston, Massachusetts, with notes on the autocompaction of peat. *Geological Society of America Bulletin,* 75, 63-80

Loeblich, A.R. Jr. & Tappan, H., 1964. Sacordina chiefly 'Thecamoebians' and Foraminiferida. In: *Treatise on Invertebrate Paleontology,* ed. R.C. Moore, part C, 1 and 2, Kansas University Press

Lohman, G.P., 1978. Abyssal benthonic foraminifera as hydrographic indicators in the western South Atlantic Ocean. *Journal of Foraminiferal Research,* 8(1), 6-34

Lutze, G.F., 1965. Zur Foraminiferen-Fauna der Ostsee. *Meyniana,* 15, 75-142

Lutze, G.F., 1968. Jahresgang der Foraminiferen-Fauna in der Bottsand-Lagune (westliche Ostsee). *Meyniana,* 18, 13-30

Lütze, G.F., 1974. Foraminiferen der Kieler Bucht (westliche Ostsee). 1. 'Hausgartengebeit' des Sonderforschungbereiches 95 der Universität Kiel. *Meyniana,* 26, 9-22

Medioli, F.S. & Scott, D.B., 1983. *Modern and early Holocene Arcellacea (Thecamoebians) from eastern Canda.* Cushman Foundation for Foraminiferal Research, Special Publication no. 21

Miller, A.A.L., Mudie, P.J. & Scott, D.B., 1982. Holocene history of Bedford Basin, Nova Scotia: Foraminifera, dinoflagellate, and pollen records. *Canadian Journal of Earth Sciences,* 19(12), 2342-2367

Mudie, P.J. & Byrne, R. 1980. Pollen evidence for historic sedimentation rates in California coastal marshes. *Estuarine and Coastal Marine Science,* 10, 305-316

Murray, J.W., 1973. *Distribution and ecology of living benthic foraminiferids.* (Crane, Russak & Co., New York)

Palmer, A.J.M. & Abbott, W.H. (this volume). Diatoms as indicators of sea-level change.

Phleger, F. B, 1960. *Ecology and distribution of recent Foraminifera.* (Johns Hopkins Press, Baltimore)

Phleger, F. B, 1966. Patterns of living marsh foraminifera in south Texas coastal lagoons. *Boletin Sociedad Geologica, Mexicana,* 28(1), 1-44

Phleger, F. B, 1967. Marsh foraminiferal patterns, Pacific coast of North America. *Anales Instituto de Biologia, Universidad Nacional Autonoma de Mexico,* 38, Serie Ciencas del Mar y Limnologia, 1, 11-38

Phleger, F. B, 1970. Foraminiferal populations and marine marsh processes. *Limnology and Oceanology,* 15(4), 522-534

Phleger, F. B and Bradshaw, J.S., 1966. Sedimentary environments in a marine marsh. *Science,* 154, 1551-1553

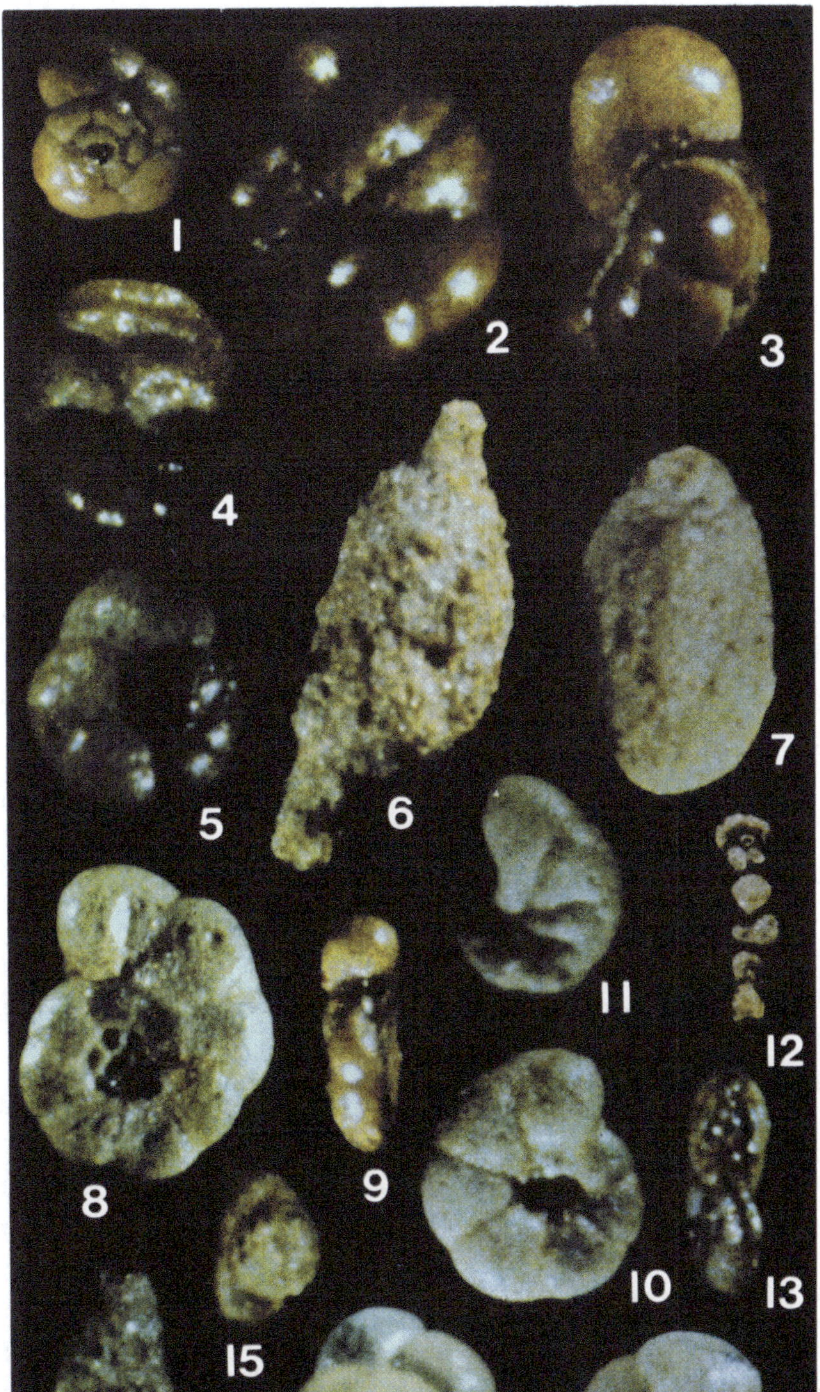
1
2
3
4
5
6
7
8
9
10
11
12
13
14
15
16
17

Photo 1

All magnifications are 20-50x; no exact magnifications are given for each specimen since sizes are variable.

1-3. *Trochammina inflata*; (1) dorsal view, (2) ventral view, (3) apertural view. Note the dark brown center in 1.

4, 5. *Tiphotrocha comprimata*; (4) ventral view, (5) dorsal view. Note the T-shaped final chamber on the ventral side and the more irregular outline of the chambers compared to *T. inflata*.

6. *Pseudothurammina limnetis*; typical side view of this single chambered species. For a better idea of variability in this species look at the type figures in Scott & Medioli (1980a).

7. *Miliammina fusca*; frontal view of the four chamber side with the aperture pointing up.

8-10. *Trochammina macrescens*; (8) dorsal view, (9) side apertural view, (10) ventral view. Note the collapsed appearance of the inner chambers of 8; this is common even in living specimens; on the apertural view note that the surface is smooth and rounded with no perforations.

11, 13. *T. macrescens f. polystoma*; (11) ventral view, showing the curvature of the sutures which are straight in the other forma. (13) side apertural view. Note the concavity of the face and you can just make out the extra apertures in the concave depression. To see the full intergradation between the two formae see Scott & Medioli (1980a).

12. *Polysaccammina ipohalina*; the aperture is at the top and clearly visible; it is usually not this clear and the specimens are often twisted and compressed. Although this species is not discussed in the text it is one of the primary species observed in the Canadian arctic marshes (Scott & Martini, 1982).

14, 15. *Ammotium salsum*; (14) side view, (15) apertural view looking down. This is a typical specimen of the species; it is distinguished from common *Ammobaculites* species by the slanted aperture at the top. Note the 'salt and pepper' appearance of the test.

16, 17. *Ammonia beccarii*; (16) dorsal view, (17) ventral view. This is the only calcareous species illustrated here; it is shown here because it is one of the few calcareous marsh species that has a world-wide occurrence. It is often cited in the literature.

Scott, D.B., 1976a. Quantitative studies of marsh foraminiferal patterns in southern California and their application to Holocene stratigraphic problems. 1st Int. Symposium on Benthonic Foraminifera of Continental Margins. Part A, Ecology and Biology. *Maritime Sediments, Special Publication*, 1, 153-170

Scott, D.B., 1976b. Brackish-water Foraminifera from southern California and description of *Polysaccammina ipohalina* n.gen., n.sp. *Journal of Foraminiferal Research*, 6(4), 312-321

Scott, D.B., 1980. Morphological changes in an estuary: a historical and stratigraphic comparison. In: *The Coastline of Canada*, ed. S.B. McCann, Geological Survey of Canada paper 80-10, 199-205

Scott, D.B., Mudie, P.J. & Bradshaw, J.S., 1976. Benthonic Foraminifera of three southern California lagoons: ecology and Recent stratigraphy. *Journal of Foraminiferal Research*, 6(1), 59-75

Scott, D.B., Medioli, F.S. & Schafer, C.T., 1977. Temporal changes in Foraminiferal distributions in Miramichi River Estuary, New Brunswick. *Canadian Journal of Earth Sciences*, 14(7), 1566-1587

Scott, D.B. & Medioli, F.S., 1978. Vertical zonations of marsh foraminifera as accurate indicators of former sea levels. *Nature*, 272(5653), 528-531

Scott, D.B., Piper, D.J.W. & Panagos, A.G., 1979. Recent salt-marsh and inter-tidal mudflat Foraminifera from the western coast of Greece. *Rivista Italiana Paleontologia*, 85(1), 243-266

Scott, D.B. & Medioli, F.S., 1980a. *Quantitative studies of marsh foraminiferal distributions in Nova Scotia: implications for sea-level studies*. Cushman Foundation for Foraminiferal Research, Special Publication no.17

Scott, D.B. & Medioli, F.S., 1980b. Living vs. total foraminiferal populations: their relative usefulness in paleoecology. *Journal of Paleontology*, 54, 814-831

Scott, D.B. & Medioli, F.S., 1980c. Post-glacial emergence curves in the Maritimes determined from marine sediments in raised basins. Canadian Coastal Conference 1980. *Proceedings, publ. by Canadian National Engineering and Science Research Council*, 428-446

Scott, D.B., Schafer, C.T. & Medioli, F.S., 1980. Eastern Canadian estuarine Foraminifera: a framework for comparison. *Journal of Foraminiferal Research*, 10, 205-234

Scott, D.B., Williamson, M.A. & Duffett, T.E., 1981. Marsh Foraminifera of Prince Edward Island: their recent distribution and application for former sea-level studies. *Maritime Sediments and Atlantic Geology*, 17, 98-124

Scott, D.B. & Martini, I.P., 1982. Marsh Foraminifera zonations in western James and Hudson Bays. *Le Naturaliste Canadien (Review of Ecological Systems)*, 109, 399-414

Streeter, S.S., 1973. Bottom water and benthonic Foraminifera in the North Atlantic: glacial-interglacial contrasts. *Quaternary Research*, 3, 131-141

Wright, H.E. Jr., 1967. A square-rod piston sampler for lake sediments. *Journal of Sedimentary Petrology*, 37, 975-976

16

Diatoms as indicators of sea-level change

A. J. M. PALMER and W. H. ABBOTT

ABSTRACT

It has long been recognized that many diatom taxa thrive in water of a particular salinity and that some diatoms are relatively indifferent to this environmental factor. In modern coastal environments, many habitats suitable for diatom growth are found and the component taxa of the modern diatom communities therein often reflect the prevailing salinity of that habitat. The distinctive siliceous valves of these microflora become incorporated in the coastal sediments through time and the microfossil assemblages thus formed can be recovered and evaluated to reconstruct the environmental history. For sea-level studies the most useful historical reconstruction is one in which a substantial change is found in the proportions of sensitive fresh, brackish, and marine diatoms. Many reconstructions have been made: fresh water to marine transitions; marine to fresh water changes and, multiple combinations of these. Where an older marine diatom assemblage is replaced by a younger fresh-water assemblage, a marine regression is usually inferred. Similarly, if a fresh-water assemblage is replaced by younger marine diatoms, a transgression is often the cause. The analyst cannot make interpretations based solely upon extensive micropaleontological observations but must also carefully scrutinize the field relationships at the study site. Diatom analysis is a powerful tool and a valid complement to other techniques in the study of sea-level change.

INTRODUCTION

Diatoms are microscopic unicellular plants which normally live in wet, naturally-illuminated environments as plankton or attached to a substratum. Many species have depth, and, in particular, salinity preferences. The life-span of diatoms is variable; a dominant living population can be replaced by another during a 'bloom' within a single season, and thereby respond to changing environmental conditions. These algae live as largely self-sufficient individuals but are sometimes united in a colony (or 'chain') for common benefit without organized roles. The two siliceous valves of the individual diatom frustule may or may not be similar, but each valve's shape and decoration is unique for each taxon. Most diatom taxa found in Pleistocene sediments are extant and thus sound paleoecological interpretations are based upon similarities between populations found in modern habitats and recovered microfossil assemblages.

Since publication of the earliest diatom literature in the 1800s, the distinction between marine and non-marine habitat preferences has been recognized. Today, within the limits of taxonomy and reliable habitat data, many diatom species can be classified according to their salt-sensitivity (Kolbe, 1927; Hustedt, 1957; Simonsen, 1962; Schrader & Schuette, 1981) as follows:

1) *Polyhalobian* (marine; greater than 30 parts per thousand salinity),
2) *Mesohalobian* (brackish water; 2-30 parts per thousand salinity),
3) *Oligohalobian* (fresh water; less than 2 parts per thousand salinity and subdivided into 'halophilic' and 'indifferent' classes), and
4) Halophobes (true fresh-water forms, intolerant of any salt).

Patrick & Reimer (1966) simplify the classification by grouping the halophilic, indifferent and halophobic classes all in the Oligohalobian category. In addition, there are *euryhaline* taxa which tolerate broad ranges of salinity (Patrick & Reimer, 1966). The reader is directed to Simonsen (1962) for a discussion of these diatoms.

Various diatom assemblages from a stratigraphic sequence can be compared and compositional differences thus detected are used to infer salinity changes in the local aquatic environment. Diatom valves which represent all the halobian classes should be anticipated in coastal sediments where changes of sea level (SL) are most effectively preserved. A coastal assemblage may consist of entirely oligohalobian or completely polyhalobian diatom valves, but most likely fossil assemblages will consist of a mixture of fresh, brackish and marine forms. The changes in dominance of a particular halobian category will be the evidence for a salinity change which in turn leads to inferences concerning changes in local SL. Because diatom populations and subsequent assemblages are often diverse (perhaps exceeding thirty taxa), it is unwise to make paleoenvironmental interpretations upon the basis of fluctuations in frequencies of a single taxon within an assemblage.

Diatom assemblages which are often found in cores of fine-grained coastal sediment usually represent nearby microfloral populations and can accurately record the changing salinity of the local environment. They are useful because:

1) diatoms are widespread in natural aquatic environments;
2) many species prefer specific salinity conditions;
3) the silica constituting the valves is relatively resistant to chemical alterations after burial; and
4) diatoms are often preserved with radiometrically dateable carbonaceous material.

DIATOMS AS SEA-LEVEL INDICATORS

1. General applications

Over the past century the diatomists of North-Europe have contributed substantially to the utilization of diatom analysis as a tool for studies of Quaternary SL changes. One of the earliest systematic diatom analyses tracing SL changes was reported by Halden in 1929. Similar work was continued in North-Europe by diatomists such as Cleve-Euler (1944), Florin (1944), Mölder (1943), Du Saar (1969) and Foged (1970). The application of diatoms as Quaternary/Holocene and possible Tertiary/latest Miocene SL indicators has been used elsewhere (eg. Abbott *in* Brooks *et al.*, 1979; and Burckle & Akiba, 1978 respectively).

Diatoms are being used also to provide basic environmental data which in turn can be applied particularly to archaeology (Jansma, 1975), an example of which is the recent report by Colquhoun *et al.* (1980) wherein changes in the elevation of the strandline (proven by Abbott's diatom analyses) in South Carolina were used to assist in the discovery of prehistoric coastal camp-sites. The salinity of the surface sea water can be altered by glaciation, glacial meltwater, coastal deformation and related phenomena. The salinity tolerances of modern diatoms (Simonsen, 1962) have also been used to reconstruct paleo-salinities of the surface waters in the Black Sea (Schrader, 1978) and Mediterranean Sea (Schrader & Gersonde, 1978).

2. Modern coastal habitats and sites of fossil accumulation

Of great importance in SL studies are *benthic* diatoms which live in association with numerous biological, geological and artificial substrates, often in sea-water depths of less than 10 m (Hendey, 1964) and probably not exceeding 100 m (Round 1971; Schrader & Schuette, 1981). The type of substrate available has considerable bearing upon the composition of associated diatom communities; the reader is directed to a broad discussion of this subject by Round (1971). Most of the benthic diatoms which live upon and within the surface sediments are elongated, or feather-shaped, forms known as Pennates. Some Pennate diatoms have the ability for vertical movement within certain substrates in response to the tidal cycle, available illumination and other factors (Pearse, 1977). The areal distribution of benthic diatom communities is controlled not only by substrate but other factors such as salinity, temperature, nutrient supply and the presence of *Spartina* marshes.

Near the sea-shore there are three zones in which diatoms may be found living: below lowest low tide (LLT), between extreme tides, and above highest high tide (HHT).

1) In the *subtidal zone* from LLT down to a depth of 10 m or more, diatoms constitute relatively simple communities.
2) In the *intertidal zone* is included the greatest variation in environmental characteristics such as: wave energy and water depth, area and composition of exposed substrate, salinity, nutrient supply and illumination. Within this zone, highly varied diatom communities develop, perhaps richly so upon muddy beaches with plentiful nutrients such as phosphates, and with much less diversity upon sandy sediments.
3) The harsh *supratidal zone* above the HHT is a suitable habitat for numerous euryhaline taxa which can tolerate the great salinity fluctuations.

Of course, within these zones the diatoms are vulnerable to disturbance and transportation as are the many substrates with which they are associated. The reader will find Hendey's (1964) useful discussion of community composition at the species level extended by other authors such as: Round (1971), Lewin & Rao (1975), Ramm (1977), and McIntire & Moore (1977).

Sea water may transport free-floating *planktic* diatoms to the coastal sites of deposition. These natural contaminants may be either *holoplanktic* forms which live and reproduce within the water column, or *meroplanktic* forms by virtue of spending part of their lives on the bottom and part as free-floaters. Most of these *planktic* diatoms are circular in outline and are thus known as Centrics.

Pennate and Centric diatoms are also transported to the coastline by flowing fresh water. Apparently, true *holoplanktic* diatoms do not normally occur in fresh-water habitats (Patrick & Reimer, 1966) whereas *meroplanktic* and *benthic* forms are common. Note that Patrick & Reimer (1966) clearly distinguish the fresh-water *epiphytic* taxa, which specifically attach themselves with mucilage, from the fresh-water *benthic* diatoms, (in contrast to the broader concept of marine *benthic* diatoms as used by Schrader & Schuette (1981), cited above). A vast amount of habitat data for fresh-water forms is located in many individual publications, but essential information will be found in Cleve-Euler (1951-1955), Hustedt (1930-1966), and Patrick & Reimer (1966).

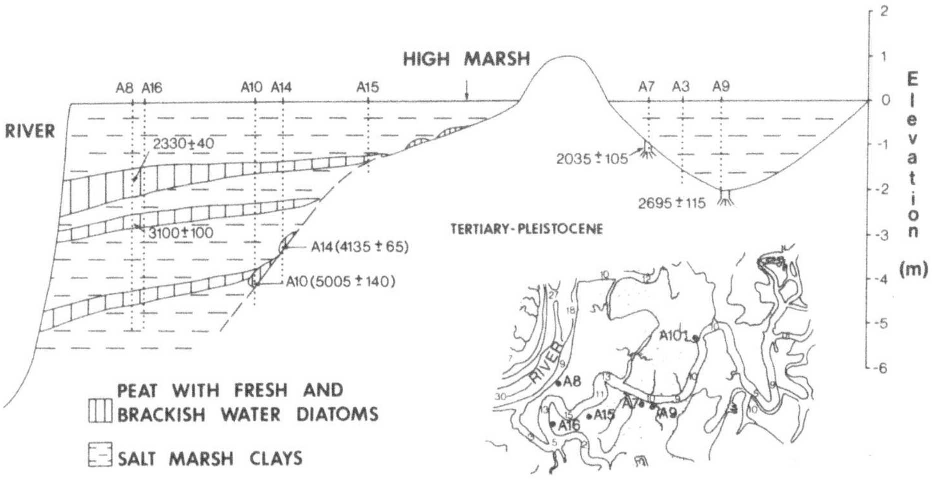

Figure 1. Stratigraphic cross-section, Cooper River-Grove Creek area. (from Brooks *et al.*, 1979, with permission, slightly modified).

In addition to the variety of habitats represented by diatoms deposited at the sea-shore are singular depositional events wherein large numbers of uni-specific communities may be deposited. The periodic *blooms* of certain diatom taxa through the growing season would seemingly contribute an inappropriate bias to the microfossil content of the sediments. As in all but the most specialized of micro-paleontological investigations, this is not a major problem because of continuous mixing of the sediments by processes such as bioturbation. This constant disturbance of sediments accumulating through time will tend to reduce the apparent importance of blooms and other seemingly ephemeral events and will result in a sediment composition analogous to the *moving average*. This continuous mixing suggests that there would be little advantage to be derived from a very close spacing of samples (measured, say, in mm), except in special cases. Nonetheless, long-term environmental changes such as movement of SL will be recorded through continuous sedimentation as new trends in the gross microfossil composition, and obviously such substantial changes are necessary to alter the proportions of halobian classes in undisturbed sediments. Theoretically, the most accurate determination of SL is possible if a sequence of shore-line sediments are preserved wherein the rise or fall of SL would be recorded as a distinct difference in continuously accumulated microfossil assemblages from the supratidal and tidal environments. Although salt-marsh deposits often are not well-preserved, they have perhaps the best potential for accurately recording the phenomenon, at least in temperate latitudes as illustrated in Figure 1. Salinity changes, that may reflect the SL changes, can be documented with high positioning accuracy if the following parameters are properly characterized:

1) Statistically valid differences in the proportions of halobian categories quantified within a cored vertical sedimentary sequence, and

2) the *elevation,* relative to a recognized datum, of the detected salinity variation within the sedimentary sequence, and its (radiometric) *age*.

3. Submerged or emerged basins

In addition to diatom assemblages incorporated into strandline sediments, fossil diatoms deposited in a basin behind a sill (Fig. 2) can be used successfully to identify local salinity changes from which changes in relative SL (or tidal range) may be inferred. It is not possible to estimate the local water depth in the basin, however; only the sill elevation is the critical value. The minimum requirements for adequate interpretations include, as before:

1) diatom-assemblage differences in halobian categories,

2) the *elevation* and radiometric *age* of the halobian category shift under investigation, and of equal importance,

3) the lowest *sill elevation* which controls the first marine effects during relative SL rise, or the last marine effects during relative SL fall.

The original elevation of the diatomaceous deposits accumulated at a strandline or behind a sill may be subsequently altered by geological processes, and these secondary events may complicate the interpretations.

4. Local regression records

A relative fall of SL is only recorded in sediments which, for adequate geological time, cease to be submerged by the sea. Regressions are well-illustrated in reports pertaining to Holocene sediments deposited behind sills in uplifted areas of northern Europe. Some useful pertinent references include the stratigraphic work of: Håkansson in Digerfeldt (1979); Eronen (1974, 1982); Foged (1972) and Glückert (1976).

The local regression at about 6000 BP in Lianvatn, Frosta, Norway is clearly represented at 42.5 m above SL (Kjemperud, 1981a). Figure 3 illustrates that the early dominance of marine-brackish diatoms was later replaced by fresh-water forms characteristic of a lacustrine environment.

5. Local transgression records

The diatomaceous sediments which record a local rise of SL typically consist of fresh-water diatoms sequentially overlain by brackish and marine taxa. Offshore, a transgressive Holocene sequence in Long Island Sound, New York, U.S.A. was cored by Schaffel (1971) as were sediments in the southeastern Baltic by Kabailiene (1974). Clearly if these

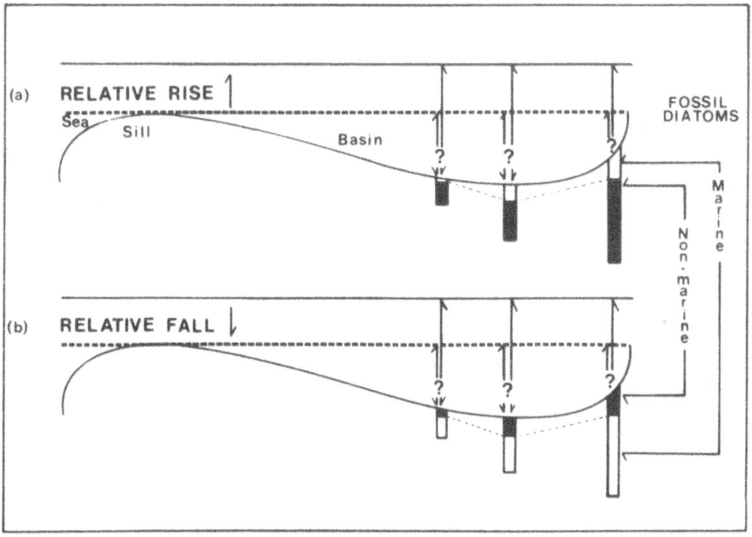

Figure 2. Illustration of changes in diatom assemblages reflecting salinity changes caused by relative changes in sea level. The depth of water column behind the sill cannot be inferred with high accuracy regardless of a relative rise (a) or a relative fall (b) in sea level; however, the sill elevation is the critical value which must be known.

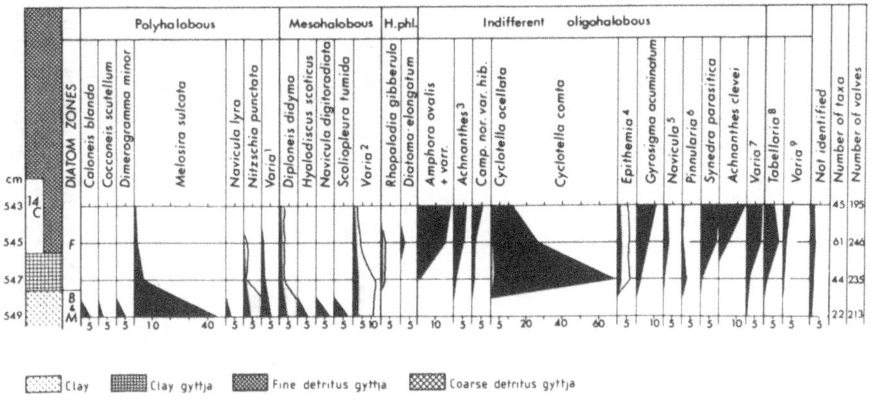

Figure 3. Illustration of marine regression shown in diatom zonation based upon salinity preferences of diatom microfossils. A fresh-water zone (F) overlies a brackish-marine zone (B&M). (From Kjemperud, 1981a, with permission.)

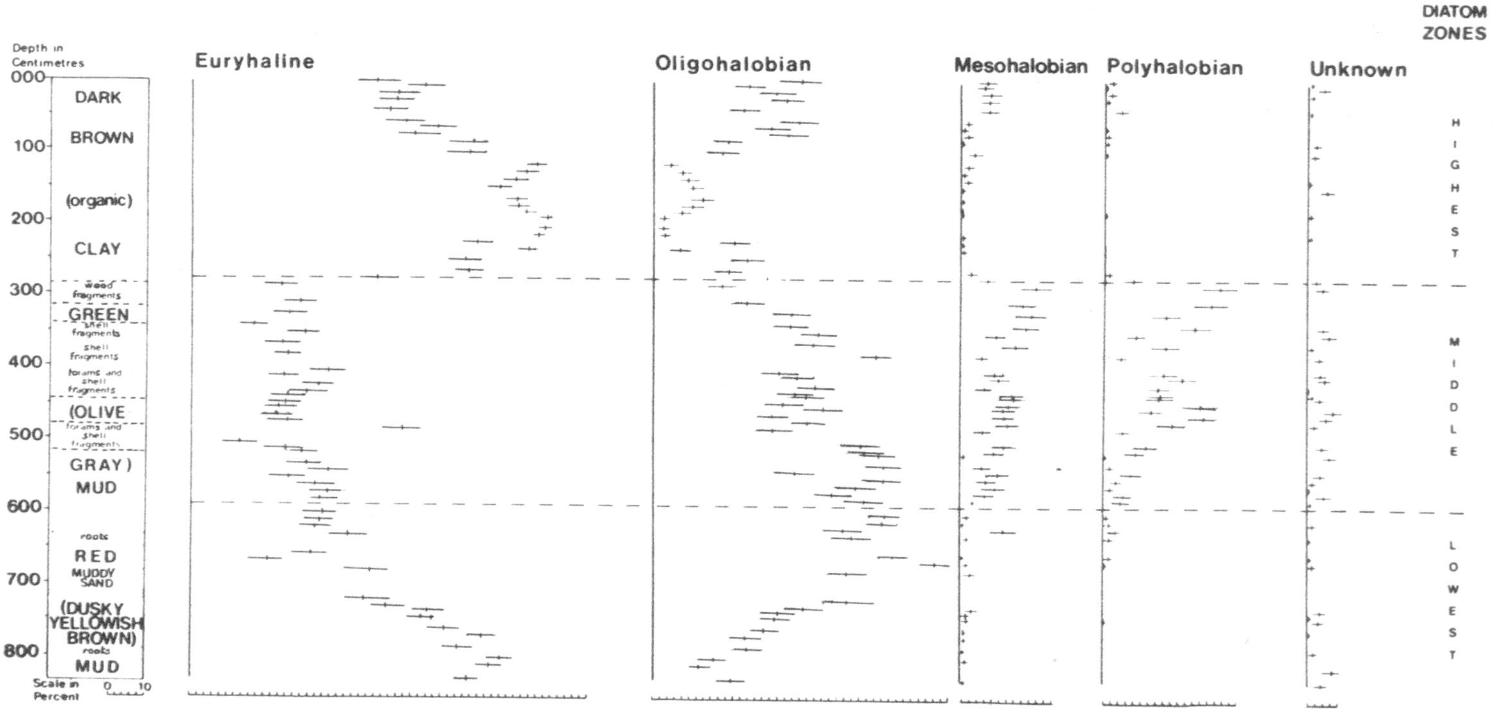

Figure 4. Distribution of components of various fossil diatom populations found at about 70 levels within core of estuarine sediments at Basin Head, Prince Edward Island. The increase in mesohalobian and polyhalobian taxa in the middle zone records a local SL rise: their decline in the upper zone does not represent a SL fall, but rather reduced salinity caused by formation of a barrier (from Palmer, 1978, with permission, slightly modified).

marine sites had been constantly under sea water a transgression could not have been recorded.

The absolute necessity of combining diatom data with all other field evidence is illustrated in the interpretation of Figure 4. At an estuarine site in Atlantic Canada local SL rose over a -1.0 m sill about 2600 BP resulting in the appearance and persistent increase of mesohalobian/polyhalobian taxa deposited over fresh-water diatoms. The continued rise of SL led, however, to formation of a nearby barrier which caused subsequent freshening of the habitat. If reliance had been placed solely upon the diatom stratigraphy, the highest (estuarine) zone may have been erroneously ascribed to a modern fall of SL[1] (which is totally contrary to other field evidence). Concerning onshore sites, some very useful references are Alhonen *et al.* (1978), Berglund (1971), Digerfeldt (1975) and Foged (1970). The reader should note that the latter (transgressive) sites are no longer submerged by sea water and have been re-exposed by geological processes. This implies subsequent regression; however parts of the record may be lost because of erosion and other factors.

6. Compound local transgression and regression records

The diatom microfossils may be adequately preserved to permit also the interpretation of multiple transgressions and regressions. Stabell (1982) for example, has reconstructed Late Quaternary diatom assemblage changes and thus the salinity variations related to complicated multiple regression-transgression sequences. Note, in Figure 5, that Stabell's interpretation of diatom analyses from Klaesvatn and Hamravatn, Norway are particularly complex (marine-lacustrine-marine-lacustrine) and by implication were difficult to analyze.

The results of analyses like those by Stabell lead to rigorous interpretations of shoreline displacement in appropriate regions. Although this latter subject is not developed here, the reader may profitably examine the applications of inferred SL changes, specifically in uplifted areas, as admirably demonstrated in important references such as: Björck & Digerfeldt (1982); Digerfeldt (1975); Florin (1946); Hyvärinen (1980); Kjemperud (1981b); Miller & Robertsson (1979); Mörner (1978); Renburg (1976); Stabell (1980); and Thomsen (1982). There are many areas of the world which have potential for fascinating interpretations of diatom analyses but few have been subjected to intensive glaciation, related crustal deformation, and great changes in relative SL as demonstrated in northern Europe. Thus, the evidence of salinity variations recorded by fossil diatoms may be much more subtle and prove to be very challenging, particularly for the analysts studying more poorly known coastal regions.

[1] This goes to show that the terms 'transgressive' and 'regressive should be used in a descriptive sense only. Transgressive and regressive sequences, or transgression and regression may, but *need not* imply rise and fall of SL.

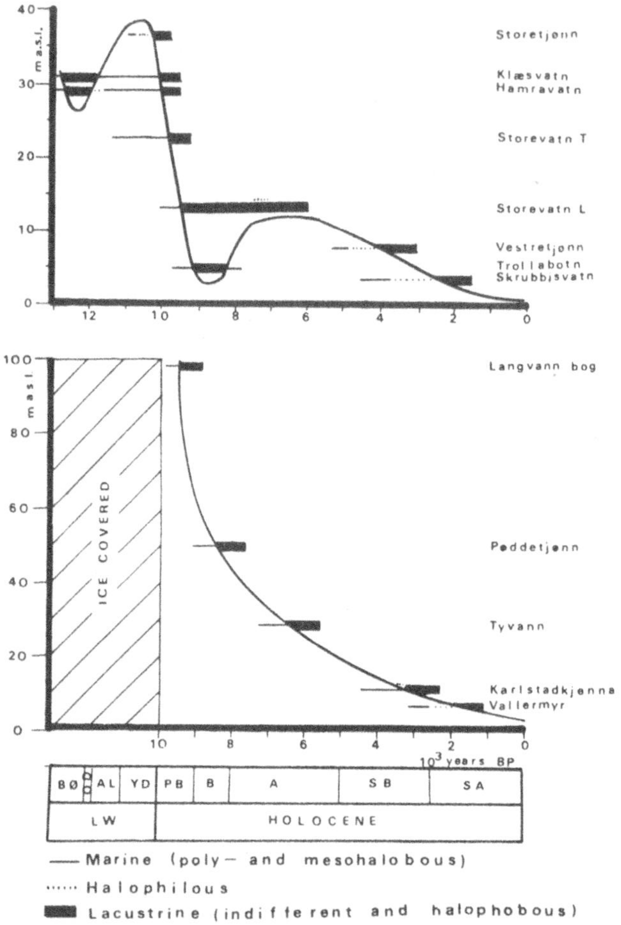

Figure 5. Illustration of *interpreted* diatom succession and a curve of local sea-level change in Norway. The simplest transitions from marine to lacustrine habitats can be seen to have occurred at Storevatn T, Langevann Bog, Pøddetjønn, and Tyvann. The most complex changes occurred at Klaesvatn and Hamravatn where there were sequences of marine-lacustrine-marine-lacustrine environments between 13 000 BP and 9000 BP (from Stabell, 1982, with permission, reduced).

PRELIMINARY MICROFOSSIL EVALUATION

1. General methodology

As in any biostratigraphic problem, reconnaissance material must be examined initially to ascertain the magnitude and direction of gross changes within the sedimentary sequence.

A suite of any convenient number of samples from a sequence is prepared for analysis; a small part of the top and bottom parts of the sequence should be included. The reconnaissance samples are then prepared and examined. The goal of this analysis is to determine the change in diatom assemblages that may reflect salinity changes. This is accomplished in several steps:

1) prepare suitable microscope slides;
2) count frequencies of each recognizable taxon and unidentifiable specimens for each sample level (about 300 individual valves);
3) assign each taxon to a halobian category and calculate the proportion of each taxa within each category; and
4) plot the frequency/core-depth relationship for each halobian category.

2. Preparation for microscopy

Potentially diatomaceous sediments are disaggregated and cleaned for analyses under the transmitted-light microscope in an oxidation process which selectively destroys most non-siliceous microfossils (Schrader, 1973). However, complete oxidation of the sediment, particularly salt-marsh peats, may be difficult or impractical. (If analyses for other microfossils are anticipated, the entire sample should not be used.) Other microfossils such as organic dinoflagellates, and siliceous forms such as possibly very abundant sponge spicules, and probably rarer radiolarians and silicoflagellates may be observed; extraneous microfossils are distinguished by ornamentation and relative size as illustrated in Figure 6.

An aliquot of the cleaned sediment suspension is transferred to a clean microscope cover slip with a disposable pipette and allowed to dry. The randomness of a fossil diatom assemblage thus attached to a microscope cover slip is a function of numerous factors including:

1) taxonomic diversity;
2) thoroughness of disaggregation of both fecal pellets and fine-grained minerals;
3) sample homogenization; and
4) the surface tension of the aqueous suspension.

Clearly, taxonomic diversity cannot be controlled, but the other factors can be affected by laboratory procedures.

The surface tension of the suspension droplets can prove to be the most vexing of all difficulties in mounting random samples. This problem can be partly resolved by altering the aqueous component. Diluted dish-washing detergent (J. Fenner, pers. comm., 1978), or photographic wetting agent can be added to reduce the surface tension and thus prevent formation of a crust or miniscus rim of diatom valves concentrated at the periphery of the subsample dried upon a cover slip. The senior author has had

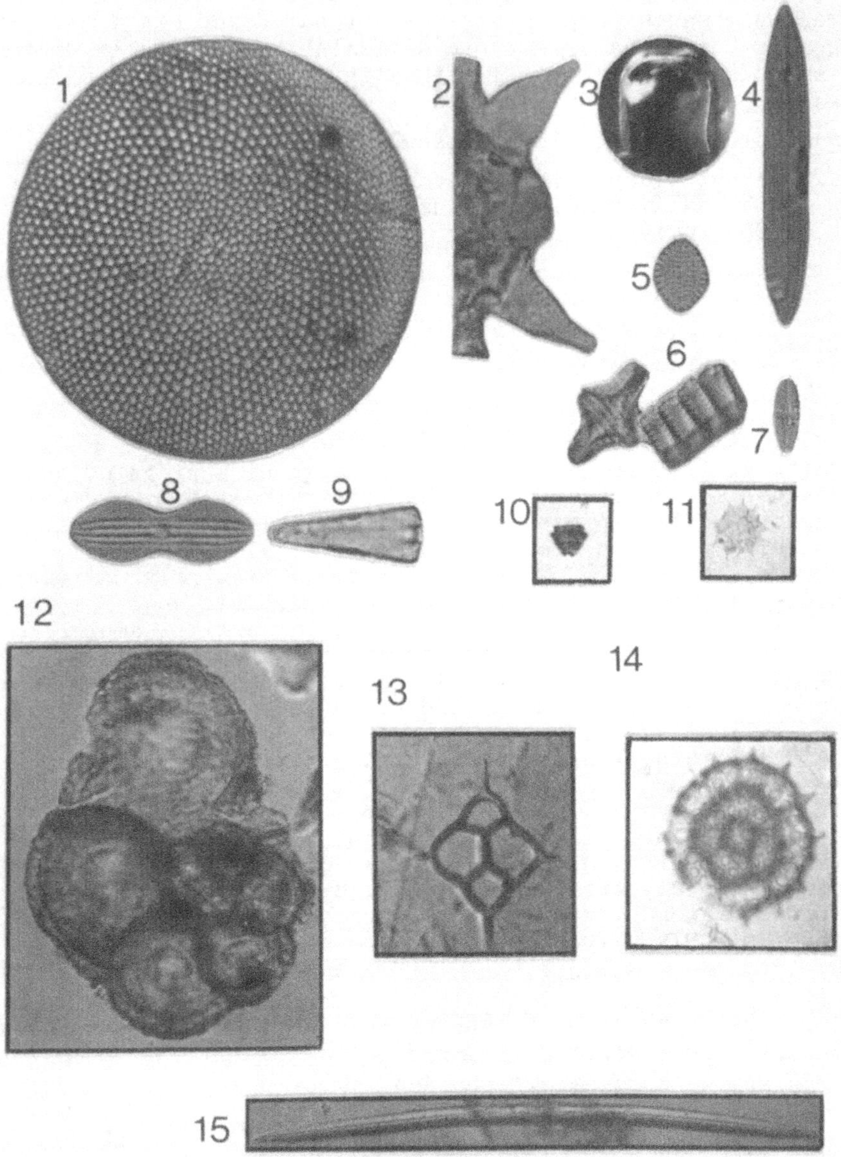

Figure 6. Illustration of various microfossils which may be found in coastal sediments.

1) 'Centric' diatom, valve view of *Coscinodiscus* sp. (500x).
2) 'Centric' diatom, girdle view of *Biddulphia aurita* (1000x).
3) 'Centric' diatom, girdle view of *Melosira* sp. (1000x).
4) 'Pennate' diatom, valve view of *Nitzschia* sp. (1000x).
5) 'Pennate' diatom, valve of *Rhaphoneis* sp. (1000x).
6) 'Pennate' diatom, valve view and girdle view of a chain, *Fragilaria* sp. (1000x).

considerable success with a mixture of 50% distilled water and 50% ethyl alcohol (Smith, 1873) in which the microfossils are suspended and also stored with no apparent deleterious effects. This proportion suitably reduces the surface tension, specific gravity and viscosity, thereby enhancing random dispersal upon a slide. Furthermore, if the well-mixed aliquot upon the warmed cover slip is immediately ignited with a toothpick taper, the resultant boiling action also contributes to: randomization, complete and immediate removal of the alcohol phase, and speedier evaporation of the distilled water.

When dry, a small drop of high refractive index mounting medium is applied to cover the inverted cover slip and representative sediment. The volatile solvent can be easily driven off by gentle heating. The cover slip can be then placed (microfossils down) upon the microscope slide and the entire unit can be heated to exploit the thermoplastic properties of the mountant which will melt and spread by capillarity to the edges of the cover slip (eg. Hyrax melts at about 130°C). The cover slip can also be picked up by a heated microscope slide.

Only after the reconnaissance samples are cleaned and mounted can the diatom microfossil content be determined. This preliminary examination should be undertaken at moderate total magnification of about 400-500x.

A qualitative examination of the prepared microscope slides will lead immediately to an assessment of the presence or absence of diatoms, concentration of microfossils, taxonomic diversity and thus overall merit of pursuing the diatom analyses. For more details on slide preparation the reader is referred to Schrader (1973).

QUANTITATIVE METHODS

1. Absolute frequency analysis

Absolute frequency analysis of diatom assemblages is used to determine the frustule or valve sedimentation rates applicable in sophisticated productivity investigations. This form of analysis is normally a supplement or corollary to SL studies rather than an integral part of them.

---

7) 'Pennate' diatom, view of raphe valve, *Acnanthes* sp. (1000x)
8) 'Pennate' diatom, valve view, *Diploneis* sp. (1000x).
9) 'Pennate' diatom, valve view of intact frustule, *Cymbella* sp. ?.
10) Tricolporate pollen grain (1000x).
11) Colonial fresh-water green algae, *Pediastrum* sp. (62·5x).
12) Foraminifera (~250x).
13) Silicoflagellate (500x).
14) Radiolarian (nasselarian type, 1000x- scanning electron micrograph).
15) Sponge spicule (1000x).

The sample preparation is quite different from that for relative frequency analysis and uses unit weights or unit volumes of diatomaceous sediment. This sediment may be subjected to: 1) 'spiking' with a known amount of exotic pollen/spores or plastic/glass beads; or 2) preliminary but detailed analyses to determine a marker taxon of consistent proportions within the assemblages suitable as a reference; or 3) high precision sub-sampling for example by the evaporation tray technique. Because the interpretations so derived go far beyond the establishment of paleoenvironmental salinity, the reader is referred to references such as Battarbee (1973), Battarbee & Kneen (1982), and Kaland & Stabell (1981) for more information.

2. Relative frequency (%) analysis

There is a variety of valid relative frequency sample sizes arising from factors such as:

1) variable concentration of microfossils in the samples under study;
2) taxonomic diversity within the assemblages;
3) unique constraints upon the time and resources of the scientists; and
4) the relative importance ascribed to *trace* microfossil occurrences by the analyst.

Regardless, no greater interpretation of paleoenvironments can be made than that substantiated by the counting data derived from samples represented by random slide preparations (Andrews, 1972).

Although several modern diatom communities have been shown to follow truncated *normal* distributions (eg. Patrick & Reimer, 1966, p 81), the structure of fossil assemblages is quite poorly understood. However, a reasonable estimate of the desirable sample size can be made by reference to a Probable Error Nomogram for normal distributions shown in Figure 7. The inflection points of the computed curves fall below a sample population of 300 individuals and there is relatively little reduction in the probable error between 300 and 1000 individuals. It is thus inferred that a sample population of about 300 randomly selected individuals (in a normal distribution) would approximate the composition of the total assemblage fairly well. A more complete approximation would be derived from counting 1000 individuals in a sample population (Setty, 1966), however, the mitigating factors enumerated above may preclude such detailed work. For more literature on this controversial subject, the reader may wish to refer to Miller (1964), Battarbee (1973), and Schrader (1973). Note that Kjemperud (Fig. 3) successfully counted 200 specimens exclusive of the colonial *Fragilaria* species.

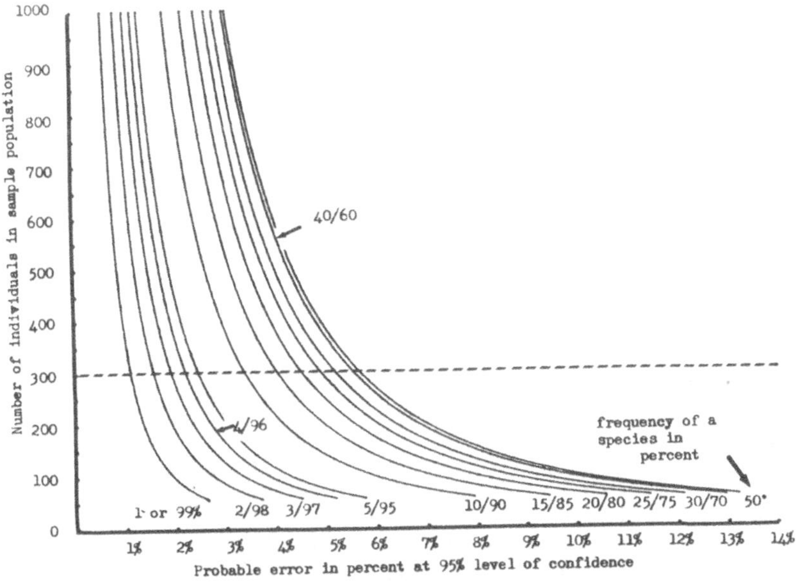

Figure 7. Probable error nomogram at 95% confidence level. Note slow decrease in probable error above a total sample population of 300 individuals. (Adapted from Galehouse, 1971.)

3. Concurrent identification and counting

A taxonomic list is begun with the first assemblage analysis and expanded as new specimens are encountered. Upon this developing list, frequencies of intact valves (or fragments greater than one-half valve) of each recognizable taxon are tabled for each stratigraphic level analyzed (along with the 'unknowns'). As the analysis proceeds, the frequencies should permit inferences about apparent variations in assemblages. Subsequent detailed sampling will be based upon these preliminary observations (there is no great need to convert the frequency counts to percentages at this stage).

Literature on diatoms commonly presents specialized data grouped for either marine and brackish forms, or freshwater taxa, thus requiring the analyst to repeatedly consult a variety of references. As a practical aid to the analyst, photomicrographs of some diatom taxa commonly found in coastal environments are presented in Plates 1& 2 and the halobian categories for those illustrated are also listed. This set of illustrations is not intended to be definitive but to assist the analyst in speedy reference to the authoritative works (eg. Cleve-Euler, 1951-1955; Greville, 1861-1866; Hendey, 1964; Hustedt, 1930-1966; Mölder & Tynni, 1967-1973; Patrick & Reimer, 1966-1975; Peragallo & Peragollo, 1897-1908; Schmidt, 1874-1979; Tynni, 1975-1978;

Van Heurck, 1896; and Van der Werff & Huls, 1957-1974) which are essential for complete descriptions of: valve shape, number of striations in 10 microns of length or circumference, growth habit, habitat preferences, etc.

Counts of individual taxa can be made on simple tally sheets or on more expensive but convenient push-button counters. The novice may not initially be able to identify each individual, but distinctive provisional names can be used. An important analytical tool at this stage is photomicrography (especially with instant photographic film) to assist in the preparation of a working file of illustrations. A complex assemblage of 30 or more taxa may be encountered in the sediments of a mixed environment with numerous unknowns. The unidentified specimens should be counted and recorded in some way such that they can be re-located for later possible identification. Three principal methods are: 1) using the mechanical grid locations if engraved upon the microscope stage; 2) using the grid location of an England Finder slide; and finally 3) circular scribing of the cover slip at the appropriate location with a diamond point. The first method is the most convenient if the use of an oil immersion (high magnification) objective is necessitated by the presence of small (3 to 50 micron diameter) diatoms.

Some practical difficulties encountered during counting include: the identification of diatoms oriented sideways in girdle view, rather than the more common valve view, and the related difficulty of identifying chains. Significantly, the major references that must be examined do not always depict the valve shape in girdle view. Where the identity of a frustule in girdle view is doubtful it is assigned to the 'unidentified' or 'unknown' category. The significance of this problem is dependent upon the frequency of frustules in girdle view.

4. Halobian category assignments

Using the information provided in the literature, each taxon is assigned to one of the halobian categories. Different varieties within given species may have different salinity preferences so classification at the highest possible level is desirable at all times. Understandably, the 'unknowns' cannot be assigned to a halobian category, nor can taxa for which habitat data are not available. These data should be combined to form a separate halobian category.

5. Graphing the results

The final treatment of numerical data is to prepare a graph of frequency versus depth for each taxon at the various stratigraphic levels. This information may be then presented in summary form by grouping the subtotals of the taxa in each halobian category as percentages and each category can then easily be graphed for various stratigraphic levels. The sedimentary sequence can then be zoned according to the proportion of each halobian category at each level.

EVALUATION OF RE-WORKED MICROFOSSILS

The best evidence for the presence of transported diatoms or bioturbation and winnowing effects is contained in:

1) the state of preservation,

2) patterns in the original abundance 'counts', and

3) sedimentary characteristics.

The abrupt appearance or disappearance of numerically important taxa will be a clue to transportation. Although a certain amount of valve breakage can be expected in sediments, if it appears to be very common, and cannot be attributed to laboratory techniques, then active re-working of sediments is a definite possibility. The reader is referred to Beyens & Denys (1982) for an appropriate discussion concerning valve disturbance. There are no criteria for recognizing disturbance; it can only be interpreted in the context of the associated sedimentary characteristics.

ACKNOWLEDGEMENTS

The authors acknowledge the scientific contributions of many individuals: We particularly thank H.B.S. Cooke, D.R. Grant and several anonymous reviewers. The plates were prepared (by A.P.) from Nomarski Differential Interference Contrast photomicrographs taken on a Zeiss Ultraphot research microscope. Figure 1 is reproduced with permission of M.J. Brooks and *The Florida Anthropologist*. Figure 3 is reproduced with consent of A. Kjemperud and *Boreas*. *The Nova Scotia Institute of Science* granted reproduction rights for Figure 4, and Figure 5 was graciously provided by B. Stabell.

We wish to thank Dr. F.J. Vernberg, Director and also the staff of Baruch Institute without whose efforts this manuscript could not have been prepared at this time. This document is contribution No.477 of the Belle W. Baruch Institute for Marine Biology and Coastal Research at the University of South Carolina.

APPENDIX 1: RECONSTRUCTION OF PALEOTIDAL RANGES

Evidently paleotides have not yet been reconstructed through the use of diatom microfossils although theoretically it should be quite possible. A procedure is outlined by which paleotides could be ascertained if suitable strandline deposits are available.

Diatom frustules are relatively resistant to post-depositional alterations leading to potential application in the reconstruction of former tidal ranges. The basic requirements are:

1) appropriate (salt marsh?) sediment in a coastal environment,
2) known underlying (bedrock?) topography for core site selection,
3) known distributions of contemporary diatom populations for direct calibration of salinity gradients recorded by the microfossil assemblages (Fig. 8), and
4) suitable coring sites from which dateable diatom assemblages can be retrieved.

The first goal is to retrieve a core (Core 1) which will contain an 'upper tidal limit' diatom assemblage (eg. 1000 BP) as illustrated in Figure 9.

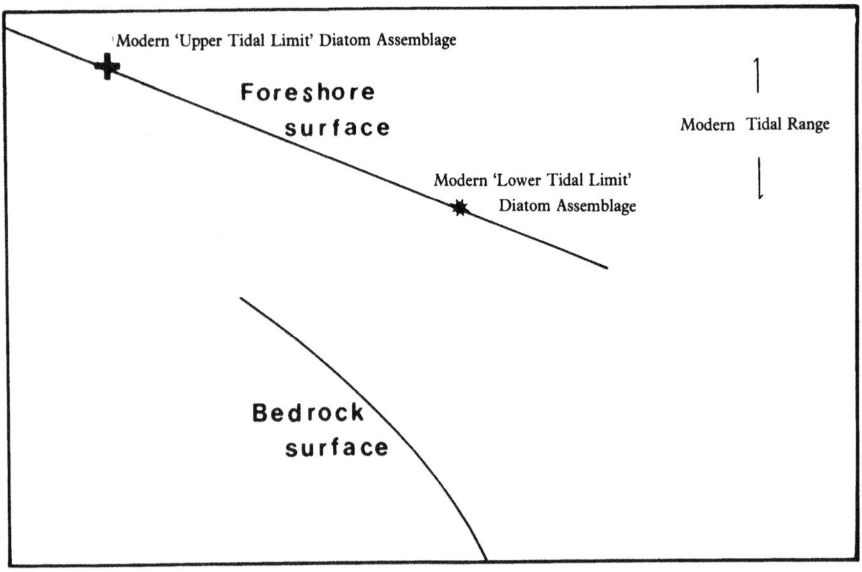

Figure 8. Modern diatom assemblages are found which distinguish the upper tidal limit (+) and the lower tidal limit (*) as shown on this schematic foreshore profile

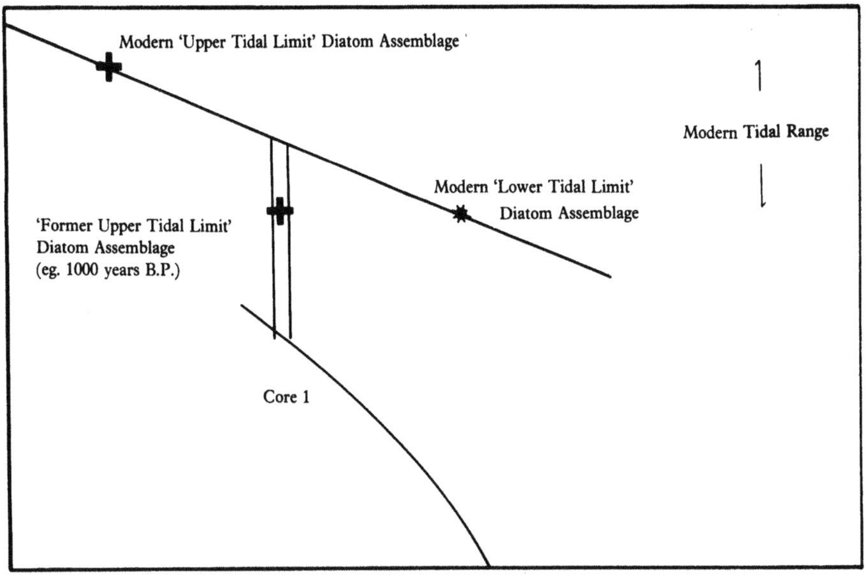

Figure 9. Core from the foreshore contains a diatom assemblage similar to that at the modern upper tidal limit (+). This assemblage is found to have an associated radiocarbon age (e.g., 1000 BP).

The 'upper tidal limit' diatom assemblage found in Core 1 (eg. 1000 BP) is then matched with a 'lower tidal limit' diatom assemblage of the same age retrieved in a stratigraphic sequence (Core 2). This comparison is illustrated in Figure 10.

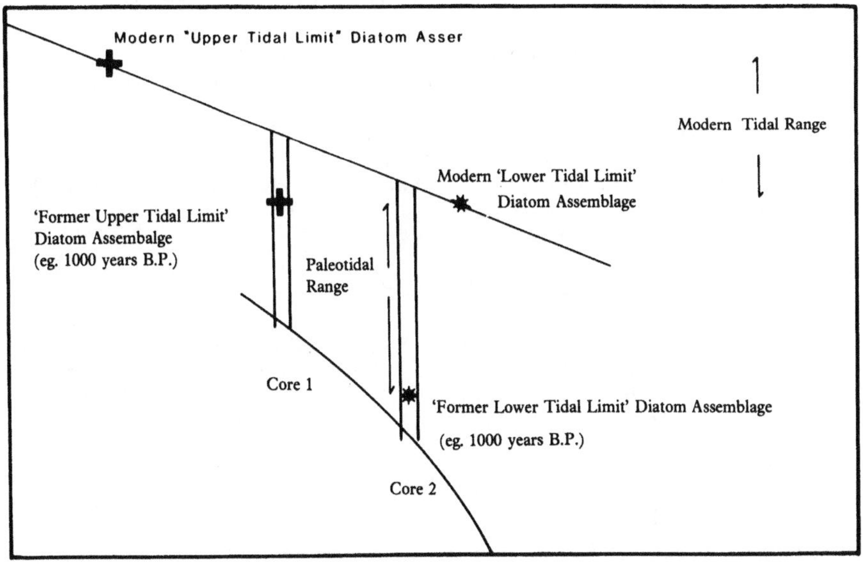

Figure 10. Core 2 from the foreshore fortuitously contains a diatom assemblage similar to that at the modern lower tidal limit (*). This assemblage has a radiocarbon age equivalent to that of the 'upper tidal limit' diatom assemblage (+) in Core 1. A paleotidal range can be established from the vertical distance between these coeval fossil assemblages.

The vertical distance between the contemporaneous diatom assemblages (here, the upper and lower tidal limit diatom assemblages of 1000 BP) is a measure of the paleotidal range for the time selected. This procedure theoretically could be repeatedly extended back in time for various ages but it is most severely constrained by the retrieval of sediments of the proper age containing the appropriate microfossil assemblages.

APPENDIX 2: AGENTS FOR MOUNTANTS AND FINDER SLIDES

I. Toluene-soluable diatom mountants

1. (CUMAR-R9; available at very low cost in large quantities only, Refractive Index > 1.60).
   Neville Chemical Company,
   Neville Island, Pittsburgh, Pennsylvania
   15225, U.S.A. and also Uithoorn, Holland

2. (Naphrax; R.I. ~ 1.72)
   Northern Biological Supplies 31 Cheltenham Avenue,
   Ipswich, Suffolk
   IPI 4LN, England

3. (Mountant; R.I. ~ 1.63)
   Asco Laboratories,
   Heaton Street, Denton, Manchester
   M34 3RG, England

II. Microscope Finder Slides

1. England Finder by Graticules Ltd. of England available as catalogue no. 20-5235 from
   Priester Scientific, 804 N.Miami Blvd.
   P.O. Box 11326, Durham, North Carolina 22703, U.S.A.

2. Field Finder, catalogue no. 12-454
   Fisher Scientific Company
   711 Forbes Avenue, Pittsburgh, Pennsylvania 15219,
   U.S.A.

SELECTED REFERENCES

Alhonen, P., Eronen, M., Nunez, M., Salomaa, R. & Uusinoka, R., 1978. A contribution to the Holocene shore displacement and environment development in Vantaa, South Finland: the stratigraphy of Lake Lammoslampi. *Bulletin of the Geological Society of Finland,* 50, 69-79

Andrews, G.W., 1972. Some fallacies of quantitative diatom paleontology. *Beihefte zur Nova Hedwigia,* 39, 285-295

Battarbee, R.W., 1973. Diatoms in lake sediments. In: *Paleohydrological Changes in the Temperate Zone in the Last 15,000 Years.* II, Specific methods, International Geological Correlation Programme Project 158, ed. B.E. Berglund, 177-225

Battarbee, R.W. & Kneen, M.J., 1982. The use of electronically counted microspheres in absolute diatom analysis. *Limnology and Oceanography,* 27(1), 184-188

Berglund, B.E., 1971. Littorina transgressions in Blekinge, South Sweden. A preliminary survey. *Geologiska Föreningens i Stockholm Förhandlingar*, 93(3), 625-652

Brooks, M.J., Colquhoun, D.J., Pardi, R.R., Newman, W.S. & Abbott, W.H., 1979. Preliminary archeological and geological evidence for Holocene sea-level fluctuations in the Lower Cooper River Valley, S.C. *The Florida Anthropologist*, 32(3), 85-103

Beyens, L. & Denys, L., 1982. Problems in diatom analysis of deposits: allochthonous valves and fragmentation. *Geologie en Mijnbouw*, 61, 159-162

Björc, S. & Digerfeldt, G., 1982. Late Weichselian shore displacement at Hunneberg, southern Sweden, indicating complex uplift. *Geologiska Föreningens i Stockholm Förhandlingar*, 104(2), 132-155

Burcle, L.H. & Akiba, F., 1978. Implications of late Neogene fresh-water sediment in the Sea of Japan. *Geology*, 6, 123-127

Cleve-Euler, A., 1944. Die Diatomeen als quartärgeologische Indikatoren. *Geologiska Föreningens i Stockholm Förhandlingar*, 66(3), 383-410

Cleve-Euler, A., 1951-1955. Die Diatomeen von Schweden und Finnland. *Kungliga Svenska Vetenskapakademiens, Handlingar*, Fjärde (4:e) Serien, 2-5
Tiel I, Band 2 (1) 1951
Tiel II, Band 4 (1) 1953
Tiel III, Band 4 (5) 1953
Tiel IV, Band 5 (4) 1955
Tiel V, Band 3 (3) 1952
(Reprinted 1968 as *Bibliotheca Phycologica*, 5, Verlag von J. Cramer)

Colquhoun, D.J., Brooks, M.J., Abbott, W.H., Stapor, F.W., Newman, W.S. & Pardi, R.R., 1980. Principles and problems in establishing a Holocene sea-level curve for South Carolina. In: *Excursions in South Eastern Geology, the Archeology-geology of the Georgia Coast*, 1980 Annual meeting, Geological Society of America, Atlanta, Georgia, eds. J.D. Howard, C.B. De Pratter & R.W. Frey, Guidebook 20, 143-159

Digerfeldt, G., 1975. A standard profile for Littorina transgressions in western Skane, South Sweden. *Boreas*, 4, 125-142

Digerfeldt, G., 1979. The highest shoreline on Hunneberg, southern Sweden. *Geologiska Föreningens i Stockholm Förhandlingar*, 101(1), 49-64

Du Saar, A., 1978. Diatom investigation of a sediment core. Downholland Moss-15. In: *Sea-level changes, North-West England during the Flandrian Stage*. M.J.Tooley (Clarendon Press, Oxford), Appendix

Eronen, M., 1974. The history of the Litorina Sea and associated Holocene events. *Commentationes Physico-Mathematicae, Societas Scientarum, Fennica*, 44, 79-195

Eronen, M., 1982. The course of shore displacement in Finland. In: *Holocene Sea-Level Fluctuations Magnitude and Causes*, ed. D.J. Colquhoun, 1981 I.G.C.P. No. 61, Annual Meeting, Department of Geology, Columbia, South Carolina, 29208, U.S.A. Manuscripts, 43-60

Florin, M.-B., 1944. En sensubarktisk transgression i trakten södra Kilsbergen enligt diatomacésuccessionen i omradets högre belägna fornsjölagerföljder. *Geologiska Föreningens i Stockholm Förhandlingar*, 66(3), 417-449

Florin, M.-B., 1946. Clypeusfloran: postglaciala fornsjölagerföljder i östra Mellansverige. *Geologiska Föreningens i Stockholm Förhandlingar*, 68(3), 429-458

Foged, N., 1970. The diatomaceous flora in a postglacial kieselguhr deposit in southwestern Norway. *Beihefte zur Nova Hedwigia*, 31, 169-202

Foged, N., 1972. The diatoms in four postglacial deposits in Greenland. *Meddelelser om Grønland*, 194, 4

Galehouse, J.S., 1971. Point counting. In: *Procedures in Sedimentary Petrology*, ed. R.E. Carver, (John Wiley & Sons Inc.), 385-407

Glückert, G., 1976. Post-glacial shore-level displacement of the Baltic, SW Finland. *Annales Academiae Scientiarum Fennicae*, Series A, III, Geologica-Geographica, 118, 1-92

Greville, R.K., 1861-1866. Descriptions of new and rare diatoms. Transactions of the Royal Microscopical Society of London, Quarterly Journal of Microscopical Science (Reprinted 1968 as *Bibliotheca Phycologia* 6 by Verlag von J. Cramer)

Halden, B.E., 1929. Kvartärgeologiska diatomacéstudier belysande den postglaciala transgressionen a svenska Västkusten. *Geologiska Föreningens i Stockholm Förhandlingar*, 51(3), 311-366

Hendey, N.I., 1964. *An introductory account of the smaller algae of British coastal waters, part V, Bacillariophyceae*. London, (Her Majesty's Stationery Office 1964), (Reprint 1976 A. Asher & Company B.U., Keizersgracht 526, Amsterdam, The Netherlands)

Hustedt, F., 1930-1966. *Die Kieselalgen Deutschlands, Österreichs und der Schweiz*, 1930 I 920 pp; 1959 II 845 pp; 1961-1966 III 816 pp, Akademische Verlagsgesellschaft, Leipzig. (Johnson Reprint Corporation, New York, London)

Hustedt, F., 1957. Die Diatomeen Flora des Fluss-systems der Weser im Gebiet der Hansestadt Bremen. *Abhandlungen Naturwissenschaftlichen Verein, Bremen*, 34, 181-440

Hyvärinen, H., 1980. Relative sea-level changes near Helsinki, southern Finland, during early Litorina times. *Bulletin of the Geological Society of Finland*, 52(2), 207-219

Jansma, M.J., 1975. Quantitative diatom analysis as an archeological aid. *Hydrobiological Bulletin*, 9(2), 87-88

Kabailiene, M., 1974. On the character of transgression and diatoms flora peculiarities of the Litorina Sea on territory of the southeastern Baltic area. *Baltica*, 5, 71-77

Kaland, P.E. & Stabell, B., 1981. Methods for absolute diatom frequency analysis and combined diatom and pollen analysis in sediments. *Nordic Journal of Botany*, 1, 697-700

Kjemperud, A., 1981a. Diatom changes in sediments at basins possessing marine/lacustrine transitions in Frosta, Nord-Trøndelag, Norway. *Boreas*, 10, 27-38

Kjemperud, A., 1981b. A shoreline displacement investigation from Frosta in Tronheims Fjorden, Nord-Trøndelag, Norway. *Norsk Geologisk Tidsskrift*, 61, 1-15

Kolbe, R.W., 1927. Zur Ökologie, Morphologie und Systematik der Brackwasser-Diatomeen. Die Kieselalgen des Sperenberger Salzgebiets. *Pflanzenforschung*, 7, 1-146

Lewin, J. & Rao, V.M.R., 1975. Blooms of surf-zone diatoms along the coast of the Olympic Peninsula, Washington VI. *Journal of Phytology*, 11, 330-338

McIntire, C.D. & Moore, W.W., 1977. Marine littoral diatoms: ecological considerations. In: *The Biology of Diatoms*, ed. D. Werner, (University of California Press, Berkeley, California), 33-371

Miller, U., 1964. Diatom floras in the Quaternary of the Gøta River Valley. *Sveriges Geologiska Undersökning*, Ser. Ca 44, 67 pp

Miller, U. & Robertsson, A.M., 1979. Biostratigraphical investigations in the Anundsjö Region, Angermanland, northern Sweden. *Early Norrland*, 12, 1-76

Mölder, K., 1943. Studien über die Ökologie und Boden-diatomeen in der Pojo-Bucht. *Suomalaisen eläin- ja kasvitieteellisen seuran vanamon julkaisuja. Helsinki (Annales Botanici Societatis zoologicae-botanicae fennicae Vanamo)*, 18

Mölder, K. & Tynni, R., 1967-1973. Über Finnlands rezente und subfossile Diatomeen
1967 I *Comptes Rendus de Géologie Finlande*, 39, 199-217
1968 II *Bulletin of the Geological Society of Finland*, 40, 151-170
1969 III *Bulletin of the Geological Society of Finland*, 41, 235-251
1970 IV *Bulletin of the Geological Society of Finland*, 42, 129-144
1971 V *Bulletin of the Geological Society of Finland*, 43, 203-220
1972 VI *Bulletin of the Geological Society of Finland*, 44, 141-159
1973 VII *Bulletin of the Geological Society of Finland*, 45, 159-179

Mörner, N.-A., 1978. Late Quaternary sea-level changes in northwestern Europe: a synthesis. *Geologiska Föreningens i Stockholm Förhandlingar*, 100, 381-400

Palmer, A.J.M., 1978. Diatom stratigraphy of Basin Head Harbour, Prince Edward Island. *Proceedings of the Nova Scotian Institute of Science*, 28, 201-215

Patrick, R. & Reimer, C.W., 1975. The diatoms of the United States 1966-volume 1, 688 pp; 1975-volume 2, 213 pp. *Monographs of the Academy of Natural Sciences of Philadelphia*

Pearse, J.G., 1977. *Radiocarbon tracing of migrating benthic diatoms*. MS thesis, Marine Science Program, University of South Carolina

Peragallo, H. & Peragallo, M., 1897-1908. *Diatomées marines de France et des districts maritimes voisins*. ed. M.J. Tempère, Micrographe-Editeur, à Grez-sur-Loing (S.-et-M.) (Réimpression A. Asher & Co., Keizersgracht 526, 1017 EK Amsterdam, The Netherlands)

Ramm, G., 1977. Structure of ephiphytic diatom population of the phytal of the Kiel Bight (western Baltic). In: Proceedings, Fourth Symposium on Recent and Fossil Marine Diatoms, Oslo 1976. *Beihefte zur Nova Hedwigia*, 54, 379-387

Renburg, J., 1976. Paleolimnological investigations in Lake Prästjön. *Early Norrland*, 9, 113-159

Round, F.E., 1971. Benthic marine diatoms. *Oceanography and Marine Biology, An Annual Review*, 9, 83-139

Schaffel, S. 1971. *Reconstruction of Late-glacial and Post-glacial Events in Long Island Sound, New York*. PhD dissertation, Geology, New York University

Schmidt, A. *et al.*, 1978. Quaternary through Neogene history Koeltz Antiquariat, Koenigstein/Taunus 1972, Otto Koeltz Science Publishers, P.O. Box 138, D-624 Koenigstein, West Germany)

Schrader, H.-J., 1973. Proposal for a standardized method of cleaning diatom bearing deep-sea and land-exposed marine sediments. *Beihefte zur Nova Hedwigia*, 45, 403-409

Schrader, H.-J., 1978. Quaternary through Neogene history of the Black Sea, deduced from the paleoecology of diatoms, silicoflagellates, ebridians and chrysomads. *Initial Reports of the Deep Sea Drilling Project*, (U.S. Government Printing Office, Washington, D.C.), 42(2), 789-901

Schrader, H.-J. & Gersonde, R., 1978. The late Messinian Mediterranean brackish to freshwater environment, diatom floral evidence. *Initial Reports of the Deep Sea Drilling Project*, (U.S. Government Printing Office, Washington, D.C.), 42(1), 761-775

Schrader,H.-J. & Schuette, G., 1981. Marine Diatoms. Chapter 28 in *The Sea*, Volume 7, ed.C. Emiliani. (John Wiley & Sons, Inc.)

Setty, M.A.P.G., 1966. Preparation and methods of study of fossil diatoms. *Micropaleontology*, 12(4), 511-514

Simonsen, R., 1962. Untersuchungen zur Systematik und Ökologie der Bodendiatomeen der westlichen Ostee. *Internationale Revue der Gesamten Hydrobiologie*, Systematische Beihefte I

Smith, H.L., 1873. Preparation of diatoms. *The Lens* (Journal of the State Microscopical Society of Illinois), ii, 209-212, in H. van Heurck, 1896. *A Treatise of the Diatomaceae*, translated by W.E. Baxter, (Wheldon & Wesley Ltd. Codicote, Herts. 1962), 67-71

Stabell, B., 1980. Holocene shorelevel displacement in Telemark, southern Norway. *Norsk Geologisk Tidsskrift*, 60, 71-81

Stabell, B., 1982. Changes in diatom floras in late Quaternary western and southern Norwegian marine and freshwater sediments: response to basin isolation from the sea. In: *The response of diatom floras during Late Quaternary shore line displacement in southern and western Norway*. PhD thesis, Department of Geology, University of Oslo

Thomsen, H., 1982. Late Weichselian shore-level displacement on Nord-Faren, southwest Norway. *Geologiska Föreningens i Stockholm Förhandlingar*, 103, 447-468

Tynni, R., 1975-1978. Über Finnlands rezente un subfossile Diatomeen, *Finland-Geologinen Tutkimuslaitus Bulletin*,
1975 VIII *Bulletin of the Geological Society of Finland*, 274
1976 IX *Bulletin of the Geological Society of Finland*, 284
1978 X *Bulletin of the Geological Society of Finland*, 296

Van Heurck, H., 1896. *A treatise on the Diatomaceae*. (William Wesley & Son, London)

Van der Werff, A. & Huls, H., 1957-1974. *Diatomeeënflora van Nederland*. Abcoude, Den Haag, (authorized Reprint 1976 by Otto Koeltz Science Publishers, P.O. Box 1380, D-624 Koenigstein/West Germany)

PLATE 1 - all illustrations at 1000x as shown on bar scale

1. *Coscinodiscus marginatus* Ehr. (P)
2. *Coscinodiscus eccentricus* Ehr. (P)
3. *Melosira juergensii* Agardh (P)
4. *Cyclotella stylorum* Brightwell (P)
5. *Stephanodiscus hantzschii* Grun. (O)
6. *Stephanodiscus astrea* (Ehr.) Grun. var. *minutula* (Kütz.) Grun. (O)
7. *Cyclotella meneghiniana* Kütz. (O)
8. *Cyclotella kutzingiana* (Thwaites) Chauvin (O)
9. *Cyclotella caspia* Grun. (M)

10a,b. *Paralia (Melosira) sulcata,* girdle view-two levels of focus (P)

11. *Biddulphia obtusa* (Kütz.) Ralfs (P)
12. *Diatoma hemiale* var. *mesodon* (Ehr.) Grun. (O)
13. *Diatoma tenue* and var. *elongatum* (Lyngbye) (O-M)
14. *Meridion circulaire* (Grev.) Agardh var. *circulaire* (O)
15. *Opephora martyi* Heribaud var. *martyi* (O)
16. *Fragilaria capucina* var. *mesolepta* Rabenhorst (O)
17. *Fragilaria vaucheriae* (Kütz.) Petersen var. *vaucheriae* (O)

18a,b. *Fragilaria leptostauron* (Ehr.) Hustedt var. *leptostauron* (O), girdle and valve views

19. *Fragilaria leptostauron* var. *rhomboides* (Grun.) Hustedt (O)
20. *Fragilaria leptostauron* var. *dubia* (Grun.) Hustedt (O)
21. *Fragilaria leptostauron* var. *leptostauron* f. *minuta* A. Cleve-Euler (O)
22. *Fragilaria pinnata* var. *hybrida* A. Cleve-Euler (O?)

23a,b. *Fragilaria construens* var. *venter* f. *typica* Mayer (O)
c. *Fragilaria construens* var. *construens* f. *typica* (O)

24. *Fragilaria pinnata* Ehr. var. *pinnata* f. *subrotunda* (O)
25. *Fragilaria construens* (Ehr.) Grun. (O) var. *construens* f. *asymetrica*
26. *Fragilaria construens* var. *binodis* (Ehr.) Grun. (O), valve view
27. *Fragilaria construens* var. *binodis,* 'chain' in girdle view

28a,b, c. *Fragilaria construens* var. *venter* f. *pusilla* (O)

29a,b, c. *Fragilaria producta* (lagerstedt) Grun. (O), valve and girdle views

30. *Fragilaria pinnata* var. *pinnata* Ehr. (O), girdle view
31. *Fragilaria pinnata* var. *minutissima* (Grun.) A. Cleve-Euler (O)
32. *Fragilaria pinnata* var. *intercedens* (Grun.) Hustedt (O)
33. *Fragilaria brevistriata* Grunow (O)
34. *Fragilaria atomus* Hustedt (O-M)
35. *Synedra fasciculata* (Agardh) Kütz. (O-M)
36. *Synedra pulchella* Ralfs *ex* Kützing (O-M)
37. *Raphoneis surirella* (Ehr.) Grun. (P)
38. *Raphoneis surirella* var. *australis* Petit (P)
39. *Grammatophora oceanica* (Ehr.) Grun. (P)

(O)=Oligohalobian; (M)=Mesohalobian; (P)=Polyhalobian

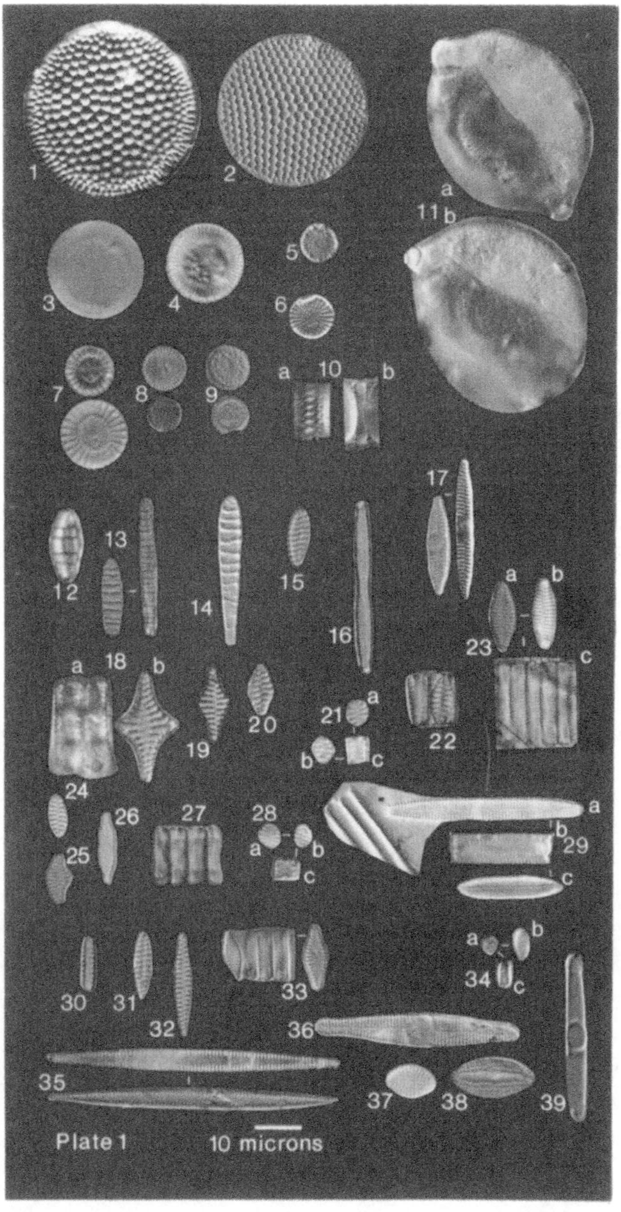
Plate 1
10 microns

PLATE 2 - all illustrations at 1000x as shown on bar scale

40,41,42. *Acnanthes lanceolata* (Bréb.) Grun., and varieties (O)
43. *Cocconeis stauroneiformis* (Van Heurck) Okuno (M)
44. *Cocconeis placentula* (O), (a) pseudoraphe valve var. *placentula?*, (b) raphe valve
45. *Rhoicosphenia curvata* (Kütz.) Grun. *ex* Rabh. (O), (a) girdle view, (b) concave raphe valve
46. *Navicula rhyncocephala* (Kütz.) var. *germainii* (Wallace) Patrick (O)
47. *Navicula cryptocephala* Kütz. (O-M)
48. *Navicula peregrina* (Ehr.) Kütz. (M)
49. *Navicula cuspidata* (Kütz.) Kütz. (O)
50. *Navicula incerta* Grun. *ex* Van Heurck (P)
51. *Navicula barbarica* Hustedt (O)
52. *Navicula salinarum* (Grun.) Cleve (M)
53. *Caloneis fusoides* Grunow (P)
54. *Pinnularia dactylus* Ehr. (O)
55. *Mastogloia smithii* Thwaites *ex* Wm. Smith (M)
56. *Diploneis interrupta* (Kütz.) Cleve var. *clancula* (Schmidt) Cleve? (P)
57. *Gomphonema acuminatum* Ehr. var. *coronata* (O)
58. *Gomphonema constrictum* Ehr. (O)
59. *Amphora coffeaefformis* Agardh (P)
60. *Amphora angusta* var. *ventricosa* (Greg.) Cleve (P)
61. *Amphora libyca* Ehr. var. *baltica* (Brander) A. Cleve-Euler (O)

(O)=Oligohalobian; (M)=Mesohalobian; (P)=Polyhalobian

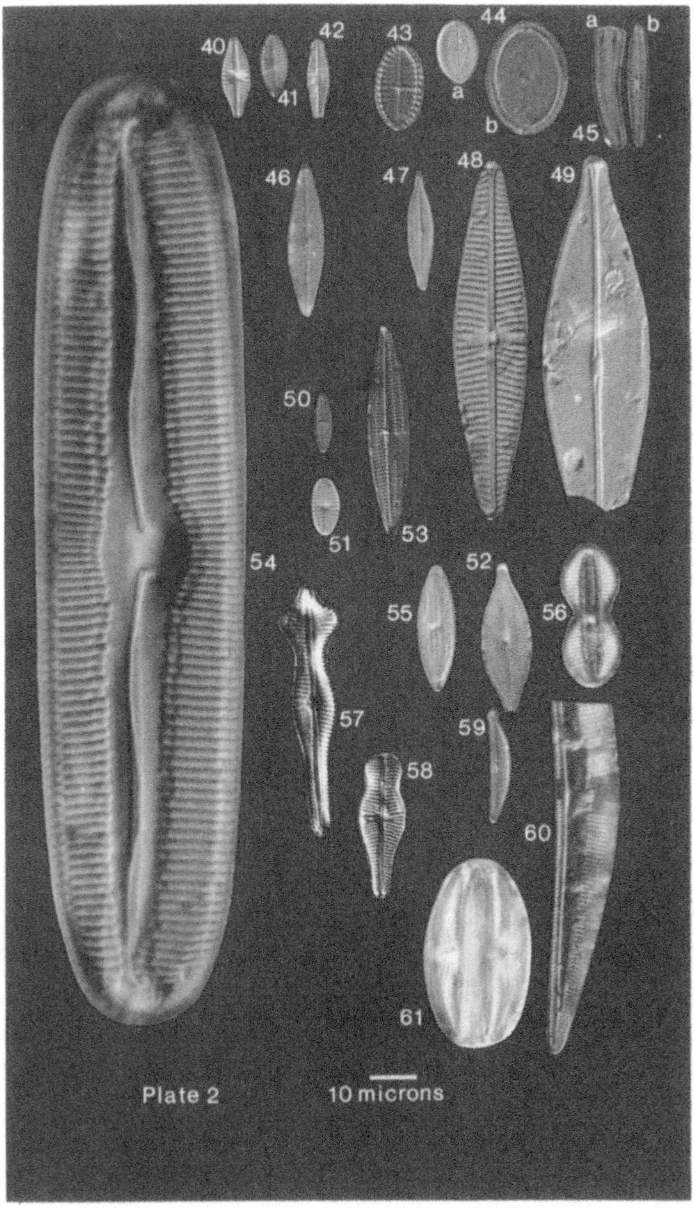
40
41
42
43
44
a
b
a
b
45
46
47
48
49
50
51
53
54
52
55
56
57
59
58
60
61
Plate 2
10 microns

PLATE 3 - all illustrations at 1000x as shown on bar scale

62. *Cymbella cistula* var. *maculata* (Kütz.) Van Heurck (O)
63. *Rhopalodia gibberula* (Ehr.) O. Müll. (O-M)
64. *Rhopalodia gibba* var. *ventricosa* (Ehr.) O. Müll. (O-M)
65. *Epithemis sorex* Kütz. (O-M)
66. *Surirella ovalis* de Brébisson (M)
67. *Surirella ovata* Kütz. (O)
68. *Surirella striatula* Turpin var. *biplicata* Grun. (M)
69. *Nitzschia closterium* (Ehr.) Wm. Smith (M)
70. *Nitzschia frustulum* var. *subsalina* Hustedt (O)
71. *Nitzschia frustulum* (Kütz.) Grun. (O)
72. *Nitzschia apiculata* (Greg.) Grun. (M)

(O)=Oligohalobian; (M)=Mesohalobian; (P)=Polyhalobian

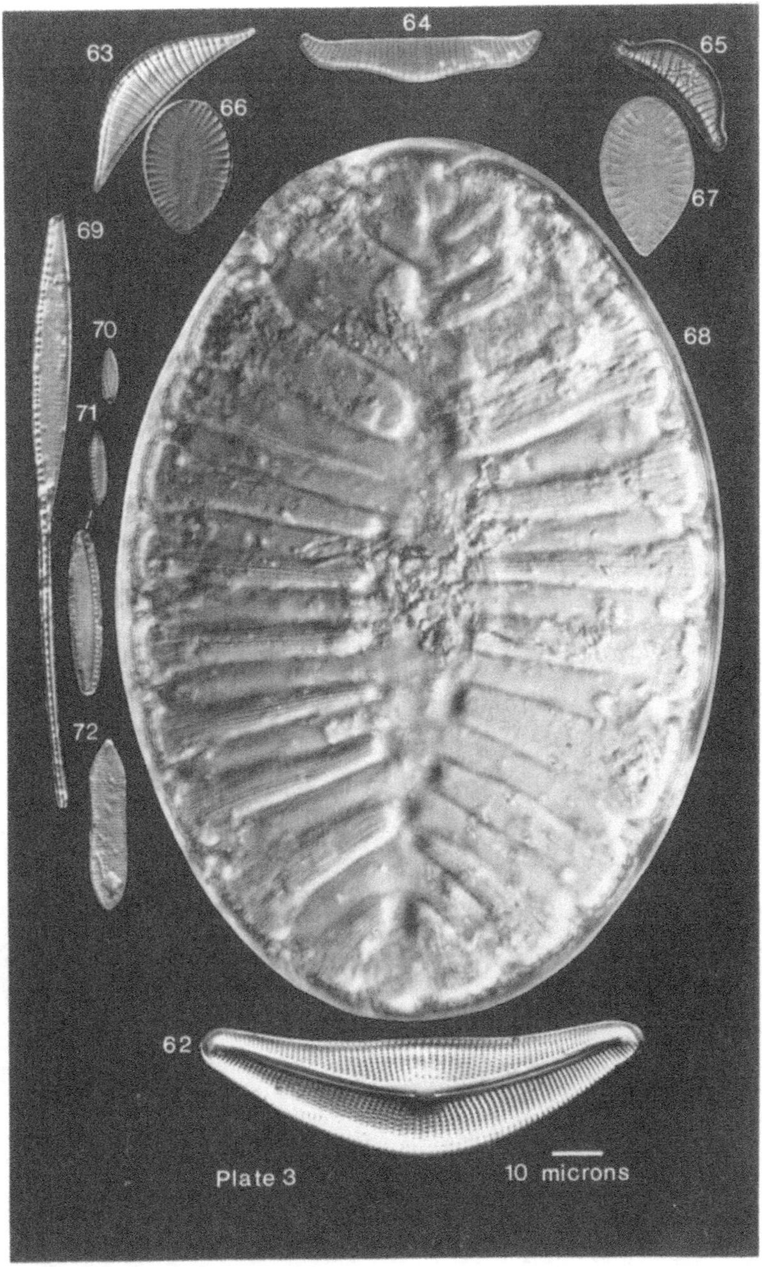
63
64
65
66
67
68
69
70
71
72
62
Plate 3
10 microns

17

Ostracode options in sea-level studies

D. VAN HARTEN

ABSTRACT

The possibilities are discussed of the application of ostracodes to problems of sea-level change. In marine sediments, ostracodes can be used to indicate depth of deposition. The current resolution is estimated to be on the order of 10^2 m for most species and genera. Significant improvements may come from future augmentation to existing data on modern ostracode depth-ranges. Paleobathymetric reconstructions based on the distribution of organic remains may be spoiled to a certain extent by *post-mortem* effects. In ostracodes, this is less of a problem than in other groups as contamination can often be recognized for what it is in ostracode faunas.

Ostracode faunas show depth-related changes in the diversity and the density. Conversely, these trends can be employed to estimate paleo-depth. Ostracodes are particularly sensitive to depth changes in the general reach of the shelf. In ancient shelf-sea sediments, the ostracode record can be utilized as a tracer for such changes. Depth decrease resulting from coastal progradation can be distinguished from shallowing due to eustatic causes.

It is relatively easy to recognize the stressed ostracode faunas that are typical of aberrant salinity conditions, whether brackish or hyperhaline. Since aberrant salinity is often associated with sea level, fossil salinity-stressed faunas may indicate this level in ancient sediments.

Some remarks about collecting and sample preparation are included in an addendum.

INTRODUCTION

Ostracodes are tiny crustaceans - the modal ostracode measures between 0.5 and 1.0 mm when adult - which characteristically possess a bivalved, calcareous shell. As in the lamellibranchs, to which the ostracodes bear a superficial resemblance the shell consists of a left and a right valve but these are not symmetrical in ostracodes. One valve is slightly larger than the other and overlaps it along parts of the margin. There may also be noticeable asymmetry in the shape of the valves.

The ostracodes, as a group, inhabit nearly all types of natural aquatic environments, both freshwater as well as marine. Single taxa specialize to a more restricted spectrum of conditions, however, and so are stenotopic to some measure. The ecologic key factor in the distribution of the ostracodes is environmental salinity. Thus the group can be divided into fresh water, brackish water and marine forms.

There seems to be a rough correlation between the salinity of the environment and such features of the ostracode shell as thickness and degree of ornamentation. Shells in fresh-water species generally are thin and smooth. Brackish-water forms mostly have thicker and somewhat more ornate shells, whereas heavily calcified and strongly ornamented shells exclusively occur in the marine ostracodes.

While there are planktonic taxa and certain forms are even able to actively swim, the overwhelming majority of the ostracodes are benthic. By and large, ostracodes are crawling about on the bottom, adhering to plants or burrowing around in soft sediments. The creatures are confined to the bottom for their entire life because they lack a planktonic stage in their ontogenetic development.

Being crustaceans, ostracodes grow by ecdysis or moulting, a process in which the young shed their old shell and secrete a new one roughly twice the size of its predecessor. In most Recent species there are 8 successive larval stages before the animals reach adulthood and sexual maturity. The shedded larval shells may accumulate in the bottom sediment if they are not destroyed or carried away by water movements.

The calcareous matter of the shell is primary calcite so that ostracodes have excellent fossilization potential. Their fossil record reaches all the way back to the Cambrian. Ostracodes, like most other groups of fossils, are geologically utilized in two ways: for dating and for the interpretation of the paleo-environment. Locating former sea levels (SLs) and tracing their changes is a special case of the latter.

For further introduction to the ostracodes the reader is referred to Pokorný (1980).

Owing to their particular attributes, ostracodes are well-suited to the purpose of environment interpretation. For SL studies two issues seem to be of specific interest: (1) *the bathymetric distribution of marine ostracodes* and (2) *the general properties of salinity-stressed ostracode faunas*. It is the aim of this paper to survey these lines of approach and to discuss a few case histories.

BATHYMETRIC DISTRIBUTION OF MARINE OSTRACODES

A change of SL may directly affect sea depth. Reliable estimates of the depth of deposition of ancient sediments are therefore crucial in attempts to locate former SL. Ostracodes are often quoted as good depth indicators but, as yet, expectations about the resolution attainable should not be set too high. Sedimentologic noise rather than an imprecise original biologic signal is frequently to blame.

Ostracodes are living in practically every marine environment, from shallow tidal pools down to depths exceeding 4000 m. Many species are reported to live, at least locally, within rather well-defined depth limits (eg. Elofson, 1941: Skagerrak; Puri *et al.*, 1964: Gulf of Naples; Puri *et al.*, 1969: Mediterranean). The actual reasons for ostracode depth-zonation are largely unknown. Most workers consider that depth-related factors such as temperature, food supply or the nature of the substratum probably play a more direct role than depth itself. Rather invariant, low temperature, for instance is believed to control the distribution of the highly specialized ostracode fauna that inhabits the floor of the world ocean. This so-called *psychrospheric* fauna is usually found in waters cooler than 10°C and is most typical in waters near 4°C. In low latitudes it occurs at greater depth than in high latitudes and in upwelling regions (Benson & Sylvester-Bradley, 1971).

With few exceptions, available data on the depth ranges of Recent marine ostracodes relate to empty shells and loose valves rather than live animals. Such data bear on what has aptly been called *taphocoenosis* (= community of the remains buried, or to be buried, together). The taphocoenoses of the modern seas are the nearest thing to what paleontologists will recover from ancient sedimentary rocks. With respect to the living community or biocoenosis the taphocoenosis contains a certain amount of sedimentologic 'noise' which results from post-mortem transport. In the sea, part of this transport is either up-slope or downslope, so that taphocoenotic depth ranges will be broader in general than the actual biotopes. The presence of faunas at depths where they do not belong is called contamination.

An extreme case in point is the depth distribution of *Loculicytheretta pavonia* (Brady), a Recent ostracode species that occurs along part of the Mediterranean coast. From Bismuth *et al.* (1978) it is reasonably certain that *L. pavonia* inhabits very shallow water only (0-4 m). The present author found valves of this species at depths down to 1166 m in the Atlantic off the Strait of Gibraltar. Finds at 620 and 700 m have been reported from the Alboran Sea by Benson & Sylvester-Bradley (1971).

To improve the precision of paleo-depth estimates, methods will have to be developed to eliminate, or at least reduce, the noise produced by contamination. In ostracodes, remedy might come from the unique fact that these animals may fossilize in several modes. Shells and loose valves, adults and larvae, and even the opposite valves of the same shell, all differ in size and shape, hence in hydrodynamic properties. As a result, these remains are apt to behave differently in a transporting medium and to ultimately become buried at different places. Little work has been done in this field but there are some encouraging experimental results by Kontrovitz & Nicolich (1979) on differential response to currents of shells and loose valves. Breman (1980) found that several ostracode species in the Adriatic show distinct geographic variation in the ratio of left and right valves, suggesting selective transport. From work in progress by the present author, it appears that larvae tend to respond differently from adults of the same species to down-slope mass transport in the Mediterranean.

The use of marine ostracodes to estimate paleo-depth depends on the actualistic comparison with the depths the Recent forms are found at. Strictly, this is valid at the species level only, which would confine proper usage to relatively young deposits. Generalizations to higher taxonomic levels, even to that of the genus, may be hazardous because the depth ranges of the various species within a higher taxon may diverge. The genus *Macrocypris*, for instance, is commonly quoted as inhabiting deep water. In the Mediterranean, it is represented by at least 3 species, all living at different depths, and the shallowest one inhabits part of the sublittoral zone. From the Bay of Biscay, 7 species of *Loxoconcha* were mentioned by Peypouquet (1980) as living at successive depths from the shore line to close to the abyssal zone; this genus is generally taken to prefer shallow and often even brackish waters. Similarly, the notorious 'genus' *Bairdia* is reported to live in very shallow as well as in very deep waters (Van Morkhoven, 1963). Defective taxonomy at the genus level might account for part of these anomalies.

It saves much time, on the other hand, to restrict oneself to genera and to refrain from specific identification In fact, the outstanding paper by Van Morkhoven (1972) on the ostracode bathymetry in the Gulf of Mexico was written on this very principle (see Fig. 1).

At the present state of the art, the resolving power of marine ostracode paleobathymetry is probably on the order of 10^2 m for most species and genera. For some of these, the shallower ones in particular, the accuracy can hopefully be refined to tens of metres by a mere increase of our insight into the Recent patterns of depth distribution. If the problem of contamination can be solved,

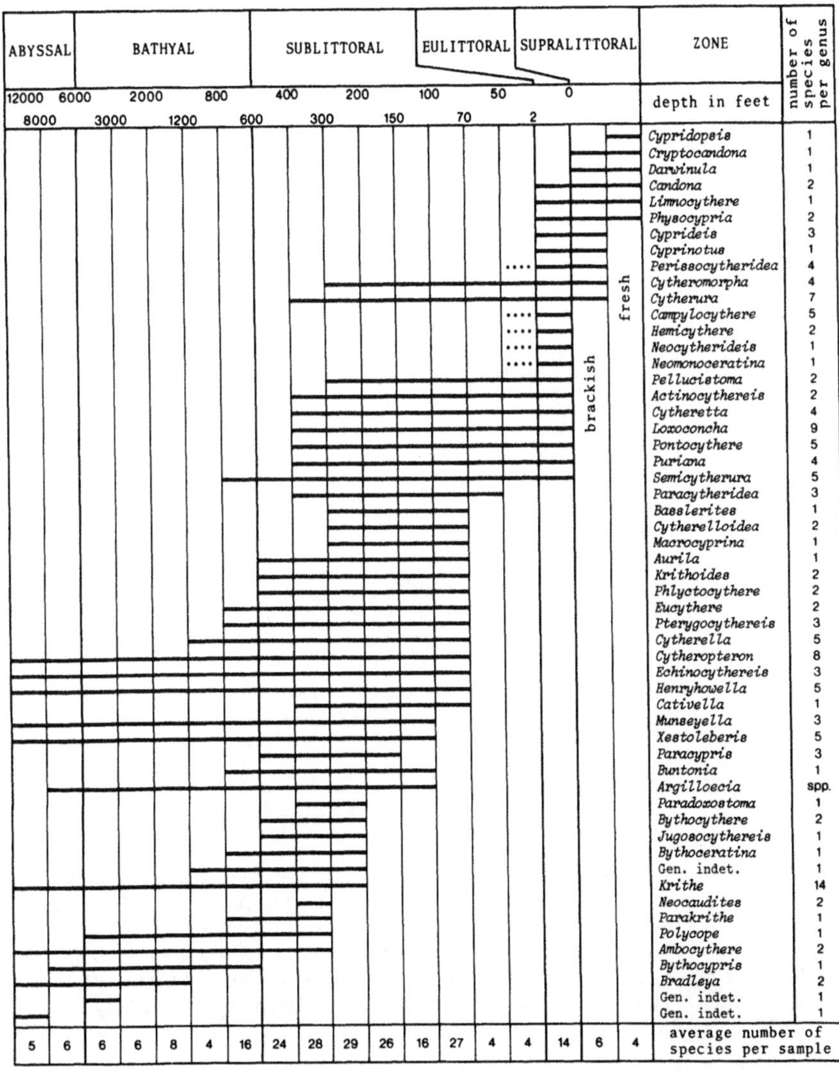

Figure 1. Depth ranges of Recent ostracode genera in the northwest Gulf of Mexico. Modified after Van Morkhoven (1972).

the metre scale comes within reach for extremely depth-restricted species such as *Loculicytheretta pavonia*.

Depth-related changes in diversity and density.

Another way to use ostracodes in paleobathymetry depends on the changes in the *diversity* (number of species) and *density* (number of specimens) of their faunas that occur with increasing depth in the sea. Pokorny (1971) postulated on the basis of fossil material that ostracode diversity increases from SL towards the base of the photic zone. Beyond that depth the diversity would decrease again. This

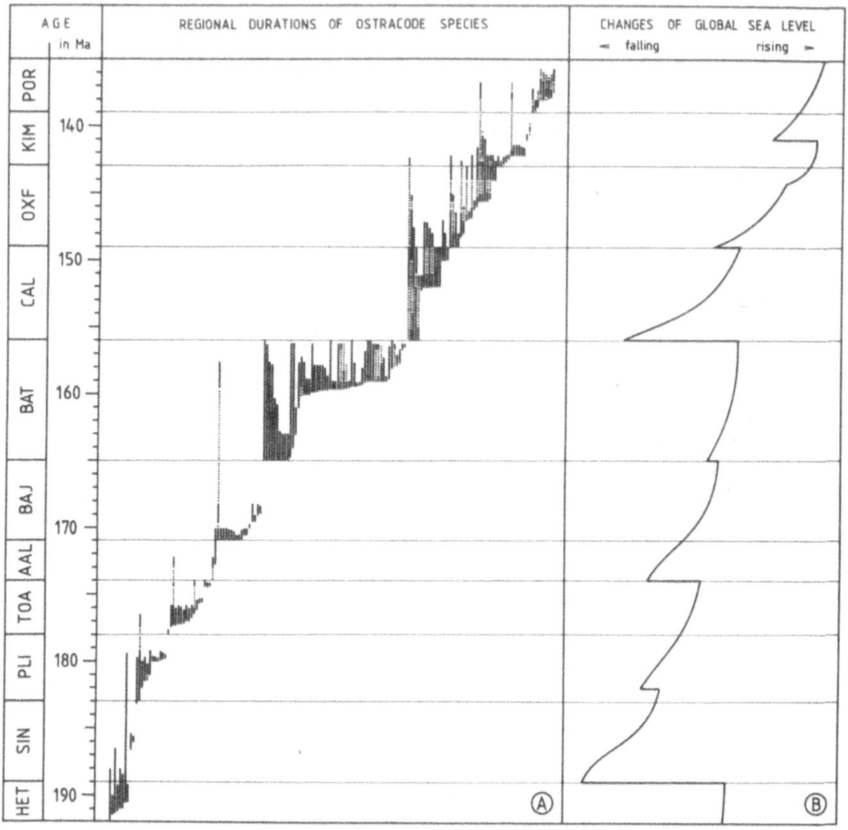

Figure 2. Duration graph of British Jurassic ostracodes (A) and eustatic curve (B). Lines in A stand for single species and are arranged according to date of entry and length of duration. Ostracode data converted from Sheppard (1978); eustatic curve modified after Vail & Mitchum (1979); time scale after Van Hinte (1976).

was confirmed for Recent ostracodes by Peypouquet (1980) who also pointed out that the eulittoral (= intertidal) zone is usually characterized by very low values for both diversity and density, whereas the supratidal environment may show low-diversity, high-density faunas. Below SL, ostracode density changes with depth in more or less the same sense as diversity. On the assumption that the diversity-density trends seen in modern ostracode faunas reflect a general principle, this method can be used to estimate paleo-depth in ancient sediments. There are no particular restrictions to young geologic ages but the results will be somewhat less precise than those based on actual species of genus bathymetry.

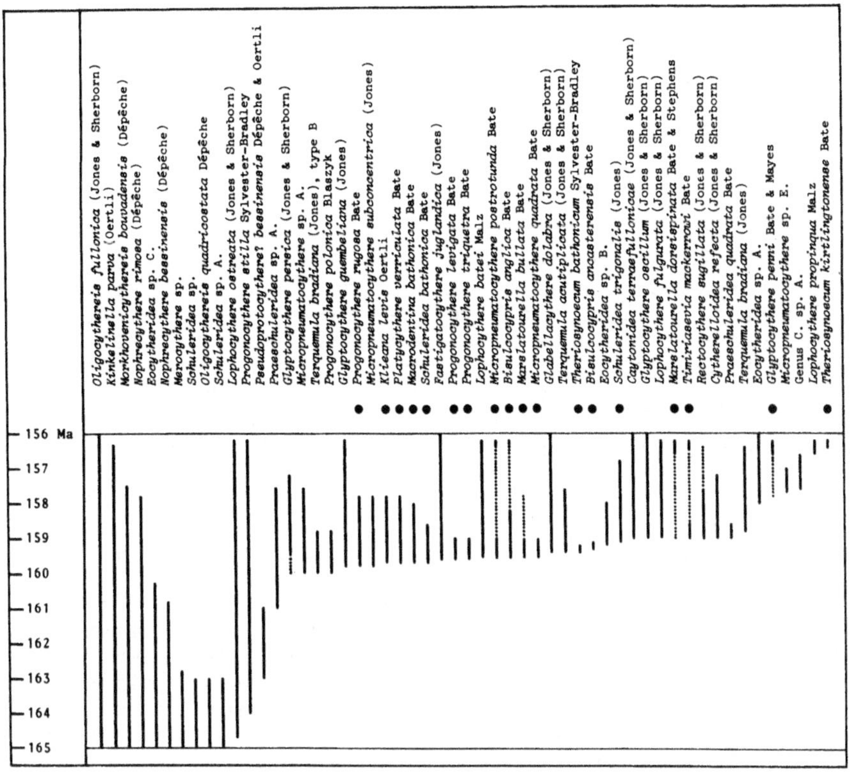

Figure 3. The Bathonian signal (compare Fig. 2). Dots indicate species reported as not, or not entirely, marine. After data in Bate (1978).

Diversity patterns in geologic time.

In most marine ostracode taphocoenoses the diversity maximum will not be found at the base of the photic zone, where its biologic origins lie, but deeper. This is due to post-mortem transport which, below wave-base, is largely down-slope. Assuming the photic zone to extend to *c.* 100 m on an average, the apparent maxima will generally occur on the lower shelf or beyond the shelf edge. For most or all of the shelf range, therefore, there will be a positive correlation between ostracode diversity and sea depth (*cf.* for instance Fig. 1: average number of species per sample).

In consequence of this, the *diversity pattern* in the ostracode record of ancient shelf seas may be expected to reflect the bathymetric history of such seas. Figure 2 illustrates this for the Jurassic of Great Britain. The graph is modified from a conventional range chart (Sheppard, 1978) by converting the chronostratic time scale to linear time (Van Harten & Van Hinte, in prep.). Each vertical line in the graph represents the regional life span, or duration, of a single ostracode species (Fig. 2A). The duration lines cluster to *diversity signals* and each such signal is considered to stand for a distinct environmental event.

These events, in view of the above correlation between ostracode diversity and sea depth on the shelf, are inferredly bathymetric. Thus the appearance of species at the base of a signal would indicate that the sea became deeper and the disappearance at the top that it shallowed again. The agreement between the diversity graph and the independently derived, eustatic curve by Vail & Mitchum (1979) supports this explanation (Fig. 2B).

On the wide and flat shelves of Jurassic northwestern Europe, eustatic changes of SL must have had considerable areal impact. There must have been marked transgressions and regressions of the sea on the land. The upper part of the Bathonian diversity signal may look somewhat suspicious in this respect. On the left, species are disappearing which suggests shallowing and regression, whereas on the right other species keep making an appearance, and this would indicate transgression. Meanwhile, according to the eustatic curve, global SL continued to rise. By taking ostracode taxonomy and ecology into account (Fig. 3), it can be seen that on the left of the Bathonian signal only marine species are present. Their gradual disappearance from around 163 Ma indicates that the sea became shallower and that regression took place. The right hand side of the signal, contrarily, to a significant degree pertains to fresh-water and brackish-water forms. The complete Bathonian signal therefore conveys that from around 160 Ma limnetic and mixohaline habitats developed while the sea was withdrawing from the region. That the record of non-marine ostracodes has been preserved at all is in itself evidence of relatively high base level and, for that matter, confirms this detail in the eustatic curve.

The ostracode diversity signals cannot but qualitatively trace SL changes. But the *addition of ecologic data makes it possible to rigidly distinguish between two fundamentally different types of regression:* those that are due to eustatic falls and those that result from coastal progradation during eustatic rises or highstands. This capability might prove very useful in future studies on SL change.

GENERAL PROPERTIES OF SALINITY-STRESSED OSTRACODE FAUNAS

In between the sea and the fresh waters of the continents lies the zone of the brackish water or *mixohalinicum*. Brackish water, in a general way, can be taken as associated with SL. The mixohalinicum can therefore be used as a rough SL indicator in ancient sediments. Brackish-water ostracode faunas are rather distinctive and can often be recognized with relative ease, even by non-specialists.

Brackish water originates from mixing of sea water and fresh water. The supplies of both may, and usually will, fluctuate and this accounts for the ecologic instability that is the most significant attribute of brackish habitats. Total salinity, ionic ratio, oxygen content and food conditions are apt to continually vary on a cyclic or non-cyclic basis. As a result, life is hard in the brackish zone and relatively few species are able to cope in the long run with the adverse conditions. Those rare species that can adapt to this particular environment will meet with little competition from other species, however, and so may thrive in unheard numbers. The well-known Recent brackish-water ostracode *Cyprideis torosa* (Jones), for instance, has been reported by Soviet authors to attain densities up to 670 000 specimens per square metre in the estuaries of the Kuban River (Pokorný, 1980). Mixohaline ostracode faunas typically tend to be numerically dominated by few, sometimes only one, species. They may even be composed of a single species (monospecific faunas).

The tendency for brackish habitats to support faunas that are hardly diverse but very dense would make them rather conspicuous as paleo-environments in ancient sediments. But overall conditions that are difficult to cope with and the ensuing lack of competition for the few successful species exist just as well in *hyperhaline* environments (salinity over 40 ‰). Hyperhaline and mixohaline faunas may therefore be very similar - they may even contain the same species - so that it is often difficult to tell them apart. This seems of little consequence in a SL context, however, as either type of aberrant salinity is most likely to develop near SL anyway.

Several ostracode taxa are quite capable of living under aberrant salinity conditions and for this reason the ostracodes can provide excellent indicators for such conditions in the geologic past. Selected species might also be used to track *haliclines,* directional changes of salinity with varying distance from a salinity source. This is because certain structural details of the shells of such species seem to correlate with the salinity level of the environment. Shells of *Cyprideis torosa,* which at higher salinities are quite smooth, develop conspicuous 'nodes' (hollow protuberances) at lower salinities (Vesper, 1975). This species also increases its size with decreasing salinity (Van Harten, 1975) and there is a correlation between salinity and the shape of pores in the shell (Rosenfeld & Vesper, 1977). Variable nodosity and size have also been observed in Tertiary brackish-water to hyperhaline ostracodes (Keen, 1982). Several ostracode species seem to vary the ornamentation and the thickness of their shell with salinity (Garbett & Maddocks, 1979). Haliclines in ancient sediments, could be useful as shore-line indicators.

Ostracode evidence has played a prominent role in the recent discovery of the Late Miocene 'Messinian Event' in the Mediterranean. In landsections around this sea, marine Miocene and Pliocene deposits are usually separated by an evaporite sequence of Messinian age. The evaporites have also been found throughout the deep basins of the western and eastern Mediterranean. They are typically associated with low-diversity ostracode faunas containing abundant *Cyprideis*. But deep-water, in part psychrospheric, ostracodes have been found in both the underlying Miocene and overlying Pliocene (Benson & Sylvester-Bradley, 1971;

Benson & Ruggieri, 1974). The *Cyprideis*-bearing deposits are known as 'lago mare' in the literature. Their paleo-environment has been described as a series of large, brackish to hyperhaline lagoons in a desert (Ruggieri, 1967; Benson & Ruggieri, 1974). This desert, if ever there was one, is now, once again, largely covered by the sea and little is known about the actual depth of deposition of lago mare. But it is currently doubted by few that dramatic changes of SL have been involved in bringing about the Mediterranean salinity crisis.

FINAL REMARKS

Ostracodes, naturally, are a specialist field. Getting optimum paleo-bathymetric resolution from the ostracode record will heavily draw on the evaluation of taxonomic and ecologic evidence. This requires a good deal of professional skills and experience. But no special skills or experience to speak of are needed to use the simple principles of depth-related changes in diversity and density and the still simpler technique of tracing diversity patterns in time. The same is true for the recognition of fossil brackish-water ostracode faunas and from that the paleo-mixohalinicum.

Though this paper is all about the application of ostracodes, it should be borne in mind that in actual SL practice the ostracode evidence is likely to be incorporated in data from other sources, such as alternative groups of fossils, lithology or sedimentology (eg. Roep & Van Harten, 1979).

Thanks to their combined biologic, ecologic and paleontologic attributes, ostracodes are currently developing into a powerful tool for studies concerned with environmental change. Though potentially dramatic in their impact, movements of the level of the sea are just another variation on this generaltheme. It is hoped that the data given and the ideas expressed in this paper may contribute to the better understanding and the sharper definition of SLs from the past.

APPENDIX I: COLLECTING AND SAMPLE PREPARATION

Collecting and sample size

Ostracodes are generally most plentiful and easiest to process in clays, claystones and shales, and marls. They may be abundant in sands but this is an exception.

Ostracodes are less frequent than foraminifers in most marine sediments. In order to be able to extract reasonably representative faunas, one has to collect fairly large samples in the field. So as to err on the safe side, a sample size of *c*.300 cc of rock should be adequate, generally.

Washing

Prior to being available for study, the ostracodes must be freed from the surrounding rock. This can be accomplished by washing in a way similar to that used for foraminifers. The aim of washing is to disintegrate the rock structure and to remove as many small-sized sediment particles as possible without actually affecting the microfossils themselves.

Ostracodes are often somewhat more fragile than foraminifers, so any hard rubbing over metal sieves should be avoided. To prevent damage, the washing can be done over a nylon cloth of suitable mesh. Closed shells and thin valves may tend to float on the surface of the washing water and care should be taken that they are not poured off.

If additional cleaning of the washed residue is required, this should be done by gently boiling in a solution of soda or a non-acid synthetic detergent. Do not apply ultra-sonic cleaning to ostracodes as this will severely damage the shells.

Picking and making observations

The picking of the dried residues is done under a stereo-microscope using low magnification. Because it is very tiring for the eyes to focus on ill-sorted material, the residues are best split into a number of sieve fractions. The often tiresome picking of sandy residues can be considerably speeded up by gently shaking them beforehand in a flat-bottomed tray. Due to mechanical sorting the microfossils will become concentrated on top of the sand so that the ostracodes can be picked off easily and rapidly.

Ostracodes are usually studied under a stereo-microscope allowing of magnifications up to *c*.200x using incident light. The smaller details of the shells can be observed with the aid of a Scanning Electron Microscope (SEM). A SEM is also needed to produce high-quality photomicrographs of ostracodes.

REFERENCES

Bate, R.H., 1978. The Jurassic Part II - Aalenian to Bathonian. In: *A Stratigraphical Index of British Ostracoda*, eds. R.H. Bate, & E. Robinson, Geological Journal Special Issue No.8, (Seel House Press, Liverpool), 213-258

Benson, R.H. & Ruggieri, G., 1974. The end of the Miocene, a time of crisis in Tethys-Mediterranean history. *Annals of the Geological Survey of Egypt*, 4, 237-250

Benson, R.H. & Sylvester-Bradley, P.C., 1971. Deep-sea ostracodes and the transformation of ocean to sea in the Tethys. *Bulletin du Centre de Recherches Pau-SNPA*, 5 suppl., 63-91

Bismuth, H., Keij, A.J., Oertli, H.J. & Szczechura, J., 1978. The genus *Loculicytheretta* (Ostracoda). *Bulletin du Centre de Recherches Exploration-Production Elf-Aquitaine*, 2, 227-263

Breman, E., 1980. Differential distribution of left and right ostracode valves in the Adriatic Sea. *Palaeogeography, Palaeoclimatology, Palaeoecology*, 32, 135-141

Elofson, O., 1941. Zur Kenntnis der marinen Ostracoden Schwedens mit besonderer Berücksichtigung des Skagerraks. *Zoologiska Bidrag fran Uppsala*, 19, 215-534

Garbett, E.C. & Maddocks, R.F., 1979. Zoogeography of Holocene cytheracean ostracodes in the bays of Texas. *Journal of Paleontology*, 53, 841-919

Harten, D. van, 1975. Size and environmental salinity in the modern euryhaline ostracod *Cyprideis torosa* (Jones, 1850), a biometrical study. *Palaeogeography, Palaeoclimatology, Palaeoecology*, 17, 35-48

Hinte, J.E. van, 1976. A Jurassic time scale. *American Association of Petroleum Geologists Bulletin*, 60, 489-497

Keen, M.C., 1982. Intraspecific variation in Tertiary ostracods. In: *Fossil and Recent Ostracods*, eds. R.H. Bate, E. Robinson & L.M. Sheppard, (British Micropalaeontological Society / Ellis Horwood Ltd, Chichester), 381-405

Kontrovitz, M. & Nicolich, M.J., 1979. On the response of ostracode valves and carapaces to water currents. In: *Taxonomy, Biostratigraphy and Distribution of Ostracodes*, ed. N. Krstić, (Serbian Geological Society), 269-272

Morkhoven, F.P.C.M. van, 1963. *Post-Palaeozoic Ostracoda, their morphology, taxonomy, and economic use*, Volume II, generic descriptions, (Elsevier, Amsterdam)

Morkhoven, F.P.C.M. van, 1972. Bathymetry of Recent marine Ostracoda in the northwest Gulf of Mexico. *Transactions, Gulf Coast Association of Geological Societies*, 22, 241-252

Peypouquet, J.-P , 1980. Les relations ostracodes-profondeur. Principes applicables pendant le Cénozoique. *Bulletin de l'Institut de Géologie du Bassin d'Aquitaine*, 28, 13-28

Pokorný, V., 1971. The diversity of fossil ostracode communities as an indicator of palaeogeographic conditions. *Bulletin du Centre de Recherches Pau-SNPA*, 5 suppl., 45-61

Pokorný, V., 1980. Ostracodes. In: *Introduction to Marine Micropaleontology*, eds. B.U. Haq & A. Boersma, (Elsevier, New York), 109-149

Puri, H.S., Bonaduce, G. & Malloy, J., 1964. Ecology of the Gulf of Naples. *Pubblicazioni della stazione Zoologica di Napoli*, 33 suppl., 87-199

Puri, H.S., Bonaduce, G. & Gervasio, A.M., 1969. Distribution of Ostracoda in the Mediterranean. In: *The Taxonomy, Morphology and Ecology of Recent Ostracoda*, ed. J.W. Neale, (Oliver and Boyd, Edinburgh), 356-411

Roep, T.B. & Harten, D. van, 1979. Sedimentological and ostracodological observations on Messinian post-evaporite deposits of some southeastern Spanish basins. *Annales Géologiques des Pays Helléniques*, Tome hors série, 1979, fasc. III, 1037-1044

Rosenfeld, A. & Vesper, B., 1977. The variability of the sieve-pores in Recent and fossil species of *Cyprideis torosa* (Jones, 1850) as an indicator for salinity and palaeosalinity. In: *Aspects of Ecology and Zoogeography of Recent and Fossil Ostracoda*, eds. H. Löffler & D. Danielopol, (Junk, The Hague), 55-67

Ruggieri, G., 1967. The Miocene and later evolution of the Mediterranean Sea. In: *Aspects of Tethyan biogeography*, eds. C.G. Adams & D.V. Ager, (Systematics Association Publication No.7), 283-290

Sheppard, L.M., 1978. The exploration application of the range tables. In: *A stratigraphical Index of British Ostracoda*, eds. R.H. Bate & E. Robinson, Geological Journal Special Issue No.8, (Seel House Press, Liverpool), 473-523

Vail, P.R. & Mitchum, R.M., 1979. Global cycles of relative changes of sea level from seismic stratigraphy. *American Association of Petroleum Geologists Memoir*, 29, 469-472

Vesper, B., 1975. To the problem of noding on *Cyprideis torosa* (Jones, 1850) *Bulletins of American Paleontology*, 65(282), 205-216

18

Shell middens as a source for additional information in Holocene shoreline and sea-level reconstruction: examples from the coast of Brazil

L. MARTIN, K. SUGUIO and J.-M. FLEXOR

ABSTRACT

Artificial accumulations of brackish-water shells and marine organisms constructed by ancient inhabitants of coastal regions may, under some conditions, be used as indicators of past sea levels. They are particularly useful in coastal areas that have experienced submergence and emergence once or several times. Assuming that the ancient inhabitants established their campsite near a suitable locality to furnish sufficient shell food and above the local high water spring tide level, it is possible to establish a horizontal relationship between shell-midden sites and former lagoonal, estuarine or marine environments. Although shell middens do not provide direct evidence of former sea-level positions, study of their stratigraphical position, internal stratigraphy, height, radiocarbon and archaeological age, fauna composition and $^{13}C/^{12}C$ isotope ratio of mollusc shells can supply valuable additional information within a more encompassing shoreline/sea-level investigation. This paper discusses several examples of the use of shell middens in the study of relative sea-level movements along the coast of Brazil.

INTRODUCTION

Artificial accumulations made up of shells of brackish-water and marine organisms are very commonly found in coastal regions around the world, as in Natal (South Africa), southern Madagascar, eastern Australia (particularly the 'New England' coast of New South Wales), Senegal, middle Atlantic coast of the United States.

In coastal Brazil, between Rio de Janeiro and Rio Grande do Sul, several hundreds of giant shell middens (known in that country as 'sambaquis' (Indian terminology) have been built up by ancient inhabitants of these areas, whose basic food was shellfish, as shown by archaeological investigations (Laming-Empéraire, 1968; Hurt, 1974; and Fairbridge, 1976). These ancient inhabitants have been named 'shellfish-eating preceramic Indians' (Fairbridge, *op.cit.*). The middens exhibit a variety of shapes and sizes, ranging from huge elongated (300 x 60 x 6m) or ellipsoidal (86 x 40 x 25 m) to small circular (10 x 1 ∿ 1.5 m) accumulations. Some of them occur as far as 40 km inland from the present coastline.

It may be assumed that site selection was decided primarily by near occurrence of shellfish in sufficient quantity to provide nourishment over a long period of time. One shell midden examined by Fairbridge (*op.cit.*) in the State of Santa Catarina was approximately 20 m high and 100 m in diameter, representing about 2.5 billion shellfish, that is almost 100 shellfish per day for a group of 100 people for 500 years. Obviously, shallow and quiet-water bay bottoms, lagoonal and estuarine zones, with muddy surface areas exposed during low tides, would have been more favourable from this standpoint than sandy areas at the open ocean margin. In the former case, biological productivity is much higher. Presently, large numbers of people pick up considerable quantities of molluscs from areas exposed during low tides as, for instance, in certain parts of the Todos os Santos Bay (State of Bahia, Brazil), without exhausting the stock. It is probable that paleo-inhabitants chose campsites above high-tide level, immediately adjacent to favourable collecting grounds, where the best conditions of comfort and safety were found. Only occasionally is there evidence that they went farther afield, in which case one may assume an important relationship between the shell midden's position and the presence of shallow-marine, lagoonal or estuarine zone in the vicinity. Ecological investigations of the dominant mollusc species found in the shell midden will indicate the paleo-environmental conditions of the surrounding area.

While it is relatively easy to establish the geographic relationship between the shell midden and a nearby ancient lagoon, estuary or bay, it is much more difficult to establish the vertical relationship between the altitude of the base of the shell midden and the position of sea level (SL) during its construction. We can only assume that initially it was above local high-water spring tide (HWST) level, a very important assumption for interpreting shell middens whose substrates are located beneath present high-tide level. A detailed investigation of the shell midden's substrate and composition may provide some criteria to eliminate the possibility of the shell middens having been built up below high-tide level. Probably, this is the case with shell middens that originated as wastedumps (mixture of shells and man-made artifacts) below houses which were built on poles, standing in shallow water during high tides. Furthermore, when the shell midden is situated near a lowland (ancient lagoonal area) we must assume, as a postulate, that its construction occurred near the mollusc collecting area. From these two assumptions, proximity of the collecting zone and construction above local HWST level, it is possible to establish a rather close relationship between geographic position of shell middens and ancient SLs.

The main purpose of this paper is to evaluate the usefulness of shell middens for SL-height/shoreline reconstruction. The method used for this evaluation is to show how information obtained from the study of shell middens correlates with SL data derived from geo(morpho)-logical and biological indicators and how that information has been used in the construction of SL curves for several sectors of the States of São Paulo and Bahia

RELATIVE SEA-LEVEL CHANGE DURING THE LAST 7000 YEARS ALONG THE BRAZILIAN COAST

Papers by Martin & Suguio, (1975, 1976a, 1978); Martin *et al.*, (1979a, 1979b, 1980); Suguio & Martin, (1976, 1978); Suguio *et al.*, (1980); and Bittencourt *et al.*, (1979), on the coastal plains of the States of São Paulo and Bahia provide very good knowledge of both the SL fluctuations during the Quaternary and the ages of associated sandy deposits. An evolutionary history of these regions during the Quaternary has been reconstructed from this information. About 120 000 years ago the relative SL (RSL) stood about 8 ± 2 m above the present level. Between the end of this transgression (Cananéia transgression) and the beginning of the following regression a first generation of sandy beach ridges was deposited. During the last glacial when SL was low, these deposits were carved by rivers and a system of valleys, some very large and deep, developed. The original wave-built terrace surfaces, frequently marked by beach-ridge alignments, have been preserved only in interfluvial zones. When, during the Holocene, SL reached its present position (about 7000 to 6500 ^{14}C years ago; Santos transgression), the low-lying parts of the ancient terraces were invaded by the sea, forming extensive lagoonal systems. About 5000 BP SL attained a maximum height, which was coincident with the maximum extension of the lagoonal areas. From that time on, RSL fell more or less regularly with two important negative fluctuations between 4100 and 3600 BP and 3000 and 2500 BP.

INFORMATION TO BE DERIVED FROM SHELL MIDDENS

1. Different types of shell-midden sites

Four distinctive types of midden sites can be recognized (modified from Fairbridge, 1976).

Type I is located on Cananéia transgression terraces, always at the margins of zones formerly, or up to the present time, occupied by lagoons. Among these we must look for the shell middens constructed during the period of greatest lagoonal extent. In general, the more inland shell middens are smaller than those situated near the present strandline. Here, at the coast, favourable living conditions for molluscs continue to exist today.

Type II shell middens are situated on Santos transgression terraces. These terraces have been deposited as beach ridges during the regressive phases that followed the SL maxima, in particular the 5100 BP maximum. Type II shell middens have necessarily been constructed after the maximum Holocene SL stands, and in some cases their position on the terraces can provide additional data on SLs.

Type III shell middens are situated on ancient lagoonal deposits in front of sandy terraces. Hence, they must have been constructed after a period of high SL, probably during the ensuing lowering of SL. One must be cautious with type III middens, because the substrate of lagoonal organic clays is liable to have been subjected to compaction, thus diminishing the height of the midden's base.

Type IV shell middens are located on rocky hills at the margins or centres of ancient lagoons or bays. Their age varies widely. Clearly, those situated more inland may have been constructed when the lagoon reached its greatest extent.

2. Faunal composition of shell middens

About fifty different species of molluscs have been identified within the shell middens but the most dominant species are very few; according to Bigarella (1949) these are:

- *Anomalocardia brasiliana* Gmelin
- *Ostrea brasiliana* sp.
- *Ostrea arborea* Chemnitz
- *Lucina jamaicensis* Chemnitz
- *Modiolus brasiliensis* Chemnitz

Except for *Ostrea arborea*, which lives fixed on aerial roots of mangrove trees, these bivalves live within sandy or clayey-sandy sediments deposited in shallow-water lagoons and bays. Shell remains of other organisms are generally negligible. Some middens are composed practically of one mollusc species, others show a mixed composition. *Ostrea* sp. and *Modiolus brasiliensis* are dominant species within more inland shell middens, and in this case *Anomalocardia* is very scarce or completely absent. On the other hand, shell middens situated nearer to the open sea are formed almost entirely of *Anomalocardia*

3. $^{12}C/^{13}C$ ratios of carbonate shells from shell middens

$\delta^{13}C$ (PDB) measurements for carbonates show a spectrum of values varying as a function of the influence of continental environmental conditions during the carbonate formation. Shells from lagoonal organisms show $\delta^{13}C$ (PDB) values between those for fresh-water organisms ($\simeq$ 13 ‰) and those for marine organisms ($\simeq$ 0 ‰). At the same time, the $\delta^{13}C$ (PDB) values for carbonate shells from a lagoon

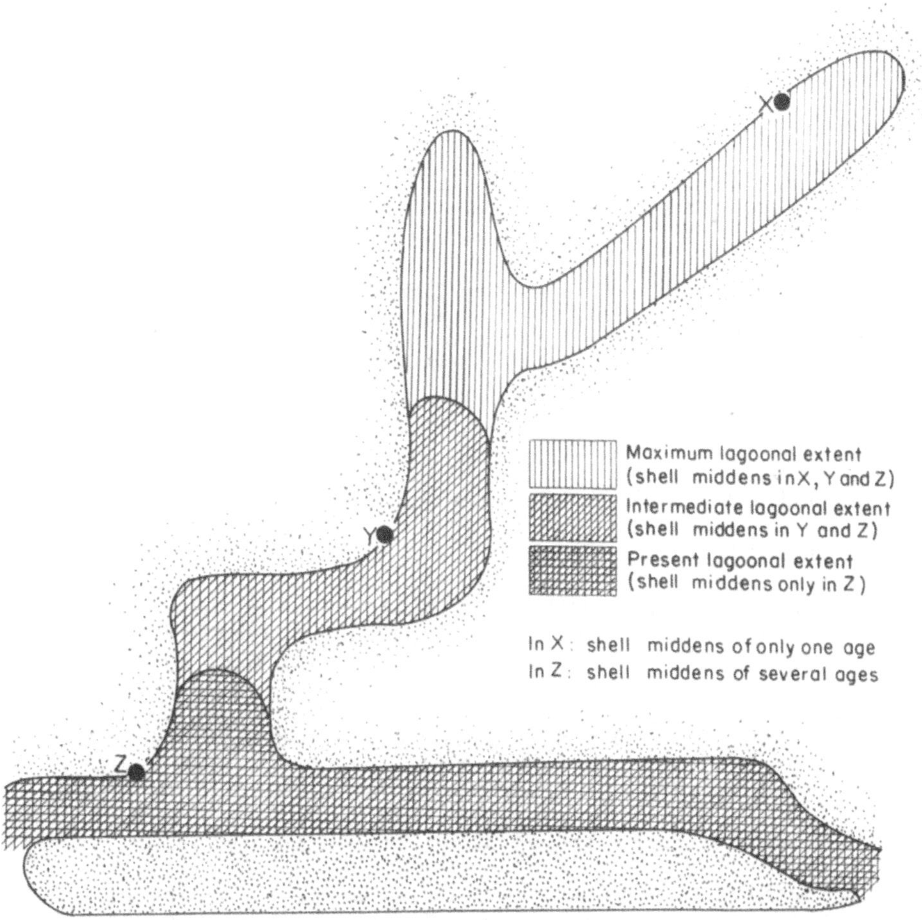

Figure 1. Age variations of the shell middens as a function of lagoonal extent.

change as a function of their geographic position within it (Flexor *et.al.*, 1979). Shells from outer zones (nearer the open sea) of a lagoon show only slightly negative $\delta^{13}C$ (PDB) values, while those from inner zones are clearly characterized by more negative values (Table 1). In the latter case, carbon, derived from decomposition of land plants, has been incorporated by the molluscs within their carbonate shells. Deeper and inner portions of the lagoons are characterized by a poor water circulation propitiating an accumulation of large quantities of plant-derived organic matter. This is independent of water salinity and explains why the carbonate shells of lagoonal organisms present $\delta^{13}C$ (PDB) values characterized by a strong land influence.

Conversely, at the same locality, but at different moments, the $\delta^{13}C$ (PDB) values change according to the increase or decrease of the lagoonal area. In this case, this parameter will be a good indicator of the lagoonal oscillations and thus indirectly of SL fluctuations (Table 2).

Taking into consideration the example in Figure 1, it is quite possible to find at 'X', 'Y' and 'Z' shell middens of similar age but with quite different $\delta^{13}C$ (PDB) values for carbonate shells: very negative in 'X' and slightly negative in 'Z'. Where, in 'Y' and 'Z' shell middens exhibit different ages, the least negative $\delta^{13}C$ (PDB) values correspond to periods of greatest lagoonal extent and vice-versa.

Generally, we do not know the paleo-RSL position with respect to the present level. However, we know if it was lower or higher and so we can deduce whether SL was (or had been) rising or falling and approximate the time that a maximum or minimum was reached.

COMPARISON OF SEA-LEVEL DATA WITH SHELL-MIDDEN DATA

1. Cananéia-Iguape region

The Cananéia-Iguape sedimentary plain has the shape of a large crescent and covers an area of about 2500 km^2. Its outer portion is presently drained by a lagoonal system and water courses subjected to tidal influence. A large part of the plain is occupied by remnants of more or less dissected Pleistocene terraces and by presently dried-out ancient lagoonal areas, at the far inland margins of which several shell middens have been found.

a) Information furnished by dating geological samples

Shell debris and wood fragments from littoral marine or lagoonal deposits have been dated, and this information, when associated with the nature of the sediments, indicates that (Fig. 2A):

a) about 6600 to 6500 BP, RSL reached approximately its present level and it was rising;

b) about 5400 BP, RSL was about 1.5 m above present level:

c) about 4400 BP, RSL was about 2 m higher than present level but it was falling. Hence, between 5400 and 4400 BP, RSL passed through a maximum height which, according to morphological data, must have been 3.5 to 4 m above the present one;

d) between 3800 and 3700 BP, RSL rose very rapidly reaching a second maximum of about 3 m above present level; between 4400 and 3700 BP it passed through a minimum;

e) after the second maximum, RSL returned more or less regularly to its present level. Unfortunately we have insufficient data for detailing this return.

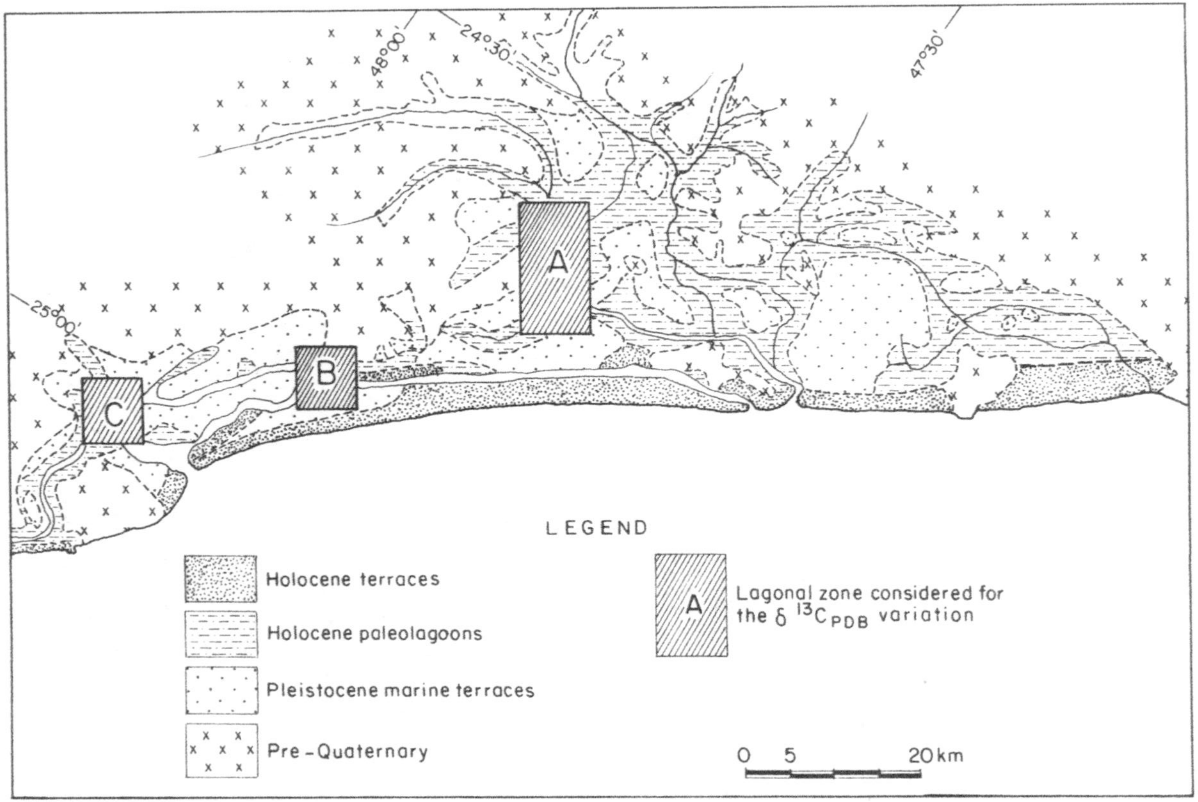

Figure 2. Reconstruction of relative sea-level fluctuation curve for the past 7000 years in the Cananéia-Iguape region (State of São Paulo, Brazil).

b) Information furnished by dating shell middens

As indicated above, between 5400 and 4400 BP RSL was always higher than at present, attaining a maximum of 3.5 to 4 m above present level. One may expect that this high SL occurred contemporaneously with the maximum lagoonal extent and that the shell middens situated more inland should date this phase (Tables 3 and 4). Analysis of the majority of these shell middens shows that only a SL higher than the present one can explain their positions (Fig. 2B). Moreover, the ages of the shell middens situated far from the present strandline lie between 5200 and 4800 BP.

Thus, we believe that the maximum lagoonal extent occurred during this period, probably at about 5000 BP. Moreover by 4100 BP RSL was probably no more than 0.5 m above the present level, and around 3800 BP it may even have been lower. This confirms the conclusion reached above on the basis of geological evidence that SL reached a minimum between 4100 and 3800 BP.

Table 5 shows that the oldest shell midden (Estaleiro = 3690 ± 80 BP) is situated on the Pleistocene terrace which may be an indication of a high lagoonal level. This hypothesis seems to be confirmed by the $\delta^{13}C$ (PDB) value of its shell carbonate. In fact, this value is much less negative than for shells of Sambaquinho, which was constructed in the same area when SL can have been at most 0.5 m above present level. Evidently, around 3700 BP RSL was well above the present level. It was stated above that since 3700 BP RSL reached a second maximum situated 3 m above the present one. The age of the Boguacu-I shell midden (3080 ± 55 BP), which was constructed when SL cannot have been more than 2.5 m above present level, shows that at about 3100 BP, the maximum of 3 m was already passed. The information from the Guarapari and Sambaquinho shell middens indicate that RSL at about 2300 and 1500 BP has not been more than 1 m and 0.5 m respectively above present level.

c) Information furnished $\delta^{13}C$ (PDB) values

As discussed above, $\delta^{13}C$ (PDB) values can be used as an index to marine versus continental influences, and as indirect evidence of SL change (Fig. 2C). Below, we discuss $\delta^{13}C$ (PDB) values obtained for shell samples from three groups of middens in the Cananéia-Iguape coastal plain (A, B and C in Fig. 3). The first two groups (A and B) concern shell middens constructed in the period 5200-4200 BP and 5200-4800 BP respectively; the middens of group C were built between 3800 and 1500 BP.

Group A We take the example of seven shell middens of region A of the Cananéia coastal plain (Fig. 3). The first six have ages ranging from 5200 to 4200 BP (Table 2). It would be interesting to have information on a shell midden constructed about 4100 BP, when SL cannot have been higher than 0.5 m above the present level. We could then compare $\delta^{13}C$ (PDB) values of carbonate shells formed in the same

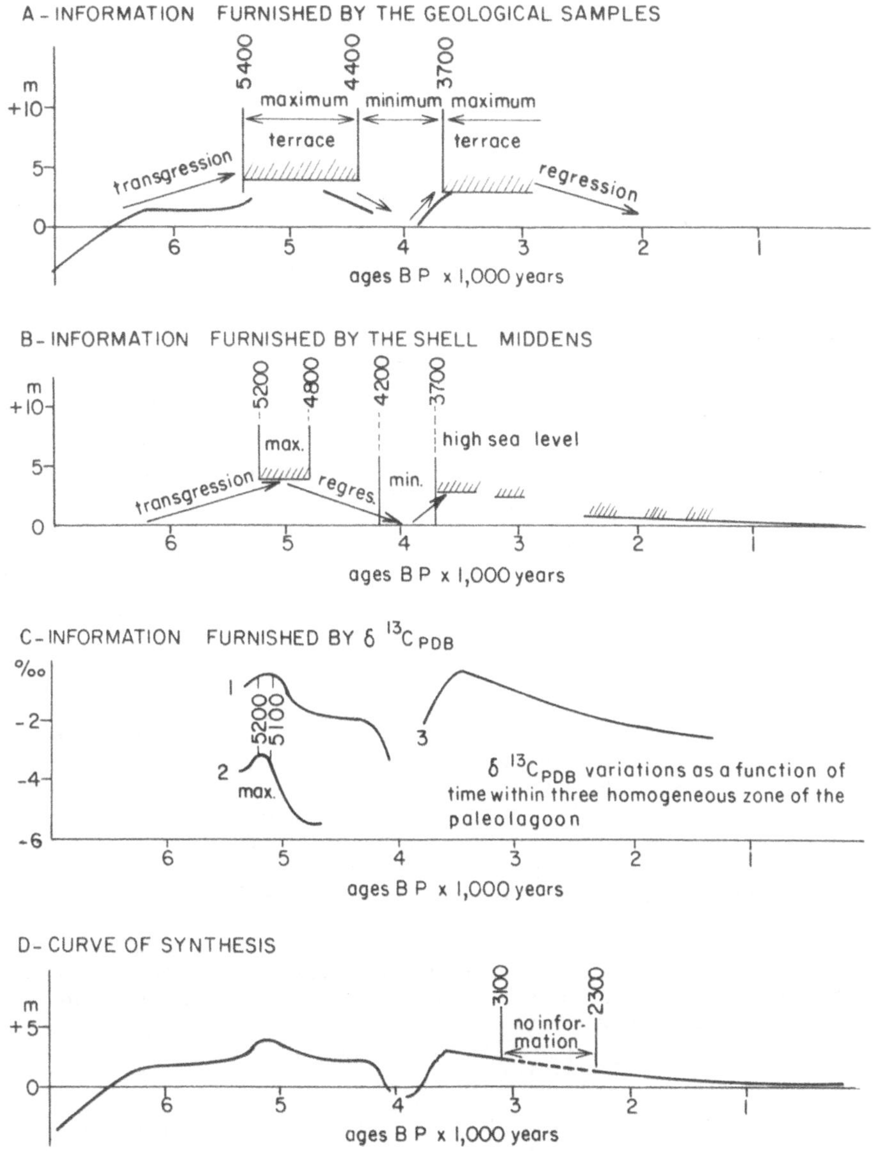

Figure 3. Cananéia-Iguape Quaternary coastal plain (State of São Paulo, Brazil).

geographic position between 5200 and 4100 BP, encompassing the first Holocene regressive phase. For want of a shell midden dated 4100 BP we have considered another about 1000 years old, when SL was quite similar to that at 4100 BP. It is assumed that $\delta^{13}C$ (PDB) values of shell carbonate from shell middens of both ages are comparable.

When plotted as a function of time (curve 1 in Fig.2C),

the $\delta^{13}C$ (PDB) values show a maximum between 5200 and 5100 BP and decrease rapidly until about 4800 BP, then slowly until 4200 BP, whereafter another rapid decrease is noticed. These $\delta^{13}C$ (PDB) differences can be interpreted both in terms of changing lagoonal extent and, assuming constant (local) tide range, in terms of changing RSL as follows;

a) the maximum lagoonal extent, and thus the SL maximum probably occurred between 5200 and 5100 BP;

b) the maximum was followed by a rapid decrease in lagoonal extent, which is indicative of a quick SL lowering between 5100 and 4800 BP;

c) lagoonal extent and hence RSL were stationary between 4800 and 4200 BP; and

d) lagoonal extent reduced rapidly after 4200 BP suggesting an equally rapid SL lowering.

Group B Here we have considered three shell middens from the same area of the inland portion of the Iguape coastal plain (area B in Fig. 3). From these data (Table 6) one may derive a curve of variation of $\delta^{13}C$ (PDB) which shows that the maximum was well established between 5200 and 5100 BP and that the fluctuation was very rapid between 5100 and 4800 BP (curve 2 in Fig. 2 C).

Group C We have considered four shell middens in the same area (area C in Fig. 3) of the Cananéia coastal plain, whose ages range from 3800 to 1500 BP (Table 7). The corresponding $\delta^{13}C$ (PDB)-variation curve is given in Figure 2 C (curve 3). Between 3800 and 3500 BP, values increase very rapidly and then decrease. Put in terms of RSL it might be that there was a rapid rise between 3800 and 3500 BP followed by a drop.

In summary, radiocarbon dating of shell debris and fossil wood fragments from both sedimentary deposits and shell middens, as well as information on $\delta^{13}C$ (PDB) of carbonate shells from shell middens has allowed us to establish with sufficient accuracy a fluctuating RSL curve for the Cananéia-Iguape region (Fig. 2 D).

2. Santos-Itanhaém region

The Santos-Itanhaém coastal plain has the same characteristics as the Cananeia-Iguape area. A lagoonal system originated during each high SL period and shell middens, some of them dated by radiocarbon method, were established at the margins of paleo-lagoons.

a) Information from geological and biological sources

A large number of radiocarbon dates of shell debris and fossil wood fragments contained in lagoonal and shallow-marine deposits, as well as of vermetid incrustations, allowed us to delineate a rather accurate curve of RSL fluctuations (Fig. 4 A). This curve is very similar to that for Cananéia-Iguape, though some differences in amplitude are apparent.

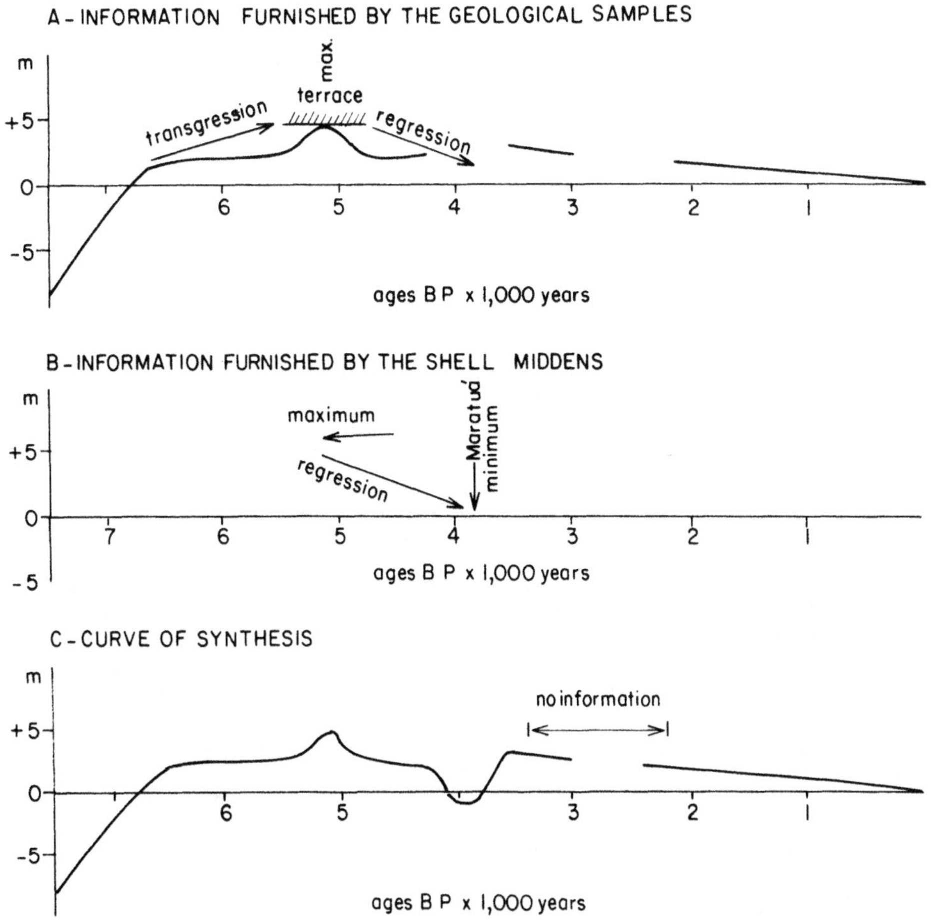

Figure 4. Reconstruction of relative sea-level fluctuation curve for the past 7000 years in the Itanhaém-Santos region (State of São Paulo, Brazil).

b) Information furnished by dating shell middens

Dated shell middens are much less numerous than in the Cananéia-Iguape area, but they are important for corroborating the information dated from geological samples (Fig. 4B; Table 8).

The Rio Preto, Araraú, and Mundo Novo shell middens are situated in an inner portion of the Itanhaém paleo-lagoon. Possibly, therefore, they were constructed during maximum lagoonal extent. However, if this hypothesis is correct, this maximum lagoonal extent could be different from that previously defined. The $\delta^{13}C$ (PDB) values of the shells from these middens are much more negative than those of the samples dating the high SL period of Cananéia-Iguape. In fact, the $\delta^{13}C$ (PDB) values of the Itanhaém shell middens are quite similar to that of the Momuna shell midden

(Table 3), which has about the same age. Thus, possibly about 4600 BP the continental influence in the inner portion of the Itanhaém paleo-lagoon was very strong, perhaps because the maximum had passed and the paleo-lagoon was beginning to dry out.

The shell middens A229 and Mar Casado (Table 8), located on a Holocene terrace, obviously were constructed after the transgression maximum, when SL was probably less than 3.5 m and 3 m respectively, above the present level.

The Maratuá shell midden, which has been completely destroyed, has furnished the most interesting information (Fig. 4B). The base of this midden was located below present SL. This fact cannot be attributed to substrate subsidence due to the shell midden's weight, because in that case the central portion would have sunk more than its margins, thereby deforming the shell midden's layers. No deformation whatever was noticed. Two samples from this shell midden dated at the beginning of radiocarbon investigations indicated ages of 7330 ± 1300 BP (Gif-15) and 7800 ± 1300 BP (Gif-16) (Laming - Empéraire, 1968). These values, now known to be in error, coincided with a period when RSL was definitely lower than today, so that initially these two ages were thought to be consistent with known SL curves. However, archaeological remains within the Maratuá shell midden suggested that its construction must have occurred more recently. Two radiocarbon ages were obtained for the same samples and indicated ages of 3925 ± 145 BP (Bah-382) and 3865 ± 95 BP (Gif-9185), which are more in keeping with modern data. It is now possible to infer that RSL about 3800 BP was lower than today, as we have also observed in the Cananéia-Iguape region.

3. Salvador region

On a sector of the coast with homogeneous conditions in the northern part of Salvador (State of Bahia), 66 reconstructions of former RSL positions were made. From these data we have delineated a very accurate curve of RSL change (Fig. 5), which shows that:

a) the present level was crossed about 7000 BP;

b) about 5100 BP, RSL was at a maximum, some 4.7 ± 0.5 m above the present level;

c) about 3900 BP, RSL reached a minimum, lower than at present;

d) about 3600 BP, RSL attained a second maximum, located more than 3 m above the present level;

e) about 2700 BP, RSL had fallen to a second minimum, near present level;

f) about 2500 BP, RSL reached a third maximum situated about 2.5 m above the present; and

g) since this time, RSL has returned progressively toward the present level.

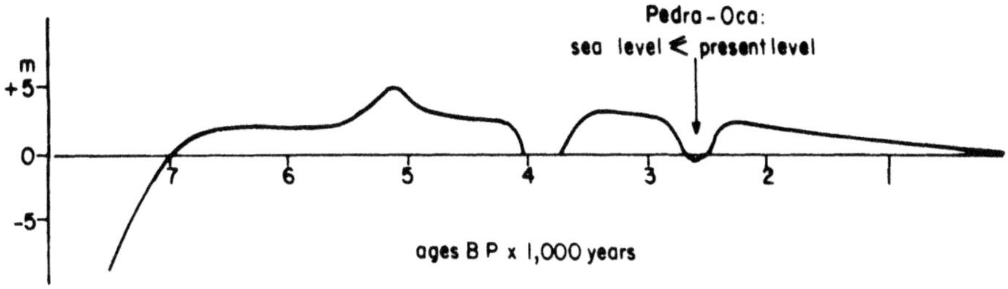

Figure 5. Relative sea-level fluctuation curve for the Salvador region (State of Bahia) with indication of the Pedra Oca shell midden

This curve presents characteristics quite similar to that of the Santos region (Fig. 4) with some small variations in amplitude.

In the Salvador region SL was low about 2700 BP, which was not evidenced in the Santos region. Perhaps, the absence of data for this period in the Santos region is indirect evidence of a lower SL. In fact, if SL has been lower, the evidence must presently be submerged.

At one time there were numerous shell middens around Todos os Santos bay, but unfortunately the great majority of them have been destroyed. It has been possible to obtain information about the Pedra Oca shell midden (Calderon, 1964) located in the sector studied by Martin *et al.* (1979a, 1980). Presently, the lower part of the non-eroded portion of this shell midden is situated 0.8 m above high-tide level (Martin *et al.*, 1979a). Two ages obtained from basal shells indicated ages of 2830 ± 130 BP (Gif-878) and 2630 ± 110 BP (Si-470). These data confirm the occurrence of a low SL position at about 2700 BP, which is also evident from other data.

FINAL CONSIDERATIONS

It is clear that shell middens are not the best tool for reconstructing, in space and time, the positions of former SLs. In fact, in practice it is not possible to establish directly the vertical relationship between the base of a shell midden and SL. The only thing of which we are more or less certain is that the base of the shell midden was above local HWST level at the beginning of its construction. If we postulate that the paleo-inhabitants established their campsites near the mollusc harvesting place, it is easy to establish the geographical relationship between the shell middens and an ancient lagoonal, estuarine or shallow-marine zone. Shell middens located further inland were probably associated with a period of maximum lagoonal extent, obviously when a SL maximum occurred. Nevertheless, it is necessary to ascertain that coastal progradation was unrelated to intense continental sedimentation, but rather to a RSL change. Plainly, only one dating is insufficient, and one must have a statistically significant series of dates in order to interpret correctly the period of maximum lagoonal extent. On the other hand, the shell middens whose substrates are located beneath present high-tide level suggest, if compaction can be ruled out, a period of lower SL. Values of $\delta^{13}C$ (PDB) for carbonate shells give us additional information about the high or low SL positions, as well as about the trend of SL change.

ACKNOWLEDGEMENT

The authors wish to express their thanks to Dr. Thomas R. Fairchild for revision of the English text.

Table 1. Variation of $\delta^{13}C$ (PDB) as a function of position of the samples within a paleo-lagoon, coastal plain of the State of São Paulo, Brazil (Flexor *et al.*, 1979).

Sample	Radiocarbon age (BP)	$\delta^{13}C_{PDB}$ ‰	Distance from the open sea (km)
Itapoã III	5245 ± 125	- 0.63	5
Jataituba	5240 + 150	- 3.26	20
Vapumauva II	5080 + 60	- 0.06	1
Pariquera Acu	5035 + 140	- 3.88	24
Vapumauva I	4680 + 115	- 0.12	2
A195	4636 + 100	- 6.17	18
Ubatuba	3870 + 100	- 0.02	?
A123	3775 + 130	- 5.69	33

Table 2. Variation of $\delta^{13}C$ (PDB) as a function of age within the same portion of a paleo-lagoon, coastal plain of the State of São Paulo, Brazil (Flexor *et al.*, 1979).

Sample	Radiocarbon age (BP)	$\delta^{13}C_{PDB}$ ‰
Itapoã III	5245 ± 125	- 0.63
Guaxixi	5110 ± 70	- 0.48
Juruvaúva I	5010 ± 115	- 0.76
Juruvaúva III	4970 ± 150	- 1.34
Juruvaúva II	4305 ± 140	- 2.07
Itapoã II	4215 ± 140	- 2.27
Vamiranga	1015 ± 70	- 3.81

Table 3. General characteristics of shell middens constructed between 5400 and 4400 BP

Shell midden	Radiocarbon age (BP)	Laboratory number	Nature of substrate	Position in the paleo-lagoon	$\delta^{13}C_{PDB}$ ‰
Itapoã III	5245 ± 125	Bah-365	P.T.	Outer	- 0.63
Jataituba	5240 ± 150	Bah-346	P.T.	Inner	- 3.26
Guaxixi	5110 ± 70	Bah-370	P.T.	Outer	- 0.48
Vapumaúva	5080 ± 60	Bah-365	P.T.	Outer	- 0.06
Pariquera Açu	5035 ± 140	Bah-295	P.T.	Inner	- 3.88
Juruvaúva I	5010 ± 115	Bah-359	P.T.	Outer	- 0.76
Juruvaúva III	4970 ± 110	Bah-361	P.T.	Outer	- 1.34
Batatal	4920 ± 100	I.9186	P.T.	Outer	----
R.das Pedras I	4860 ± 100	Bah-343	C.B.	Inner	- 3.17
Momuna	4790 ± 100	Bah-308	P.T.	Inner	- 5.39
R.das Pedras III	4750 ± 110	Gif-3641	P.T.	Inner	- 4.67
	4710 ± 145	Bah-300			
Rio Comprido	4560 ± 110	Gif-3646	C.B.	Outer	----
Cananéia	4340 ± 110	Gif-3435	P.T.	Outer	- 1.28
	4300 ± 140	Bah-302			

(P.T.= Pleistocene Terrace; C.B. = Crystalline Basement; Bah = Laboratorio de Física Nuclear Aplicada, Universidade Federal da Bahia, Brazil; Gif = Laboratoire des Faibles Radioactivités, Gif-sur-Yvette, France and I = Isotopes, USA)

Table 4. Shell middens constructed between 5000 and 3700 BP on Holocene terraces and lagoonal deposits

Shell midden	Radiocarbon age (BP)	Laboratory number	Position of sea level	$\delta^{13}C_{PDB}$ ‰
Vapumaúva II	4680 ± 110	Bah-362	< + 3.5m	- 0.72
Nóbrega	4380 ± 160	SPC-21	< + 3.5m	----
Boguaçu II	4160 ± 100	Bah-303	< + 1.5m	- 1.39
	4120 ± 100	Gif-3436		
Ararapira II	4175 ± 100	Bah-290	< + 0.5m	- 1.67
Ubatuba	3870 ± 100	Bah-294	< + 2.0m	- 0.02
Ararapira I	3790 ± 110	Gif-3437	< 0m	----

(SPC = Centro de Pesquisa Geocronológicas do Instituto de Geociências, Universidade de São Paulo, Brazil)

Table 5. Shell-middens constructed since 3700 BP.

Shell midden	Radiocarbon age (BP)	Laboratory number	Nature of substrate	Position of sea level	$\delta^{13}C_{PDB}$ ‰
Estaleiro	3690 ± 80	Bah-367	P.T.	> 0m	- 0.65
Fosfasa I	3350 ± 135	Bah-340	P.T.	> 0m	- 1.80
Pereirinha	3330 ± 125	Bah-286	H.T.	> 0m	- 0.09
Boguaçu I	3080 ± 55	Bah-285	H.T.	< + 2.5m	- 3.57
Boguaçu III	3220 ± 90	Bah-307	H.T.	< + 3.0m	- 3.25
	3090 ± 110	Gif-3645			
Pindu	3090 ± 120	Bah-348	C.B.	> 0m	- 4.31
Guarapari	2285 ± 45	Bah-368	L.	< + 1m	- 2.24
R. das Minas	1850 ± 100	Gif-3643	H.T.	< + 0.5m	----
Sambaquinho	1500 ± 120	Bah-292	L.	< + 0.5m	- 2.61
Itapitangui	1490 ± 120	Bah-293	H.T.	< + 1.5m	- 4.45
Vamiranga	1015 ± 70	Bah-369	P.T.	indefinite	- 3.81

(H.T. = Holocene Terrace and L. = Lagoonal).

Table 6. $\delta^{13}C_{PDB}$ values of carbonate shells from three shell middens constructed in the same inner zone of the Iguape coastal plain

Shell midden	Radiocarbon age (BP)	$\delta^{13}C_{PDB}$ ‰
Jataituba	5250 ± 150	- 3.26
Pariquera Açu	5040 ± 140	- 3.88
Momuna	4790 ± 115	- 5.81

Table 7. $\delta^{13}C_{PDB}$ values of carbonate shells from four shell middens of the Cananéia coastal plain

Shell midden	Radiocarbon age (BP)	$\delta^{13}C_{PDB}$ ‰
Fosfasa II	3790 ± 110	- 2.63
Estaleiro	3490 ± 60	- 0.67
Guarapari	2285 ± 45	- 2.24
Sambaquinho	1500 ± 120	- 2.61

Table 8. General characteristics of the dated shell middens of the Santos-Itanhaém coastal plain

Shell midden	Radiocarbon age (BP)	Laboratory number	Nature of substrate	Position of sea level	$\delta^{13}C_{PDB}$ ‰
Piaçagüera	4930 ± 100	I-4491	C.B.	> 0m	----
Rio Preto	4635 ± 100	Bah-331	P.T.	> 0m	- 6.07
Ararаú	4630 ± 130	Bah-296	C.B.	> 0m	- 5.29
Mundo Novo	4575 ± 110	Bah-446	C.B.	> 0m	- 8.91
A229	4520 ± 130	Bah-328	P.T.	< + 3.5m	- 2.98
Mar Casado	4400 ± 130	Gif-1194	H.T.	< + 3.5m	----
Maratuá	3925 ± 145	Bah-382	L.	< 0m	----
	3865 ± 95	I-9185			

REFERENCES

Bigarella, J.J., 1949. Nota prévia sobre a composicão dos Sambaquis do Paraná e Santa Catarina. *Boletim de Arquivos de Biologia e Tecnologia*, 4, 95-106

Bittencourt, A.C.S.P., Martin, L., Vilas-Boas, G.S. & Flexor, J.-M., 1979. Quaternary marine formations of the coast of the State of Bahia, Brazil. *Proceedings 1978 International Symposium on Coastal Evolution in the Quaternary*, eds. K. Suguio, T.R. Fairchild, L. Martin & J.-M.Flexor, (São Paulo), 232-253

Calderon, V., 1964. *O sambaqui de Pedra Oca. Instituto de Ciências Sociais*. Universidade Federal da Bahia, 2

Fairbridge, R.W., 1976. Shellfish-eating preceramic indians in coastal Brazil. *Science*, 191, 353-359

Flexor, J.-M., Martin, L. & Suguio, K., 1979. Utilisation du rapport isotopique $^{13}C/^{12}C$ comme indicateur d'oscillations lagunaires. *Proceedings 1978 International Symposium on Coastal Evolution in the Quaternary*, eds K. Suguio, T.R. Fairchild, L. Martin & J.-M. Flexor, (São Paulo), 356-375

Hurt, W.R., 1974. The interrelationships between the natural environment and four sambaquis, coast of Santa Catarina, Brazil. *Occasional Papers and Monographs*, 1, (Indiana University Museum, Bloomington)

Laming-Empéraire, A., 1968. Missions archéologiques françaises au Chili et au Brésil meridional. *Journal de la Société des Américanistes*, LVII, 77-99

Martin, L. & Suguio, K., 1975. The State of São Paulo coastal marine Quaternary geology: The ancient strandlines, International Symposium on the Quaternary, (Curitiba). *Anais Academia Brasileira de Ciências,* 47, (Rio de Janeiro), 249-263

Martin, L. & Suguio, K., 1976a. Etude préliminaire du Quaternaire marin: Comparaison de São Paulo et Salvador, Brésil. *Cahier O.R.S.T.O.M.,* Série Géologie VIII, 33-47

Martin, L. & Suguio, K., 1976b. Les variations du niveau de la mer au Quaternaire récent dans le sud de l'état de São Paulo: Utilisation des 'sambaquis' dans la détermination des anciennes lignes de rivage holocènes. *Actes du XLII Congrès International des Américanistes,* IX, 73-83

Martin, L. & Suguio, K., 1978. *Excursion route along the coastline between the town of Cananéia (State of São Paulo) and Guaratiba outlet (State of Rio de Janeiro).* 1978 International Symposium on Coastal Evolution in the Quaternary, Special Publication no.2, 95 pp (São Paulo)

Martin, L., Flexor, J.-M., Villas-Boas, G.S., Bittencourt, A.C.S.P. & Guimaraẽs, M.M.M., 1979a. Courbe de variation du niveau relatif de la mer au cours des 7000 dernières années sur un secteur homogène du littoral brésilien (nord de Salvador, Bahia). *Proceedings 1978 International Symposium on Coastal Evolution in the Quaternary,* eds. K. Suguio, T.R. Fairchild, L. Martin & J.-M. Flexor, (São Paulo), 264-274

Martin, L., Suguio, K. & Flexor, J.-M., 1979b. Le Quaternaire marin du littoral brésilien entre Cananéia (SP) et Barra de Guaratiba (RJ). *Proceedings 1978 International Symposium on Coastal Evolution in the Quaternary,* eds K. Suguio, T.R. Fairchild, L. Martin & J.-M. Flexor, (São Paulo), 296-331

Martin, L., Suguio, K., Flexor, J.-M., Bittencourt, A.C.S.P. & Vilas-Boas, G.S., 1980. Le Quaternaire marin brésilien (littoral pauliste, sud-fluminense et habianais). *Cahier O.R.S.T.O.M.,* Série Géologie, IX(1), 95-124

Suguio, K. & Martin, L., 1976. Brazilian coastline Quaternary formations: The States of São Paulo and Bahia littoral zone evolutive schemes. International Symposium on Continental Margins of Atlantic Type, São Paulo, 1975. *Anais Academia Brasileira de Ciências,* 48, (Rio de Janeiro), 325-334

Suguio, K. & Martin, L., 1978. *Quaternary marine formations of the States of São Paulo and southern Rio de Janeiro.* 1978 International Symposium on Coastal Evolution in the Quaternary, Special Publication, No.1, 55 pp, (Sao Paulo)

Suguio, K., Martin, L. & Flexor, J.-M., 1980. Sea-level fluctuations during the past 6000 years along the coast of the State of São Paulo. In: *Earth Rheology, Isostasy and Eustasy,* ed. N.-A. Mörner, (John Wiley & Sons, Chichester), 471-486

Editor's note

Other publications worthwhile consulting are:

Bailey, G.N., 1977. Shell mounds, shell middens, and raised beaches in the Cape York Peninsula. *Mankind*, 11, 132-143

Bailey, G.N., 1978. Shell middens as indicators of post-glacial economies: a territorial perspective. In: *The Early Postglacial Settlement of Northern Europe*, ed. P. Mellars, New approaches in archaeology (Duckworth, London), 37-63

Bailey, G.N., 1983. Problems of site formation and the interpretation of spatial and temporal discontinuities in the distribution of coastal middens. In: *Quaternary Coastlines and Marine Archaeology*, eds. P.M. Masters & N.C. Flemming, (Academic Press, London), 559-582

Bailey, G.N., Deith, M.R. & Shackleton, N.J., 1983. Oxygen isotope analysis and seasonality determinations: limits and potential of a new technique. *American Antiquity*, 48, 390-398

Colquhoun, D.J., Brooks, M.J., Michie, J., Abbott, W.B., Stapor, F.W., Newman, W.S. & Pardi, R.R., 1981. Location of archaeological sites with respect to sea level in the Southeastern United States. In: *Florilegum Florinis Dedicatum*, eds. L.-K. Königsson & K. Paabo, *Striae*, 16, 144-150

Shackleton, N., 1970. Stable isotope study of the palaeo-environment of the neolithic site of Nea Nikomedeia, Greece. *Nature*, 227, 5261, 943-944

AGE

19

Radiocarbon dating

W. G. MOOK and O. VAN DE PLASSCHE

ABSTRACT

The paper discusses radiocarbon dating in terms of inherent and incidental problems and uncertainties, both in general and with reference to sea-level studies. Intrinsic problems and uncertainties are treated in terms of physical, geophysical and geochemical aspects. Incidental difficulties may arise from contamination (chemical, mechanical and/or botanical) and from geological processes that disturb the age relationship between sample and indicator or event to be dated. Recommendations are given for sample collection and handling.

INTRODUCTION

Radiocarbon dating is the current major instrument for providing the timescale of sea-level (SL) studies concerning the last 15 000 to 40 000 years. Even as a well-established and powerful technique, it has certain intrinsic as well as incidental problems and uncertainties, which may affect sample collection, data evaluation and interpretation. This paper discusses the principles of the method as well as its limitations. Procedures are recommended to avoid or minimize the effects of various problems.

1. Instrinsic problems and uncertainties.

These can be grouped in terms of disciplinary aspects, under physical, geophysical and geochemical. Radioactive decay is a statistical process. The inherent *physical* uncertainty in the radioactivity measurement results in a standard deviation conventionally quoted with ^{14}C ages.

Geophysical and *geochemical* aspects determine the ^{14}C content of the environment in which the sample material has been formed. Variations in the natural ^{14}C concentration of atmospheric CO_2 - the source of all samples - may cause time periods resulting from ^{14}C dates to be significantly smaller or larger than the true periods. This relationship between the ^{14}C timescale and the astronomical timescale is a matter of current *geophysical* interest.

Various *geochemical* processes may also affect the original ^{14}C content of the sample material. These concern isotope fractionation during CO_2 assimilation and carbonate formation, and carbon isotope exchange between sample and environment.

2. Incidental problems and uncertainties.

These may arise from *chemical, mechanical* and/or *botanical* contamination of the sample material, and from *geological* events which complicate or entirely wipe out the relationship between the (age of the) sample and the phenomenon, event or process to be dated.

The many complications may suggest that radiocarbon dating has too many pitfalls to be a successful method. Yet, the reproducibility of the vast majority of ^{14}C dates and the consistency obtained for archaeological, geological and palynological chronologies are proof of its validity. Moreover, for many problems adequate physical or chemical solutions have been found. The main advantage of radiometric dating over other dating techniques is the fact that the rate of radioactive decay of a specific isotope does not depend on environmental factors and has been constant throughout.

PRINCIPLES

^{14}C is continuously produced by the interaction of cosmic ray neutrons with nitrogen atoms in the atmosphere. This radioactive carbon isotope enters the global carbon cycle and finally decays. A stationary state of production, distribution and decay is inherent in a constant ^{14}C concentration in atmospheric CO_2. This also applies to dissolved inorganic carbon in the mixed layer (ie. the upper 100 m) of the oceans, exchanging rapidly with atmospheric CO_2. Deviations of this constant ^{14}C concentration, known as natural ^{14}C variations, will be discussed below.

Living plants and animals incorporating this carbon have the same ^{14}C concentration (apart from isotope fractionation, see page 536). Once carbon has become fixed in the vegetable or animal tissue, or in carbonate crystals, ^{14}C can only disappear by radioactive decay at a known rate. Thus, the age of carbon-containing matter can be determined by measuring ^{14}C left in the sample at present.

The *conventional* ^{14}C *age* reported by all laboratories is defined by using the following assumptions and internationally agreed conventions:

(1) during geological times, the ^{14}C activity of carbon-containing material during its formation has always been the same,

(2) this ^{14}C activity is defined by standard oxalic acid distributed by the US National Bureau of Standards (NBS),

(3) ^{14}C activities of the carbon content of dated samples are to be corrected for isotope fractionation, based on the measured $^{13}C/^{12}C$ ratio,

(4) the ^{14}C half-life ($T_{\frac{1}{2}}$) used is 5568 years; this is known to be in error by 3%; the better value of 5730 years should not be used in order to avoid confusion with earlier reported dates; moreover, larger discrepancies are caused by the natural ^{14}C variations (see page 525)

(5) ^{14}C ages are given in years BP (Before Present), ie. before AD 1950.

1. Physical aspects

The age (T) of a sample is calculated from the measured activity A^{14} by:

$$T = - \frac{T_{\frac{1}{2}}}{\ln 2} \ln A^{14}/A_0^{14} = -8033 \ln A^{14}/A_0^{14} \qquad (1)$$

(detailed treatment of physical aspects by Mook & Streurman, 1983) where ln stands for log to the base e, and A_0^{14} is the original ^{14}C activity as defined by the NBS oxalic acid, which equal 95% of the ^{14}C activity of old oxalic acid with $\delta^{13}C = -19‰$ (Godwin, 1962), or 74.6% of the ^{14}C activity of new oxalic acid from NBS with $\delta^{13}C = -25‰$ (Mann, 1983).

Use of the activity ratio A^{14}/A_0^{14} is preferred over the absolute (specific) ^{14}C activity (in disintegrations per gram of carbon per unit of time). This ratio

$$A^{14}/A_0^{14} = a^{14} \ (\times 100\%) \qquad (2)$$

is always used in geochemical applications where, for instance, recent variations are being studied. Thus, dealing with ages, a recent sample contains 100% ^{14}C, 50% refers to a sample which is 5568 years old etc.

The one physical uncertainty, intrinsic to the phenomenon of radioactivity, is the statistical (random) nature of the radioactive decay. It causes the results of repeated measurements to spread around a 'true' value. One measurement thus differs from this true value by an amount indicated by the standard deviation (σ). The chance that the true age (activity) is between certain limits from the measured age (activity) is:

68 % chance between $T + \sigma$ and $T - \sigma$
95 % chance between $T + 2\sigma$ and $T - 2\sigma$
99.7% chance between $T + 3\sigma$ and $T - 3\sigma$

With the conventional age ^{14}C laboratories should always give one sigma errors. At high ages (>30 000 years) the statistical age distribution becomes significantly asymmetrical: $+\sigma$ becomes larger than $-\sigma$ because of the log scale (Fig. 1).

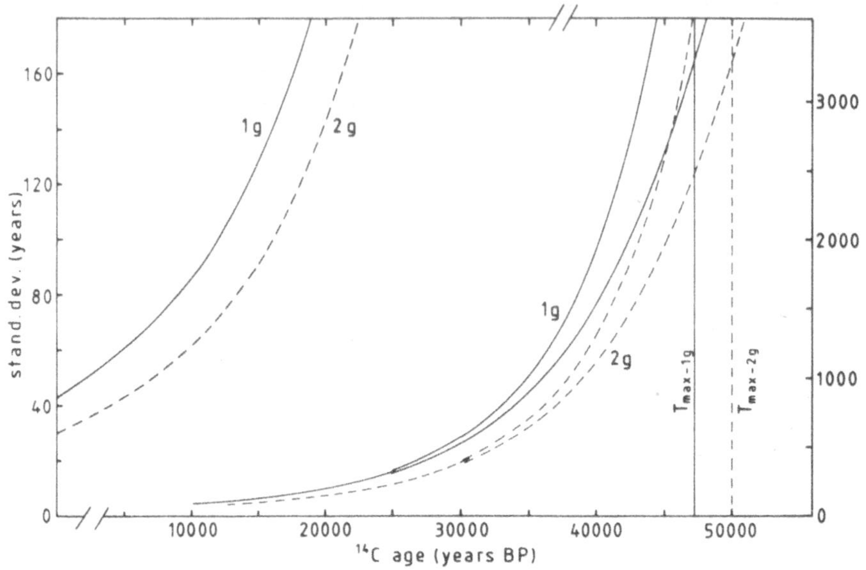

Figure 1. Standard deviation in age obtained by dating ^{14}C samples of different ages. Curves are presented for two different-size counters: one which contains the equivalent of 1 g, the other 2 g of carbon (dashed lines). The dating limits are shown by the vertical lines of T_{max}.

a) Factors determining σ

In this stage of our discussion we should pay attention to some aspects of the amount of sample available for dating. The fact is that the radioactivity measurement is more accurate at higher activities. In other words, the value of the standard deviation depends on:

(1) the amount of sample available for dating; many laboratories have different-size counters in operation, from which the larger produce more accurate dates, but require more sample (see also page 555).

(2) the activity and thus the age of the sample; Figure 1 shows the relation between age and σ for two counter sizes; one which needs 1 gram of carbon (as CO_2, C_6H_6, or any other counting medium), the other twice that amount;

(3) the duration of activity measurement, generally two days.

As a rule an increase of sample size or measuring time by a factor of 2 reduces σ by a factor of $\sqrt{2}$. Similarly concerning the age, an activity reduction by a factor of 2 (equivalent to an age of one half-life) increases σ by $\sqrt{2}$. Table 1 presents σ values resulting from analysing various amounts of carbon of different ages. Although the numbers specifically refer to the counters available in the Groningen laboratory, they are generally valid.

Table 1. Standard deviations resulting from analysing various amounts of carbon of various ages. Although each column refers to a specific counter of the Groningen laboratory, the data are generally applicable. Counting time is routinely 2 days (2700 minutes). However, the numbers for detectors 10 (5 equal mini-counters) are based on counting one sample during one month.

Detector nr.:	10	8	7	1	2	6	9
grams of C:	0.025	0.75	1.0	1.5	3.0	4.0	25.0
Age yrs BP							
0	± 70	± 50	± 45	± 35	± 25	± 20	± 9
5000	100	70	60	50	35	30	12
10000	150	100	90	70	50	45	16
15000	220	150	130	110	80	70	25
20000	330	230	200	160	120	100	35
25000	550	370	320	270	190	160	50
30000	950	650	550	450	320	270	80
35000	+1800 -1500	+1200 -1000	+1000 -900	800	550	480	130
40000	+3600 -2500	+2300 -1800	+2000 -1600	+1600 -1300	+1100 -900	850	220
45000	>38000	+4900 -3000	+4000 -2700	+3100 -2200	+2100 -1600	+1800 -1400	400
50000		>42000	>43000	>44000	+4400 -2800	+3600 -2500	700
55000					>47000	>48000	+1400 -1200
60000							+2900 -2100
65000							>59000
	>43000	>46000	>47100	>48700	>51500	>52700	>64100

b) Age limit

If samples are very old, the ^{14}C activity can no longer be distinguished from that formed by cosmic radiation background penetrating even the heavy shielding around the actual ^{14}C counter. It is always one of the main efforts of laboratories to reduce the background as much as possible by various measures. An age (activity) beyond (below) the highest measurable age (lowest measurable activity) is presented as, for instance, T > 45 000 years BP (equivalent to $a^{14} < 0.37\%$) (vertical lines in Fig 1).

c) Average values

In general it is not judicious to calculate the average of a series of ages, because 'duplicate' samples are seldomly 'identical'. In case an average is required and the ages are not too far apart, it should be weighted according to:

$$\bar{T} = \frac{T_1/\sigma_1^2 + T_2/\sigma_2^2 + \ldots}{1/\sigma_1^2 + 1\sigma_2^2 + \ldots} \quad (3)$$

with a standard deviation of the average:

$$\bar{\alpha} = \frac{1}{\sqrt{(1/\sigma_1^2 + 1/\sigma_2^2 + \ldots)}} \quad (4)$$

where $T_1 \pm \sigma_1$, $T_2 \pm \sigma_2$, are the various ages.

d) Errors

As stated before, the statistical nature of radioactive decay is the only intrinsic physical uncertainty of the method. All other physical deviations from the 'truth' should be referred to as errors, and these are necessarily incidental, that is not inherent to the method.

Errors can be systematical or restricted to a single event, so-called blunders. Several authors (Pardi & Marcus, 1977; Baxter, 1983; Scott, Baxter & Aitchison, 1983) have tried to quantify non-counting errors. The latter two papers as well as the international intercomparison of the old and new oxalic standards (Mann, 1983) and various cross-checks between high-precision laboratories on tree-ring sequences, showed that errors are not evenly spread over laboratories. These non-counting errors are completely incomparable with counting standard deviations and probably do not obey laws of statistics. Part of the discrepancies may be caused by different procedures of chemical treatment for removal of contamination (page 547).

Our conclusion is that with ^{14}C ages only 1σ standard deviation should be quoted, contrary to suggestions to use 2σ or even 3σ values. Any scientist is able to multiply by 2 or 3 if the need is felt, but this may be only justified to some laboratories' data and would cause a lot of additional uncertainty and confusion.

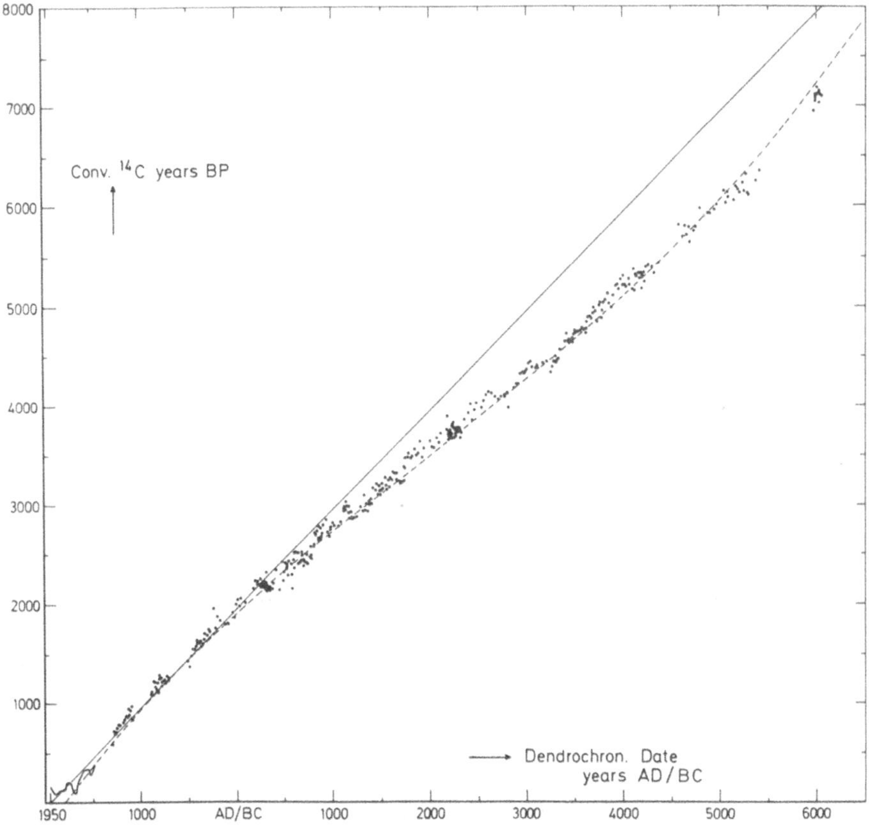

Figure 2. Calibration measurements by Suess (1978) on the Bristlecone Pine. The dashed line represents the third order curve by Ralph *et al.* (1973). The data are corrected according to Suess' indication on a standard offset.

2. Geophysical aspects

One of the assumptions underlying ^{14}C dating is a constant ^{14}C activity of atmospheric CO_2 (page 526). ^{14}C measurements of dendrochronologically dated tree rings have shown that this is not entirely true. Many hundreds of data have been published on Bristlecone Pine wood (Suess, 1970, 1978; Damon, Long & Wallick, 1972; Ralph, Michael & Han, 1973; Klein, Lerman, Damon & Ralph, 1982). In more recent years high-precision data have become available on German Oak (De Jong, Mook & Becker, 1979; Bruns, Münnich & Becker, 1980; Pearson, 1980; Stuiver, 1982). Now that we have a better insight in the shape of the curve relating ^{14}C with true ages we can distinguish three types of variations:

- long-term trend. A sinusoidal curve with a period of almost 9000 years, which can be accounted for by an equivalent variation in the earth magnetic field (Bucha, 1970). Ralph *et al.* approximate this trend by the equation

$$T_{conv} = 1907 - 0.892T_D - 6.97 \times 10^{-5}T_D^2 - 1.15 \times 10^{-8}T_D^3 \quad (5)$$

(adjusted to apply to conventional ages T_{conv}; T_D = dendrochronological date AD or BC, where the years BC count negative) (Fig. 2). For converting conventional into true ages Wendland & Donley (1971) reported an equation which we modified to apply to conventional ages:

$$T_{hist} = +1838 - 0.71\,T_{conv} - 1.61 \times 10^{-4}T_{conv}^2 + 1.5 \times 10^{-8}T_{conv}^3 \quad (6)$$

where T_{hist} is given in AD(1)/BC(-) ages.

- Medium-term variations. The 'Suess wiggles' (De Jong *et al.*, 1979). These may amount to an apparent change in ^{14}C ages of a few hundred years within a historical period of less than half a century. These variations can be explained by changes in the ^{14}C production rate due to changes in solar activity (De Jong & Mook, 1980). Especially these variations have to be discussed in more detail (Fig. 5).

- Short-term fluctuations. These small variations of about 3‰ or less, correlated with the 11-year sunspot cycle (Stuiver, 1978; Tans *et al.*, 1979), are caused by low ^{14}C production during periods of high solar activity, causing increased geomagnetic field, shielding the atmosphere against penetration of cosmic radiation.

a) Calibration of ^{14}C ages

Generally there is no need for converting ^{14}C ages into true or astronomical ages (calibration). Radiocarbon provides a timescale which is commonly used for establishing archaeological or geological chronologies.

In SL studies it may be relevant, however, to carry out the age conversion. The long-term-trend correction is simply applied by Equation 6. The medium-term corrections are more complicated. If we apply frequency diagrams of ^{14}C ages (Geyh & Streif, 1970; Roeleveld, 1974; Zagwijn, 1983) for indicating periods of beginning or ending peat growth, we have to realise that also horizontal parts of the calibration curve (Figure 3) cause clustering of ^{14}C dates (De Jong & Mook, 1981). Furthermore, a non-linear relation between conventional ^{14}C ages and true ages may show up in the SL curve as non-realistic irregularities or may change the shape of or even remove real irregularities, ie. changes in rate of SL rise (Figure 4). It is important to realise that an irregularity or the lack of one may also result from differences in indicative meaning of the dated samples. For this reason data sets should not be considered in the light of natural ^{14}C variations alone, but also against the background of their indicative meaning and available geological and paleoecological information.

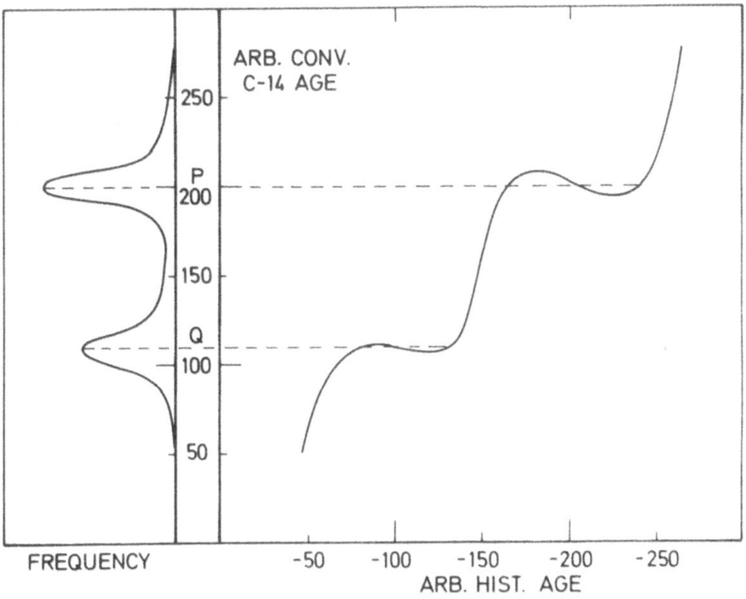

Figure 3. Effect of natural ^{14}C variations on the clustering of ^{14}C dates. The peaks in the histogram (left) are only caused by the non-linearity of the calibration curve.

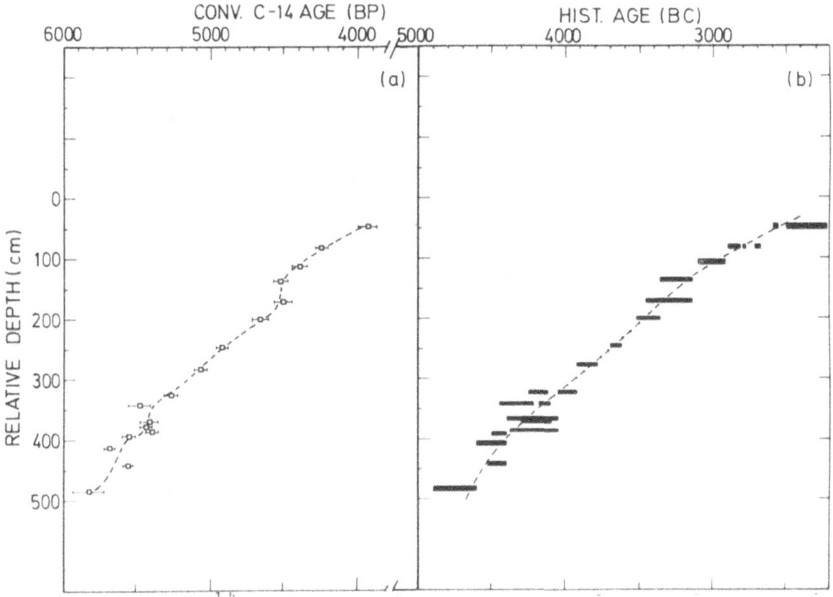

Figure 4. Set of ^{14}C ages from a sequence of peat layers (a) converted into historical ages (b). It is seen that the irregularities in curve (a) are less or less obvious in curve (b).

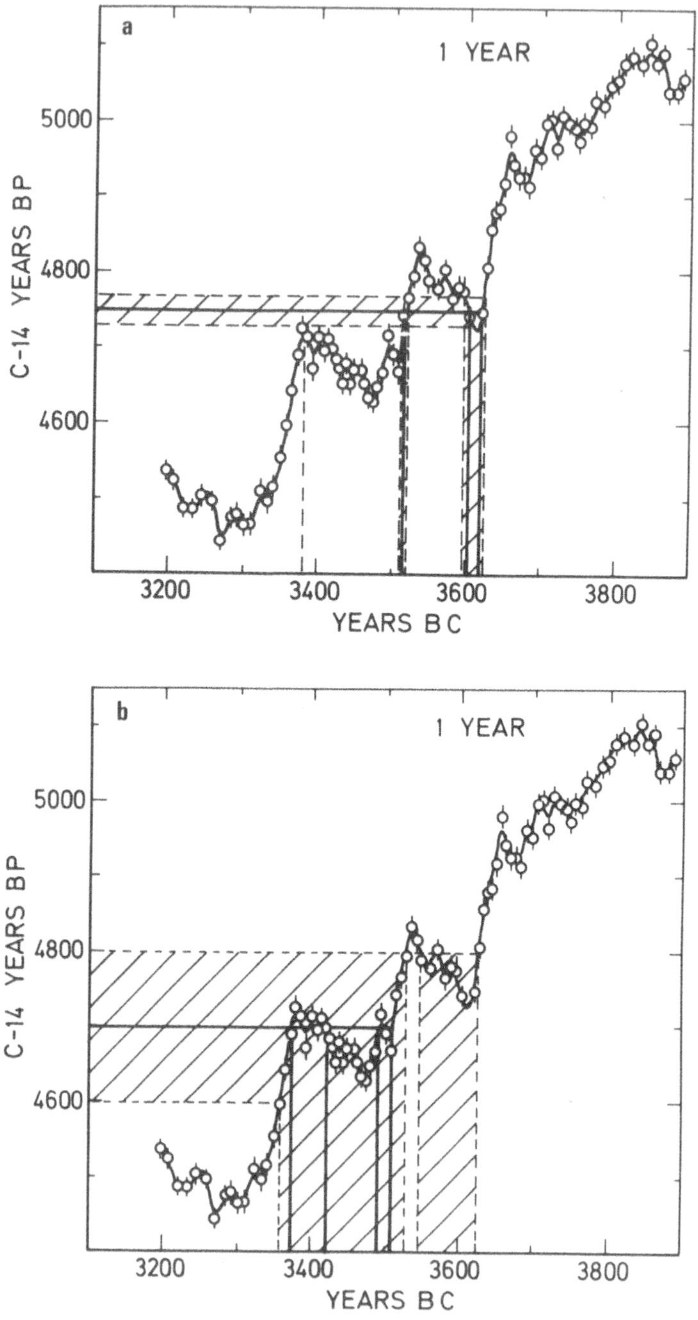

Figure 5. Part of the ^{14}C calibration curve from high-precision analyses of single tree-rings (De Jong *et al.*, 1979). The lines drawn in Figures 5a and 5c are spline fits obtained by computer through the single data.

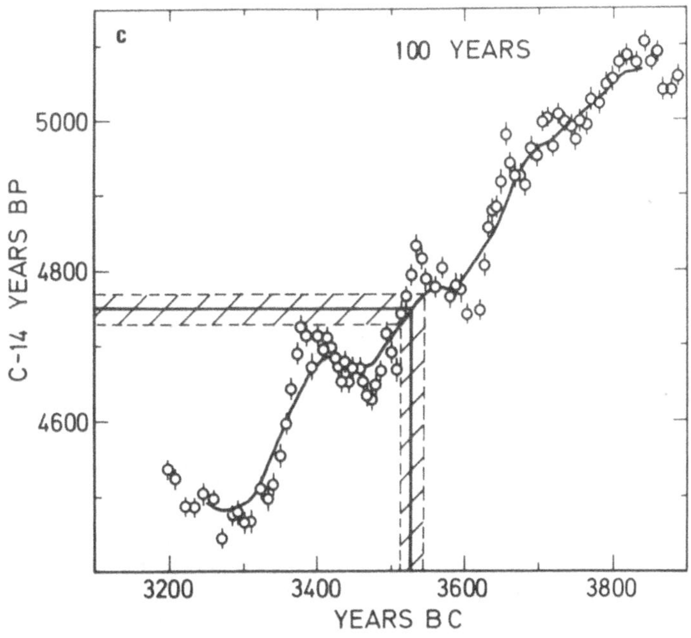

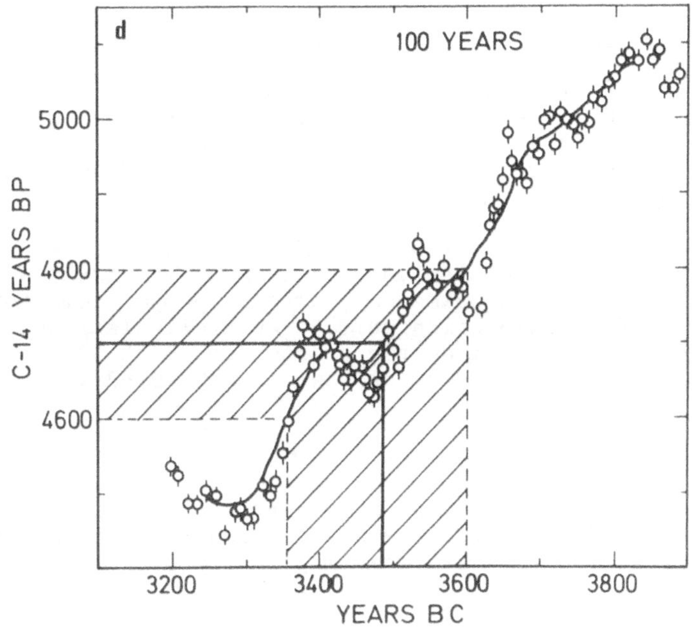

Figures 5b and 5d are calculated as moving averages over 100 years of the preceding spline data (Mook *et al.*, 1979). As an illustration the calibration is shown for conventional dates of 4750 ± 20 BP (a and b) and 4700 ± 100 BP (c and d).

An example of calibration is presented in Figure 5, showing part of the timescale analysed with high precision. ^{14}C dates of 4750 ± 20 BP and 4700 ± 100 BP are converted into true ages. It is seen that the medium-term variations may cause considerable uncertainty about the true ages (Mook, de Jong & Geertsema, 1979). Carrying out the calibration in Figure 5b results in a *dating range* rather than a corrected date with a standard deviation: 3510-3545 BC* with * referring to the paper providing the calibration curve. Quoting a standard deviation is not justified because the uncertainty is asymmetrical (non-linear curve).

b) Time width of ^{14}C samples

For the period shown in Figure 5 two different calibration curves have been derived from the same set of data. The curve denoted by '1 YEAR' is a cubic spline fitted through all data. The '100 YEARS' curve has been obtained by a moving-average procedure: each point of this curve is the average over 100 single-year data of the 1 YEAR curve. The lines of Figures 5b and 5d show less detailed variations than Figures 5a and 5c. The reason for showing these two calibration curves is to make clear that each sample should be calibrated using the curve obtained with the same number of tree rings per sample as years represented by the ^{14}C sample: the ^{14}C sample and the calibration data should have the same time width (growth period).

There seems to be less ambiguity in the derived true ages of samples with large time width. Of course this is only partly true. The fact is that, by dating 100-year samples, there is less certainty about the origin of the material within the growth period: not every year will be equally represented in the sample; in the case of peat, because of varying accumulation rates or even the occurrence of a hiatus (Mook, 1983).

Once the single calibration data, whether one-year (De Jong & Mook, 1980) or ten-year data (Stuiver, 1982) are available, one can produce any curve by proper computer programs. The calibration curve, now available with reasonable precision from the present to about AD 1, is expected to be extended by another two thousand years in 1985.

3. Geochemical aspects

Also because of geochemical phenomena, the original ^{14}C activity may differ from the assumed constant value. We will separately deal with some major effects.

a) Isotope fractionation

The phenomenon that during photosynthesis $^{12}CO_2$ is taken up preferentially over $^{14}CO_2$ is called isotope fractionation. The effect occurs in principle in all chemical and phase transitions. For instance, $H_2{}^{16}O$ evaporates easier than $H_2{}^{18}O$, precipitating carbonate contains a higher percentage of ^{13}C than the dissolved HCO_3^- ions in solution, etc. This fractionation differs for different plant species and even for each single specimen or part thereof. This means that plants growing at the same time, using the same atmospheric CO_2, contain different concentrations of ^{14}C. This is also true for ^{13}C while a special relationship exists between the fractionation for ^{14}C and that for ^{13}C. Because ^{13}C is stable, the $^{13}C/^{12}C$ abundance ratio did not change in time. This presents the possibility of making a fractionation correction. If $^{13}C/^{12}C$ is corrected to a certain agreed value, the ^{14}C activity changes proportionally. Isotopic abundances of ^{13}C are given as:

$$\delta^{13}C = \frac{(^{13}C/^{12}C)_{sample}}{(^{13}C/^{12}C)_{PDB\ standard}} - 1\ (\times\ 10^3 ‰) \qquad (7)$$

All ^{14}C samples have to be corrected to $\delta^{13}C = -25‰$. On theoretical grounds, the correction for ^{14}C activities amounts to:

$$\frac{a_c^{14}}{a_m^{14}} = \left(\frac{1 - (-25‰)}{1 + \delta^{13}C_m}\right)^2 \qquad (8)$$

where c and m refer to the corrected and measured values, respectively. The age correction is approximately:

$$\Delta T = (\delta^{13}C + 25) \times 16 \text{ years} \qquad (9)$$

or 16 years per one per mil deviation from -25‰.

Figure 6 shows some ranges of $\delta^{13}C$ values in natural compounds and the equivalent age deviations if no correction were applied. Especially large are the corrections for marine carbonates and for C3 plants. The latter group of plants (for instance, corn, cane, semi-arid grasses) have a special photosynthesis cycle resulting in $\delta^{13}C$ values as high as -10‰, equivalent to a deviation of 240 years too young if not corrected for.

A special type of sample are marine carbonates. These have $\delta^{13}C$ values of about 0‰, requiring a fractionation correction of about +400 years. This, however, is eventually about equal and opposite to the apparent age correction (see (*b*) below). Therefore, many ^{14}C ages have been published in literature where no corrections for either of these two effects have been applied. This matter deserves special attention, because some laboratories apply only the fractionation correction, so that the conventional ages reported are 400 years too old.

b) Apparent age of marine carbonates

^{14}C ages obtained from marine carbonates are generally 400 years too old. This is because of an apparent age of surface ocean water (Berger, Taylor & Libby, 1966; Mangerud, 1972; Mangerud & Gulliksen, 1975; Taylor & Slota, 1979; Olsson, 1983). The reason for this apparent age is that upwelling of deep ocean water containing (up to 20%) less

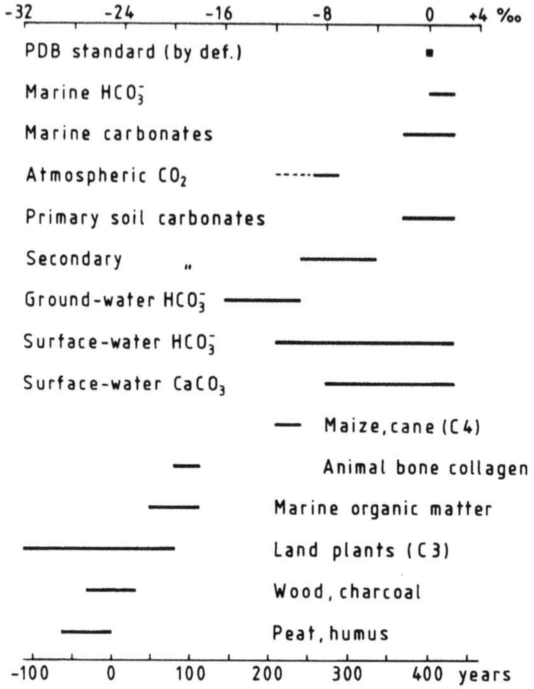

Figure 6.

$\delta^{13}C$ values for different natural compounds. The lower scale shows the equivalent changes in $\delta^{14}C$ age, if the fractionation correction is applied. The ranges indicated show only the most common variations.

^{14}C than in equilibrium with the atmosphere has reduced the ^{14}C activity of dissolved carbon in the mixed layer (upper 100 m) by about 5%, equivalent to 400 years. This means that the ^{14}C content of dissolved inorganic carbon of surface ocean water - and thus also the carbonate formed in it - is smaller than expected, as if the water had some age. This phenomenon is sometimes referred to as the reservoir effect. Although it has not been established for the entire ocean surface water, it is assumed to be a general phenomenon (Gillespie & Polach, 1979). We have to check carefully whether published data have been corrected for this apparent age or not (see *(a)* above).

c) Apparent age of fresh-water carbonates

^{14}C ages obtained from fresh-water carbonates may be affected by an apparent age of the ground or surface water in which they were formed (Mook, 1980). This is caused by the fact that part of dissolved carbon in ground water originates from dissolution of soil limestones of considerable age and, thus, low ^{14}C activity. The original ^{14}C activity of ground water varies between 65% and 85%, so that the apparent age correction can range between 3500 and 1300 years. In that case one has to expect $\delta^{13}C$ values as low as -10‰. This phenomenon is by some authors referred to as the hard-water effect. Apart from this apparent age, the ground-water carbon, precipitated as carbonate, may have some real age, adding to the former. This makes the application of fresh-water carbonates less useful.

In shallow surface water the original ^{14}C activity is relatively quickly restored through carbon exchange between the dissolved inorganic fraction and atmospheric CO_2. In that case there is no ageing effect on carbonates formed in these waters, while $\delta^{13}C$ is expected to be about 0‰ or slightly positive. However, the problem remains that it is in general impossible to distinguish between these secondary carbonates and old carbonates eroded and possibly washed into the basin.

Apart from the above, there is a possibility of formation of secondary carbonate in carbonate deposits (page 544).

d) Apparent age of brackish-water carbonates

Marine-shell carbonate shows $\delta^{13}C$ values in excess of 0‰. Negative $\delta^{13}C$ values point to the presence of a fresh-water component in the water in which the shell lived. The salinity of the water in which they grew, or rather the fractional contribution of fresh water with relatively low ^{14}C content, is to some extent indicated by the $\delta^{13}C$ value (Mook, 1971; Eisma, Mook & Das, 1976; Eisma, Mook & Laban, 1981). If this is true a correction may be applied, based on assumptions of $\delta^{13}C$ values of purely marine and fresh-water carbonate of $\delta^{13}_m = +2‰$ and $\delta^{13}_f = -10‰$ respectively, and ^{14}C contents of $a^{14}_m = 100\%$ and, for instance, $a^{14}_f = 85\%$, respectively:

$$\frac{a^{14}_o - a^{14}_m}{a^{14}_f - a^{14}_m} = \frac{\delta^{13} - \delta^{13}_m}{\delta^{13}_f - \delta^{13}_m}$$

where a^{14}_o and δ^{13} denote the original values during formation. During prolonged time δ^{13} does not change, while a^{14} diminishes only by decay. This equation is simply based on conservative mixing of fresh and sea water. Then:

$$a^{14}_o = a^{14}_m + \frac{\delta^{13} - \delta^{13}_m}{\delta^{13}_f - \delta^{13}_m}(a^{14}_f - a^{14}_m) \qquad (10)$$

and the age is:

$$T = -8033 \ln a^{14}/a^{14}_o$$

where a^{14} is the measured value. Applying this method, the resulting ^{14}C ages are corrected for isotope fractionation. Although this correction has been successful in some cases, we can not yet state that it is generally applicable. Furthermore, the problem remains estimating the ^{14}C content of the fresh-water component (a^{14}_f)

e) Isotopic exchange

This process involves no chemical changes, merely exchange of isotopes between the sample material and the environment. Samples consisting of organic carbon do not show exchange. Carbonates especially, are in principle able to exchange carbon isotopes with the environmental solution. Again this isotopic exchange involves no chemical dissolution or precipitation of carbonates. It consists of exchange of carbonate ions (CO_3^{2-}) between the solid crystals and the ground water. This process is very slow and probably extends only to the first surface layer of CO_3^{2-} ions in the crystal (Thilo & Münnich, 1970). It can therefore only be significant if the carbonate consists of very fine grains, where the relative surface area is large, and if the water temperature is well above normal (hydrothermal regions).

Generally speaking, isotopic exchange is less likely than re-crystallisation (see below). Generally, any proof for the absence of recrystallisation can serve as proof that isotopic exchange has not occurred (Mangerud, 1972).

f) Recrystallisation

During recrystallisation, that is, change of the crystal structure, chemical exchange with the environment and thus isotopic changes may occur. For instance, metastable high-Mg calcite and aragonite, when brought into contact with Mg-free water (rain, ground water), change into stable low-Mg calcite. In this process, where dissolution and reprecipitation take place on a micro scale, isotopes as ^{14}C and ^{13}C may be gained from or lost to the surrounding solution. Consequently, the ^{14}C age may be influenced. The process is particularly important when dealing with corals (100% aragonite) (Chappell & Polach, 1972), and with shells partly (eg. *Mytilus edulis*) or entirely (eg. *Cerastoderma edule*) consisting of aragonite.

A large degree of recrystallisation of primary marine carbonate in a fresh-water environment can be recognised by $\delta^{13}C$. Values lower than 0‰ are suspect. Another way to check recrystallisation is to determine the crystal structure by X-ray analysis. If shells still consist only of the metastable mineral, recrystallisation and isotopic changes have not occurred.

A mixture of low- and high-Mg calcite does not necessarily point to recrystallisation. Beachrock cement, for instance, consisting of aragonite or high-Mg calcite may also contain low-Mg calcite as a primary product. Calcite cement on aragonite cement from which needles are missing, however, suggest replacement through dissolution and reprecipitation (Bathurst, 1976).

For ^{14}C dating shells, hard and non-porous marine specimens with well-preserved textures can be used with great confidence. It is not possible to obtain meaningful ages on carbonate that has been exchanged or recrystallised (Evin, 1983).

CONTAMINATION

With contamination we refer to any compound now present in the sample that was not formed together with the sample material. In the previous section we dealt with the *sample*

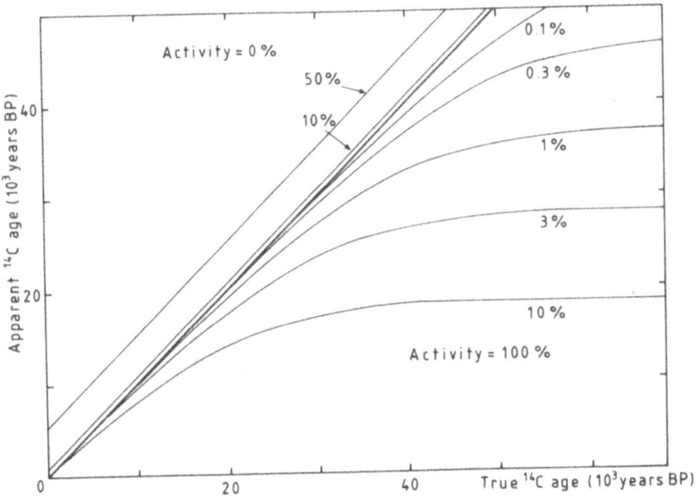

Figure 7. Apparent ages resulting from contamination in the true sample (with true ages). Contamination with carbon containing no ^{14}C (0%) results in errors not depending on the age (50% contamination with fossil carbon results in ages which are 5568 years too large). The curved lines give apparent ages by contamination with 0.1 to 10% of recent carbon (100% activity).

carbon; here we will discuss the presence of *foreign* carbon, which may be chemically different, in other cases identical to the sample carbon.

The main problems for the sample submitter and the ^{14}C specialist are the recognition of contaminations, in the field, and their removal in the laboratory.

Essential for the effect of contamination on the age of a ^{14}C sample are the (fractional) amount of contaminants and their ^{14}C content. If these are denoted by x and a_c^{14} and the uncontaminated and contaminated sample activity by a_s^{14} and a^{14}, respectively,

$$a^{14} = xa_c^{14} + (1 - x)a_s^{14} \qquad (11)$$

or

$$a_s^{14} = (a^{14} - xa_c^{14})/(1 - x) \qquad (12)$$

If instead of activities, a^{14}, ages or at least estimates of ages are known, they first have to be converted to activities by:

$$a^{14}(\text{in }\%)/10^2 = e^{-T/8033} \qquad (13)$$

Figure 7 shows the effect of various degrees of recent (a_c^{14} = 100%) or old (a_c^{14} = 0%) contamination on samples of varying ages. It is clear that the main problems are presented by small and difficult to recognise contamination rather than the easily detected and those that completely distort the age. Let us present an example of this type of calculation. The ^{14}C age of a peat sample with an expected age of 4200 years BP came out as 2500 years BP. Could this discrepancy have been caused by contamination? According to the observed profile, roots from the living surface vegetation penetrated the peat. What amount of roots is required to cause the 1700 year shift? According to Equation 13: a_s^{14}= 59.3% and a^{14} = 73.3%. while a_c^{14} is probably 100%. From Equation 11 the fraction of contamination is deduced to be:

$$x = (a^{14} - a_s^{14})/(a_c^{14} - a_s^{14}) \qquad (14)$$

or in our case x = 34.3%. This means that about one-third of the sample should have consisted of recent roots, which seems unlikely, or, if possible, the sample should not have been used for dating.

Table 2. Survey of possibilities of contamination of chemical, botanical or geological origin. The letters refer to the sections in the text. The type of carbon to be dated is given in the first column. The second distinction made is between contaminating processes taht occur before or during sample formation and after.

NATURAL, PRE-SAMPLING CONTAMINATION

	Pre- or during-formation	Post-formation
Organic sample	a. Fossil organic matter b. Fossil carbonate	c. Organic (humic) infiltration d. Root penetration e. Secondary carbonate f. Admixture (by animals)
Carbonate sample	g. Fossil carbonate h. Fossil organic matter	i. Secondary carbonate

HUMAN, DURING-/POST-SAMPLING CONTAMINATION

	During-sampling	Post-sampling
all types	j. Admixture during collection	k. Preservatives l. Packing materials

Table 2 presents a listing of various types of contamination grouped in categories. Of course the boundaries between them are not always strict. The letters in Table 2 refer to the following sections, where each type will be discussed.

Some of the potential sources of contamination in this long list are no problem in practice, because the contaminant can easily be removed in the laboratory. Another group requires somewhat more elaborate and sophisticated chemical treatment, whilst some contaminants cannot be removed at all. It is therefore of the utmost importance that the sample collector in the field forms his opinion on probability of various contaminations. The forms, provided by the laboratory and to be completed by the sample submitter, ask very specific questions about this. It is important that the collector is aware of these questions during sampling.

1. Contamination of organic samples pre- or during formation

a) Fossil organic matter.

Contamination with older organic matter can occur, for instance with peat or samples of vegetation horizons (fossil AO soil horizons in clay deposits; Schoute 1984). The sediment on which a temporary vegetation cover developed may already have contained organic matter. This observation was made by Schoute *et al.* (1981), dating vegetation horizons. The organic content of the deposit consists of resistant organic debris transported by rivers and is of considerable age (up to 6000 BP). Since this organic content is very low (in the order of 1% by weight) only soils low in organic matter will suffer by this contamination (vegetation horizons contain 3% of carbon or less).

Clearly, the contaminating organic matter is impossible to recognise and remove. However, in general the contaminant merely consists of highly resistant material, so that dating a less resistant fraction (alkali soluble humic acid) may in some cases provide a better age (Schoute, Mook & Streurman, 1983).

During peat growth smaller or larger pieces of eroded older peat (or gyttja, wood, etc.) may drift or wash into the peat-forming vegetation. Careful inspection of the structure and contents of the peat cores may sometimes lead to detection of allochthonous particles. If the peat does not consist of wood remains, bits of wood should be removed from the sample.

b) Fossil carbonate

As discussed above, the soil may, prior to vegetation, have contained carbonate, deposited as a fraction of the sediment, or transported by wind. Any carbonate in an organic ^{14}C sample is easily and routinely removed by treatment with acid.

2. Contamination of organic samples post-formation

c) Organic (humic) infiltration.

One of the main problems in dealing with contamination is the infiltration of organic carbon. Primarily younger carbon from soils at lower depth may have been transported in solution or in colloidal form by infiltrating rain water to the layer to be dated where the organic molecules were adsorbed. The acid soluble fulvic acid and alkali soluble humic acid are removed by chemical treatment, provided sufficient resistant material (lignin, cellulose, humine) is contained by the sample. This is no problem in charcoal, wood and good-quality peat. In sample types such as bone and soils the success of chemical treatment is less certain. From old samples (T > 10000 BP) often the alkali extract is dated as well. This fraction should contain the infiltrated, younger contamination. If the dates on residue and extract are about equal, the chance that the residue age is still too low because of contamination is small.

d) Root penetration

Also a major problem, especially in dating peat, is the penetration of roots. Kaye & Barghoorn (1964), for example, observed that rootlets of living *Spartina* grasses reach a depth of over 30 cm beneath the rhizomes, and that the entire section (over 4 m) of marsh peat studied was interlaced by vertical rootlets. Indeed, they speak of a 'constant and unavoidable source of contamination'.

The rejuvenating effect of root penetration during peat growth depends on the rate of organic accumulation and on the depth and volume of the root system. As for root penetration after formation of a peat bed, one should look at the bed from a stratigraphic-genetic viewpoint. For instance, if *Phragmites* peat is overlain by fen-wood peat, it is clear that woody roots and rootlets, if present in a sample of the reed peat, have to be removed.

The effect of root contamination in peat samples can be reduced by removing all roots and rootlets recognizable under the binocular microscope.(Streif, 1971; Van de Plassche, 1979, 1980, 1982). This is, however, a very laborious procedure. As an alternative it is suggested for peat samples to remove all rhizomes (usually in a horizontal position), (sub)vertically positioned roots (Figure 8), and bits of wood visible with the naked eye.

e) Secondary carbonate

Small calcareous particles can be washed or blown into the peat layer, exposed to the atmosphere or to surface water for some time after its formation. However, the carbonate is easily removed.

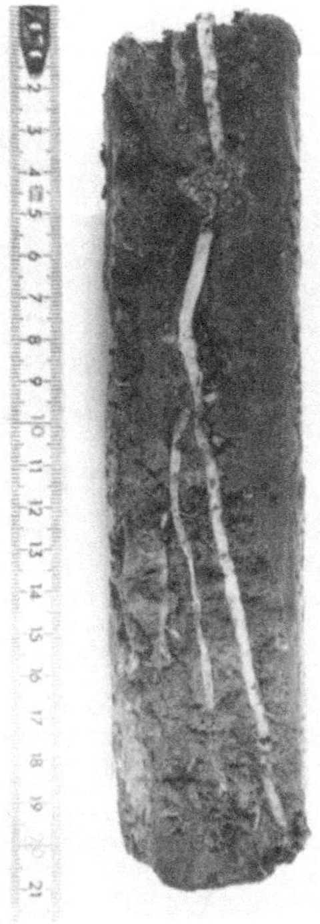

Figure 8.

Peat core showing roots descending from higher up in the peat bed. The necessity of their removal is obvious.

f) Admixture of foreign material

Admixture of material of different age may result from, for example, soil organism activity or falling trees. It is important to try to recognise this phenomenon in the field. Dark, amorphous peat must be suspected since this has probably been subject to oxidation and mineralization due to the activity of soil organisms. Peat may be very dark as a result of oxidation; however, if the plant remains are still distinguishable, it may be regarded as suitable for dating.

Foreign carbon may also enter a sample through root channels or broken-off plant stems.

3. <u>Contamination of carbonate samples pre- or during formation</u>

g) Fossil carbonate

Carbonate formation (chemical, biological) and sedimentation may have occurred in a marine environment where also continental material was deposited containing eroded fossil limestone. For this reason ages obtained on carbonates of this origin often give maximum ages.

(h) Fossil organic matter.

Formation and deposition of carbonate can hardly take place without the presence or simultaneous formation of organic matter. Chemical treatment with acid easily provides CO_2 from the carbonate for ^{14}C analysis without affecting the organic carbon.

4. Contamination of carbonate samples post-formation

i) Secondary carbonate.

Apart from the processes of isotopic exchange or recrystallisation, the ^{14}C content of a primary, *in situ* formed carbonate (corals, shells, coralline algae) may be disturbed by infiltration of sea or ground water and subsequent formation of secondary carbonate on (encrustation) or in the sample carbonate by evaporation of the water or by loss of CO_2 (Salomons & Mook, 1985). The different origins of the primary and secondary carbonate may be recognised by differences in colour, crystal structure (X-ray analysis) or structure (electron microprobe analysis). Wherever possibilities exist for the presence of secondary carbonate in the sample the ^{14}C ages are to be considered as minimum ages.

5. During or post-sampling contamination

j) Admixture during collection.

Great care is required that during sampling no material from other layers than the one to be dated becomes mixed with the sample. For example, when collecting basal peat samples, part of the AO-horizon of the underlying soil (if present) could be easily included. If transitions between mineral and organic beds are very gradual, it is important to define criteria in order to sample as consistently as possible (for the purpose of comparability). Peat cores, taken with open tube augers should be 'peeled', ie. the outermost rind should be removed. A final example worth mentioning is the case where an algal crust has been covered by several younger generations of encrusting algae. If this is suspected, the biological structure (evolution) of the entire crust should be studied prior to sample selection.

k) Preservatives.

Especially with bones that have not been collected by the submitter the possibility exists that they have been treated with preservatives. These should be removed, prior to the standard chemical treatment.

l) Packing materials.

Contamination by packing material such as paper, cloth, wool, cardboard, *etc.* can easily be avoided. The sample should be kept and sent in glass, plastic, or aluminium foil, preferably dry, dark, and cool especially if the material is wet.

The type and intensity of chemical pretreatment for removal of contamination depend on the type, quality and quantity of sample. Each sample is carefully examined for the occurrence of contamination: plant roots are removed, pieces of charcoal are hand-picked or sieved out of the soil sample, the outer layer of bones, wood and shells is removed mechanically or chemically. The chemical treatment is described for the types of material which we deal with most frequently and which are likely to be contaminated. The experimental procedures on the chemical treatment are those used in our laboratory, but are quite common. They are extensions of the original work of Hl. de Vries & J.C. Vogel, unless stated otherwise (Mook & Streurman, 1983).

The amount of sample required for physical reasons is given in Table 3. The more sample is lost by the chemical treatment, the more sample is needed. In combination with Table 1, the reader is able to estimate personally the amount of sample needed by the laboratory to obtain the required precision.

Table 3. Usual carbon content of various types of ^{14}C sample and amount of dry sample in grams needed to obtain 1 gram of carbon for ^{14}C analysis without or with chemical pretreatment. The chemical procedures for removal of contamination are according to Mook & Streurman (1983).

	% C	Treatment		
		without	A	AAA
Wood	50	2	2.5	4.0
Charcoal	70	2	2.5	3.5
Peat (wet)	5	20	25	40
Peat (dry)	50	2	2.5	4.0
Humic soil	0-5	40	50	120
Bone	0-5	-	50	-
Charred bone	0-5	-	40	40
Carbonates	12	10	15	-
Deep-sea cores	0-5	40	50	-

A = extraction with acid (HCl)
AAA = extraction with acid (HCl), alkali (NaOH) and acid (HCl).

1. Wood

Recent wood consists of various fractions: resins, sugars, lipids, hemicellulose, lignin (up to 30%) and cellulose (up to 70%). The latter two are the most stable compounds. The others are more or less soluble and mobile and may infiltrate radially into the tree.

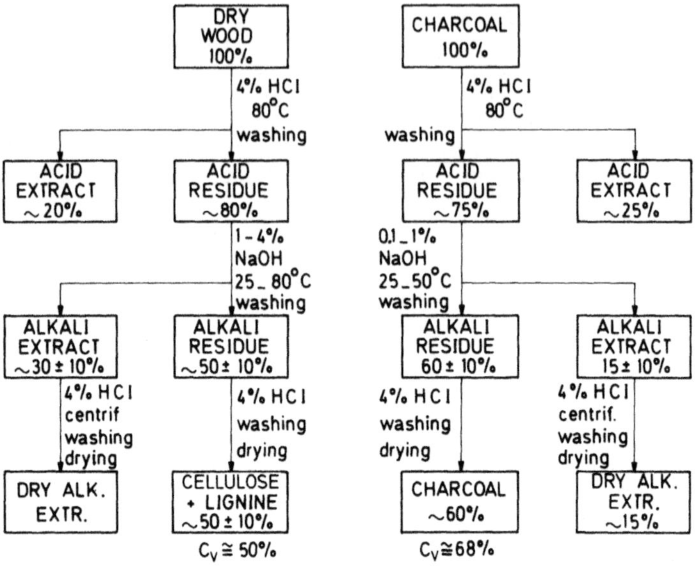

Figure 9. Diagrams showing chemical (AAA) treatment of wood and charcoal for ^{14}C dating. The percentages refer to the fraction left after the respective step. C_V is the percentage of carbon in the actual residue or extract.

The routine treatment of wood samples consists of the following extraction steps (Fig. 9)(see also Olsson 1980):

(i) with 4% HCl solutions at 80°C during 24 hours to remove resins, sugars, soil carbonate and infiltrated humic (fulvic) acids;

(ii) with 0.5 to 4% NaOH solution at 20 to 80°C during at least 24 hours to remove infiltrated tannic acids and part of the lignine;

(iii) with 4% HCl solution at 80°C during several hours to remove any CO_2 absorbed during step (ii).

Treatment by the single step (i) will be referred to as the A treatment, (i) to (iii) as the AAA treatment. The extracts are separated from the residue. The acid or alkali are removed from the residues by washing with demineralized water. The alkali extract can be precipitated by acidification, washing and drying. After the AAA treatment only the cellulose and part of the lignin are left as a residue. Generally the acid extract from step (i) contains about 20% of the original wood. The percentage of alkali extract varies greatly, depending on the quality of the wood sample: generally between 10% and 30%, sometimes up to 50%. The overall yield after AAA treatment is in general 40% to 50%.

The final carbon content (C_V) of the organic residue after AAA treatment is 50 ± 6%. This is to be compared with the carbon percentages of, for instance lignin (C_V = 44%) and resins (C_V > 70%).

^{14}C calibration studies, on tree rings in particular, may require the preparation of pure cellulose. This procedure consists of a number of steps (De Jong, 1981):

(i) extractions with 4% HCl solution at 80°C during 24 hours to remove resinous compounds and to make the wooden chips more accessible to the further bleaching reagents;

(ii) bleaching of the residue with $NaClO_2$ and acetic acid (Green, 1953), performed by the subsequent addition, over 12 hours, of ten 15g portions of $NaClO_2$ powder and 1ml of 100% acetic acid per 200g of wooden chips in 2 litres of water at 70°C;

(iii) extraction of the residue with 4% NaOH solution at 80°C during 24 hours, thus isolating the (α) cellulose;

(iv) treatment with 4% HCl solution at 80°C during several hours to remove CO_2 possibly absorbed during step (iii).

The steps (ii) to (iv) are separated by washing with demineralized water to pH ≃ 7. The final yield is 25 to 40% of cellulose (Tans, 1978), compared to a yield after AAA treatment of about 45%.

2. Charcoal

After physical removal of roots, clay, sand, etc. by hand-picking and sieving, charcoal is subjected to the same procedure as wood, generally the AAA treatment. The concentration of the alkali solution is taken 0.1 to 1% depending on the age and the available quantity of sample. The alkali extract generally contains 15 ± 10% of the original sample, if the extraction is carried out at room temperature; at higher temperatures this may amount to 45%. Rootlets which survived this treatment may be removed by a 12% H_2SO_4 + $NaClO_2$ solution. However, part of the charcoal also dissolves in this case.

The C_V of charcoal generally is 68 ± 5%, which is more or less indicative for the degree of completeness obtained in the removal of contamination (Figure 9). The ash content is generally below 5%, when sand, etc. have been removed effectively.

3. Peat

In peat the remains of separate plant species can still be recognised. Recent peat primarily consists of lignin, resins, carbohydrates, lipids and cellulose. During the decomposition of peat several kinds of chemical compounds are formed. Generally groups of compounds are distinguished on the basis of their solubility in acid and alkali solutions. The most stable chemicals are the fulvic acids, humic acids

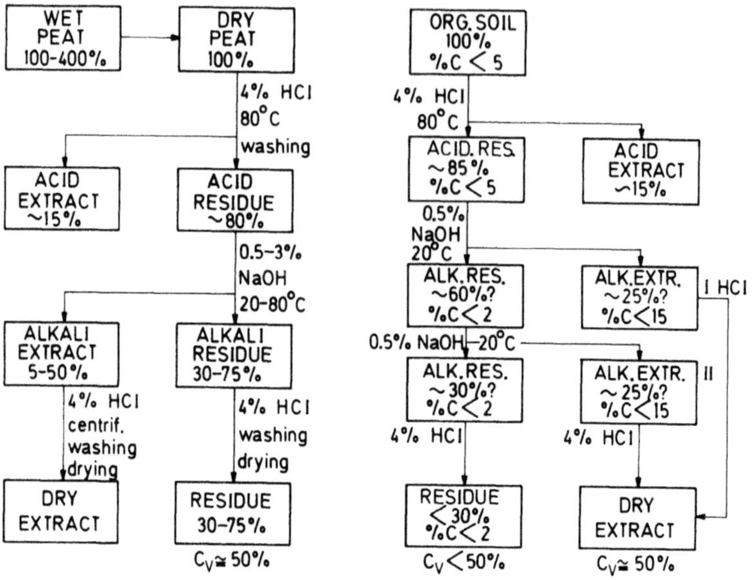

Figure 10. Diagrams showing chemical treatment of peat and organic deposits (*cf.* Fig. 9).

and humine, of which the first two are the most mobile in ground water. Their mobility is determined by the acidity (pH) of the water. The precipitation of the mobile fractions in lower stratigraphical layers causes an apparent rejuvenation of ^{14}C samples in these layers.

The pre-treatment of peat samples for ^{14}C dating consists of the removal of sand and clay, and possibly of younger roots, followed by some chemical extractions, depending on the age and state of preservation of the peat (Fig. 10).

(i) extraction with 4% HCl solution at 80°C during 24 hours to remove carbonates and fulvic acids;

(ii) extraction with 0.5% NaOH solution at room temperature during 24 hours; only very old samples are treated with 1 to 3% NaOH solution above 50°C;

(iii) acidification with 4% HCl solution at 80°C and separating the residue.

During step (i) about 15% of fulvic acids are removed. The alkali extract obtained at room temperature varies greatly between 5% and 50%, depending slightly on the alkali concentration. If high ages are to be determined the alkali extract is precipitated by acidification and dated as well. These extract ages may be significantly smaller than the residue ages.

The carbon percentage, C_V, of peat is generally 50 ± 5%. The real carbon content of the samples varies with the actual clay and sand content.

Contamination by recent roots in particular is a serious problem, because these can no longer be distinguished from the original peat constituents in the laboratory. In the field, on the other hand, recent roots are often clearly visible because of their white appearance.

4. Organic deposits

Here we are dealing with different types of soil containing organic matter, in which no plant remains are recognizable. The mineral matrix may be sand, clay or löess type deposits. The organic content consists mainly of fulvic and humic acids and humine. The contribution of contaminants to the apparent ^{14}C age can be considerable, because the carbon content in these samples may be very low (often below 3%). The special example of vegetation horizons is treated by Schoute *et al.* (1983).

The possible origins of the organic content of the sample are:

- Autochthonous carbon, formed during deposition: this is generally the material being dated.
- Resistant (allochthonous) carbon which was part of the deposited matrix prior to formation of the autochthonous carbon (see Schoute, Mook & Streurman, 1983: organic content of deposited marine clay); this carbon causes higher ages.
- Reworked or eroded allochthonous organic matter which might even be of tertiary origin, and which is redeposited in the dated layer.
- Younger carbon from stratigraphically lower depth, as fulvic and humic acids.
- Roots from higher levels.

It is extremely difficult to obtain meaningful ages from samples of this type. After possible removal of roots, carbonates and part of the organic contamination are removed by treatment with 4% HCl solution at 80°C during 24 hours (Fig. 10). The 0.5% alkali extraction at room temperature during 24 hours sometimes dissolves all remaining carbon. Generally the alkali extract gives a minimum age. However, as with the vegetation horizons in clay deposits the residue may be too old, whereas the alkali extract can be shown to present a more consistent age.

The carbon content of the alkali extract (C_V) is often well above 50%.

5. Bones

Animal bones contain inorganic as well as organic carbon. The first fraction consists of hydroxyapatite ($Ca_3(PO_4)_2Ca(OH)_2$) and calcium carbonate. Isotopic carbon exchange with the ground-water carbonate as well as precipitation of secondary carbonates renders it completely unsuitable for ^{14}C dating.

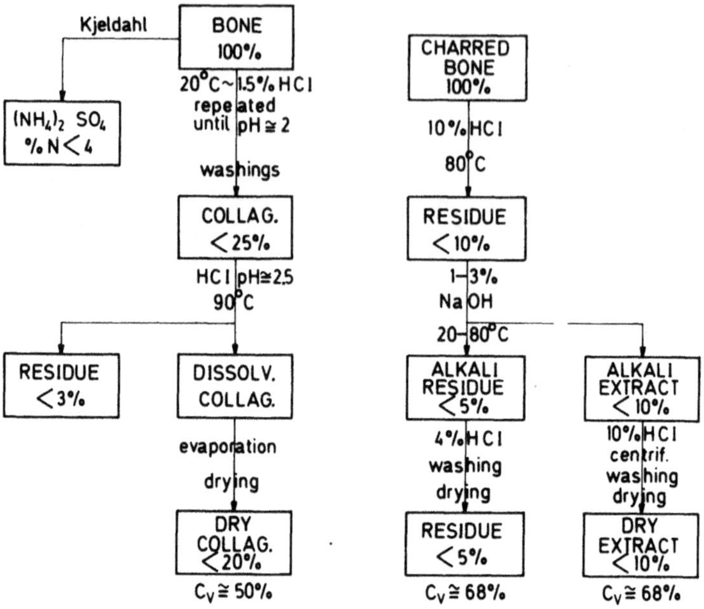

Figure 11. Diagrams showing chemical treatment of bone and charred bone (*cf.* Fig. 9). %C values refer to the carbon content of the original sample submitted, whereas C_V refers to the carbon content of the fraction combusted to CO_2 (= sample - ash content).

The organic bone matrix, collagen, is a complex combination of aminoacids, which is remarkably resistant to high temperatures and soil acids. However, in (sub)tropical soils bones decompose more rapidly through bacterial activity. The remaining fraction of collagen can be determined by the Kjeldahl method. In fresh bone collagen, the N_2 content is about 18 mol%, whereas the collagen content is 20% to 25% by weight.

Bone samples impregnated with preservative are treated with acetone until the addition of water to the acetone fraction no longer causes a white suspension. The surfaces of selected pieces of hard, non-porous bones are rinsed with 10% HCl solution during 30 minutes after roots, clay and marrow have been removed.

The procedure for separating the collagen fraction is essentially according to Longin (1970) (see also: Olsson *et al.*, 1974) (Fig. 11):

(i) addition of aliquots of 1 to 2% HCl solution until pH remains about 2; this may take a few days; 10 to 50% of the collagen is decomposed into soluble aminoacids;

(ii) thorough washing with demineralized water to remove Ca^{2+} ions, which might form Ca-humates;

(iii) dissolution of the collagen in slightly acid (pH 2 to 3) demineralized water at 90°C during 5 to 18 hours;

(iv) removal of the insoluble humic contamination by centrifuging;

(v) evaporation at 120°C to dry.

The following criteria are used for estimating the reliability of the chemical purification of collagen:

(1) C_V should be 45 to 50%;

(2) δ^{13}should be -18 to 22‰ (mainly -19 to -21‰);

(3) the ash fraction after combustion to CO_2 should be below 5%;

(4) the insoluble fraction removed by step (iv) should be below 10%.

Impurities remaining after step (v) generally result in lower $\delta^{13}C$ values; the insoluble compounds have $\delta^{13}C$ values of -22 to -29‰. Bone samples from Australia, Indonesia, South America and South Africa have indicated $\delta^{13}C$ values up to -10‰, depending on the $^{13}C/^{12}C$ ratio of the vegetation (Vogel & Van der Merwe, 1977).

6. Charred bones

In charred bones the contribution of charred collagen to the total carbon content is very low (about 0.5 to 3%). This increases the effect of a possible humic contamination on the ^{14}C age. The chemical treatment is somewhat different from that discussed for uncharred bones (Fig. 11).

(i) treatment with 10% HCl solution at 80°C to dissolve the phosphate and carbonate and soluble contaminants;

(ii) physical removal (for instance by sieving) of clay, and possible roots;

(iii) treatment with a 1 to 3% alkali solution at 20 to 80°C, depending on the quantity of sample available.

In general both the residue and the alkali extract are dated. As criteria for the reliability of the treatment we accept that:

(1) C_V should be 65 to 70% (comparable to charcoal);

(2) the ash fraction after combustion should be smaller than 10%;

(3) $\delta^{13}C$ should be -22 ± 2‰;

(4) there should be consistency between the residue and alkali extract ages.

7. Shells

The outer layer of shells may contain secondary carbonates, precipitated in the soil. This carbonate is removed by rinsing with a dilute HCl solution, during which about 30% of the sample may be lost. This first fraction however, can be collected and dated to detect the presence of secondary carbonate. The samples are subsequently heated in demineralized water at 90°C during 24 hours to remove organic particles.

GEOLOGICAL ASPECTS

Even if a given sample of dateable material is perfectly reliable, certain geological and geomorphological processes may have acted on the material which need to be recognised in order to avoid or reduce errors.

Our original aim is to date SL indicators or processes and events (possibly) related to SL changes. This goal is simply achieved if the indicator consists of dateable material. However, if this is not the case the age relationship or 'association' between datable material and indicator needs to be established.

1. Reworking

One requirement is that the datable material occurs in place. A clear example of a wasted radiocarbon assay is the case where pieces of driftwood, taken from a swashmark deposit in Subboreal beach sands, was dated over 10 000 BP. The requirement of in-place position, however, is not applicable in all cases. For example young bivalved molluscs washed onto a beach during storm and covered by sand soon after can be used to date paleo-tide levels derived from sedimentary structures in the beach sands. Had the molluscs been dead for some time, their ligaments would have decayed and the shell halves separated during transportation.

In arctic regions the age relationship between shells and raised beach deposits may have been severely disturbed. Donner & Jungner (1975), Donner, Erronen & Jungner (1977) and Ten Brink (1974) have shown that one should be very critical in dating raised beaches by means of shells from a mixed death assemblage in the deposit. Due to wave action or freeze and thaw processes (Olsson, Stenberg & Göksu, 1967) shells of different species and age may be brought together. It appeared that only the shell species from the intertidal zone were contemporaneous with the beach formation. Thus, Ten Brink (1974) concluded that samples including the barnacle *Balanus balanus* and/or *Mytilus edulis*, which usually both live in the intertidal zone, could be used to date the position of former SL. Other shells, not restricted to this zone, appeared to be unsuitable for this purpose. In this connection the reader may also refer to Barr (1971), Kelly (1973), Blake (1975) and Aloïsi *et al.* (1973).

2. Erosion

If the top of a peat bed is sampled, it should be ascertained that it is not eroded. Usually this is evident from the more or less gradual transition to the overlying clay.

A peat bed may contain an intercalated clay layer characterized by very sharp upper and lower boundaries, absence of roots from the peat above and fine lamination. Such beds originate when clay is deposited in between slabs of peat ripped apart vertically, of which the upper part floats during extreme high water levels. Clearly, it is useless to date either the peat immediately above or below the secondary clay bed.

3. Compaction

Another process that may be mentioned is compaction. When sampling peat cores, sample width should be adapted to the degree of compaction. The greater the sample interval, the larger will be the 'rejuvenating' effect of this source of error. A procedure sometimes followed in SL studies is to relate the age of a sample from the base of a peat bed, that has been lowered due to compaction of underlying deposits, to the altitude of the bed where it is found to rest on non- or much less compactable deposits. This procedure assumes that the base of the peat bed is isochronous. However, even over short distances the base of a peat bed may be strongly diachronous.

RECOMMENDATIONS

- Before separating a ^{14}C sample from a profile or a core, study the stratigraphical conditions of the sample (peat erosion, *in situ* position, possibilities for contamination); be aware of the questions asked on the submission forms. In general, provide oneself with a basis for evaluating dating results, not only by carefully inspecting sites and cores, but also by sampling in series, dating different kinds of material, two fractions, etc.

- Collect more samples or a greater amount of sample than required for one dating (later check may be required, sample may get lost), estimate whether the amount of sample available is sufficient to provide the required precision in age.

- Pack the samples in plastic or aluminium foil or in glass bottles and write name of site, sample number, depth and horizontal position and name of collector on each packed sample immediately.

- If possible, send sample to the ^{14}C laboratory as originally packed, together with complete information requested on the laboratory's information form and about any post-sampling treatment.

- If samples have to be stored before submission to the laboratory, keep them at a cool, dark and dry place; drying the samples in an oven at 50°C (avoid contamination) has the disadvantage that roots in peat can not easily be recognised and removed later.

- At a certain stage of this continuity the character of the sample may have to be determined. For instance, tree species (wood, charcoal), type of carbonate (calcitic secondary carbonate in aragonitic shells or coral samples), origin of bones.

- Do not change conventions agreed to by the international ^{14}C community at their regular meetings:
Quote conventional ^{14}C ages, based on the 5568 years half-life, in years BP (ie. before AD 1950) with one sigma standard deviation, together with the laboratory's code and number. If corrected ages are to be used, indicate this clearly by an asterisk and mention the calibration curve or table used (use the experience of your ^{14}C laboratory!), for instance:

T_{conv} = 1400 ± 40 BP (GrN-x) corected to:

600 - 660 AD*

*Stuiver, M., 1982. Radiocarbon 24: 1-27

- For obtaining reliable ages on marine shells, one should:
  (1) select the hard, non-porous specimens, preferentially in life-position (not reworked),
  (2) remove the outer part (in the laboratory) mechanically and/or chemically,
  (3) in some cases obtain more than one date on one batch of shells (for instance, inner and outer part separately),
  (4) determine $^{13}C/^{12}C$ ratio and crystal structure for establishing the occurrence of isotopic exchange, recrystallisation (for instance, coral should be 100% aragonite) and the marine origin of the shells; dates on shell carbonate are only reliable if $\delta^{13}C$ is 0‰ or positive,
  (5) check carefully whether conventional ages are reported (corrected to $\delta^{13}C = -25‰$) and whether apparent-age correction has been applied.

REFERENCES

Aloïsi, J.-Cl., Monaco, A., Thommeret, J. & Thommeret, Y., 1978. The Holocene transgression in the Golfe du Lion, Southeastern France: paleogeographic and paleobotanical evolution. *Géographie Physique et Quaternaire*, XXXII(2), 145-162

Barr, W., 1971. Postglacial isostatic movement in north-eastern Devon Island: a reappraisal. *Arctic*, 24, 249-268

Bathurst, R.G.C., 1976. Carbonate sediments and their diagenesis. *Developments in Sedimentology*, 12, (Elsevier, Amsterdam)

Baxter, M.S., 1983. An international tree ring replicate study. *Proceedings of the Groningen Conference on ^{14}C and Archaeology*, eds. W.G. Mook & H.Tj. Waterbolk, PACT Publication 8, 123-131

Berger, R., Taylor, R.E. & Libby, W.F., 1966. Radiocarbon content of marine shells from the Californian and Mexican West Coast. *Science*, 153, 864-866

Blake, W., Jr., 1975. Radiocarbon age determinations and postglacial emergence at Cape Storm, southern Ellesmere Island, Arctic Canada. *Geografiska Annaler*, 57, Series A

Brink, N.W. ten, 1974. Glacio-isostasy: new data from West Greenland and geophysical implications. *Bulletin of the Geological Society of America*, 85, 219-228

Bruns, M., Münnich, K.O. & Becker, B., 1980. Natural radiocarbon variations from AD 200 to 800. *Radiocarbon*, 22, 273-277

Bucha, V., 1970. Influence of the earth's magnetic field on radiocarbon dating. In: *Radiocarbon Variations and Absolute Chronology*, ed. I.U. Olsson (John Wiley, New York), 501-510

Chappell, J. & Polach, H.A., 1972. Some effects of partial recrystallization on ^{14}C dating of later Pleistocene corals and molluscs. *Quaternary Research*, 2, 244-252

Damon, P.E., Long, A. & Wallick, E.I., 1972. Dendrochronological calibration of the carbon-14 time scale. *Proceedings of the 8th International Conference on Radiocarbon Dating*, (New Zealand), 28-71

Donner, J. & Jungner, H., 1975. Radiocarbon dating of shells from marine Holocene deposits in the Disko Bugt area, West Greenland. *Boreas*, 4, 25-45

Donner, J., Erronen, M. & Jungner, H., 1977. The dating of the Holocene relative sea-level change in Finnmark, North Norway. *Norsk Geografisk Tiddskrift*, 31,103-128

Eisma, D., Mook, W.G. & Das, H.A., 1976. Shell characteristics, isotopic composition and trace element contents of some euryhaline molluscs as indicators of salinity. *Palaeogeography, Palaeoclimatology, Palaeoecology*, 19, 39-62

Eisma, D., Mook, W.G. & Laban, C., 1981. An early Holocene tidal flat in the Southern Bight. *Special Publication of the International Association of Sedimentologists*, 5, 229-237

Evin, J.C., 1983. Les matériaux d'origine terrestre utilisés pour les datations par le radiocarbone. *Proceedings of the Groningen Conference on ^{14}C and Archaeology*, eds. W.G. Mook & H.Tj. Waterbolk, PACT Publication 8, 235-275

Geyh, M.A. & Streif, H., 1970. Studies on coastal movements and sea-level changes by means of the statistical evaluation of ^{14}C data. *Report on the Symposium on Coastal Geodesy* (Munich), ed. R. Sigl, 599-611

Gillespie, R. & Polach, H.A., 1979. The suitability of marine shells for radiocarbon dating of Australian Prehistory. *Proceedings of the 9th International Conference on Radiocarbon Dating* (Los Angeles, La Jolla), 404-421

Godwin, H., 1962. Radiocarbon dating. *Nature*, 195, 943-945

Green, J.W., 1963. Wood cellulose. In: *Methods in Carbohydrate Chemistry*, ed. R.L. Whistler, (Academic Press, New York), volume 3, 10-14

Jong, A.F.M. de, 1981. *Natural ^{14}C variations*. Ph.D. thesis, University of Groningen

Jong, A.F.M. de & Mook, W.G., 1980. Medium-term atmospheric ^{14}C variations. *Radiocarbon*, 22, 267-272

Jong, A.F.M. de, Mook, W.G. & Becker, B., 1979. Confirmation of the Suess wiggles: 3200-3700 BC. *Nature*, 280, 48-49

Jong, A.F.M. de & Mook, W.G., 1981. Natural ^{14}C variations and consequences for sea-level fluctuations and frequency analysis of periods of peat growth. *Geologie en Mijnbouw*, 60, 331-336

Kaye, C.A. & Barghoorn, E.S., 1964. Late Quaternary sea-level change and crustal rise at Boston, Massachusetts, with notes on the autocompaction of peat. *Bulletin of the Geological Society of America*, 75, 63-80

Kelly, M., 1973. Radiocarbon dated shell samples from Nordre Strømfjord, West Greenland. *Rapport Grønlands Geologiske Undersøgelse*, 59

Klein, J., Lerman, J.C., Damon, P.E. & Ralph, E.K., 1982. Calibration of radiocarbon dates: tables based on the consensus data of the workshop on Calibrating the Radiocarbon Time Scale. *Radiocarbon*, 24, 103-150

Longin, R., 1970. *Extraction du collagène des os fossiles pour leur datation par la méthode du carbone 14*. Thesis, University of Lyon

Mangerud, J., 1972. Radiocarbon dating of marine shells, including a discussion of apparent age of recent shells from Norway. *Boreas*, 1(2), 143-172

Mangerud, J. & Gulliksen, S., 1975. Apparent radiocarbon age of recent marine shells from Norway, Svalbard and Ellesmere Island. *Quaternary Research*, 5, 263-273

Mann, W.B., 1983. An international reference material for radiocarbon dating. *Radiocarbon*, 25, 519-527

Mook, W.G., 1971. Paleotemperatures and chlorinities from stable carbon and oxygen isotopes in shell carbonate. *Palaeogeography, Palaeoclimatology, Palaeoecology*, 9, 245-263

Mook, W.G., 1980. Carbon-14 in hydrogeological studies. In: *Handbook of Environmental Isotope Geochemistry*, eds. P. Fritz & J.C. Fontes, (Elsevier, Amsterdam), volume I, 49-74

Mook, W.G., 1983. ^{14}C calibration and the time-width of ^{14}C samples. *Proceedings of the Groningen Conference on ^{14}C and Archaeology*, eds. W.G. Mook & H.Tj. Waterbolk, PACT Publication 8, 517-525

Mook, W.G. & Streurman, H.J., 1983. Physical and chemical aspects of radiocarbon dating. *Proceedings of the Groningen Conference on ^{14}C and Archaeology*, eds. W.G. Mook & H.Tj. Waterbolk, PACT Publication, 8, 31-55

Mook, W.G., Jong, A.F.M. de & Geertsema, H., 1979. Archaeological implications of natural carbon-14 variations. *Palaeohistoria*, 21, 10-18

Olsson, I.U., 1980. ^{14}C in extractives from wood. *Radiocarbon*, 22, 515-524

Olsson, I.U., 1983. Dating non-terrestrial materials. *Proceedings of the Groningen Conference on ^{14}C and Archaeology*, eds. W.G. Mook & H.Tj. Waterbolk, PACT Publication 8, 277-293

Olsson, I.U., Stenberg, A. & Göksu, Y., 1967. Uppsala natural radiocarbon measurements VII. *Radiocarbon*, 9, 454-470

Olsson, I.U., El-Daoushy, M.F.A.F., Abd-El-Mageed, A.I. & Klasson, M., 1974. A comparison of different methods for pretreatment of bones. *Geologiska Föreningens i Stockholm Förhandlingar*, 96, 171-181

Pardi, R. & Marcus, L., 1977. Non-counting errors in ^{14}C dating. *Annals of the New York Academy of Sciences*, 288, 174-180

Pearson, G.W., 1980. High-precision radiocarbon dating by liquid scintillation counting applied to radiocarbon timescale variation. *Radiocarbon*, 22, 337-345

Plassche, O. van de, 1979. Reducing the effect of root contamination of peat samples. *Océanis*, 5, 215-218

Plassche, O. van de, 1980. Compaction and other sources of error in obtaining sea-level data; some results and consequences. *Eiszeitalter und Gegenwart*, 30, 171-181

Plassche, O. van de, 1982. Sea-level change and water-level movements in the Netherlands during the Holocene. *Mededelingen Rijks Geologische Dienst*, 36(1), 1-93

Ralph, E.K., Michael, N.H. & Han, M.C., 1973. Radiocarbon dates and reality. *Masca Newsletter*, 9, 1-20

Roeleveld, W., 1974. *The Groningen Coastal Area*. Ph.D. Thesis, Free University, Amsterdam

Salomons, W. & Mook, W.G., 1985. Isotope geochemistry of carbonates in the weathering zone. In: *Handbook of Environmental Isotope Geochemistry*, Volume 2, eds. P. Fritz & J.C. Contes, (Elsevier, Amsterdam, Oxford, New York), in press

Schoute, J.F.Th., 1984. *Vegetation horizons and related phenomena. A palaeoecological-micromorphological study in the younger coastal Holocene of the northern Netherlands (Schildmeer area)*. Ph. D. thesis. Free University, Amsterdam, (Strauss & Cramer, Hirschberg)

Schoute, J.F.Th., Mook, W.G. & Streurman, H.J., 1983. Radiocarbon dating of vegetation horizons: methods and preliminary results. *Proceedings of the Groningen Conference on ^{14}C and Archeology*, eds. W.G. Mook & H.Tj. Waterbolk, PACT Publication 8, 295-311

Schoute, J.F.Th., Griede, J.W., Mook, W.G. & Roeleveld, W., 1981. Radiocarbon dating of vegetation horizons, illustrated by an example from the Holocene coastal plain in the northern Netherlands. *Geologie en Mijnbouw*, 60, 453-459

Scott, E.M., Baxter, M.S. & Aitchison, T.C., 1983. ^{14}C dating reproducibility: evidence from a combined experimental and statistical program. *Proceedings of the Groningen Conference on ^{14}C and Archaeology*, eds. W.G. Mook & H.Tj. Waterbolk, PACT Publication 8, 133-145

Streif, H., 1971. Stratigraphie und Fazies Entwicklung im Küstengebiet von Woltzeten in Ostfriesland. *Beiheft Geologisches Jahrbuch*, 119, 1-59

Stuiver, M., 1978. Radiocarbon timescale tested against magnetic and other dating methods. *Nature*, 273, 271-274

Stuiver, M., 1982. A high-precision calibration of the AD radiocarbon time scale. *Radiocarbon*, 24, 1-26

Suess, H.E., 1970. Bristlecone pine calibration of the radiocarbon time scale 5200 BC to the present. In: *Radiocarbon Variations and Absolute Chronology*, ed. I.U. Olsson, (John Wiley, New York), 303-311

Suess, H.E., 1978. La Jolla measurements of radiocarbon in tree-ring dated wood. *Radiocarbon*, 20, 1-18

Tans, P.P., 1978. *Carbon 13 and carbon 14 in trees and the atmospheric CO_2 increase*. Ph.D. thesis, University of Groningen

Tans, P.P., Jong, A.F.M. de & Mook, W.G., 1979. Natural atmospheric ^{14}C variation and the Suess effect. *Nature*, 280, 826-828

Taylor, R.E. & Slota, P.S., 1979. Fraction studies on marine shell and bone samples for radiocarbon analyses. *Proceedings of the 9th International Conference on Radiocarbon Dating*, (Los Angeles, La Jolla), 422-432

Thilo, L. & Münnich, K.O., 1970. Reliability of carbon-14 dating of ground water: effect of carbonate exchange. *Proceedings of the IAEA Conference on Isotopes in Hydrology* (Vienna), 259-269

Vogel, J.C. & Merwe, N.J. van der, 1977. Isotopic evidence for early Maize cultivation in New York State. *American Antiquity*, 42, 238-242

Wendland, W.M. & Donley, D.L., 1971. Radiocarbon calendar age relationship. *Earth and Planetary Science Letters*, 11, 135-139

Zagwijn, W.H., 1983. Geological aspects of carbon-14 dating. *Proceedings of the Groningen Conference on ^{14}C and Archaeology*, eds. W.G. Mook & H.Tj. Waterbolk, PACT Publication 8, 71-90

20

Dendrochronological dating

A. HEYWORTH

ABSTRACT

A brief discussion is presented of methods of sampling trees, measuring ring widths, matching ring-width curves, and using them to establish the pattern of water-table changes.

INTRODUCTION

The relationship of submerged forest trees to tidal levels is discussed in Chapter 13. If considerable numbers of such trees are present in an exposure, it is clearly desirable to date a significant proportion of them, in order to determine their age range, and whether differences in altitude can be related to differences in age. Large numbers of radiocarbon dates may be impracticable on the grounds of expense and furthermore, the resolution (perhaps no better than ± 150 yrs; International Study Group, 1982) may be insufficient to reveal the pattern of change.

Matching of ring-width variations, however, will fix the relative ages of trees to within one year, and a chronological framework for the growth and death of the forest can thus be built up. One radiocarbon date may then be sufficient to link this chronology to others.

DENDROCHRONOLOGICAL CORRELATION

It is not necessary here to go into the details of the principles and theory of dendrochronology. These have been discussed very fully by, for example, Fritts (1976).

Matching of different trees by their ring widths is, of course, possible only if the growth of annual rings is influenced by some common external factor. A very important factor at most sites is the variation in rainfall, soil moisture, and associated transpiration stress. Submerged forest trees, however, grew so near to the water table that it is difficult to see how water shortage could ever have affected them. Asphyxiation of part of the root system by a rise in the water table seems more likely to have been the cause of narrow annual rings. Sunshine and temperature are other important factors, although variations in the latter are less severe at coastal sites than inland. The more equable environment at such sites is, no doubt, responsible for the fact that, in general, submerged forest trees show a more complacent ring-width pattern than trees from inland and high-altitude sites. Nevertheless, ring-width matching of submerged forest trees is possible, and extremely good correlations between trees are often found.

It should be mentioned here that it is often necessary to allow for longer-term trends, especially the growth-trend, ie. the tendency for a tree's early rings to be wider than those formed later in its life, Ring-width *indices,* as opposed to actual ring-widths. are then used in the correlation of different trees. In this paper, however, the examples given will, in the interests of simplicity, be confined to the measured widths.

PREPARATION OF SAMPLES FOR RING-WIDTH MEASUREMENT

Most submerged forest trees are very well preserved, since they have never dried out since they died. When exposed, on the beach, they are, of course, much more easily accessible than when buried. The best method of sampling is to saw complete discs, at least 10 cm thick, from the stumps or trunks. As the wood is often extremely hard, a chainsaw is usually necessary, but great care should be taken to avoid the ingress of sand and salt water into the mechanism.

By allowing these thick slices to dry out gradually, it is usually possible to avoid the splitting and warping which occurs with thinner slices. When dry (which may take several weeks) the wood may be sawn into thinner pieces, preferably using a bandsaw. This gives clean, flat faces, and it is frequently possible to measure the ring widths without further preparation. Sanding of the surface may be required, although this sometimes gives a fluffy finish, which obscures the ring boundaries. A very useful tool in these circumstances is an electric plane, but a very sharp hand plane can be used, although some skill and effort are required in the notoriously difficult task of obtaining a good finish by planing across the end grain of hard oak.

An alternative method of preparation is to use the still wet, relatively soft, wood, and to take a thin slice off the surface with a sharp razor blade. This cuts cleanly through the vessels and fibres and the ring boundaries are usually exposed very clearly. Some wood is too hard for this treatment, even when freshly collected.

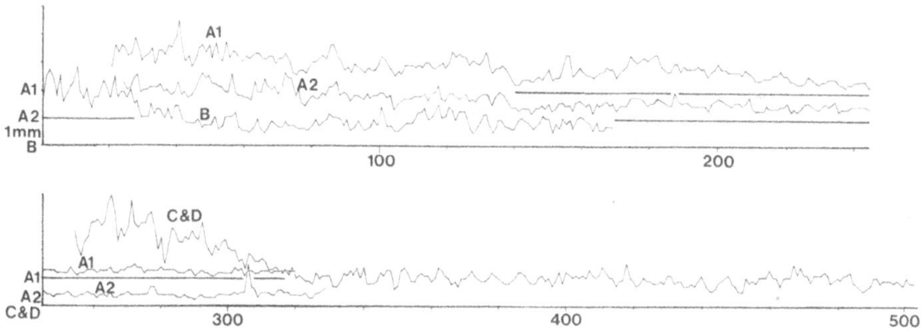

Figure 1. Ring-width curves from four trees, A, B, C and D, in the submerged forest in Bridgwater Bay, Bristol Channel plotted in their matching positions. Two radii from tree A are shown, and the mean ring-widths of the matching pair of trees C and D. Different base lines are used, as indicated. A total of 506 years is covered by the four trees.

As soon as possible after collection a note should be made as to whether the centre of the tree and the bark are intact. It may be necessary to measure the rings in the sapwood when wet, as this may shrink considerably and disintegrate when dried.

The ring widths can be measured by methods varying in sophistication from a simple magnifier with a built-in scale, such as a map measurer, to a large photogrammetric plotting instrument. It is usually necessary to measure the ring widths only to the nearest 0.1 mm, although where several consecutive rings are less than, say, 0.5 mm wide, it is important that the increases and decreases in width are correctly recorded.

Ring boundaries in oak, which has very large spring vessels, are easily identified. They are usually clearly visible in pine, although this, of course, has no vessels, by changes in density and colour. Oak is the most reliable and commonly used species in dendrochronological studies, since it produces one ring every year. Other species, including pine, sometimes produce more than one ring per year, or may fail to produce any ring in some years. Useful results have, however, been obtained from pine, although it is sometimes difficult to find matching patterns even in trees known to be of the same date.

When the rings have been counted and measured, and the figures checked, the widths are plotted on tracing paper, in the manner of Figure 1. It is common practice to use graph paper with a logarithmic vertical scale. This is useful when there is a great variation in ring widths, but a linear scale is used in Figure 1. At least two radii from each tree should be measured, as there is often considerable variation around the tree. These different radii should first be checked against one another. The tracing paper plots can be superimposed, and slid along until a good visual match is obtained. In the case of two radii from the same tree, the matching position is, of course, known, but the closeness of the fit can be judged.

In trees with a relatively complacent ring-width response, a less subjective indication of correlation is needed. A very useful and reliable assessment of the match can be obtained by using the computer program devised for this purpose by Baillie & Pilcher (1973). This FORTRAN program is simple to use routinely with large numbers of trees. The program, in effect, slides one series of ring widths (corrected for growth trend) along a second, and calculates the correlation coefficient for each position of overlap, in steps of one year. The correlation is based not simply on the percentage of agreement between increases and decreases in the ring widths in the first and second trees, but also on the magnitude of the variations. The Student's t value for each position of overlap is then calculated, and the results tabulated, with a summary of positions with a t value greater than an arbitrary figure of 3.5, indicating positions of high correlation. In practice, this program is very useful in suggesting possible matching positions, which can then be checked visually. In many cases the t value at one position is so high as to leave no reasonable doubt that this is the correct match.

When checking radii from the same tree against one another, the visual and computer checks will show whether there is good agreement. If the correlation between these radii is not extremely good then that tree cannot be considered reliable for matching with other trees, as it is clearly not responding normally to external factors. In some cases, however, a tree may produce a normal pattern on one side of the trunk, whilst the other side, where, for example, the roots have been injured, has an anomalous pattern. In such cases it is worthwhile trying to match each radius of the tree with other trees. This may show that the anomalous behaviour is confined to one radius. Trees in coastal situations frequently show such asymmetrical patterns, as a result of the effects of salt water, which might, for example, fill a drainage channel and kill the roots on one side of an adjacent tree whilst the roots on the other side are unaffected. The effects of the prevailing wind may also cause asymmetrical growth.

When the various radii in each tree have been checked against one another, then the ring widths from these radii can be combined to produce an average ring-width pattern for each tree. Ring-width plots from adjacent trees can then be compared, and if two or more show good agreement, these values can be combined to produce an average ring-width series for that group of trees. This can be extended to produce a complete floating chronology for the whole submerged forest, and the differences in the dates of germination and death of the trees can be related to altitude and position. An estimate of the number of missing outer rings may have to be made if the bark is not preserved in some trees. The establishment of such a chronology means that one radiocarbon sample can be taken from the largest or best-preserved tree, whilst the altitude can be carefully worked out from the overall chronology of a group of trees. Most submerged forest wood is so well-preserved that it provides an ideal source of material for radiocarbon dating, with no danger of contamination. Submerged forest wood has been used extensively in the calibration of radiocarbon dates and in investigating possible variations in past levels of radiocarbon in the atmosphere (Campbell & Baxter, 1979; Heyworth & Kidson, 1982; International Study Group, 1982). There is also the possibility of dating the submerged forests directly in calendar years by matching the floating chronology with an established absolute tree-ring chronology.

RING-WIDTH MATCHING: AN EXAMPLE

In the submerged forest in Bridgwater Bay, in the Bristol Channel, trees occur over a vertical range of several metres, on a sloping beach. Within this range there are distinct horizontal forest belts, where large stumps and fallen trunks are incorporated into alder carr peat. Many trees can be sampled at the same altitude. In order to establish the course of sea-level (SL) rise at the site it was necessary to know whether the horizontal belts of trees were killed by a series of rapid rises in SL, with intervening stands or falls in level, or whether a gradual rise was responsible.

One of the those groups of trees gave ^{14}C dates of 4500-5000 BP. Figure 1 shows ring-width plots from four oak trees in this group. Tree A was the first to be measured and two radii, A1 and A2 are shown. These matched at the correct position, with a t value of 10, at 296 years of overlap, that is, the two radii were virtually indistinguishable. The two radii were then averaged, though, for simplicity, this curve is not shown. A pair of radii were similarly measured in tree B, and the average of these two is shown in Figure 1. Tree B matched with tree A, with a t value of 7.6 at 150 years of overlap ('overlap' here means the number of years the shorter plot is moved, from left to right along the longer plot, to reach the matching position). Trees C and D were similarly matched, the highest t value of several pairs of radii being 13.5. These two trees each covered almost exactly the same period, both trees having germinated in the same year. The average plot of C and D matched that of A with a t value of 3.56 at 485 years of overlap. At this position, tree A only coincides with the C and D average over a period of 76 years. About 100 years of coincidence is normally regarded as the acceptable minimum for significant matching, but in this case, individual radii of A correlated with individual radii of C and of D, with higher t values, so that the match was regarded as correct. The visual matching of the A curve with the C and D curve is obscured in Figure 1, by the growth trend of C and D, the tree having wide rings in the early part of its life, and narrower ones as it ages.

A total of 506 years is covered by these trees, a period during which the water table must have been practically constant. The trees, when alive, could not have been more than 50-100 m from the storm beach, of which the 4500-5000 BP position can still be identified. A very gradual rise of SL over the 506 year period seems, therefore, to be demonstrated, with the later trees probably producing roots at a slightly higher level than the older ones. Clearly, this period could not have been a time of rapid SL rise.

At this particular site, the underlying topography is such that it is very unlikely that factors other than SL could have brought about waterlogging and the preservation of trees. It must be borne in mind that, at some sites, trees may have been killed and preserved as a result of local wetness well before the effects of SL were felt at the site. In such a case, only the most recent trees would be useful as SL indicators.

REFERENCES

Baillie, M. & Pilcher, J., 1973. A simple cross-dating program for tree-ring research. *Tree-ring Bulletin*, 33, 7-14

Campbell, J. & Baxter, M.S., 1979. Radiocarbon measurements on submerged forest floating chronologies. *Nature*, 278, 409-413

Fritts, H.C., 1976. *Tree-rings and Climate*. (Academic Press, London)

Heyworth, A. & Kidson, C., 1982. Sea-level changes in south-west England and Wales. *Proceedings of the Geological Association*, 93(1), 91-111

International Study Group, 1982. An inter-laboratory comparison of radiocarbon measurements in tree-rings. *Nature*, 298, 609-623

ALTITUDE

21

Determination of altitude

W. G. JARDINE

ABSTRACT

Ideally, the altitude of a sea-level indicator is determined initially in relation to a nearby temporary bench mark (TBM) and, thereafter, *via* surveyed bench marks (subsidiary datum-levels) in relation to a reference datum-level (which in certain cases will be a national reference datum-level). For purposes of correlation between reference datum-levels in areas remote from each other, it may be necessary for the altitude of a datum-level recorded on a TBM located near to a sea-level indicator to be determined in relation to a 'local tidal datum-level'. Examples of local tidal datum-levels that have been used are: high water mark of ordinary spring tide at a chosen site or sites on a chosen day or days; high water mark during calm conditions, as indicated by the upper limit of drifted *Enteromorpha* spp.; mean level of lower low water spring tide; the level of the lowermost reach of a river possessing a gentle gradient; the upper limit of *Fucus vesiculosus*; the upper limit of barnacle growth. Tide tables, giving predicted levels of the sea at sites devoid of permanent bench marks, also may be used in correlation between one area and another; problems encountered in Kenya, together with possible solutions, provide examples. Comparison of several national reference datum-levels, eg. in NW Europe, may be made by use of a correlation table. Measurement of the vertical distance (altitude) above or below a reference datum-level should be made by surveyor's level and staff. Altitude determination at off-shore sites presents special problems; a standard procedure to resolve these problems is recommended, but practical issues may necessitate the use of a simplified (and less reliable) procedure.

INTRODUCTION

The altitude ('height' in geodetic terminology) of a given point, eg. a sea-level indicator (SLI) such as is of relevance in this volume, may be defined as the vertical distance of the point above (or below) a reference datum-level. For the above definition to be of use, the term 'reference datum-level' must be defined and, in order that altitude may be determined, the selection of reference datum-levels and the measurement of vertical distances must be considered. In addition, so that altitudes determined in one geographical area may be compared with those determined in other geographical areas, methods of comparing and correlating reference datum-levels are discussed below.

It must be emphasised that the methods of determination of altitude discussed in this chapter are those that have been practised, and are capable of being practised, frequently in remote areas with the minimum of equipment and limited assitance, by earth scientists who investigate changes of SL and of shoreline position. Certain fundamental elements of geodetic theory and surveying practice are included in the discussion below, but no more than are required to ensure that altitudes may be determined satisfactorily in the field and that the principles involved in comparison of altitudes may be understood.

Important principles and details of practical geodesy are given in Torge (1980) and Vaniček *et al.* (1980). Explanations of field surveying practices are to be found in handbooks such as *Mapping for field scientists* (Ritchie *et al.*, 1977).

As an essential preliminary, the nature of Mean Sea Level and Mean Tide Level is considered. It should be noted that, throughout the text (except in quotations from other authors), the terms Mean Sea Level (MSL) and Mean Tide Level (MTL) are used in this form for consistency and clarity. In the literature the terms Mean Sea Level and mean sea level are both used, in certain cases interchangeably. It should *not* be thought that the term Mean Sea Level refers solely to the surface level of the sea on a global scale (see section below) or to a level of the sea determined over a long period of time or in the course of an 'official' national survey, whereas the term mean sea level refers to a level of the sea determined over a short period of time or in the course of an 'unofficial' survey, such as that in which a user of this manual might be engaged.

SURFACE LEVELS OF THE SEA

Independently of the effects of wind (eg. waves) and other 'superficial' agents, the level of the surface of the sea at a given location changes continuously throughout the tidal cycle, in broad terms ranging between the level of high water and that of low water, two levels which vary between 'spring' and 'neap' values depending on the position of the

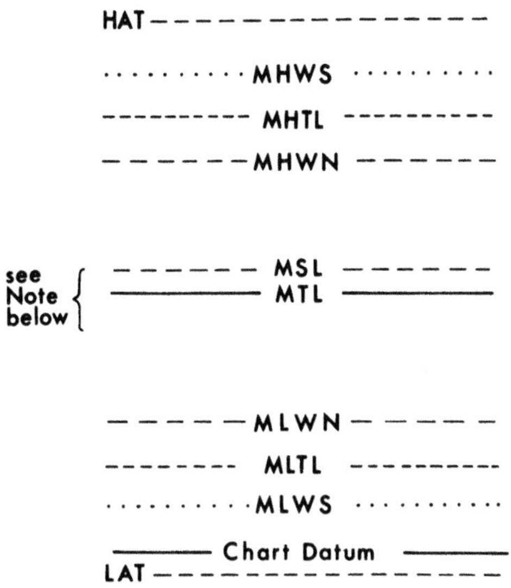

Figure 1. Diagrammatic representation of positions of the surface level of the sea at a location where the tidal cycle is not perfectly symmetrical (together with Chart Datum, which is a level to which soundings on a published chart are reduced, and above which tidal predictions and tidal levels are given in tide tables). Abbreviations used: LAT and HAT, lowest and highest astronomical tide levels; MLWS and MHWS, mean low and mean high water levels of spring tides; MLTL and MHTL, mean low and mean high tide levels; MLWN and MHWN, mean low and mean high water levels of neap tides; MTL, Mean Tide Level; MSL, Mean Sea Level. Note: Mean Sea Level may be lower, rather than higher, than Mean Tide Level (see text). Vertical scale varies, depending on the location. Adapted from various sources.

moon in relation to the earth. Various positions of the surface level of the sea are shown in Figure 1. It should be noted that, ecologically, mean high tide level (MHTL) is an important 'SLI', in practice being used more frequently than mean high water of spring tides (MHWS) or mean high water of neap tides (MHWN) (H. Streif, pers. comm.).

In theory, MSL is the surface level which the oceans and seas of the world would adopt if tidal forces ceased to operate (Doodson & Warburg, 1941, p 100). In practice, MSL, as zero reference datum is the average level of the surface of the sea as calculated from a large number of observations taken at equal intervals of time (commonly one hour) over a period of several years; eg. for Denmark, MSL is the average level of the sea recorded in Danish mareographs between AD 1885 and 1905, whilst for the mainland of Britain, Ordnance Datum, Newlyn, is the mean value of SL determined at Newlyn (Cornwall) during the period AD 1915 to 1921.

In detail, MSL shows appreciable variations from day to day, from month to month, and even from year to year. Thus, for the year AD 1974, local MSL at Newlyn was 0.086 m above 'standard' Ordnance Datum, Newlyn. 'The reliability of a value of Mean Sea Level will depend on the length of period of observation taken to determine it. A value of sorts [ie. a rather unreliable value] may be determined by meaning 25 hourly observations; a more reliable result will be obtained from 39 hourly observations ... Better still will be the results obtained by meaning hourly heights of water level observed over a period of 29 days. The best results cannot be obtained from observations which do not extend over a period of at least 18.6 years (one cycle of the moon's nodes). ... In non-tidal waters a long period of observations is no less important [than in tidal waters]' (Hatfield, 1969, p 13; see also Clark, 1958, pp 544-546, for further details and examples based on data collected at Takoradi, Ghana, West Africa, and at Galveston, Texas coast of the U.S.A.).

MTL at a given location is the average value of all the levels of high water and low water at that location, taken over a period, preferably long rather than short. At a given location the values of MSL and MTL will agree only if the tidal curve is perfectly symmetrical (Hatfield, 1969, p 13); in practice, MTL differs (either positively or negatively) from MSL by an amount which is approximately constant, but the 'constant' differs widely from one location to another, depending on the shallow water distortion of the tidal wave (*cf.* Gordon, 1957, Fig. 1; Suthons, 1960, p 68). Information on the detailed relationship between MTL and MSL is minimal but, on the basis of the above statements, presumably where the tidal curve is 'positively skewed' the relationship between MTL and MSL will be the opposite from that where the tidal curve is 'negatively skewed'. It follows that, in an area where the nature of the tidal curve is known and the relationship between MTL and MSL (MTL higher or lower than MSL) is known at one location, the relationship between MTL and MSL at other locations within the same tidal area may be deduced. Along the coast of The Netherlands a relationship of this type appears to exist (O. van de Plassche, pers. comm.).

As a datum-plane for surveying, 'mean sea level is the best datum to use as it is not so variable as the high or low water marks or the mean level of high and low tides' (Clark, 1958, p 544). It is worth noting, however, that, 'Mean Tide Level may be obtained in the field more easily than Mean Sea Level, and from fewer observations' (Hatfield, 1969, p 13) and thus there are advantages on certain occasions in using MTL rather than MSL as a reference datum-level. An example of such an occasion is when the altitudes of present coastal forms are being compared with the altitudes of former coastal forms. 'It should be understood that Mean Sea Level, being a calculated value based on measurements made at regular intervals in the course of no longer than approximately the last 150 years, is a level of which there is no long-term record in sedimentary deposits or landforms. In contrast, Mean Tide Level ... is a level whose (approximate) former value may be determined from the sedimentary and geomorphological record of former high tide, low tide and tidal range values. For practical purposes, eg. the construction of sea-level curves, values of former Mean Tide Level rather than of former Mean Sea Level are used (although those who construct such curves may not be aware, or may not state, that this is the case). In effect, former Mean Tide Level and former Mean Sea Level at a given site at a given time are taken to be equivalent in value.' (Jardine, 1982, p 26). These points should be borne in mind when field procedures for relating a temporary datum-level to a local reference datum-level are being considered.

REFERENCE DATUM LEVEL

1. Nature and definition

To understand the nature of, and to define, a reference datum-level it is useful first to consider briefly the concept of the *geoid*.

In geodetic and surveying practice, the principal limitations related to geometrical measurement of the earth's shape can be circumvented if all measurements are referred to surfaces on which the *geopotential*, the combined potential of the gravitational attraction and the rotational acceleration of the earth, is constant. 'The normals to such surfaces are readily found in practice, for they are the directions of the local vertical in which the plumb-line hangs. Alternatively, they are the normals to the horizontal planes (tangent planes to the equipotential surfaces) as defined by the spirit level. All the instruments that a surveyor uses are set up with respect to the tangent plane to the equipotential surface that passes through the point of observation. ... heights are derived from measurements of differences of potential between points.' (Cook, 1972, p 14; see also Fig. 2).

'The *geoid* is the particular surface of constant geopotential which, over the seas, coincides with mean sea level[1]. Since measurements of heights or potential differences on land begin from a reference station at sea level, values of height or potential obtained ... are values above the height or potential of the geoid. While this is a strictly correct statement, its interpretation needs some care, for while the geoid is accessible over the three-quarters of the surface of the earth covered by the oceans, over the remaining area it lies generally below the surface of the ground and so, while no less a physical reality, is inaccessible to direct observation' (Cook, 1972, p 15).

A reference datum-level may be defined as either a (short) horizontal line or a horizontal surface (of limited extent), commonly located near SL, that is used as a reference level above or below which the altitude of a given SLI is measured. From the above consideration of the geoid, it may be seen that, certainly in theory, the most useful

[1] +1 m. See Torge, 1980, p 44

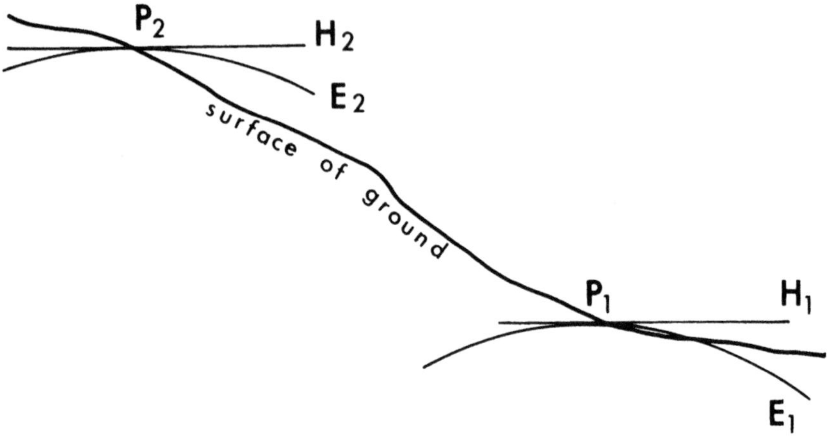

Figure 2. Diagrammatic representation of the relationships between equipotential surfaces, E_1 and E_2, at two points, P_1 and P_2, on the surface of the earth and horizontal planes, H_1 and H_2, tangential to E_1 and E_2. Not to scale. For the purposes of discussion in the text, P_1 is to be regarded as being located at (local) MSL. Adapted, with permission, from Cook (1972, Fig. 1).

reference datum-level that can be chosen in any given limited geographical area is the relevant part of the geoid or its near equivalent, local MSL. In practice, commonly a national reference datum-level is chosen to correspond as closely as possible with MSL at the location that has been selected as the national reference station. For example, for the mainland of England, Wales and Scotland, the reference datum-level, OD, ie. Ordnance Datum, Newlyn, coincides with MSL calculated from hourly SL readings at Newlyn between 1st May AD 1915 and 30th April AD 1921. Examples of other national reference datum-levels in north-western Europe are: NN, ie. Normaalnullpunkt (Federal Republic of Germany); DNN (Denmark); NAP, ie. Normaal Amsterdams Peil (The Netherlands); NGF, ie. Nivellement général de la France (France). The present altitudes of these reference datum-levels in relation to local MSL are indicated in Table 1.

Examples of *subsidiary* datum-levels are 'bench marks' (whose altitudes above the national reference datum-level have been determined by spirit levelling) placed by the Ordnance Survey of Great Britain at selected locations throughout the United Kingdom, and comparable indicators placed similarly by government or other authorised surveyors in various countries throughout the world. An essential feature of such datum-levels is that, within any given national territory, the altitude of each of the individual datum-levels is determined in relation to a national reference datum-level at (comparatively) frequent intervals (*c*. 20 to 30 years). It is of interest to note that, for

Table 1. National reference datum-levels of the North Sea and Irish Sea areas in relation to local Mean Sea Level and Normaal Amsterdams Peil (NAP). Based on information provided by J. Mangerud (Norway), H. Streif (Federal Republic of Germany), H. Krog (Denmark), O. van de Plassche (The Netherlands), L. Jones (Belgium), M. Ters (France) and the Ordnance Survey (U.K.) and on data included in Jacobsen (1964) and Rossiter (1967).

Area	Reference datum-level	Present altitude of reference datum-level in relation to local mean sea level	Altitude of reference level in relation to Normaal Amsterdams Peil (NAP)
Sweden (west coast)	Zero of precision levelling system	?	? 0.007 m below NAP
Norway (north)	Nordnorsk Null (MSL at Narvik, 1948-56, & MSL at Evenskjaer, (1945-56)	Zero	? 0.010 m below NAP
Norway (south)	Normal Null, 1954 (MSL at Tregde)	Zero (±0.04 m)	? 0.012 m below NAP
Federal Republic of Germany	Normalnullpunkt	'Nearly identical with MSL of North Sea'	? 0.016 m above NAP
Denmark	DNN (MSL, Danish mareographs, 1885-1905)	Zero	0.124 m below NAP
The Netherlands	Normaal Amsterdams Peil	*c.* 0.1 m above MSL	Zero
Belgium	Provisional datum	2.012 m below MSL at Ostend	2.323 m below NAP
	Recent datum	Zero	0.311 m below NAP
France	Nivellement général de la France (NGF)	Zero	? 0.128 m below NAP
Mainland of England, Scotland and Wales	Ordnance Datum, Newlyn (MSL at Newlyn, Cornwall, 1915-1921)	Zero	0.203 m below NAP
Republic of Ireland	Ordnance Datum, Dublin (level to which the tide fell in Dublin Bay on 8 April 1837)	*c.* 2.5 m below MSL at Dublin	?
Northern Ireland	Ordnance Datum, Belfast (MSL at Belfast): new datum, not yet tied to other reference datum-levels	Zero	?

practical reasons, the United Kingdom national reference datum-level is *not* recorded physically but is taken as the level at 4.750 m (15.588 feet) below a bench mark established at the Tidal Observatory situated on the south pier at Newlyn, Cornwall. Similarly, in the Federal Republic of Germany the national reference datum-level (NN) is situated 37.000 m below a 'standard' bench mark which was established at the (former) Berlin Observatory in 1879. The standard bench mark of 1879 was replaced in 1912, because of the demolition of the observatory, by an underground mark approximately 40 km to the east of Berlin. This 'standard bench mark of 1912' (*cf.* Torge, 1980, p 224) at Berlin-Hoppegarten lies 54.638 m below NN. In the United States of America, precise levelling networks are tied to 26 tidal stations. Adjusted heights establish the National Geodetic Vertical Datum of 1929 (formerly termed the 'Sea Level Datum of 1912') (*cf.* Torge, 1980, p 225).

A third type of datum-level is the 'temporary bench mark' (TBM) that may be established, by the field scientist, within a (horizontal) distance of a few metres or tens of metres of the site or sites at which relevant SLIs are located. The TBM may take the form of *either* a small paint mark or other readily-recognisable (and durable) mark placed on a fixed object such as an exposure of solid rock or a (semi-) permanent artificial structure, *or* a mark on a peg fixed firmly in stable ground. In the latter case it may be necessary to protect the TBM from disturbance by animals.

A TBM may be used as a temporary datum-level for the duration of the field season in which it is established, but its altitude in relation to a more permanent reference datum-level (eg. a surveyed bench mark, where possible) should be determined at, or shortly after, the time that the TBM is established. The altitude of the TBM also should be checked (in relation to a more permanent datum-level) before the end of the field season in which the TBM was established and, if it is desired to use the TBM for a further season or seasons, its altitude should be checked at the beginning and end of such a season or seasons.

2. Selection and use

National reference datum-levels, subsidiary datum-levels (surveyed bench marks) and temporary bench marks (TBMs) have been considered above. Clearly, sites at which relevant SLIs are located seldom are situated sufficiently close to a national reference datum-level for that datum-level to be used directly in altitude determination. In contrast, in many field situations surveyed bench marks may be used in the direct measurement of the altitudes of SLIs, and in almost every field situation it is possible (and advantageous) to establish TBMs in relation to which the altitudes of SLIs may be determined. Broadly, in this context two sets of sites of SLIs exist: sites located within surveying distance of a surveyed bench mark; sites not located within surveying distance of a bench mark.

It is difficult to define precisely what constitutes 'within surveying distance of a bench mark' because the maximum distance over which it is possible and advisable to survey depends on several factors whose importance varies from place to place within a limited geographical area and also on a global scale: eg. type and relief of terrain over which surveying would require to be made, clarity of the atmosphere at the time of surveying, type of surveying instrument used. In what must be regarded as very general terms, an acceptable surveying distance would be that in which the 'closing error' given by traverses in opposite directions between the relevant bench mark and the relevant SLI (or a TBM, where the altitude of that datum-level is being established) is small enough to be ignored; this value also depends on several factors, but mostly on the degree of accuracy required, 0.150 m probably being the maximum acceptable closing-error value in any circumstances (*cf.* Gray, 1975, p 14). In practice, 1 km would be the maximum possible surveying distance in a majority of cases but, in suitable flat terrain, up to 5 km might be possible on occasions.

a) Sites located within surveying distance of a bench mark

In certain cases the sites of SLIs may be located sufficiently close (eg. up to *c.* 50 m in terrain where the range of relief is less than 3 m) to a surveyed bench mark for that bench mark to be used directly as a local datum-level. More commonly, however, it is necessary to establish a TBM within a distance of a few metres or tens of metres of the site or sites, as discussed above (page 576), and the altitudes of the SLIs in relation to the TBM then to be determined by accurate vertical measurement (see below, pages 583-585).

An advantage of using either a single TBM or a single surveyed bench mark as datum-level for several adjacent sites is that comparison of altitudes of relevant horizons at these sites can be made with ease. In certain cases, however, it may be advantageous to save time initially by using several TBMs and/or permanent bench marks as datum-levels although, eventually, the altitudes of the datum-levels used in this way must be determined in relation to the national reference datum-level for the geographical area concerned.

b) Sites not located within surveying distance of a bench mark

The determination of the altitudes of SLIs at sites located at remote distances from a surveyed bench mark presents special problems. In such situations it is not sufficient to establish a temporary bench mark or bench marks in the vicinity of the sites concerned (as indicated in *(a)* above) although, in almost every case, this will be an essential preliminary part of the procedure.

If more than one TBM is established, the individual TBMs will require to be correlated with each other, preferably by the selection of one of the TBMs as a 'Key TBM' and by the altitudes of each of the other TBMs being measured thereafter in relation to the datum-level represented by the Key TBM. Altitudes of SLIs at relevant sites then may be determined in relation to adjacent TBMs, and so to the Key TBM.

A further requirement is that the altitude of the datum-level recorded at the TBM be determined in relation to a datum-level (termed here the 'local tidal datum-level' for convenience of discussion) that can be correlated with reference datum-levels in other areas. For this to be achieved, the practice is for a local tidal datum-level to be selected on the basis of its equivalence to a position of the surface of the sea. For reasons discussed above, ideally the position chosen is (local) MSL. Field studies in remote areas, however, seldom extend over periods sufficiently long for satisfactory determinations of MSL to be made (see pages 570-573). In practice, therefore, visually recognisable positions of the surface of the sea (and occasionally other markers), rather than a calculated level, have been used. Some examples are these:

1. High water of ordinary spring tide at a chosen site or sites on a chosen day or days (McCann, 1964, p 8 and p 16).
2. High water mark during calm conditions; taken as the upper limit of drifted *Enteromorpha* spp. (Synge & Stephens, 1966, p 105). It should be noted that this method has been questioned on the grounds that the limit of drifted *Enteromorpha* is affected by factors such as the gradient of the beach and its exposure (Sissons, 1967, p 164).
3. Mean level of lower low water spring tide. The establishment of a common tidal datum is a special problem associated with coral reef structures. In the case of the northern Great Barrier Reef of Queensland, Australia, all levelled profiles were measured with respect to low water datum (mean level of lower low waters at spring tides), and all reductions to datum were made by observing a still water level at a known time on a given profile and relating this to a curve of predicted tide levels (McLean *et al.*, 1978, p 168).
4. The level of the lowermost reach of the Sénégal River, in conjunction with a dense network of bench marks surveyed in relation to a local datum-level (Faure *et al.*, 1980). The slope of the Sénégal River varies annually between 0.001 and 0.002 to 0.003 percent over the 270 km reach extending from the mouth of the river inland to Podor, and near the mouth of the river the average tide level is *c.* 0.05 m above local datum-level, which corresponds with MTL at Dakar between 1942 and 1945.
5. The upper growth limit of the seaweed *Fucus vesiculosus* (Sollid *et al.*, 1973, pp 237-238). It should be noted that, in Finnmark, Sollid *et al.* (p 237) checked the

observation that, 'It has commonly been assumed that the height of the seaweed limit above mean tide level is constant from place to place', and concluded that a 'mean figure of 0.6 m above mean tide level should be fairly correct for Varangerhalvøya and for the rest of Finnmark, where climate and tidal range, etc. are similar. In the area, Porsangerhalvøya-Bergsfjordhalvøya, the upper growth limit of seaweed was found to lie 0.3 ± 0.2m above mean tide level at 55 observation points. The height of the seaweed limit is, however, less certain at the heads of fjords which are commonly frozen over during the whole winter. Observations indicate that it occasionally reaches no higher than mean tide level' (Sollid *et al.*, 1973, p 238).

6. The upper limit of barnacles (*cf.* Donner, 1959, p 9; Donner, 1963, p 5). It should be noted that two barnacles used in this method, *Balanus balanoides* and *Chthamalus stellatus*, indicate approximately the upper limit of the mid-littoral zone of rocky shores in the British Isles, but the boundaries of zones shown by organisms, such as barnacles, do not correspond precisely with levels of the tide (*cf.* Donner, 1959, p 9; Gray, 1975, p 7).

Clearly, the usefulness of each of the methods listed above is dependent upon the difference in altitude between the relevant local tidal datum-level and the corresponding position of MSL or MTL being known, because, as discussed above, MSL is the best reference datum-level that can be selected, whilst MTL is an acceptable alternative in certain circumstances.

The use of tidal information in the recording of the altitudes of surfaces related to former SLs in areas where permanent bench marks are not available is illustrated by two contrasting examples. In the first, from Varangerfjorden, northern Norway, altitudes were measured in relation to an arbitrary datum-level, the altitude of which was obtained in relation to MSL for northern Norway by recording the level of the sea at known times and comparing these levels with appropriate predicted values at ports around Varangerfjorden, given in (British) Admiralty Tide Tables (Rose, 1978, p 123). This method may be applied in many parts of the world where bench marks are not available, because Admiralty Tide Tables (which include detailed instructions on, and examples of how to use, the tables) give tidal information for hundreds of ports throughout the world (*cf.* use of British Admiralty tidal data to derive local MSL in western Malaysia and the Straits of Malacca; Streif, 1979, p 557).

The second example is taken from Kenya, on the eastern coast of Africa. 'To refer the levellings to datum (lowest low tide as calculated from astronomical observations), tide tables published by the East African Harbours Corporation were used. These tables, however, only include the time and height of low water and high water. Thus a diagram allowing for interpolation of sea-level at various tidal ranges at different times was constructed' (Åse, 1978, p 213).

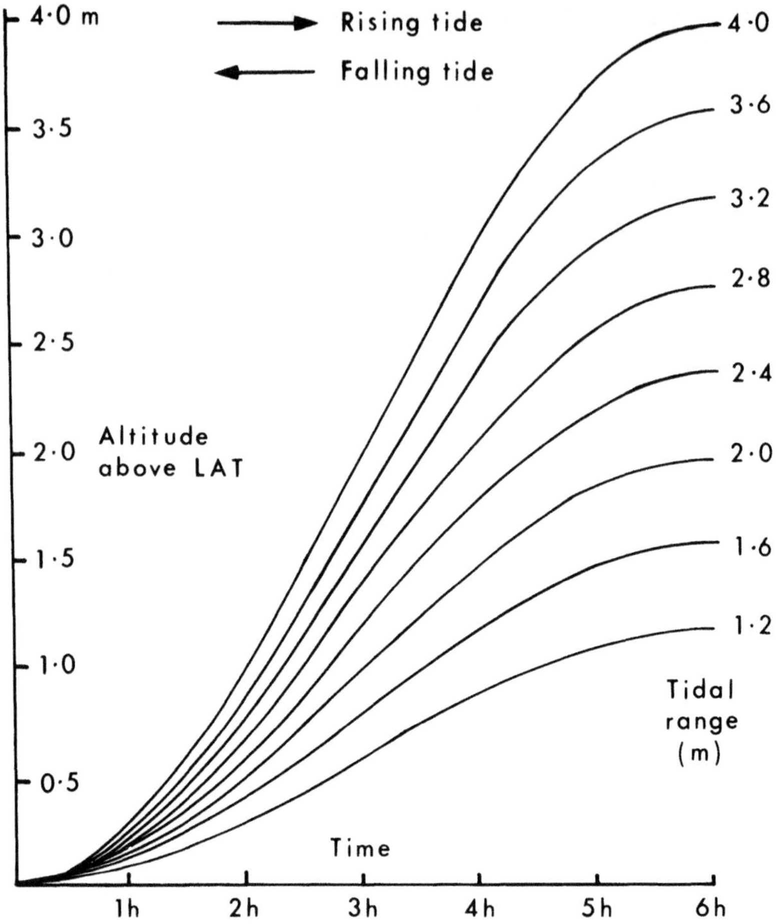

Figure 3. Modified version of Figure 11 in Åse (1978), adapted with permission. For use of the diagram see pages 579-581.

A modified version of the diagram is reproduced in Figure 3. As an example of its use, at a location included in the tide tables, where on a given day the predicted tidal range was 3.2 m and the time of measurement of the surface of the sea was 4 hours after or 4 hours before the predicted time of lowest position of the tidal waters, the position of low water (equivalent to the chosen datum-level) would be approximately 2.5 m below the observed level of the surface of the sea.

It is suggested that, in the cases considered by Åse (1978), for purposes of comparison with other areas it would have been at least as useful, and probably more useful, if the datum-level selected had been MTL rather than that of the lowest astronomical tide (LAT). An approximate value of (predicted) MTL in the area concerned may be obtained from the information included in Åse's graph. On the assumption that MTL occurs at a level midway between that of successive high and low tide levels (a reasonable assumption where flood-tide and ebb-tide intervals are equal at six hours), in the example given above where tidal range on the relevant day was taken to be 3.2 m, the (predicted) position of MTL would be 1.6 m above low tide level. In the same example, Åse's graph suggested that, 4 hours before or 4 hours after the predicted time of low tide, the observed level of the surface of the sea would be *c*. 2.5 m above low tide level. It follows that MTL would be *c*. 0.9 m below the observed level of the surface of the sea at the time of observation.

In the more general case of relating a datum-level at a Key TBM to local MTL, the question arises as to what is the best procedure. If biological tide-level markers (such as *Fucus vesiculosus*) are present, it is tempting to use these, despite the doubts attached to their precise relationships with positions of the tide. Ideally, the installation of a short-term tide-gauge station in the study area will allow at least reliable calculation of local MTL and, perhaps, reasonably reliable calculation of local MSL on the basis of observations extending throughout the length of a field season (see above, pages 571-573). If it is possible to install a suitable tide gauge, this method certainly is recommended. If the installation of a tide gauge is impossible, the following procedure is suggested.

At a chosen shore location situated close to the Key TBM, successive positions of high tide and low tide marks in relation to the datum-level at the Key TBM are recorded over as long a period as possible, preferably over a minimum of 15 consecutive days (so that measurements obtained in both spring and neap tidal conditions are included). Averaging of the values so obtained gives an acceptable indication of the altitude of the datum-level at the Key TBM above local MTL.

The methods adopted in the recording of the positions of high tide and low tide will vary from one location to another, depending on a number of factors, including the nature of the shore and the ingenuity of the operator: paint marks may be suitable on rocky shores, lines of pegs (possibly quite short along the line of high tide, but certainly markedly longer along the line of low tide) on sandy shores. Experience shows that it is more difficult to record and measure positions of low tide than positions of high tide. This is because, while measurements are being made in the former situation, the water level is rising and markers showing the level of low tide either are being covered or their stability is being threatened. Also, in the case of (concave) crescentic shores, the low water line will be shorter (frequently much shorter) than the high water line. In such cases, for any individual low tide or high tide event, it may be possible to average measurements obtained from as few as three points on the low water line whereas, on the same shore, it may be possible to average measurements obtained from as many as thirty (or more) points on the high water line.

In the case-study in Kenya, described by Åse (1978), the procedure described two paragraphs above could have been adopted without the necessity of published tide tables being used. A disadvantage of this procedure compared with that adopted by Åse is that the procedure given above would have required observation of the level of the surface of the sea at successive intervals of approximately six hours over a period of at least 15 days (a very demanding commitment in a field situation) whereas Åse's method allowed observations of the level of the surface of the sea to be made at any convenient time. An advantage of the method described above is that it may be applied in virtually any coastal situation without the necessity of having access to tide tables for the area concerned. It has the additional advantage of being based on real rather than predicted positions of the level of the surface of the sea. Because this method may be applied in most coastal situations and because it relates datum-levels located in a given area to a 'standard' reference datum-level, MTL, it allows observations made in one area to be compared with those made in other areas. Obviously, if tide gauges are used and it therefore is possible to relate the relevant datum-levels to local MSL, comparisons with other areas will be on an even more reliable basis than where local MTL is the reference datum-level.

COMPARISON OF NATIONAL REFERENCE DATUM-LEVELS

In the above discussion of selection and use of reference datum-levels, two alternative situations were considered. In one situation, where relevant sites were not located close to a surveyed bench mark, it was recommended that altitudes of SLIs be measured in relation to temporary bench marks the altitudes of which, in turn, be measured in relation to local MTL, which is approximately equivalent to local MSL. In the other situation, where relevant sites were located within surveying distance of a 'permanent' bench mark, it was recommended that altitudes of SLIs be measured in relation to temporary bench marks, hence to a permanent bench mark and, in turn, to the relevant national reference datum-level. To place the results obtained by either of these recommended methods within an international context, the relationships between individual national reference datum-levels and the corresponding local MSLs, and the relationships among various national reference datum-levels are considered briefly.

Taking northwestern Europe as an example, in the case of the majority of the national reference datum-levels, the present altitude corresponds exactly with the relevant local MSL (Table 1). In theory, therefore, altitudes determined in any one country of this region may be compared directly with altitudes determined in another country of this region. As discussed elsewhere, however (Jardine, 1976, p 79), the mean surfaces of the North Sea and the Irish Sea do not conform precisely to the geoid and, for this reason, it may be preferable to use another suitable reference datum-level, such as the international geodetic reference 'plane' for

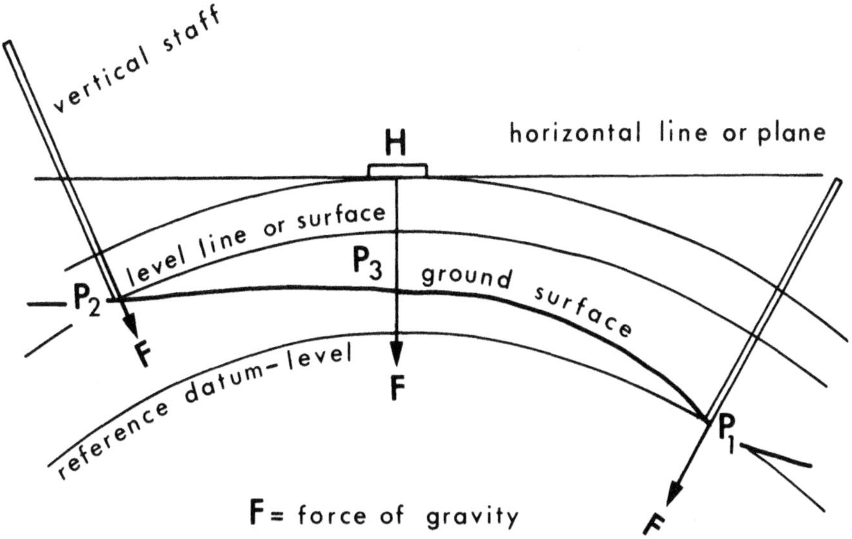

Figure 4. Diagrammatic representation of the relationships between a horizontal line or plane and 'level' lines or surfaces at points on the surface of the earth. P_1 is the position of a (local) reference datum-level. Horizontality is defined by spirit level = surveyor's level (H), located vertically above point P_3. Vertical scale, and therefore curvature of the surface of the earth, greatly exaggerated. Adapted, with permission, from Ritchie *et al.* (1977, Fig. 66).

Europe, Normaal Amsterdams Peil (NAP), for comparisons of altitudes determined in more than one national territory. As shown in Table 1, the altitudes of NW European national reference datum-levels vary considerably in relation to NAP. If NAP is to be used as the reference level, altitudes determined in countries other than The Netherlands should be recalculated in relation to NAP before comparisons are made.

VERTICAL DISTANCE, ABOVE OR BELOW A REFERENCE DATUM-LEVEL

The essential problem involved in the determination of altitude is that of measuring the vertical distance between two points (in the present context, one located on a reference datum-level, the other located on a SLI) that are not situated so that one point lies directly vertically above the other. The problem is illustrated in terms of geodetic theory in Figure 2 and in simpler, surveying terms in Figure 4. From the latter figure it may be seen that a horizontal line or plane as defined by a surveyor's level at a point, although at right angles to the vertical, does not coincide with a 'level' line or surface (equivalent to an equipotential surface in Fig. 2), which 'is one at a constant distance from the centre of the earth'

(Ritchie *et al.*, 1977, p 111). The altitude of a point such as P_2 in Figures 2 and 4 may be defined *either* as the vertical difference between tangent planes to the equipotential surfaces E_1 and E_2 (Fig. 2) *or* the vertical distance between the 'level' line or surface passing through P_2 and the reference datum-level passing through P_1 (Fig. 4).

Several different types of instrument have been used in the field to measure the altitudes of SLIs, notably: the aneroid barometer; the hand-held (Abney) level; the rod-supported (Abney) level in conjunction with a graduated surveying staff; the surveyor's level and staff (*cf.* Gray, 1975, pp 11-15). Of these, the aneroid barometer and the hand-held level do not give sufficiently accurate results for their use to be considered seriously in modern determinations of altitudes in studies concerned with changes of SL. The rod-supported level used in conjunction with a surveying staff gives somewhat more accurate results but this method is not recommended either. The only method that is sufficiently reliable to be recommended here is the use of a surveyor's level (preferably of the 'automatic' type) in combination with a surveying staff graduated in metres, tenths of a metre (100 mm) and two-hundredths of a metre (5 mm).

The methods involved in using the combination of surveyor's level and staff in terrain of either regular or irregular topography, and the precautions to be taken to minimise operational errors, are discussed in detail in standard text-books such as Ritchie *et al.*, (1977, especially pp 110-129). It should be noted that, to obviate the problem mentioned above in the first paragraph of this section - non-coincidence of a horizontal line or plane and a 'level' line or surface (see also Fig. 4) - in field practice the distance between surveyor's level and staff should never exceed 150 m and normally should be markedly less than 150 m. It also should be noted that, 'all traverses should be closed with an error of less than 0.15 m ... Traverses with errors larger than this should be repeated; a task which can be minimised by the creation of frequent temporary benchmarks during the original survey, so that in the case of a re-survey only these need be checked and the section of the traverse with the error located.' (Rose, 1981, p 327; *cf.* Ritchie *et al.*, 1977, p 126).

Additional problems are encountered in the determination of the altitude of a (former) SLI where the indicator is buried below ground level at the relevant location. In such cases, errors in the determined altitude that are too great to be ignored may occur, especially should the SLI be located on a steeply-sloping (buried) surface (Fig. 5). Generally, buried SLIs have been discovered by coring operations. The method commonly employed, therefore, in the determination of the depth of the SLI below ground level is to measure the length of the part of the coring device that lies below the ground surface. Clearly, great care must be taken to ensure that the coring device is

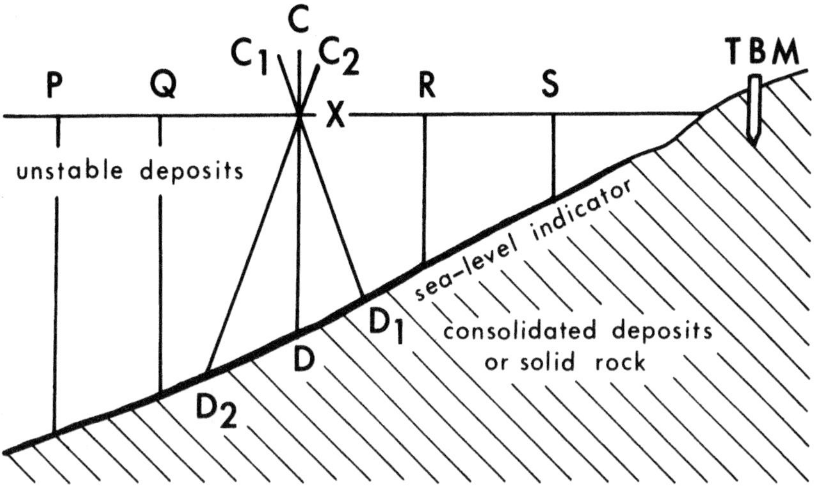

Figure 5. Diagrammatic representation of problems related to accurate determination of the vertical distance (altitude) of a sea-level indicator located on a sloping surface buried below ground level. If the coring device, CXD, is inserted vertically, the distance XD gives the true depth of the indicator below the ground surface at X. If the coring device, C_1XD_1 or C_2XD_2, is inserted at an angle to the vertical (tilt greatly exaggerated in the diagram), the distance XD_1 or XD_2 gives an incorrect value for the depth of the sea-level indicator below the ground surface at location X. The approximate gradient of the surface on which the sea-level indicator lies may be given by information obtained in boreholes sunk at locations such as P, Q, R and S. Measurements of altitude in an area such as that illustrated should be referred initially to a temporary bench mark (TBM) established on the stable ground shown on the right of the diagram. Not to scale.

aligned vertically rather than tilted; even a small angle of deviation from the vertical will produce a marked error in the assessment of vertical distance should the depth of the base of the coring device below ground level be greater than a few metres. Machine-operated coring devices encased in rigid metal casings of 150 mm diameter or greater may be established and retained in a vertical position much more readily than hand-operated, non-encased metal augers and plastic coring tubes. If it is suspected that the coring device has not been inserted precisely vertically, and the angle of tilt can be determined (even approximately) by measurement above the ground surface, it may be possible to calculate trigonometrically the probable error involved, should the (approximate) slope of the SLI be known on the basis of information derived from a series of (truly) vertical boreholes (Fig. 5).

In passing, it may be noted that in a situation such as that illustrated in Figure 5, where the consistency of the deposits overlying the former SLI may be such that the surface level at the borehole site is liable to change in the course of boring operations, it is advisable to establish a temporary bench mark (TBM) on the stable ground adjacent to the area of operation at the outset of the investigations. All measurements of altitude at the borehole site should be referred to this TBM until such time as they may be referred to a more permanent reference level (see above, page 576).

ALTITUDE DETERMINATION FROM A RESEARCH VESSEL

In the discussion above it has been assumed that the sites of relevant SLIs are located on land, or offshore at a distance less than the recommended maximum between surveyor's level and surveyor's staff (ie. 150 m, see last section above). In the case of a more-distant site, where the (former) SLI is located on the present sea floor or below the sea floor, in addition to the problems associated with accurate position-fixing of the site (not considered here), several problems arise in the accurate measurement of the altitude of the SLI at the site.

In practice the relevant SLI may be located in an offshore borehole, in which case the level of the indicator would be determined initially by combining two individual procedures:

1. Determination of water depth below a temporary reference plane by means of echo sounding (accuracy *c.* ± 2 dm) or, in rare cases, by the use of a plumb line.
2. Measurement of the length of the sedimentary core below the sea floor (accuracy *c.* ± 1 dm).

A further problem requires to be considered. The procedures mentioned above give the position of the former SLI below a reference plane or datum-level that is normally located on a floating vessel. A major problem is the determination of the altitudinal relationship between this datum-level (which is changing its vertical position continuously throughout the tidal cycle) and a stationary reference datum-level on land. The reference datum-level on land most commonly used is Chart Datum (Hatfield, 1969, p 14) or the slightly higher datum, (local) mean low water of spring tides (MLWS) (H. Streif, pers. comm.).

The standard procedure to be followed in the conversion of a depth value obtained in relation to an offshore 'floating' datum-level to an altitude related to a reference datum-level on land is too lengthy to be given in detail here. (It is to be found in handbooks such as Hatfield, 1969, pp 25-32.) Also, seldom is it possible for the investigator of former SLs to follow the standard procedure fully because frequent depth sounding may be impracticable (H. Streiff, pers. comm.). It is recommended, however, that whenever possible the standard procedure be followed so that the determined altitudes of offshore former SLIs

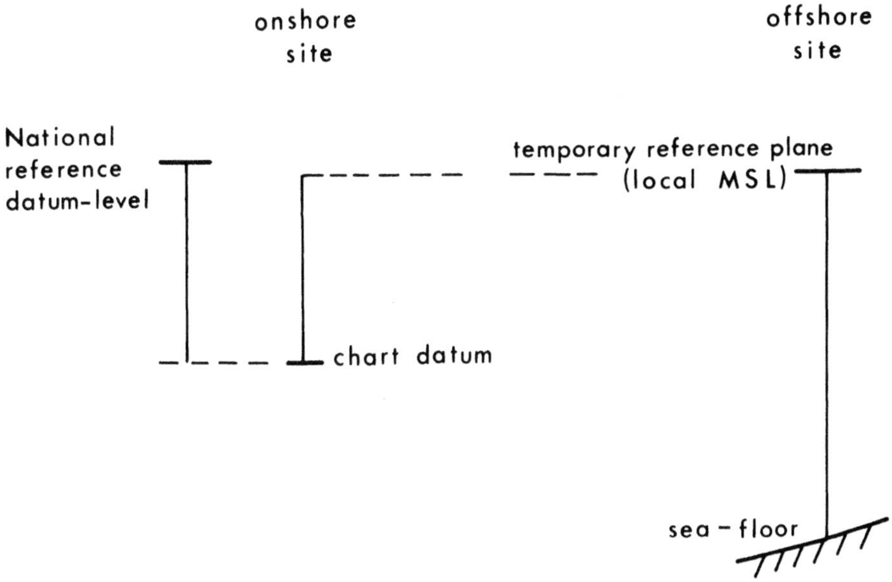

Figure 6. Greatly simplified representation of relationships between levels involved in measurement of altitudes of sites located offshore. Vertical scale greatly exaggerated. It should be noted that: 1. Generally the chart datum used by a survey or research vessel is that of the port from which the vessel operates. 2. With few exceptions, each national reference datum-level originally was equivalent to MSL in the vicinity of the national reference site.

may be as reliable as other conditions allow. In summary, the procedure involves the following operations (*cf.* Fig. 6):

1. Frequent depth-sounding at the water surface directly above the offshore site for at least the period from one high water to an adjacent low water (or, preferably, the period covering two consecutive high water and two adjacent low water positions).
2. Simultaneously with the observations made in 1 above, depth observations at an onshore site where a chart datum has been established previously.
3. Plotting of results on graph paper, leading to altitude determination in relation to the relevant chart datum.
4. Conversion of altitudes relative to the chart datum to altitudes relative to the appropriate national reference datum-level.
5. Where appropriate, conversion of altitudes to those relative to NAP (Table 1) or another international reference datum-level.

A considerably simplified (and correspondingly less accurate) procedure than the standard procedure is for the water depth at the relevant offshore site to be determined at the time when (by consulting tide tables) the level of the surface of the sea at that site is considered to coincide with the level of MLWS at the location on land where the relevant reference datum-level is sited. The sum of the water depth thus obtained and the depth of the SLI in the sedimentary core below the sea-floor gives the approximate altitude of the indicator below MLWS at the relevant site on land and, by further calculation (see above, pages 578-582), the altitude of the SLI in relation to a standard reference datum-level may be obtained.

Where altitudes have been determined at sea, the methods used in depth measurement at sea, together with the method used in conversion to altitude related to a land-sited reference datum-level, should always be stated.

ACKNOWLEDGEMENTS

Constructive comments by W.S. Newman, O. van de Plassche and H. Streif were much appreciated and have led to considerable improvement of the text. P. Vaníček's comments and B.D.F. Methley's advice on geodetic aspects have saved the author from embarrassment. A. Morrison kindly read several versions of the text, and Mrs. S. Hall drew the diagrams. Permission to modify and adapt previously published diagrams is gratefully acknowledged (see individual figure captions).

REFERENCES

Åse, L.-E., 1978. Preliminary report on studies of shore displacement at the southern coast of Kenya. *Geografiska Annaler,* 60A, 209-221

Clark, D., 1958. *Plane and geodetic surveying for engineers. volume one. plane surveying,* 5th edition, (Constable, London)

Cook, A.H., 1972. The geoid. *Earth-Science Reviews,*8, 13-44

Donner, J.J., 1959. The Late- and Post-glacial raised beaches in Scotland. *Annales Academiae Scientiarum Fennicae,* A, III, 53, 25 pp

Donner, J.J., 1963. The Late- and Post-glacial raised beaches in Scotland II. *Annales Academiae Scientiarum Fennicae,* A, III, 68, 13 pp

Doodson, A.T. & Warburg, H.D., 1941. *Admiralty manual of tides.* (His Majesty's Stationery Office, London)

Faure, H., Fontes, H.C., Hebrard, L., Monteillet, J. & Pirazzoli, P.A., 1980. Geoidal change and shore-level tilt along Holocene estuaries: Sénégal River area, West Africa. *Science,* 210, 421-423

Gordon, D.L., 1957. Mean Sea Level around the British Isles. *The Chartered Surveyor,* 90, 28-33

Gray, J.M., 1975. *Measurement and analysis of Scottish raised shoreline altitudes.* (Department of Geography, Queen Mary College, London, Occasional Papers No.2)

Hatfield, H.R., 1969. Tides and tidal streams, Chapter 2 in *Admiralty Manual of Hydrographic Surveying, Volume Two,* (The Hydrographer of the Navy, Taunton)

Jacobsen, N.K., 1964. Traek af Tøndermarskens naturgeografi med saerligt henblik på morfogenesen. *Folia Geographica Danica,* 7, 350 pp

Jardine, W.G., 1976. Some problems in plotting the mean surface level of the North Sea and the Irish Sea during the last 15,000 years. *Geologiska Föreningens i Stockholm Förhandlingar,* 98, 78-82

Jardine, W.G., 1982. Sea-level changes in Scotland during the last 18,000 years. *Proceedings of the Geologists' Association,* 93, 25-41

McCann, S.B., 1964. The raised beaches of north-east Islay and western Jura, Argyll. *Transactions of the Institute of British Geographers,* 35, 1-16

McLean, R.F., Stoddart, D.R., Hopley, D. & Polach, H., 1978. Sea-level change in the Holocene on the northern Great Barrier Reef. *Philosophical Transactions of the Royal Society of London,* A, 291, 167-186

Ritchie, W., Tait, D.A., Wood, M. & Wright, R., 1977. *Mapping for field scientists.* (David & Charles, Newton Abbot)

Rose, J., 1978. Glaciation and sea-level change at Bugöy-fjord, south Varangerfjord, north Norway. *Norsk Geografisk Tidsskrift,* 32, 121-135

Rose, J., 1981. Raised shorelines. In: *Geomorphological Techniques,* ed. A. Goudie, (George Allen & Unwin, London), 327-341

Rossiter, J.R., 1967. An analysis of annual sea-level variations in European waters. *Geophysical Journal of the Royal Astronomical Society,* 12, 259-299

Sissons, J.B., 1967. Comments on the paper by F.M. Synge and N. Stephens in Transactions No. 39. *Transactions of the Institute of British Geographers,* 42, 163-168

Sollid, J.L., Andersen, S., Hamre, N., Kjeldsen, O., Salvigsen, O., Sturød, S., Tveitå, T. & Wilhelmsen, A., 1973. Deglaciation of Finnmark, north Norway. *Norsk Geografisk Tidsskrift,* 27, 233-325

Streif, H., 1979. Holocene sea level changes in the Strait of Malacca. In: *Proceedings of the '1978 International Symposium on Coastal Evolution in the Quaternary',* eds K. Suguio, T.R. Fairchild, L. Martin & J.-M. Flexor, (Sao Paulo), 552-572

Suthons, C.T., 1960. The computation of Mean Sea Level. *Bulletin Géodésique,* 55, 65-68

Synge, F.M. & Stephens, N., 1966. Late- and Post-glacial shorelines, and ice limits in Argyll and north-east Ulster. *Transactions of the Institute of British Geographers,* 39, 101-125

Torge, W., 1980. *Geodesy.* (Walter de Gruyter, Berlin)

Vaniček, P., Castle, R.O. & Balazs, E.L., 1980. Geodetic leveling and its applications. *Reviews of Geophysics and Space Physics,* 18, 505-524

22

Compaction and consolidation

J. T. GREENSMITH and E. V. TUCKER

ABSTRACT

Gravitational compaction and consolidation factors are briefly reviewed and discussed in relation to subsidence and estimating sea level. Most coastal Holocene regions show uneven subsidence, this being controlled by variability in the thickness, lithofacies and diagenetic and burial history of the successions. Tectonism is also influential. Broad quantification of compaction and subsidence is possible in uniform successions but only over limited geographical areas.

INTRODUCTION

In all work connected with sedimentary successions, whether dealing with the foundations of engineering structures, reclamation projects or more academically oriented considerations of relative sea-level (RSL) change, the assessment of the magnitude and effect of compaction and consolidation is of prime importance. It has been suggested (Boswell, 1961) that certain peat seams can progressively reduce to 10% of their original thickness as a consequence of the continuance of compaction processes, and clayey muds to 11-25%, sands to 66-75% and calcite muds to 50%, but these are maximum figures unlikely to be attained in coastal situations over a geologically short period of several thousand years. When the clay percentage in a sediment is high, eg. 50-60%, and in the presence of organic matter, which tends to increase with clay content, the reduction in thickness of a given layer by compaction may be as much as 60% over a period as short as 100 years, but this is only likely to occur under particular conditions. De Glopper (1973) records such values in some drained topsoils on the former site of the Zuiderzee. In the absence of human interference where thick beds of clayey mud are concerned, say 30 m thick, and even with an overburden subsequently amounting to 300 m, the reduction may only amount to about 30%. Reductions of 50-60% normally require overburdens greater by a factor of 2-3 (x300 m), something unattained in Holocene coastal areas (Weller, 1959; Brown, 1975).

The net result of thickness reduction, the bulk of which probably occurs within the first few thousand years after deposition, is expressed by subsidence or surface lowering. In general, over a given region, this type of subsidence is likely to be uneven or differential being controlled, for example, by sediment type and thickness, natural and man-induced drainage (fluid withdrawal) and altitude and form of the substrate. Subsidence due to drainage is frequently observed in reclaimed ground protected by seawalls and can amount, over a period of 200-400 years to several metres, carrying downwards below contemporary mean high water (MHW) mark a deposit laid down originally at or slightly above MHW mark. This is subsidence essentially due to a drying-out series of processes and is much faster than when a sediment remains permanently wet as it may well have done, and even continue to be, in some coastal situations.

In assessing the original altitude of dated and undated marker levels eg. peat seams, cheniers in Holocene successions and relating such to contemporary SLs, subsidence and its contributory factors should be taken into account. Yet, there is no field of knowledge where the problems of this assessment seem so difficult, almost intangible and beset by innumerable variable factors that will significantly influence conclusions drawn. There is a considerable body of work on the compaction and consolidation of argillaceous sediments, particularly in moderately thick, relatively uniform successions (see Rieke & Chilingarian, 1974, for extensive bibliography). The quantitative work of Skempton (1970) on Quaternary and older clays is a prime example as is the more critical appraisal by De Glopper (1973) of near-surface Holocene deposits in The Netherlands. There appears to be less on the compaction of mixed lithological sequences, eg. sand-silt-clay such as commonly occur in present coastal areas (Allen, Chilingar, Mayuga & Sawabini, 1971; Chilingarian & Wolf, 1975). Despite all this work which includes attempts to analyze and deduce subsidence rates due to compaction (eg. Terzaghi & Peck, 1967; Sandhu & Wilson, 1969), it is clear that, until more field data are collected and laboratory experiments carried out for a wide range of lithological successions, all assessments should be regarded as tentative and approximate (Marsden & Davis, 1967).

It is generally recognised that the term consolidation, as used by civil engineers, is synonymous with the physical process of gravitational compaction (usually abbreviated to compaction) as used by geologists (Gill & Lang, 1977). In this article compaction, as understood by geologists is preferred and is defined as the slow expulsion of pore fluids and the reduction of voids within a sediment prism as a result of stresses due to overburden load. The ensuing reduction in volume of the prism almost wholly takes place in a vertical direction. Void reduction is dependent on a range of factors including lithology, mineral content, particle size and shape, water content, pore-water pressure, liquid limits, drainage, rate of loading, maximum effective stresses and initial thickness, and occurs more readily in organic-rich argillaceous sediments than in sands and sandy gravels.

Consolidation, in the geologist's sense, refers to a variety of processes including not only compaction but also cementation and recrystallisation. The last two involve chemical precipitation and mineral reorganisation by the activities of migratory interstitial fluids, often under enhanced pressure and temperature conditions. Certain clay muds, for example, may be cemented relatively quickly by aragonite, calcite and amorphous silica and become much less susceptible to compaction than hitherto.

In this non-mathematical article a brief review of the main factors which need to be taken into account when establishing the rate and total amount of natural compaction in Holocene sedimentary successions is given (see Terzaghi & Peck, 1967 and Ragavan & Miller, 1975, for mathematically-based discussions). Compaction due to human activities, such as occurs during the artifical drainage of reclaimed ground and during the extraction of water, gas and oil from underground reservoirs (Geertsma, 1973), is clearly of considerable local, if not regional, importance and is subsumed in the general discussion on natural processes. In assessing the degree of subsidence in any area or region compaction is only one factor and, indeed, it may be subordinate to tectonic (including earthquake) and isostatic factors, though this is difficult to demonstrate in practice (Greensmith & Tucker, 1980).

The instruments used for determining compaction, such as consolidometers, oedometers, filter presses, piston and triaxial apparatus are the subject of many texts (eg. Capper & Cassie, 1969) and are not discussed here.

FACTORS RELEVANT TO COMPACTION AND CONSOLIDATION

1. Sediment thickness and variability

The sub-Holocene topography of most marine basins is commonly irregular, a situation inherited from events occurring within Pleistocene times or even earlier (Greensmith & Tucker, 1976; Ente, 1976). Marked changes in base-level, and periodic glacial and fluvial incursions in certain regions, have together contributed to the uneven relief, a condition which frequently persists through to the present day. The total thickness of Holocene successions is thus very variable, changing laterally from a 'feather-edge' thickness, where they abut more ancient deposits, to several tens of metres (>100 m in some basins, especially at the sites of buried channels). Onto this lateral variation in thickness is superimposed lithological variation, a function of the changes in the spatial distribution and migration of lithofacies with the passage of time. Regressive sequences of sand, silt, clay and peat are interspersed with transgressive sequences of peat, shelly sand, silt and sand both on a local and regional scale. Moreover, individual layers of sediment show, in detail, considerable internal variability, such as in calcareous shell content, proportion of degraded plant matter (organic C), clay mineral species, cementing agents and post-depositional disturbance by organisms. The amount of compaction and consolidation will therefore depend not only on the total thickness of the succession but also on the detailed physical character of the sediments comprising the succession, especially the proportion of more readily compactible materials.

In a succession dominated by clay-deficient quartz sand deposits, resting on a substrate of Pleistocene sands and gravels, the amount of subsidence due to compaction is almost certainly low. Van Straaten (1965) discussed coastal barrier sand sequences with thin intercalations of silty clay near Scheveningen and IJmuiden (The Netherlands) and estimated lowering by compaction to be of the order of 0.2-0.4 m. Providing estimates such as these are accurate then reasonably accurate SL determinations are feasible. But, if little account is taken of the substrate then the estimates may be misleading. Van de Plassche (1980) discarded results from the same area on the basis that subsidence, at least locally where sand deposits rested on several thick clay beds, may have exceeded 0.6 m.

2. Discontinuities in sedimentary successions

The processes of sediment accumulation in any area may ensure either regular, unbroken deposition or irregular, interrupted deposition. The latter is the more usual geological situation. Depositional rates may be as little as 2 m per 1000 years, as in some estuarine areas, or as much as 120 m per 1000 years in some deltaic mud areas (McClelland, 1967; Skempton, 1970, Table 1). When deposition is slow the chances are that the sediment, providing it is uniform in texture and composition, will compact normally and evenly. When deposition is irregular the rates of accumulation are likely to be variable and deposition in certain areas may be matched by erosion or non-deposition in others nearby. Compaction is likely to be uneven. Discontinuities in Holocene marine successions can be marked by sharp changes in lithology, such as washover shell sands resting discordantly on marsh silts, or subtidal pebbly sand lenses resting on tidal flat muds. Peat or plant-rich layers and human occupation surfaces can also be regarded as discontinuities due to a change in sub-environmental circumstances (Hacquebord, 1976; Greensmith & Tucker, 1980). If periods of subaerial exposure are initiated synchronously with the interrupted deposition it is usual for physico-chemical changes to begin to affect the upper layers of the exposed sediment. Desiccation, leaching and cementation are three such changes, all capable of altering the compaction characteristics of that particular sediment. For example, the shear strength (cohesion) of soft plastic clay in the Thames estuary, England is normally 0.14-0.20 kg/cm^2, but frequently rises to about 0.70 kg/cm^2 as a consequence of these processes (the local consolidated Eocene London Clay has values of 1.08-1.62 kg/cm^2). In coastal and fluviatile marsh environments with well developed creek systems it is usual for levee deposits to desiccate and drain much more readily than adjacent deposits. Compaction of levee deposits in the Lake IJssel region extends to depths of 2-3 m below the surface (Pons & Zonneveld, 1965; Ente, 1971). If the levee or other subaerially compacted deposits are subsequently incorporated into the succession then levels of differential compaction are created, and their effect is difficult to assess with any degree of refinement when calculating overall compaction (Greensmith & Tucker, 1971). Moreover, the situation could then arise whereby dateable (human occupation) material located on a levee surface actually correlates with a contemporaneously produced peat seam, now located at a markedly lower altitude.

Some fine silts and clay-rich sediments lose strength and compact readily on re-wetting, something frequently likely to occur in a coastal situation either due to short-term storm surges or long-term events, and the consequent subsidence caused by this hydrocompaction mechanism can amount to 4-6 m per 30-40 m thickness of strata. Any inter-granular strength of the sediments due, for example, to incipient calcite cementation, can also be lost during the re-wetting.

3. Effects of loading

Natural overburden pressures are imposed by the weight of the sediment column. In clays, effective overburden pressure is determined by measuring pore pressures using piezometers and establishing sedimentation compression curves, which take into account whether the sediment is normally-consolidated (no load pressures greater than those existing have been applied) or over- or under-consolidated (existing pressures less than or greater than those previously applied)(see Terzaghi, 1941 and Skempton, 1970 for details).

The chronology and time-interval of successive events affecting any sediment accumulation must, however, influence the reaction of that accumulation to loading. Deposition may be fast, uniform and continuous, in which case it may be possible to crudely calculate the amount of subsidence due to loading over a designated period. Results obtained during laboratory tests indicate that, under artificially applied pressures of 0.54-1.62 kg/cm^2, Holocene coastal silty clays compact much more, by a factor of 5-6, than freer draining silty sands. At these predetermined pressures, and assessed over a duration of one hour only, the settlement of clays can amount to 0.7 mm, whereas with sands it can amount to 0.15 mm. Unfortunately, the various formulae applied in soil mechanics are mostly derived from short timescale observations either from laboratory tests (a few hours) or in the field (1-3 years), and therefore look unreliable. Whether extrapolation of these formulae to periods of several hundreds or thousands of years is feasible is unknown, or even whether the formulae are valid for layers some tens of metres thick.

In general, deposition in most coastal regions over periods of more than a few thousand years is likely to be irregular, non-uniform and discontinuous, with periods of non-loading interspersed with periods of loading. Changes in lithology and loading characteristics will ensue. Unloading during erosive phases could result in elastic rebound, though this is probably insignificant quantitatively in unconsolidated or poorly consolidated materials. When a 4 m thick layer of sand was removed from over a compressible layer in the IJsselmeerpolder area rebound was assessed at 2-3 cm, small compared with the previous compaction of almost 70 cm in 3 years.

4. Sediment characteristics

The physical and chemical character of a sediment determines the way in which it will behave when loaded. A few general examples only need be quoted. A siliciclastic mud will react differently to loading than will a carbonate mud, because of basic differences in the physical and chemical properties of the component minerals and the ways in which those minerals react to stress conditions. Although siliciclastic grains (quartz, feldspar, clay minerals) are susceptible to solution and reprecipitation processes under loading, they are much less so in these respects than aragonite, calcite and dolomite. The fabric of siliciclastic deposits is much more likely to be affected by physical re-orientation and closer packing of component particles in early diagenesis than carbonate minerals, in which solution and cementation processes are relatively more prominent and frequently faster acting. Clay-rich sands will generally compact much more readily than clay-deficient, 'clean-washed' sands because the clay particles, though inelastic, can be squeezed, bent and physically relocated in between the sand grains (Meade, 1966). On the other hand, it has to be appreciated that a small amount of plastic fine material can increase the strength and reduce the rate of compaction by increasing the cohesion of the sand. As a further contrast, well-sorted, 'clean-washed' sand may be cemented much more readily and quickly than less permeable muddy sand and become less compressible and subject to liquefaction (Ingles & Grant, 1975).

Early diagenetic cement precipitation is not normally considered an important factor in assessing the consolidation of Holocene sediments, particularly those of argillaceous type, yet the geological record indicates that early precipitation of a range of inter-granular carbonates, silica in various guises, and iron oxides, hydroxides and sulphides is the rule, even in argillaceous rocks (Ho &

Coleman, 1969; Sangrey, 1972). The location of these chemical precipitates within the sediments depends on factors such as the microstructure, grain orientation, grain size and sorting, the porosity and permeability, and the geochemistry and activity of the interstitial or pore waters.

a) Clay In clay-dominant sediments the proportion and type of clay mineral and the primary microtexture of the clay-mineral aggregates, eg. honeycomb texture in kaolinite-rich sediments, can have a very marked effect on the progress of compaction. Most grains are capable of adsorbing and absorbing water and other materials, as particularly well expressed by the montmorillonite (smectite) group, and most are capable of ionic exchange. Water adsorbed around the edges of discrete particles can be removed readily by compaction and desiccation, whereas absorbed water between unit cells cannot and requires higher pressures and temperatures for removal. In montmorillonites with Na^+ as the exchangeable ion, the interlayer water appears to be about 3 molecular layers thick; in the Ca^+ montmorillonites the thickness is 4 molecular layers. This suggests that any clay sediments carrying clay minerals dominated by montmorillonites will follow a different compaction and consolidation path from those dominated by illites and kaolinites, which generally have fewer water interlayers. It has been demonstrated by Lomtadze (1954) that a bentonitic clay can retain more than 20% water when a pressure of 3000 kg/cm^2 is applied, whereas a kaolinitic clay has a water retention measurable in single figures. Clays of such small moisture content become very complex systems and their behaviour under compression is difficult to assess or explain in terms of simple bonding mechanisms (Williamson, 1974).

The moisture content of sediments is very diverse. On deposition in a marine environment the content in the upper levels of the sediment may be as high as, or higher than, 200%, with void ratios of 5.7. In argillaceous deposits of estuarine and intertidal origins the moisture content in some instances approximately equals the liquid limit. The withdrawal of that moisture either as a consequence of squeezing by loading or by ground-water pumping, with a lowering of the water table, will usually lead to significant changes in porosities and void ratios, and subsidence will ensue. In coastal Essex, England, Holocene plastic clays down to a depth of 15 m sustain relatively high void ratios of 1.5-2.0, these comparing with associated silty sands at about 0.9. In general, for a given uniform lithology, the porosity and void ratio will diminish with depth. Void ratios in the silty clays of Essex vary between 0.7 and 2.1 at depths up to 6 m, decreasing to 0.4-1.1 at depths between 15-20 m. Poorly sorted silts tend to have lower void ratios than sands. In contrast, well-sorted silts have larger void ratios than sands, because the grains in silts are usually more angular and do not interlock as tightly as sand-size particles.

The behaviour of marine and fresh-water, post-glacial, quick or sensitive clays, which lose strength on disturbance (shock loading) cannot be completely ignored in evaluating consolidation in successions peripheral to intensely glaciated regions, though none are known in Britain. In the rapid transformation from firm solids to viscous liquids there appears to be a collapse and marked re-organisation of the microfabric of the clay and quartz particles, which significantly alters pore space dimensions and void ratios (Pusch, 1966; Cabrera & Smalley, 1973; Moon, 1978).

b) Peat The accumulation of plant debris in quantity during sedimentation and its ultimate incorporation and degradation within the succession to form intercalated peat seams, has inevitable repercussions on compaction calculations. The texture of natural peats is extremely variable from coarse fibrous at one extreme to amorphous granular at the other (Berry & Poskitt, 1972; Chappell & Burton, 1975). This probably reflects differences in the nature of the plant constituents, environments of accumulation (eg. bog peat, drifted peat, *Phragmites* coastal marsh peat), rates of accumulation and rates of plant degradation both subaerially and subsequently, after incorporation into the succession. Coarse fibrous peat has an open structure and natural water content as high as 1000%. It is highly compressible and the particles are subject to viscous creep and deformation under loading. The pore water is free, not absorbed, and tends to move from the micropores into the macropores as compression increases and permeability decreases. Moreover, the water may be compressible because of the presence of dissolved gases. Granular peat has a much greater resistance to compression and has a similar rheological behaviour to clay. Primary compaction is by drainage of pore water, which is usually adsorbed around particles, from the macropores; in the later stages there is a very slow drainage of water out of the micropores.

The thickness of peat after compaction will thus be a function not only of its original thickness but also of initial and progressive changes in its internal constitution. It is generally stated that peats can be subject to a 80-90% change in thickness by compaction, and this is probably a reasonable assessment for fibrous peat but, clearly, in view of the known variability in constitution and structure and response to compaction, it would be unwise to assume that such a high figure for reduction in thickness is invariably achieved. It would be wiser to closely examine and deduce the nature of the peat before making such assessments.

It is sometimes argued that if a basal peat has been subject to loading by a thick cover of quartz sands for a period of a few thousand years, then the base of a younger peat formed on top of the sands will practically have its original altitude, since the basal peat has been compacted to such an extent that only a little extra subsidence will follow after limited additional loading. This is almost certainly correct, providing the sands intervening between the two peats have a low clay and silt content and are uncemented; their compaction is likely to be of the order of 2-6% over such short intervals. With 'clean' carbonate sand, because of early cementation, the percentage compaction may be even less.

The time span involved in the accumulation of even a moderately thick peat is considerable. The thickest peats commonly reflect (a) *in situ* sedimentation of plant materials keeping pace with subsidence and (b) the maintenance of high water tables. Radiocarbon dating of the lower and upper levels of such individual peat seams in the Thames estuary of England, approximately one metre in thickness, indicates time spans of 2300 years or more (Devoy, 1979). Unless the percentage decrease in thickness can be reasonably quantified it is probable that the altitude of the lowest dated, preferably basal, level in any given seam, providing it originated at or close to MHW mark, is a more accurate datum for judging overall subsidence within an area and locating contemporaneous SLs. This is particularly so if the peat rests directly on sands and gravels, which are less readily compactible. Where peats are very thin, a few centimetres only, the possibilitity of them reflecting a long time span is markedly reduced. It may prove impossible to date the top and the bottom, so that one date has to suffice.

CONCLUSIONS

The preceding sections have outlined briefly the real problems which beset all attempts to quantify compaction and consolidation, especially those relevant to the objectives of deducing rates of subsidence and SL change during Holocene times. In successions comprising a uniform lithology, particularly of sand type, it is probable that a reasonably broad quantification can be achieved. Even in argillaceous successions a fair assessment might be made, albeit only suitable for short-term civil engineering purposes. But the assessments can only be approximate and, for workers on SL problems thinking in terms of a time span measurable in thousands of years, essentially unsatisfactory. Mixed successions of strata with a wide range of physical attributes are more frequent in coastal marine areas and this variability ensures that a high degree of accuracy is unlikely to be attained.

Despite these negative comments, the Holocene SL worker must try to combat such difficulties and attempt to quantify compaction and consolidation using all the available evidence from the geological, geotechnical, geomorphological, biological and archaeological fields. As mentioned above, in quartz and lime sand sequences the problems are probably surmountable to give a reasonable order of accuracy to any assessment. In other sequences, it is incumbent on the workers to adopt a collaborative multi-disciplinary approach and not depend on one particular set of facts. Compaction cannot be ignored, but it is of little value to SL studies if assessments are based either on one or even a number of assumptions. It is potentially dangerous to assume, because peat seams in adjacent successions are horizontal, that the overall compaction can be ignored, unless there is strong support from a range of compatible data. It is also unwise to assess compaction over a wide area using arguments solely based on the correlation of the altitude, constitution and age of a peat seam or shell body. Plant and shell materials can be swept into deeper or shallower water than that in which they accumulated. They can form on irregular as well as even surfaces and, indeed, may themselves be the cause of the irregularity. But, if the nature of the substrate and overburden is fully comprehended and a particular interpretable lithological sequence is recognised, which has areally common physical, and hence correlatable, characteristics, then the assessment of compaction may have greater validity, although the *strong diachroneity* of such sequences has to be recognised.

It also needs to be stressed that the problems attendant on compaction and consolidation cannot be divorced from tectonic effects. Tectonic instability is widespread and even in areas assumed to be stable its influence may still be pervasive, although of low magnitude. The subsidence due to tectonic downwarping is additive to that caused by compaction and consolidation. The North Sea Basin is a prime example of a region where subsidence due to plate tectonism appears to have been persistent from at least Tertiary times through to the present day. Certain marginal areas of the Basin, eg. the outer Thames, southeastern England show probable evidence of continuing differential tectonic downwarping (D'Olier, 1972; Greensmith & Tucker, 1980). In areas such as this, the problem of what should be used as a datum point for calculating compaction, consolidation and subsidence is exacerbated by the tectonism; the altitude of surface features is bound to change. Criteria that might be used, though with great caution, to help establish a datum at particular points in Holocene time might be the base of (a) certain peat seams, (b) shelly sand marsh bars and (c) certain types of chenier all deduced, on present day distribution, to have originated close to MHW mark. Even so, these criteria are only likely to be of some value over a restricted geographical area because of the variability in the altitude of MHW over relatively short distances.

REFERENCES

Allen, D.R., Chilingar, G.V., Mayuga, M.N. & Sawabini, C.T., 1971. Studio e previsione della subsidenza. *Enciclopedia della Scienza e della Tecnica*.

Berry, P.L. & Poskitt, T.J., 1972. The consolidation of peat. *Géotechnique*, 22, 27-52

Boswell, P.G.H., 1961. *Muddy sediments*. (Heffer, Cambridge)

Brown, L.F., 1975. Role of sediment compaction in determining geometry and distribution of fluvial and deltaic sandstones. In: *Compaction of Coarse-grained Sediments, I*, eds. G.V. Chilingarian & K.H. Wolf, *Developments in Sedimentology* 18A, (Elsevier, Oxford), 247-292

Cabrera, J.G. & Smalley, I.J., 1973. Quickclays as products of glacial action: a new approach to their nature, geology, distribution and Geotechnical properties. *Engineering Geology*, 7, 115-153

Capper, P.L. & Cassie, W.F., 1969. *The mechanics of engineering soils*. (Spon, London)

Chappell, B.A. & Burton, P.L., 1975. Consolidation of fibrous peat. *Journal of Geotechnology, Engineering Division*, 100, 741-753

Chilingarian, G.V. & K.H. Wolf (eds), 1975. *Compaction of coarse-grained sediments, I*. Developments in Sedimentology 18A, (Elsevier, Oxford)

Devoy, R.J.N., 1979. Flandrian sea-level changes and vegetational history of the lower Thames estuary. *Philosophical Transactions of the Royal Society, London*, B, 285, 355-410

D'Olier, B., 1972. Subsidence and sea-level rise in the Thames Estuary. *Philosophical Transactions of the Royal Society, London*, A, 272, 121-130

Ente, P.J., 1971. Sedimentary geology of the Holocene in Lake IJssel region. *Geologie en Mijnbouw*, 50, 373-382

Ente, P.J., 1976. The geology of the northern part of Flevoland in relation to the human occupation in the Atlantic time. *Helinium*, 16, 15-35

Geertsma, J., 1973. Land subsidence above compacting oil and gas reservoirs. *Journal of Petroleum Technology*, 25, 734-744

Gill, E.D. & Lang, J.G., 1977. Estimation of compaction in marine geological formations from engineering data commonly available. *Marine Geology*, 25, M1-M4

Glopper, R.J. de, 1973. Subsidence after drainage of the deposits in the former Zuyder Zee and in the brackish and marine forelands in The Netherlands. *Van Zee tot Land*, 50, 1-205

Greensmith, J.T. & Tucker, E.V., 1971. Overconsolidation in some fine-grained sediments: its nature, genesis and value in interpreting the history of certain English Quaternary deposits. *Geologie en Mijnbouw*, 52, 193-202

Greensmith, J.T. & Tucker, E.V., 1976. Major Flandrian transgressive cycles, sedimentation and palaeogeography in the coastal zone of Essex, England. *Geologie en Mijnbouw*, 55, 131-146

Greensmith, J.T. & Tucker, E.V., 1980. Evidence for differential subsidence on the Essex coast. *Proceedings of the Geologists' Association, London*, 91, 169-175

Hacquebord, L., 1976. Holocene geology and palaeogeography of the environment of the levee sites near Swifterbant. *Helinium*, 16, 36-41

Ho, C. & Coleman, J.M., 1969. Consolidation and cementation of recent sediments in the Atchafalaya Basin. *Geological Society of America Bulletin*, 80, 183-191

Ingles, O.G. & Grant, K., 1975. The effect of compaction on various properties of coarse-grained sediments. In: *Compaction of coarse-grained sediments, I*, eds. G.V. Chilingarian & K.H. Wolf, *Developments in Sedimentology*, 18A, (Elsevier, Oxford), 293-348

Lomtadze, V.D., 1954. About the role of compaction processes of clayey deposits in the formation of underground waters. (trans.) *Doklady Akademii Nauk S.S.S.R.*, 98, 451-454

Marsden, S.S. & Davis, S.N., 1967. Geological subsidence. *Scientific American*, 216, 93-100

McClelland, B., 1967. Progress of consolidation in delta front and prodelta clays of the Mississippi River. In: *Marine Géotechnique*. ed. A.F. Richards, (Illinois Press, Urbana), 22-40

Meade, R.H., 1966. Factors influencing the early stages of the compaction of clays and sands - a review. *Journal of Sedimentary Petrology*, 36, 1085-1101

Moon, C.F., 1978. High voltage electron microscopy as an adjunct to scanning electron microscopy in the study of fine sedimentary particles. In: *Scanning Electron Microscopy in the Study of Sediments*, ed. W.B. Whalley, (Geo Abstracts, Norwich), 71-80

Plassche, O. van de, 1980. Compaction and other sources of error in obtaining sea-level data: some results and consequences. *Eiszeitalter und Gegenwart*, 30, 171-181

Pons, L.J. & Zonneveldt, I.S., 1965. Soil ripening and classification. *International Institute for Land Reclamation and Improvement, Publication*, 13, 128 pp

Pusch, R., 1966. Quick-clay microstructure. *Engineering Geology*, 1, 433-443

Raghavan, R. & Miller, F.G., 1975. Mathematical analysis of sand compaction. In: *Compaction of Coarse-grained Sediments, I*, eds.G.V. Chilingarian & K.H. Wolf, *Developments in Sedimentology*, 18A, (Elsevier, Oxford), 403-524

Rieke, H.H. & Chilingarian, G.V., 1974. *Compaction of Argillaceous Deposits*. (Elsevier, Oxford)

Sangrey, D.A., 1972. Naturally cemented sensitive soils. *Géotechnique*, 22, 139-152

Skempton, A.W., 1970. The consolidation of clays by gravitational compaction. *Quarterly Journal of the Geological Society, London*, 125, 373-408

Sandhu , R. S. & Wilson, E.L., 1969. Finite element analysis of land subsidence. In: *Land Subsidence*. I.A.S.H. - Unesco Publication No.89, AIHS 2, 393-400

Straaten, L.M.J.U. van, 1965. Coastal barrier deposits in South- and North-Holland. *Mededelingen Geologische Stichting*, N.S. 17, 41-75

Terzaghi, K., 1941. Undisturbed clay samples and undisturbed clays. *Journal of the Boston Society of Civil Engineers*, 28, 211-231

Terzaghi, K. & Peck, R.B., 1967. *Soil Mechanics in Engineering Practice*. (John Wiley & Sons, London), 2nd edition

Weller, J.M., 1959. Compaction of sediments. *American Association of Petroleum Geologists Bulletin*, 43, 273-310

Williamson, W.O., 1974. The strength of clays at small moisture contents. *Minerals Science and Engineering*, 6, 3-18

Index

LIST OF CONTRIBUTORS

W.H. Abbott Mobil Exploration and Producing Services Inc., Applied Stratigraphy, P.O. Box 900, DALLAS, Texas 75221, USA.

W.H. Adey Marine Systems Laboratory, National Museum of Natural History, Smithsonian Institution, WASHINGTON, DC 20560, USA.

J.T. Andrews Institute of Arctic and Alpine Research and Department of Geological Sciences, University of Colorado, BOULDER, Colorado 80309, USA

K.-E. Behre Niedersächsisches Landesamt für Marschen- und Wurtenforschung, Viktoriastrasse 26/28, D-2940 WILHELMSHAVEN, Federal Republic of Germany

J. Coudray U.E.R. - Faculté des Sciences, Université de la Réunion, Université Francaise de l'Océan Indien, B.P. 5 - 97490 SAINTE - CLOTILDE, France D.O.M.

J.-M. Flexor Observatorio Nacional - CNPq, Departamento de Geofisica, rua General Bruce 586, 20921 RIO DE JANEIRO, Brazil

J.T. Greensmith Department of Geology, University College London, Gower Street, LONDON WC1E 6BT, UK

D. van Harten Geologisch Instituut, Universiteit van Amsterdam, Nieuwe Prinsengracht 130, 1018 VZ AMSTERDAM, Netherlands

A. Heyworth Department of Geography, The University College of Wales, Llandinam Building, Penglais, ABERYSTWYTH, Dyfed SY23 3DB, UK

A.C. Hine Department of Marine Science, University of South Florida, 140 Seventh Avenue South, ST. PETERSBURG, Florida 33701, USA

D. Hopley Department of Geography, James Cook University of North Queensland, TOWNSVILLE, 4811 Australia

W.G. Jardine Department of Geology, The University, GLASGOW G12 8QQ, UK

C. Kidson Department of Geography, The University College of Wales, Llandinam Building, Penglais, ABERYSTWYTH, Dyfed SY23 3DB, UK

L.R. Kump Department of Marine Science, University of South Florida, 140 Seventh Avenue South, ST. PETERSBURG, Florida 33701, USA

J. Laborel Station Marine d'Endoume, Rue de la Batterie-des-Lions, 13007 MARSEILLE, France

L. Martin Observatorio Nacional - CNP 9, Departamento de Geofisica, rua General Bruce 586, 20921 RIO DE JANEIRO, Brazil (ORSTOM)

F.S. Medioli Department of Geology, Life Sciences Centre, Dalhousie University, HALIFAX, N.S. B3H 3J5, Canada

L.F. Montaggioni U.E.R. - Faculté des Sciences, Université de la Réunion, Université Française de l'Océan Indien, B.P. 5 - 97490 SAINTE - CLOTILDE, France D.O.M.

W.G. Mook Laboratorium voor Isotopen Fysica, Rijksuniversiteit Groningen, Westersingel 34, 9718 CM GRONINGEN, Netherlands

A. Palmer Belle W. Baruch Institute for Marine Biology and Coastal Research, University of South Carolina, COLUMBIA, S.C. 29208, USA

K.S. Petersen Geological Survey of Denmark, 31 Thoravej, DK-2400 COPENHAGEN NV, Denmark

P.A. Pirazzoli Former Address:- Laboratoire de Géomorphologie, Ecole Pratique des Hautes Etudes, 92120 MONTROUGE, France
Current Address:- C.N.R.S. 'INTERGEO', 191 Rue Saint- Jacques, 75005 PARIS, France

O. van de Plassche Instituut voor Aardwetenschappen, Vrye Universiteit, Postbus 7161, 1007 MC AMSTERDAM, Netherlands

Th. B. Roep Former Address:- Geologisch Instituut, Universiteit van Amsterdam, Nieuwe Prinsengracht 130, 1018 VZ AMSTERDAM, Netherlands
Current Address:- Instituut voor Aardwetenschappen, Vrye Universitei Postbus 7161, 1007 MC AMSTERDAM, Netherlands

D.B. Scott Department of Geology, Life Sciences Centre, Dalhousie University, HALIFAX, N.S. B3H 3J5, Canada

K. Suguio Instituto de Geosciências, Universidade de Sao Paolo, Caixa Postal 20899, 01000-SAO PAOLO-5P, Brasil

E.V. Tucker Applied Earth Science, Department of Geography and Earth Science, Queen Mary College, University of London, Mile End Road, LONDON E1 4NS, UK

GPSR Compliance

The European Union's (EU) General Product Safety Regulation (GPSR) is a set of rules that requires consumer products to be safe and our obligations to ensure this.

If you have any concerns about our products, you can contact us on ProductSafety@springernature.com

In case Publisher is established outside the EU, the EU authorized representative is:

Springer Nature Customer Service Center GmbH
Europaplatz 3
69115 Heidelberg, Germany

Batch number: 10823811

Printed by Printforce, the Netherlands